Plant Genomic Prediction

AI-Enhanced Models for Plant Genomics and Crop Improvement

Plant Genomic Prediction

AI-Enhanced Models for Plant Genomics and Crop Improvement

Edited by

Jen-Tsung Chen
National University of Kaohsiung, Taiwan

CABI

CABI is a trading name of CAB International

CABI
Nosworthy Way
Wallingford
Oxfordshire OX10 8DE
UK

Tel: +44 (0)1491 832111
E-mail: info@cabi.org
Website: www.cabi.org

CABI
200 Portland Street
Boston
MA 02114
USA

T: +1 (617)682-9015
E-mail: cabi-nao@cabi.org

A catalogue record for this book is available from the British Library, London, UK.

ISBN-13: 9781836991304 (hardback)
9781836991311 (ePDF)
9781836991328 (ePub)

DOI: 10.1079/9781836991328.0000

Commissioning Editor: Ward Cooper
Editorial Assistant: Emma McCann
Production Editor: James Bishop

Typeset by Exeter Premedia Services Pvt Ltd, Chennai, India
Printed and bound in the UK by CPI Group (UK) Ltd, Croydon, CR0 4YY

Table of Contents

Contributors

Atman Adiba, Regional Agricultural Research Center of Tadla, National Institute of Agricultural Research, Avenue Ennasr, P.O. Box 415, Rabat 10090, Morocco; Email: atman.adiba@inra.ma

Sarfraz Ahmad, Department of Genetics and Plant Breeding, College of Agriculture Mandawa, Swami Keshwanand Rajasthan Agriculture University, Bikaner, Rajasthan, India; Email: sarfraz_firoz@rediffmail.com

Mohd Ali, Post Graduate Department of Zoology, Khalsa College Charitable Society, Amritsar, Punjab, India; Email: xyaansari774@gmail.com

T. Anam, Department of Genetics and Plant Breeding, Gazipur Agricultural University, Gazipur 1706, Bangladesh; Email: tasfia.priya1998@gmail.com

Shreyas Padmanabha Sharma Beedubail, BioAtlantis Ltd., Clash Industrial Estate, Tralee, Co. Kerry, V92 RWV5, Ireland; Email: reach.shreyas01@gmail.com

Sourav Bhattacharjee, Department of Botany, Pandit Deendayal Upadhyaya Adarsha Mahavidyalaya Tulungia, Bongaigaon, Assam, India; Email: souravbh263@gmail.com

Sudip Bhattacharya, Department of Biological Science, Midnapore City College, Midnapore-721129, West Bengal, India; Email: sudipbhattacharya234@gmail.com

Abdellatif Boutagayout, Environment and Valorization of Microbial and Plant Resources Unit, Faculty of Sciences, Moulay Ismail University, Meknes, Morocco; Email: hamdanianasse@gmail.com

Debdeep Chattopadhyay, Post-Graduate & Research Department of Biotechnology, St. Xavier's College (Autonomous), 30, Mother Teresa Sarani, Kolkata-700016, West Bengal, India; Email: chattopadhyay.debdeep777@gmail.com

Anamika Das, Department of Genetics and Plant Breeding, Bidhan Chandra Krishi Viswavidyalaya, Mohanpur-745212, Nadia, West Bengal, India; Email: anamikabbsr@gmail.com

Ankan Das, Palli Siksha Bhavana (Institute of Agriculture), Visva-Bharati, Sriniketan, West Bengal-731236, India; Email: rickdas.official@gmail.com

Shreya Dave, Department of Biotechnology, Brainware University, Kolkata-700125, West Bengal, India; Email: sddave08@gmail.com

Sandip Debnath, Department of Genetics and Plant Breeding, Palli Siksha Bhavana (Institute of Agriculture), Visva-Bharati, Sriniketan, West Bengal-731236, India; Email: sandip.debnath@visva-bharati.ac.in

Alka Devi, Dhiti Lifesciences, Pvt. Ltd., New Delhi 110020, India; Email: alkamondela@gmail.com

Dhivyapriya Dharmaraj, Department of Genetics and Plant Breeding, Amrita School of Agricultural Sciences, Amrita Vishwa Vidyapeetham, Coimbatore-642109; Email: d_dhivyapriya@cb.amrita.edu

Shreyasi Dutta, Department of Botany (UG & PG), Raja Narendralal Khan Women's College (Autonomous), Gope Palace, Midnapur 721 102, India; Email: shreyasidutta@rnlkwc.ac.in

Mariam El Harsal, National Institute of Agricultural Research (INRA), BO 415, Rabat, Morocco; Email: mariam.elharsal@inra.ma

Abdelaaziz Farhaoui, Equipe de Recherche en Didactique des Sciences et Sciences de l'Education et TICE. (DSSE_TICE), Centre Régional des Métiers de l'Education et de la Formation de Fès Meknès, Annexe Meknès, Maroc; Phytopathology Unit, Department of Plant Protection, Ecole National of Agriculture Meknes, Km10, Rte Haj Kaddour, BP S/40, Meknes 50001, Morocco; Email: farhaoui.aziz22@gmail.com

Umer Fayaz, Accelerated Crop Improvement, Pigeonpea Breeding, International Crops Research Institute for the Semi-Arid Tropics, Hyderabad, India; Email: umerfayaz1999@gmail.com

Rahul Goyal, ICAR-National Research Centre on Equines, Hisar, Haryana 125001, India; Email: rahul0001goyal@gmail.com

Payel Guha, Department of Biotechnology, Brainware University, Kolkata-700125, West Bengal, India; Email: payalguha381@gmail.com

Anas Hamdani, National Agricultural Research Institute, BP 578, Meknes, Morocco; Email: hamdanianasse@gmail.com

A.K.M. Aminul Islam, Department of Genetics and Plant Breeding, Gazipur Agricultural University, Gazipur 1706, Bangladesh; Email: aminulgpb@bsmrau.edu.bd

Indrajit Kalita, Department of Botany, Gauhati University, Guwahati, 781014, Assam, India; Email: kalitaindrajit7@gmail.com

Sharmistha Sarma Kalita, Department of Botany, Gauhati University, Guwahati, 781014, Assam, India; Department of Botany, Salbari College, Salbari, Baksa, Assam, India; Email: sarmakalita004@gmail.com

Srinivasan Kameswaran, Department of Botany, Vikrama Simhapuri University College, Kavali-524201, SPSR Nellore District, Andhra Pradesh, India; Email: skameswaran@vsu.ac.in

Mohamed Kouighat, National Institute of Agricultural Research (INRA), BO 415, Rabat, Morocco; Email: kouighat84@gmail.com

Arti Kumari, Nalanda College of Horticulture, Noorsarai, Nalanda 803113, Bihar, India; Email: artikumarijha07@gmail.com

Prerna Kumari, University Department of Botany, Ranchi University, Ranchi, Jharkhand, India-834008; Email: prekmri1999@gmail.com

Atharva Anand Mahajan, Advanced Centre for Treatment, Research and Education in Cancer (ACTREC), Tata Memorial Centre, Mumbai, Maharashtra, India; Email: atharva66mahajan@gmail.com

Binod Kumar Mahto, University Department of Botany, Ranchi University, Ranchi, Jharkhand, India-834008; Email: binodkrmahto@ranchiuniversity.ac.in

Aqsa Majgaonkar, Senior Education Officer, Conservation Education Centre, Bombay Natural History Society, Mumbai 400 063, India; Email: aqsa.majgaonkar@xaviers.edu.in

Manonmani Karthikeyan, Department of Genetics and Plant Breeding, Amrita School of Agricultural Sciences, Amrita Vishwa Vidyapeetham, Coimbatore-642109; Email: k_manonmani@cb.amrita.edu

Roberto de la Rosa Santa Maria, Colegio de Postgraduados (CP), Campus Tabasco, Producción Agroalimentaria en el Trópico, H. Cárdenas, Tabasco, México; Email: robdelarosas@colpos.mx

Osval A. Montesinos-López, Facultad de Telemática, Universidad de Colima, Colima, Colima, 28040, México; Email: oamontes1@ucol.mx

Rashmi Mukherjee, Department of Botany (UG & PG), Raja Narendralal Khan Women's College (Autonomous), Gope Palace, Midnapur 721 102, India; Email: rashmimukherjee@rnlkwc.ac.in

Kalirajan Murugasridevi, Department of Agricultural Entomology, Amrita School of Agricultural Sciences, Amrita Vishwa Vidyapeetham, Coimbatore-642109; Email: k_murugasridevi@cb.amrita.edu

Puja Ohri, Department of Zoology, Guru Nanak Dev University, Amritsar, Punjab 143005, India; Email: puja.zoology@gndu.ac.in

Brahim Ouahzizi, Biochemistry of Natural Resources Team, Faculty of Sciences and Techniques, Moulay Ismail University, Errachidia, Morocco; Email: b.ouahzizi@edu.umi.ac.ma

Parkirti, Department of Zoology, Guru Nanak Dev University, Amritsar, Punjab 143005, India; Email: parkirti209@gmail.com

Ishrat Parween, University Department of Botany, Ranchi University, Ranchi, Jharkhand, India-834008; Email: parween8789@gmail.com

Dhaval Patel, Research and Development Division, Latambarcem Brewers Private Limited, Bicholim, Goa 403 503, India; Email: dhaval8294patel@gmail.com

Krutika Patil, Division of Microbiology, Indian Agricultural Research Institute, New Delhi 110012, India; Email: krutikapatil43432@gmail.com

Indrani Paul, Department of Biotechnology, Brainware University, Kolkata-700125, West Bengal, India; Email: paulindrani54@yahoo.com

José Elías Peregrina-Chavarría, Facultad de Telemática, Universidad de Colima, Colima, Colima, 28040, México; Email: eliasperegrinachava@gmail.com

Murali Shankar Perumal, Department of Plant Pathology, Pushkaram College of Agricultural Sciences, Tamil Nadu Agricultural University, Pudukottai; Email: muralishankar.tnau@gmail.com

Biswajit Pramanik, Biotechnology Unit, Crop Improvement Division, ICAR-Central Research Institute for Jute and Allied Fibres, Barrackpore, West Bengal-700121, India; Email: biswajit1996pramanik@gmail.com

Sourish Pramanik, Department of Plant Pathology, Punjab Agricultural University, Ludhiana 141004, India; Email: sourishpramanik2002@gmail.com

Mohammad Ghani Raghib, Department of Horticulture, Faculty of Agriculture, Nangarhar University, Jalalabad, Nangarhar, Afghanistan; Email: mohammadghaniraghib@gmail.com

Mandala Ramakrishna, Department of Botany, Vikrama Simhapuri University College, Kavali-524201, SPSR Nellore District, Andhra Pradesh, India; Email: mramakrishna@vsu.ac.in

Harshdeep Rana, BRIC-National Agri-Food and Biomanufacturing Institute (BRIC-NABI), Knowledge city, S.A.S. Nagar, Sector 81, Mohali-140306; Email: Ranadeeepharsh@gmail.com

Subhajit Roy, Department of Biotechnology, Brainware University, Kolkata-700125, West Bengal, India; Email: subhajitroy013@gmail.com

Aryadeep Roychoudhury, Discipline of Life Sciences, School of Sciences, Indira Gandhi National Open University, Maidan Garhi, New Delhi –110068, India; Email: aryadeep.rc@gmail.com

Arup Kr. Saha, Department of Genetics and Plant Breeding, School of Agriculture, Seacom Skills University, Kendradangal, Birbhum-731236, West Bengal, India; Email: arups838@gmail.com

Saniya, Department of Horticulture, ICAR-National Research Centre for Grapes, Pune, Maharashtra, India; Email: saniyacg.17@gmail.com

Rahele Ghanbari Moheb Seraj, Department of Horticultural Science and Engineering, Faculty of Agriculture, University of Kurdistan, Sanandaj, 66177-15175, Iran; Email: r.ghanbari@uma.ac.ir

Alisha Sharma, Department of Zoology, Guru Nanak Dev University, Amritsar, Punjab 143005, India; Email: sharmaalisha405@gmail.com

Roohi Sharma, Department of Zoology, Guru Nanak Dev University, Amritsar, Punjab 143005, India; Email: sharmaroohi77.rs@gmail.com

F. A. Shraboni, Department of Genetics and Plant Breeding, Gazipur Agricultural University, Gazipur 1706, Bangladesh; Email: shraboniagri@gmail.com

Aditya Pratap Singh, Department of Genetics and Plant Breeding, Bidhan Chandra Krishi Viswavidyalaya, Mohanpur-745212, Nadia, West Bengal, India; Email: adityapratapbckv@gmail.com

Anchita Singh, Department of Zoology, Guru Nanak Dev University, Amritsar, Punjab 143005, India; Email: anchitasinghthakur@gmail.com

Arun Dev Singh, Department of Botanical and Environmental Sciences, Guru Nanak Dev University, Amritsar, Punjab, India; Email: arundevsingh1727@gmail.com

Jaswinder Singh, Post Graduate Department of Zoology, Khalsa College Charitable Society, Amritsar, Punjab, India; Email: singhjassi75@yahoo.co.in

Dwaipayan Sinha, Department of Botany, Government General Degree College, Mohanpur, Paschim Medinipur, West Bengal 721 436, India; Email: dwaipayansinha@hotmail.com

Sneha Kali, Department of Biotechnology, Brainware University, Kolkata-700125, West Bengal, India; Email: sneha.kali.2002@gmail.com

Battana Swapna, Department of Botany, Vikrama Simhapuri University College, Kavali-524201, SPSR Nellore District, Andhra Pradesh, India; Email: swapnaivsr@gmail.com

Bhaben Tanti, Department of Botany, Gauhati University, Guwahati, 781014, Assam, India; Email: btanti@gauhati.ac.in

Jaishree Tanwar, Department of Genetics and Plant Breeding, Agriculture University, Jodhpur-342304, Rajasthan, India; Email: jaishreetanwar13@gmail.com

Vavar Vafaee, Department of Horticultural Science and Engineering, Faculty of Agriculture, University of Kurdistan, Sanandaj, 66177-15175, Iran; Medicinal Plants Breeding & Development Research Institute, University of Kurdistan, Sanandaj, 66177-15175, Iran; Email: y.vafaee@uok.ac.ir

Showkat A. Waza, Department of Genetics and Plant Breeding, Sher-i-Kashmir University of Agricultural Sciences and Technology of Kashmir, India; Email: sawgpb@skuastkashmir.ac.in

Preface

The world has faced multiple serious crises for decades, primarily caused by the increasing human population, which has led to environmental pollution, shortages of natural resources, and food scarcity. The situation can be exacerbated by global climate change, such as rising temperatures worldwide, which leads to multiple environmental stresses that significantly decrease agricultural nutrition and food production. One of the promising solutions is to develop so-called 'climate-smart' or 'stress-resilient' future crops, which largely depends on advancing crop breeding technologies. Among the approaches, 'genomic selection' undoubtedly plays a key role, holding promise for accelerating crop improvement.

A main component of genomic selection, i.e., genomic prediction, can accelerate genetic gains in crop breeding programs by building models. Genomic prediction has been developed over recent decades, and currently, it has opportunities to be greatly advanced by cutting-edge artificial intelligence (AI), particularly deep learning algorithms, coupled with high-throughput omics approaches such as metabolomics, proteomics, and transcriptomics.

This book explores AI-enhanced genomic prediction by integrating high-throughput technology, such as smart phenotyping, and fusing multiple omics data of metabolomics, proteomics, and transcriptomics with AI-based models to advance their coverage and accuracy. Updated methods for multi-environment prediction and building multi-trait models are comprehensively summarized in this book. Combinatory strategies for accelerating crop improvement are proposed involving AI-enhanced genomic prediction together with CRISPR/Cas9-mediated genome editing technology. The knowledge benefits the development of plant stress tolerance by combating critical stressors, including heat, cold, salt, drought, pests, and pathogens in major crops, which supports global nutrition and food security.

Altogether, the knowledge in this book supports the vision of climate change-ready, disease-resilient, and sustainable agriculture, which meets the sustainable development goals (SDGs) by advancing next-generation breeding tools. It is an ideal reference for readers in all aspects of plant sciences and agricultural biotechnology, particularly plant stress physiology and crop breeding.

Editor's biography

Dr Jen-Tsung Chen is a professor of cell biology at the National University of Kaohsiung in Taiwan. He teaches cell biology, genomics, proteomics, plant physiology, and plant biotechnology. Dr Chen's research interests include bioactive compounds, chromatography techniques, plant molecular biology, plant biotechnology, bioinformatics, and systems pharmacology. He is an active editor of academic books and international journals to advance the exploration of multidisciplinary knowledge involving plant physiology, plant biotechnology, nanotechnology, materials science, ethnopharmacology, and systems biology. He serves as an associate editor, editorial board member, and guest editor in reputed journals. Dr Chen has published books in collaboration with international publishers and is currently handling book projects on diverse topics, including AI technologies, drug discovery, drug development, herbal medicine, medicinal biotechnology, nanotechnology, bioengineering, plant functional genomics, plant speed breeding, epigenetics, functional RNAs, and CRISPR-based genome editing. Dr Chen is a productive author in academic publications and was recognized as one of the World's Top 2% Scientists between 2023 to 2025 by Elsevier and Stanford University. Dr Chen received the Springer Nature Editor of Distinction Award in 2025.

1 An Introductory Overview of Plant Genomic Prediction

Sarfraz Ahmad[1]*, Saniya[2] and Mohammad Ghani Raghib[3]
[1]Department of Genetics and Plant Breeding, College of Agriculture Mandawa, Swami Keshwanand Rajasthan Agriculture University, Bikaner, Rajasthan, India; [2]Department of Horticulture, ICAR-National Research Centre for Grapes, Pune, Maharashtra, India; [3]Department of Horticulture, Faculty of Agriculture, Nangarhar University, Jalalabad, Nangarhar, Afghanistan

Abstract

Genomic prediction (GP), also known as genomic selection, has transformed plant breeding by enabling the estimation of breeding values using genome-wide markers. Since its introduction, GP has evolved into a central methodology for accelerating crop improvement by leveraging high-dimensional molecular and phenotypic data. Advances in statistical models, machine learning, and deep learning have significantly enhanced predictive accuracy for complex traits, including yield, stress tolerance, and quality. Recent integration of multi-omics datasets like genomics, transcriptomics, metabolomics, and phenomics further strengthens GP by capturing biological complexity and genotype–environment interactions. Additionally, high-throughput phenotyping and speed breeding approaches synergize with GP to reduce breeding cycle time and increase genetic gain. Despite challenges such as computational demands, limited training populations, and uneven adoption across breeding programs, GP continues to expand in scope and utility. Emerging frontiers include cloud-based federated models, AI-enhanced algorithms, and predictive pipelines tailored for climate resilience and sustainable agriculture. This chapter reviews the theoretical foundations, methodological advances, and diverse applications of GP in plants, highlighting its potential to reshape global crop breeding strategies in the coming decades.

Keywords: Genomic prediction, Genotype × environment interaction, Machine learning, Multi-omics integration, Plant breeding

1.1 Introduction

The global demand for food is projected to increase by more than 50% by 2050, driven by population growth, dietary shifts, and climate change (FAO, 2021). Conventional plant breeding, while historically successful, is time-consuming and resource-intensive, often requiring several years to release improved cultivars. To accelerate genetic gains, the integration of molecular breeding tools has become essential. One of the most transformative innovations in this space is genomic prediction (GP), which uses genome-wide molecular markers to estimate the breeding values of individuals before phenotypic evaluation (Meuwissen *et al.*, 2001).

Early adoption of GP was most prominent in animal breeding but has rapidly spread

*Corresponding author: sarfraz_firoz@rediffmail.com

 Plant Genomic Prediction: AI-Enhanced Models for Plant Genomics and Crop Improvement (ed. Jen-Tsung Chen)
DOI: 10.1079/9781836991328.0001

into plant breeding since the 2010s, fueled by decreasing genotyping costs and the availability of dense marker datasets (Heffner *et al.*, 2009). Unlike traditional marker-assisted selection that targets a few major loci, GP captures the aggregate effect of thousands of small-effect loci across the genome, making it particularly powerful for polygenic traits such as yield and drought tolerance (Crossa *et al.*, 2017; Juliana *et al.*, 2019).

The past decade has witnessed remarkable progress in GP methodologies. Statistical approaches such as genomic best linear unbiased prediction (GBLUP) and Bayesian regressions remain foundational, but machine learning and deep learning models are increasingly adopted to capture nonlinear interactions and improve predictive accuracy (Montesinos-López *et al.*, 2018; Azodi *et al.*, 2019). Recent reviews emphasize that the integration of big data, multi-omics, and high-throughput phenotyping represents the next frontier in predictive breeding (Crossa *et al.*, 2025).

Applications of GP in major crops, including wheat, maize, rice, soybean, and barley, have demonstrated its potential to increase selection accuracy and shorten breeding cycles (Sinha *et al.*, 2023). For instance, genomic-enabled prediction combined with speed-breeding techniques can reduce the time to develop new varieties by half, a critical advantage under rapidly changing climates (Watson *et al.*, 2018). Furthermore, GP has been extended beyond yield traits to include abiotic stress tolerance, disease resistance, and nutritional quality, making it an indispensable tool for sustainable agriculture (Singh *et al.*, 2021).

Despite these advances, GP faces limitations. Challenges include the computational intensity of large-scale models, the need for diverse and representative training populations, and the difficulty of modeling genotype × environment interactions (G × E) (Jarquin *et al.*, 2020). Moreover, disparities in adoption persist, with resource-rich breeding programs advancing faster than those in low- and middle-income countries. Addressing these bottlenecks requires the integration of novel statistical models, AI-driven approaches, and international collaboration.

This chapter provides a comprehensive overview of GP in plants, beginning with its theoretical foundations, followed by methodological advances, key applications, limitations, and future directions. By consolidating knowledge across statistical, machine learning, and deep learning approaches, the chapter underscores GP's role in transforming plant breeding for global food and nutritional security.

1.2 Theoretical Foundations of Genomic Prediction

Genomic prediction (GP) is rooted in the principles of quantitative genetics and statistical learning, designed to capture the aggregate effect of genome-wide molecular markers on complex traits. Unlike marker-assisted selection, which focuses on a limited number of loci with large effects, GP uses thousands of single nucleotide polymorphisms (SNPs) across the genome to estimate genomic estimated breeding values (GEBVs). These values serve as predictors of an individual's genetic potential, enabling breeders to make early and informed selection decisions.

1.2.1 Quantitative genetics basis

The foundation of GP lies in the infinitesimal model of quantitative genetics, which assumes that complex traits are influenced by numerous loci of small additive effects. In the breeder's equation ($\Delta G = i \times h^2 \times \sigma P / L$), genetic gain ($\Delta G$) per cycle depends on selection intensity (i), trait heritability (h^2), phenotypic variance (σP), and cycle length (L). Genomic prediction accelerates breeding by improving selection accuracy (h^2) and reducing cycle length (L) through early prediction of breeding values (Lorenz *et al.*, 2011; Cobb *et al.*, 2019).

Heritability plays a critical role: traits with moderate to high heritability are generally predicted with higher accuracy. However, GP has also proven useful for low-heritability traits, especially when multi-environment trials and large training populations are employed (Gill *et al.*, 2021). This ability to account for small-effect loci makes GP particularly advantageous for polygenic traits such as yield, drought tolerance, and nutritional content.

1.2.2 Statistical frameworks of genomic prediction

The statistical premise of GP is to fit a model that relates phenotypic values (y) to marker genotypes (X) through regression:

$$y = X\beta + \epsilon$$

where X is the genotype matrix, β represents marker effects, and ε is the error term. Since the number of markers (p) usually exceeds the number of observations (n), traditional regression fails due to overfitting and multicollinearity. To address this, GP relies on regularization techniques and shrinkage methods.

The most commonly used models include:

- **RR-BLUP (ridge regression BLUP):** Assumes equal variance for all marker effects, shrinking them toward zero. It is computationally efficient and widely applied in practice.
- **GBLUP (genomic BLUP):** A reformulation of RR-BLUP that uses a genomic relationship matrix (GRM) to capture genetic similarity between individuals.
- **Bayesian models (BayesA, BayesB, BayesCπ, Bayesian ridge regression):** These models allow for heterogeneous marker variances, accommodating scenarios where some loci contribute more strongly than others (Meuwissen *et al.*, 2001; Gianola, 2013).

Bayesian approaches often outperform linear models when trait architecture involves a few major loci along with many minor loci (Azodi *et al.*, 2019). Recent developments include hierarchical Bayesian frameworks that incorporate prior biological knowledge, such as candidate genes or pathways, thereby improving interpretability (Sun *et al.*, 2023).

1.2.3 Genotype × environment interactions (G×E)

One of the central theoretical challenges in GP is capturing genotype × environment (G × E) interactions. Plant performance depends not only on genetic makeup but also on environmental variability. Classical GP models assumed stability across environments, which often underestimated predictive accuracy in multi-environment trials. To address this, mixed models and reaction norm models have been developed to explicitly account for G × E (Jarquin *et al.*, 2020).

In addition, multi-environment deep learning models can capture nonlinear responses to environmental covariates, improving prediction for traits such as drought tolerance and flowering time (Montesinos-López *et al.*, 2018). Recent advances also integrate climate variables and soil data with genomic markers, enabling breeders to forecast performance under future climate scenarios.

1.2.4 Marker density and linkage disequilibrium

The effectiveness of GP depends on marker density and the extent of linkage disequilibrium (LD) between markers and quantitative trait loci (QTLs). High-density SNP arrays and genotyping-by-sequencing (GBS) technologies have made it feasible to achieve sufficient LD coverage across diverse crop genomes (Poland and Rife, 2012). However, beyond a certain threshold, increasing marker density yields diminishing returns in predictive accuracy.

Moreover, the persistence of LD across breeding populations influences the transferability of prediction models. When prediction models are applied to unrelated populations or across breeding cycles, accuracy declines due to LD decay. Strategies such as updating training populations, incorporating functional genomic information, and leveraging haplotype-based approaches are being explored to maintain prediction reliability.

1.2.5 Training population design

The predictive ability of GP models largely hinges on the size, diversity, and representativeness of the training population. Larger training sets typically enhance prediction accuracy, but representativeness is equally critical to ensure that the genomic relationships between training and validation populations are well captured (Lorenz and Smith, 2015). Optimal training population design involves balancing diversity with genetic relatedness to maximize predictive power across breeding generations.

Recent strategies include dynamic training sets that are iteratively updated with new phenotypic and genotypic data. Machine learning algorithms are also being employed to optimize training population selection by identifying subsets of individuals that maximize information gain (Hamdia *et al.*, 2021).

To operationalize the theoretical frameworks of genomic prediction, plant breeding programs follow a structured workflow that links data generation to decision making. This process begins with germplasm collection and extends through genotyping, phenotyping, and data integration, culminating in model training and the prediction of genomic estimated breeding values. The resulting predictions inform selection strategies and variety deployment. This stepwise pipeline provides the foundation upon which diverse methodological approaches can be applied, as illustrated in Fig. 1.1, and will be discussed in detail in Section 1.3 (Methods of Genomic Prediction).

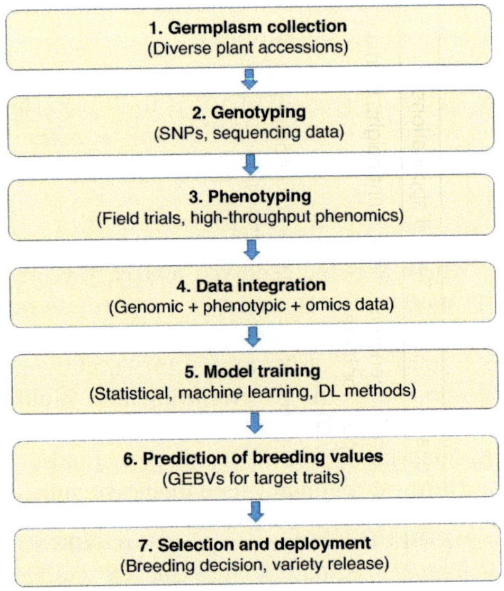

Fig. 1.1. Genomic prediction pipeline in plants. The sequential workflow of genomic prediction (GP) begins with germplasm collection, followed by genotyping through SNP arrays or sequencing and phenotyping via field trials or high-throughput platforms. The resulting genomic and phenotypic datasets are integrated to train predictive models (statistical, machine learning, or deep learning). These models generate genomic estimated breeding values, which guide selection decisions and enable the release of improved cultivars (Author's own image).

1.3 Methods of Genomic Prediction

The choice of GP method directly influences prediction accuracy, model interpretability, and applicability across traits and environments. GP methods can be broadly classified into statistical models, machine learning approaches, deep learning methods, multi-omics integrative frameworks, and phenomic-assisted approaches. Each category has its strengths and limitations, making it crucial to select the right model for the breeding context.

Before discussing these methods in detail, Table 1.1 provides a comparative summary of the major GP models, their features, advantages, limitations, and typical applications.

1.3.1 Statistical models

Statistical models form the backbone of genomic prediction. Their strength lies in their theoretical grounding in quantitative genetics, computational efficiency, and interpretability.

- **RR-BLUP (ridge regression best linear unbiased prediction):** One of the earliest GP models, RR-BLUP assumes equal variance for all markers, shrinking estimates toward zero. It is well-suited for highly polygenic traits but may underperform when a few loci of large effect dominate trait architecture (Meuwissen *et al.*, 2001; Lorenz *et al.*, 2011).
- **GBLUP (genomic BLUP):** A variant of RR-BLUP that operates on a genomic relationship matrix (GRM). GBLUP captures kinship among individuals, making it useful in structured populations such as biparental crosses (Crossa *et al.*, 2017).

Table 1.1. Summary of genomic prediction models and their characteristics (Author's own table)

Model/approach	Type	Key features	Strengths	Limitations	Typical applications
RR-BLUP	Statistical	Ridge regression assuming equal marker effects	Computationally efficient, simple, suitable for polygenic traits	Assumes homogeneous variance, poor for traits with major QTLs	Yield, plant height, biomass
GBLUP	Statistical	Uses genomic relationship matrix (GRM)	Accounts for kinship structure, stable across populations	Limited handling of nonlinear effects, sensitive to population structure	Biparental populations, pedigree-informed breeding
Bayesian models (BayesA, BayesB, BayesCπ)	Statistical	Heterogeneous marker variances with different priors	Flexible, effective for traits with both large- and small-effect loci	Computationally intensive, requires prior specification	Grain quality, disease resistance
LASSO/ Elastic net	Statistical/ Penalized regression	Variable selection with sparsity	Identifies influential markers, handles high-dimensional data	May overlook polygenic background, tuning required	Drought tolerance, root traits
Random forest (RF)	Machine learning	Ensemble of decision trees	Captures nonlinear effects, robust to noise, interpretable feature importance	Can overfit, less efficient on massive SNP data	Biomass, stress tolerance
Support vector machine (SVM)	Machine learning	Kernel-based classification/regression	Effective for nonlinear associations, adaptable	High computational cost, sensitive to kernel choice	Flowering time, disease scoring
RKHS regression	Machine learning	Nonlinear regression using kernel functions	Captures epistasis and G × E interactions	Parameter tuning needed, may not scale well	Complex adaptive traits
Gradient boosting machines (GBMs)	Machine learning	Iterative boosting of weak learners	High accuracy, handles unbalanced datasets	Computationally expensive, risk of overfitting	Yield stability under stress
Convolutional neural networks (CNNs)	Deep learning	Multi-layer filters capturing local patterns	Learns haplotype blocks, strong for epistasis	Requires large training data, black-box model	Genomic regions controlling yield

Continued

Table 1.1. Continued

Model/approach	Type	Key features	Strengths	Limitations	Typical applications
Recurrent neural networks (RNNs/LSTM)	Deep learning	Sequential modeling of data	Effective for time-series and longitudinal omics data	Computationally intensive, prone to vanishing gradients	Developmental stages, growth dynamics
Transformer models	Deep learning	Attention mechanisms capturing long-range dependencies	Scales well, state-of-the-art accuracy	Requires massive datasets and GPUs	Multi-trait GP, multi-omics integration
Hybrid models (e.g. CNN-GBLUP)	Deep learning + Statistical	Combines interpretability with nonlinear learning	Improved balance between accuracy and interpretability	Still developing, computational demand	Large-scale breeding programs
Multi-omics GP	Integrative	Combines genomics, transcriptomics, metabolomics, proteomics	Captures biological pathways, reduces missing heritability	High data integration complexity	Stress tolerance, quality traits
Phenomic-assisted GP	Integrative	Combines high-throughput imaging and sensor data with genomics	Improves $G \times E$ modeling, bridges genotype-phenotype gap	Expensive infrastructure, data handling challenges	Climate resilience, disease resistance

- **Bayesian models:** These include BayesA, BayesB, BayesCπ, and Bayesian Ridge Regression. Unlike RR-BLUP, Bayesian approaches allow heterogeneous marker variances, enabling flexible modeling of trait architectures. For example, BayesB is particularly effective for traits with a mix of large-effect and small-effect QTLs (Gianola, 2013).
- **Penalized regression methods:** Approaches like LASSO (least absolute shrinkage and selection operator) and Elastic Net extend RR-BLUP by introducing sparsity, making them efficient for scenarios where only a subset of markers contributes significantly (Azodi *et al.*, 2019).

Despite their robustness, these statistical methods may struggle with nonlinear interactions, epistasis, and complex genotype × environment (G × E) effects, motivating the rise of machine learning and AI-based methods.

1.3.2 Machine learning approaches

Machine learning (ML) offers flexible frameworks to capture nonlinearities and complex interactions between markers, traits, and environments.

- **Random forests (RFs):** RF models are ensemble-based decision tree methods that handle high-dimensional marker data well. They are non-parametric and robust to noise but can be computationally demanding for very large datasets. RF has shown competitive performance in predicting complex traits such as biomass and disease resistance (Bombrun *et al.*, 2020).
- **Support vector machines (SVMs):** SVMs use kernel functions to map input data into higher-dimensional spaces, effectively capturing nonlinear marker–trait associations. However, SVMs require careful kernel selection and parameter tuning, which can limit scalability (Azodi *et al.*, 2019).
- **Kernel methods (RKHS – reproducing kernel Hilbert space):** RKHS regression uses kernel functions to capture nonlinear and epistatic interactions. It often outperforms linear models in traits influenced by non-additive effects (De Los Campos *et al.*, 2009).
- **Gradient boosting machines (GBMs):** GBMs iteratively improve prediction accuracy by focusing on errors from previous iterations. They have been increasingly applied in GP pipelines due to their efficiency in handling unbalanced datasets (Li *et al.*, 2018).

Overall, ML methods excel in flexibility and adaptability but may lack the strong genetic interpretability of statistical models.

1.3.3 Deep learning and AI-enhanced genomic prediction

Deep learning (DL) has emerged as a powerful extension of machine learning, leveraging multi-layered neural networks to automatically learn complex feature hierarchies.

- **Convolutional neural networks (CNNs):** CNNs can model spatial and sequential relationships among markers, capturing local haplotype structures. Recent studies have shown that CNNs outperform linear models in traits with strong epistatic effects (Montesinos-López *et al.*, 2018).
- **Recurrent neural networks (RNNs) and LSTMs:** RNN-based models are particularly effective for time-series phenotypic data and sequential omics datasets. Long short-term memory (LSTM) architectures capture long-range dependencies, useful for modeling developmental traits (Montesinos-López *et al.*, 2021).
- **Transformer architectures:** Inspired by natural language processing, transformers have been adapted for genomic data, where markers are treated as 'tokens.' These models capture long-range genomic dependencies more efficiently than RNNs (Choi and Lee, 2023).
- **Hybrid models:** Combinations of DL with statistical frameworks (e.g. CNN-GBLUP) enhance interpretability while retaining nonlinear modeling capacity. Such hybrid approaches are gaining popularity in large-scale breeding programs (Galli *et al.*, 2022).

While DL models often achieve superior accuracy, they require large training datasets, extensive computational resources, and careful hyperparameter optimization, making them less accessible for resource-limited breeding programs.

1.3.4 Multi-omics and integrative models

Traditional GP relies solely on genomic markers, but recent advances incorporate multi-omics data, including transcriptomics, metabolomics, epigenomics, and proteomics alongside genomic and phenotypic data.

- **Transcriptomic-assisted GP:** Expression QTLs (eQTLs) provide functional insights that improve prediction of gene-regulated traits.
- **Metabolomic-assisted GP:** Metabolite profiles serve as intermediate phenotypes, improving prediction accuracy for stress and quality traits (Ahmad *et al.*, 2025b).
- **Integrative Bayesian and ML frameworks:** Multi-kernel learning and multimodal deep learning architectures enable the joint modeling of diverse omics layers (Ballard *et al.*, 2024).

These integrative approaches address missing heritability and enhance the biological relevance of predictions. For example, integrating genomics with metabolomics has improved salt tolerance prediction in rice (Ullah *et al.*, 2022).

1.3.5 High-throughput phenotyping and genomic prediction

Advances in digital phenotyping have revolutionized data collection, complementing genomic datasets with high-resolution phenotypic measurements.

- **Imaging-based phenomics:** UAVs (drones), hyperspectral imaging, and LiDAR provide high-throughput trait data such as canopy temperature, biomass, and disease indices (Ahmad *et al.*, 2025a).
- **Sensor networks:** Soil moisture, nutrient sensors, and thermal imaging enable real-time monitoring of crop–environment interactions.
- **Integration with GP models:** High-dimensional phenomic data are increasingly combined with genomic markers using ML and DL frameworks, improving prediction accuracy under variable environments (Juliana *et al.*, 2019).

The integration of phenomics with GP bridges the gap between genotype and phenotype, enabling more accurate modeling of G × E interactions and accelerating selection for climate resilience.

1.4 Applications of Genomic Prediction in Plants

Genomic prediction has transitioned from theoretical promise to practical application in diverse crop breeding programs. By integrating high-density marker data, advanced phenotyping, and predictive models, GP accelerates genetic gain, reduces breeding cycle time, and enhances the precision of trait improvement. Applications span a wide range of crop species and trait categories, each presenting unique opportunities and challenges. This section discusses major applications in cereal crops, legumes, horticultural crops, and industrial crops (Table 1.2), followed in Section 1.4.5 by trait-focused examples (Table 1.3) that illustrate the breadth of GP in modern agriculture.

1.4.1 Applications in major cereal crops

Wheat

Wheat has been one of the pioneering crops for GP applications. The genomic complexity of bread wheat, with its large hexaploid genome, necessitates advanced prediction models. GP has been applied to accelerate improvement in yield stability, disease resistance, and abiotic stress tolerance. For example, Juliana *et al.* (2019) demonstrated that integrating high-throughput phenotyping with GP significantly improved

Table 1.2. Summary of genomic prediction applications across major crops and traits (Author's own table)

Crop	Key traits targeted	Application outcomes of GP	References
Wheat	Yield stability, drought tolerance, rust resistance, protein content	Improved prediction accuracy under stress environments; integration with HTP enhanced resilience breeding	Juliana *et al.*, 2019; Azizinia *et al.*, 2020
Rice	Yield, grain quality (amylose, gelatinization), blast resistance, salinity tolerance	Multi-omics-assisted GP improved accuracy for stress tolerance and quality traits	Iqbal *et al.*, 2021
Maize	Hybrid performance, kernel composition, drought tolerance, G × E adaptation	Accurate prediction of hybrid yield and stress tolerance; ML models optimized training sets	Fritsche-Neto *et al.*, 2021; Mora-Poblete *et al.*, 2023
Barley	Malting quality (β-glucan, protein), disease resistance	GP accelerated early selection for brewing-related traits	Taheri-Kafrani *et al.*, 2021
Chickpea	Yield, drought tolerance	Accurate yield prediction in semi-arid conditions	Roorkiwal *et al.*, 2016
Apple	Fruit firmness, sugar content, disease resistance	Accurate prediction of fruit quality traits in perennial crops	Muranty *et al.*, 2015
Sugarcane	Cane yield, sucrose content, fiber content	GP feasible in highly polyploid genomes for industry traits	Gouy *et al.*, 2013

prediction accuracy for grain yield under drought conditions. More recently, GP has been used to predict resistance to rust diseases by integrating genomic markers with environmental covariates, enhancing model transferability across environments (Azizinia *et al.*, 2020).

Rice

Rice breeding programs have leveraged GP to accelerate improvement in both agronomic and quality traits. Studies show that GP models can accurately predict yield, amylose content, and blast resistance, especially when combined with multi-omics data (Iqbal *et al.*, 2021). Additionally, genomic prediction for salt tolerance has been enhanced by integrating metabolomic and transcriptomic data, providing actionable insights for developing resilient cultivars.

Maize

Maize has served as a model crop for testing GP models due to its genetic diversity and strong private-sector investment. GP has been widely used to predict hybrid performance, kernel composition traits, and drought tolerance (Fritsche-Neto *et al.*, 2021). Multi-environment genomic prediction models have shown high accuracy in predicting G × E interactions, enabling targeted breeding strategies across diverse environments (Mora-Poblete *et al.*, 2023).

Barley and other cereals

Barley breeding has benefited from GP in the improvement of malting quality traits, such as beta-glucan content and protein concentration. In sorghum, GP has been applied to model biomass yield and drought adaptation traits, supporting breeding for bioenergy applications.

1.4.2 Applications in legume crops

Legumes present unique challenges due to their often lower levels of research investment compared with cereals. Nevertheless, GP is gaining momentum in major species.

- **Soybean:** GP has been applied to predict seed protein and oil content, flowering time, and resistance to soybean cyst nematode. Integrating GP with speed breeding

Table 1.3. Trait-wise applications of genomic prediction in plants (Author's own table)

Trait category	Crops applied	Key outcomes of GP applications	References
Yield and yield components	Wheat, rice, maize, barley, chickpea	Improved prediction of grain yield and harvest index in polygenic contexts; multi-environment GP enhanced yield stability	Roorkiwal *et al.*, 2016; Juliana *et al.*, 2019; Fritsche-Neto *et al.*, 2021; Iqbal *et al.*, 2021; Taheri-Kafrani *et al.*, 2021
Disease resistance	Wheat (rusts), rice (blast)	GP enabled early prediction of resistance; integration with environmental data improved durability of resistance deployment	Azizinia *et al.*, 2020; Iqbal *et al.*, 2021
Abiotic stress tolerance	Wheat, maize, rice	GP with environmental covariates improved drought tolerance prediction; multi-omics integration enhanced salinity tolerance prediction in rice	Azizinia *et al.*, 2020; Iqbal *et al.*, 2021; Mora-Poblete *et al.*, 2023
Quality traits	Wheat (protein, micronutrients), rice (amylose, gelatinization temperature), apple (firmness, sugar), barley (malting quality)	Accurate prediction of nutritional and processing quality traits, aiding biofortification and food industry traits	Muranty *et al.*, 2015; Juliana *et al.*, 2019; Taheri-Kafrani *et al.*, 2021; Iqbal *et al.*, 2021
Hybrid performance	Maize	Reliable prediction of hybrid vigor and yield, reducing field testing requirements; ML-based optimization of training sets	Fritsche-Neto *et al.*, 2021

approaches accelerates trait fixation in soybean improvement programs.

- **Chickpea:** Studies have demonstrated that GP can predict yield and drought tolerance, traits crucial for semi-arid regions (Roorkiwal *et al.*, 2016).
- **Common bean:** GP models are increasingly used to accelerate genetic gain for disease resistance, including angular leaf spot and anthracnose.

1.4.3 Applications in horticultural crops

Fruits

GP is being increasingly explored in perennial fruit crops where long breeding cycles limit progress. In apple, GP has been applied for predicting fruit quality traits such as firmness, sugar content, and disease resistance (Muranty *et al.*, 2015). In grapevine, GP has been used to predict aroma compounds and disease resistance, helping improve both yield and quality traits in viticulture.

Vegetables

Tomato and pepper breeding programs have applied GP to predict yield, fruit weight, and disease resistance. In cucumber, GP has facilitated early selection for resistance against downy mildew, demonstrating its utility in accelerating breeding of short-cycle horticultural crops.

1.4.4 Applications in industrial and specialty crops

GP is also gaining ground in industrial crops such as sugarcane, cotton, and bioenergy grasses.

- **Sugarcane:** Given its complex polyploid genome, GP has shown promise in predicting traits such as cane yield, fiber content, and sucrose concentration (Gouy *et al.*, 2013).

- **Cotton:** GP is being tested for fiber quality and yield stability across multiple environments.
- **Bioenergy crops:** In switchgrass and miscanthus, GP has been applied to predict biomass yield, lignin content, and drought resilience, supporting biofuel production systems.

1.4.5 Trait-focused applications of GP

Yield and yield components

GP is widely applied to predict grain yield, harvest index, and related traits. Yield is highly polygenic and environment-dependent, making GP superior to traditional marker-assisted selection for yield-related breeding.

Disease resistance

GP has been successfully used to predict resistance against fungal, bacterial, and viral pathogens. In wheat, genomic prediction of stripe rust resistance has improved durability of resistance deployment (Azizinia *et al.*, 2020). In rice, GP enhances blast resistance prediction, especially when multi-environment data are considered.

Abiotic stress tolerance

Abiotic stresses such as drought, salinity, and heat are critical targets for GP. In maize and wheat, GP models incorporating environmental covariates have improved prediction accuracy for drought tolerance (Azizinia *et al.*, 2020; Mora-Poblete *et al.*, 2023). Salt tolerance in rice has been predicted with greater accuracy using multi-omics-assisted GP frameworks (Iqbal *et al.*, 2021).

Quality traits

Quality traits, including nutritional content and processing attributes, are key breeding targets. In wheat, GP predicts grain protein and micronutrient concentrations, while in rice, GP has been applied to predict cooking and eating quality traits such as amylose and gelatinization temperature.

Hybrid performance

In crops like maize, GP has been widely applied for predicting hybrid performance, reducing the need for extensive field testing and lowering breeding costs (Fritsche-Neto *et al.*, 2021).

1.4.6 Summary and outlook

Applications of genomic prediction span a wide diversity of crops and traits, highlighting its versatility and impact in modern plant breeding. While GP is already mainstream in crops such as maize, wheat, and rice, its expansion into legumes, horticultural species, and industrial crops shows its broader potential. Trait-wise, GP is particularly powerful for predicting complex, polygenic, and environmentally sensitive traits such as yield, disease resistance, and stress tolerance. With the increasing availability of multi-omics and phenomics data, GP applications are poised to expand further, delivering robust predictions that can accelerate crop improvement across global agricultural systems.

1.5 Challenges and Future Perspectives in Plant Genomic Prediction

Despite significant progress, the widespread deployment of genomic prediction in plant breeding faces multiple challenges. These arise from data limitations, model constraints, biological complexities, and issues of adoption and equity. At the same time, rapid technological innovation and integration of multi-omics and artificial intelligence (AI) approaches hold promise for overcoming current limitations. This section critically evaluates the challenges of GP implementation and outlines future perspectives for its advancement.

1.5.1 Technical and computational challenges

One of the most pressing challenges in GP lies in the computational demand of handling increasingly large datasets. With modern breeding programs generating millions of markers across thousands of individuals, conventional statistical models often struggle with scalability (Tempelman, 2015). Although machine learning (ML) and deep learning (DL) methods offer solutions, they introduce their own challenges in terms of hyperparameter optimization, overfitting, and interpretability (Montesinos-López *et al.*, 2021).

Additionally, data heterogeneity complicates model development. Genotyping platforms, phenotyping methods, and environmental covariates often vary across breeding programs, limiting model transferability across populations and environments. Standardized pipelines for data integration remain underdeveloped, posing a significant bottleneck (Davis-Turak *et al.*, 2017).

1.5.2 Biological complexities

The genetic architecture of most agronomic traits poses another major challenge. Traits such as yield and drought tolerance are highly polygenic, with small-effect loci scattered across the genome. Predictive accuracy often declines when models are applied to unrelated populations or new environments, highlighting issues with genotype ×environment (G×E) interactions (Crossa *et al.*, 2022).

Furthermore, the epigenetic and gene regulatory networks underlying complex traits are poorly captured by marker-based GP models. The integration of epigenomic, transcriptomic, and metabolomic data is still in its infancy, and, while promising, requires substantial methodological innovation to be practically implemented (Larue *et al.*, 2024).

1.5.3 Phenotyping bottlenecks

While sequencing costs have plummeted, phenotyping remains a major bottleneck in GP. Accurate and high-throughput phenotyping (HTP) is essential for training reliable models, but field-based phenotyping is expensive, labor-intensive, and prone to environmental variability. Even with drone-based imaging and sensor platforms, issues of data standardization, processing, and integration persist (Araus and Kefauver, 2018).

In perennial crops like fruit trees, long juvenile phases delay the availability of phenotypic data, limiting the pace of model development. Novel phenotyping strategies that balance cost, speed, and accuracy are urgently needed to complement genotyping advances.

1.5.4 Economic and institutional barriers

The successful implementation of GP requires substantial investment in infrastructure, expertise, and data management. Public-sector breeding programs, particularly in developing countries, often lack the resources to establish high-throughput sequencing and phenotyping platforms (Baenziger, 2019).

Moreover, intellectual property (IP) concerns surrounding genotyping platforms, software, and predictive algorithms create barriers to open access and equitable adoption. Breeding institutions may face difficulties in sharing data across borders due to legal and regulatory frameworks, further limiting collaborative progress.

1.5.5 Ethical, social, and policy considerations

Although GP does not involve transgenesis, its reliance on genomic data raises concerns about data ownership, privacy, and benefit-sharing, particularly in low- and middle-income countries where genetic resources originate. Equitable access to predictive breeding technologies remains a central ethical challenge (FAO, 2022).

Furthermore, policymakers and funding agencies often prioritize short-term yield gains, while long-term traits such as sustainability and climate adaptation receive less attention. Embedding GP within holistic agricultural policies will be crucial for maximizing its impact on food security.

1.5.6 Future perspectives

Integration of multi-omics and systems biology

The incorporation of transcriptomics, proteomics, metabolomics, and epigenomics promises to enhance the resolution of GP models by capturing regulatory layers beyond DNA polymorphisms. Early studies in rice and maize have demonstrated the potential of multi-omics-assisted GP to improve accuracy for stress tolerance traits (Fan *et al.*, 2025). Developing scalable pipelines for multi-omics integration will be a key future direction.

Advances in artificial intelligence and deep learning

Deep learning models, particularly convolutional and graph neural networks, offer unique advantages for modeling non-linear relationships and high-dimensional genomic data. Coupled with explainable AI techniques, these models could improve both predictive accuracy and biological interpretability (Charles *et al.*, 2024).

Climate-smart and resilient breeding

Future applications of GP will increasingly focus on breeding for climate resilience, including tolerance to drought, heat, salinity, and emerging diseases. Incorporating environmental covariates and crop growth models into GP frameworks can better capture G × E interactions and guide region-specific breeding strategies (Naqvi *et al.*, 2022).

Digital phenotyping and big data integration

Advances in drone imaging, remote sensing, and sensor networks will provide rich phenomic datasets to complement genotypic data. Integrating these into GP pipelines will require big data frameworks, cloud computing, and automated data processing pipelines (Araus and Kefauver, 2018).

Global collaboration and open science

To ensure equitable benefits, future GP research must embrace open-access databases, shared computational tools, and international partnerships. Initiatives such as the Integrated Breeding Platform (IBP) and Excellence in Breeding (EiB) already provide models for collaborative GP deployment in developing regions (Xu *et al.*, 2022).

1.6 Conclusion

Genomic prediction represents a paradigm shift in plant breeding, offering powerful tools to accelerate the selection of superior genotypes. By combining statistical models, machine learning, and deep learning with high-throughput genotyping and phenotyping, GP has demonstrated significant success in predicting complex traits such as yield, disease resistance, abiotic stress tolerance, and hybrid performance across major crops. Nevertheless, limitations persist, including data scarcity, computational demands, genotype × environment interactions, and the challenge of integrating multi-omics and phenomics data. Addressing these requires scalable algorithms, improved data-sharing frameworks, and AI-driven solutions to enhance model robustness and interpretability. Looking ahead, the convergence of GP with genome editing, explainable AI, and climate-resilient breeding approaches will expand its impact. By fostering innovation and equitable access, GP has the potential to drive sustainable crop improvement and contribute meaningfully to global food security.

References

Ahmad, S., Shah, M.H., Raghib, M.G., Belwal, V., Ali, S. *et al.* (2025a) Technical advancements in plant high-throughput phenotyping and phenomics: An overview. In: *Plant High-Throughput Phenotyping and Functional Phenomics*. CRC Press, pp. 246–259.

Ahmad, S., Shah, M.H., Saniya, Raghib, M.G., Belwal, V. *et al.* (2025b) High-throughput metabolomics for plant phenotyping. In: *High-Throughput Plant Metabolomics*. CAB International, Wallingford, pp. 236–251.

Araus, J.L. and Kefauver, S.C. (2018) Breeding to adapt agriculture to climate change: Affordable phenotyping solutions. *Current Opinion in Plant Biology* 45, 237–247.

Azizinia, S., Bariana, H., Kolmer, J., Pasam, R., Bhavani, S. *et al.* (2020) Genomic prediction of rust resistance in tetraploid wheat under field and controlled environment conditions. *Agronomy* 10(11), 1843.

Azodi, C.B., Bolger, E., McCarren, A., Roantree, M., de Los Campos, G. *et al.* (2019) Benchmarking parametric and machine learning models for genomic prediction of complex traits. *G3: Genes, Genomes, Genetics* 9(11), 3691–3702.

Baenziger, P.S. (2019) Modern plant breeding: A perspective from the public sector in the United States. In: *Sustaining Global Food Security: The Nexus of Science and Policy*. CSIRO Publishing, Melbourne, pp. 148–161.

Ballard, J.L., Wang, Z., Li, W., Shen, L. and Long, Q. (2024) Deep learning-based approaches for multi-omics data integration and analysis. *BioData Mining* 17(1), 38.

Bombrun, M., Dash, J.P., Pont, D., Watt, M.S., Pearse, G.D. *et al.* (2020) Forest-scale phenotyping: Productivity characterisation through machine learning. *Frontiers in Plant Science* 11, 99.

Charles, S., Subeesh, A., Dhanya, V.G. and Malathi, V.M. (2024) Opportunities and prospects of Artificial Intelligence in plant genomics. In: *Unraveling New Frontiers and Advances in Bioinformatics*. Springer Nature Singapore, Singapore, pp. 93–113.

Choi, S.R. and Lee, M. (2023) Transformer architecture and attention mechanisms in genome data analysis: A comprehensive review. *Biology* 12(7), 1033.

Cobb, J.N., Juma, R.U., Biswas, P.S., Arbelaez, J.D., Rutkoski, J. *et al.* (2019) Enhancing the rate of genetic gain in public-sector plant breeding programs: Lessons from the breeder's equation. *Theoretical and Applied Genetics* 132(3), 627–645.

Crossa, J., Pérez-Rodríguez, P., Cuevas, J., Montesinos-López, O., Jarquín, D. *et al.* (2017) Genomic selection in plant breeding: Methods, models, and perspectives. *Trends in Plant Science* 22(11), 961–975.

Crossa, J., Montesinos-López, O.A., Pérez-Rodríguez, P., Costa-Neto, G., Fritsche-Neto, R. *et al.* (2022) Genome and environment based prediction models and methods of complex traits incorporating genotype × environment interaction. In: Ahmadi, N. (ed.) *Genomic Prediction of Complex Traits: Methods and Protocols*. Springer, pp. 245–283.

Crossa, J., Martini, J.W., Vitale, P., Pérez-Rodríguez, P., Costa-Neto, G. *et al.* (2025) Expanding genomic prediction in plant breeding: Harnessing big data, machine learning, and advanced software. *Trends in Plant Science* 30(7), 756–774.

Davis-Turak, J., Courtney, S.M., Hazard, E.S., Glen Jr, W.B., da Silveira, W.A. *et al.* (2017) Genomics pipelines and data integration: Challenges and opportunities in the research setting. *Expert Review of Molecular Diagnostics* 17(3), 225–237.

De Los Campos, G., Naya, H., Gianola, D., Crossa, J., Legarra, A. *et al.* (2009) Predicting quantitative traits with regression models for dense molecular markers and pedigree. *Genetics* 182(1), 375–385.

Fan, B.L., Chen, L.H., Chen, L.L. and Guo, H. (2025) Integrative multi-omics approaches for identifying and characterizing biological elements in crop traits: Current progress and future prospects. *International Journal of Molecular Sciences* 26(4), 1466.

FAO (2021) *The State of Food and Agriculture 2021*. Food and Agriculture Organization of the United Nations, Rome.

FAO (2022) *Harnessing Agricultural Data for Equitable Innovation: Policy guidelines on digital agriculture*. Food and Agriculture Organization of the United Nations, Rome.

Fritsche-Neto, R., Galli, G., Borges, K.L.R., Costa-Neto, G., Alves, F.C. *et al.* (2021) Optimizing genomic-enabled prediction in small-scale maize hybrid breeding programs: A roadmap review. *Frontiers in Plant Science* 12, 658267.

Galli, G., Sabadin, F., Yassue, R.M., Galves, C., Carvalho, H.F. *et al.* (2022) Automated machine learning: A case study of genomic 'image-based' prediction in maize hybrids. *Frontiers in Plant Science* 13, 845524.

Gianola, D. (2013) Priors in whole-genome regression: The Bayesian alphabet returns. *Genetics* 194(3), 573–596.

Gill, H.S., Halder, J., Zhang, J., Brar, N.K., Rai, T.S. *et al.* (2021) Multi-trait multi-environment genomic prediction of agronomic traits in advanced breeding lines of winter wheat. *Frontiers in Plant Science* 12, 709545.

Gouy, M., Rousselle, Y., Bastianelli, D., Lecomte, P., Bonnal, L. *et al.* (2013) Experimental assessment of the accuracy of genomic selection in sugarcane. *Theoretical and Applied Genetics* 126(10), 2575–2586.

Hamdia, K.M., Zhuang, X. and Rabczuk, T. (2021) An efficient optimization approach for designing machine learning models based on genetic algorithm. *Neural Computing and Applications* 33(6), 1923–1933.

Heffner, E.L., Sorrells, M.E. and Jannink, J.L. (2009) Genomic selection for crop improvement. *Crop Science* 49(1), 1–12.

Iqbal, Z., Iqbal, M.S., Khan, M.I.R. and Ansari, M.I. (2021) Toward integrated multi-omics intervention: Rice trait improvement and stress management. *Frontiers in Plant Science* 12, 741419.

Jarquin, D., Howard, R., Crossa, J., Beyene, Y., Gowda, M. *et al.* (2020) Genomic prediction enhanced sparse testing for multi-environment trials. *G3: Genes, Genomes, Genetics* 10(8), 2725–2739.

Juliana, P., Crossa, J., Montesinos-López, O.A., Poland, J., Huerta-Espino, J. *et al.* (2019) Integrating genomic-enabled prediction and high-throughput phenotyping in breeding for climate-resilient bread wheat. *Theoretical and Applied Genetics* 132, 177–194.

Larue, F., Rouan, L., Pot, D., Rami, J.F., Luquet, D. *et al.* (2024) Linking genetic markers and crop model parameters using neural networks to enhance genomic prediction of integrative traits. *Frontiers in Plant Science* 15, 1393965.

Li, B., Zhang, N., Wang, Y.G., George, A.W., Reverter, A. *et al.* (2018) Genomic prediction of breeding values using a subset of snps identified by three machine learning methods. *Frontiers in Genetics* 9, 237.

Lorenz, A.J. and Smith, K.P. (2015) Adding genetically distant individuals to training populations reduces genomic prediction accuracy in barley. *Crop Science* 55(6), 2657–2667.

Lorenz, A.J., Chao, S., Asoro, F.G., Heffner, E.L., Hayashi, T. *et al.* (2011) Genomic selection in plant breeding: Knowledge and prospects. *Advances in Agronomy* 110, 77–123.

Meuwissen, T.H.E., Hayes, B.J. and Goddard, M.E. (2001) Prediction of total genetic value using genome-wide dense marker maps. *Genetics* 157(4), 1819–1829.

Montesinos-López, O.A., Montesinos-López, A., Crossa, J., Gianola, D., Hernández-Suárez, C.M. *et al.* (2018) Multi-trait, multi-environment deep learning modeling for genomic-enabled prediction of plant traits. *G3: Genes, Genomes, Genetics* 8(12), 3829–3840.

Montesinos-López, O.A., Montesinos-López, A., Pérez-Rodríguez, P., Barrón-López, J.A., Martini, J.W. *et al.* (2021) A review of deep learning applications for genomic selection. *BMC Genomics* 22(1), 19.

Mora-Poblete, F., Maldonado, C., Henrique, L., Uhdre, R., Scapim, C.A. *et al.* (2023) Multi-trait and multi-environment genomic prediction for flowering traits in maize: A deep learning approach. *Frontiers in Plant Science* 14, 1153040.

Muranty, H., Troggio, M., Sadok, I.B., Rifai, M.A., Auwerkerken, A. *et al.* (2015) Accuracy and responses of genomic selection on key traits in apple breeding. *Horticulture Research* 2, 15060.

Naqvi, R.Z., Siddiqui, H.A., Mahmood, M.A., Najeebullah, S., Ehsan, A. *et al.* (2022) Smart breeding approaches in post-genomics era for developing climate-resilient food crops. *Frontiers in Plant Science* 13, 972164.

Poland, J. and Rife, T. (2012) Genotyping-by-sequencing for plant breeding and genetics. *The Plant Genome* 5(3), 92–102.

Roorkiwal, M., Rathore, A., Das, R.R., Singh, M.K., Jain, A. *et al.* (2016) Genome-enabled prediction models for yield related traits in chickpea. *Frontiers in Plant Science* 7, 1666.

Singh, R.K., Sood, P., Prasad, A. and Prasad, M. (2021) Advances in omics technology for improving crop yield and stress resilience. *Plant Breeding* 140(5), 719–731.

Sinha, D., Maurya, A.K., Abdi, G., Majeed, M., Agarwal, R. *et al.* (2023) Integrated genomic selection for accelerating breeding programs of climate-smart cereals. *Genes* 14(7), 1484.

Sun, N., Wang, Y., Chu, J., Han, Q. and Shen, Y. (2023) Bayesian approaches in exploring gene–environment and gene–gene interactions: A comprehensive review. *Cancer Genomics & Proteomics* 20(6 suppl), 669–678.

Taheri-Kafrani, A., Kharazmi, S., Nasrollahzadeh, M., Soozanipour, A., Ejeian, F. *et al*. (2021) Recent developments in enzyme immobilization technology for high-throughput processing in food industries. *Critical Reviews in Food Science and Nutrition* 61(19), 3160–3196.

Tempelman, R.J. (2015) Statistical and computational challenges in whole genome prediction and genome-wide association analyses for plant and animal breeding. *Journal of Agricultural, Biological, and Environmental Statistics* 20(4), 442–466.

Ullah, M.A., Abdullah-Zawawi, M.R., Zainal-Abidin, R.A., Sukiran, N.L., Uddin, M.I. *et al*. (2022) A review of integrative omic approaches for understanding rice salt response mechanisms. *Plants* 11(11), 1430.

Watson, A., Hickey, L.T. and Jahnke, S. (2018) Speed breeding is a powerful tool to accelerate crop research and breeding. *Nature Plants* 4(1), 23–29.

Xu, Y., Zhang, X., Li, H., Zheng, H., Zhang, J. *et al*. (2022) Smart breeding driven by big data, artificial intelligence, and integrated genomic–enviromic prediction. *Molecular Plant* 15(11), 1664–1695.

2 AI-Based Multi-omics for Predicting Plant Complex Traits

Krutika Patil[1], Alka Devi[2*], Arti Kumari[3] and Rahul Goyal[4]
[1]Division of Microbiology, Indian Agricultural Research Institute, New Delhi, India; [2]Dhiti Lifesciences, New Delhi, India; [3]Nalanda College of Horticulture, Noorsarai, Nalanda, Bihar, India; [4]ICAR-National Research Centre on Equines, Hisar, Haryana, India

Abstract

Artificial intelligence (AI) is transforming plant sciences by facilitating accurate species identification, early detection of diseases, plant yield prediction, and the optimization of agricultural practices. Leveraging machine learning and image recognition, AI supports biodiversity protection, and environmental and ecological research efforts. It is pivotal in plant breeding, expediting the creation of resilient, high-yield crop varieties by identifying desirable plant characteristics. AI-driven models, informed by soil and climate data, advance sustainable farming and enhance global food security. AI-driven phenotyping automates the assessment of plant traits, deepening insights into growth and adaptation mechanisms. Current research focuses on enhancing the reliability and transparency of AI models while addressing ethical concerns such as data privacy and algorithmic bias. Interdisciplinary collaboration among scientists, data experts, and engineers is key to unlocking AI's full potential in shaping a sustainable and secure agricultural future.

Keywords: Crop yield, Food security, Machine learning, Plant breeding

2.1 Introduction

Artificial intelligence (AI) encompasses the machine-based emulation of human cognitive functions, most notably learning (the acquisition of data and governing principles), reasoning (the application of those principles to derive conclusions), and self-correction (the iterative refinement of performance) (Cembrowska-Lech *et al.*, 2023). As a groundbreaking technology, AI is reshaping numerous fields, and its application within plant sciences promises to transform our approaches to understanding, cultivating, and sustainably managing plant life (Singh *et al.*, 2021). In particular, AI-driven techniques are being employed for tasks such as plant identification, disease detection, yield forecasting, high-throughput phenotyping, and the implementation of precision agriculture (Garske *et al.*, 2021; Javaid *et al.*, 2023). Machine learning models integrated with image recognition have revolutionized plant species identification, enabling swift and precise classification that bolsters ecological studies and supports biodiversity preservation. In plant phenotyping, AI-based systems automate the capture and quantify the morphological and physiological traits, granting researchers deeper insights into growth

*Corresponding author: alkamondela@gmail.com

 Plant Genomic Prediction: AI-Enhanced Models for Plant Genomics and Crop Improvement (ed. Jen-Tsung Chen)
DOI: 10.1079/9781836991328.0002

dynamics and adaptive strategies. Continuous, real-time surveillance of plant development and health not only fosters cutting-edge research but also enhances crop management protocols (Roper *et al.*, 2021). Moreover, AI-powered diagnostic platforms equip plant pathologists and agronomists with the ability to detect diseases and pest infestations at their onset, allowing for prompt remedial action and substantially reducing yield losses (Shoaib *et al.*, 2023). In plant breeding, AI plays an essential part in developing robust, high-yield crop varieties. By processing extensive genomic datasets, machine learning algorithms can swiftly pinpoint desirable genetic traits, thereby shortening breeding cycles (Yoosefzadeh Najafabadi *et al.*, 2023). Additionally, AI-based predictive models that integrate climatic and soil parameters deliver actionable strategies for optimizing crop management and minimizing environmental impacts (Singh *et al.*, 2020). Together, these innovations drive sustainable agricultural practices and strengthen global food security in the face of climate change and rising population demands (Talaviya *et al.*, 2020).

Despite AI's profound impact on plant sciences, researchers are actively working to refine these models for greater robustness and transparency (Shah and Konda, 2021; Rudin *et al.*, 2022). Concurrently, there is a growing emphasis on ethical issues such as data privacy, algorithmic bias, and the reliable deployment of AI in agriculture. Moreover, harnessing AI's full potential requires close partnership among plant scientists, data scientists, and engineers to drive innovation and ensure practical, sustainable solutions (Ryan *et al.*, 2023).

Over the past few years, plant science has been fundamentally reshaped by the rise of big data analytics coupled with AI-driven methodologies (Javaid *et al.*, 2023). This synergy has unlocked novel avenues for probing and enhancing plant biology, agricultural practices, and crop productivity. In this context, big data analytics refers to the deployment of sophisticated computational approaches to sift through and interpret vast, multifaceted datasets derived from sources such as genomics, transcriptomics, proteomics, phenomics, metabolomics, and a variety of environmental monitoring systems (Javaid *et al.*, 2022). Big data analytics methodologies are crucial for distilling actionable insights from extensive plant science datasets (Khiarak *et al.*, 2018). These approaches leverage statistical modeling, machine learning, and AI-driven techniques to uncover underlying patterns, establish correlations, and build predictive frameworks. Commonly used algorithms, including random forests (RF), support vector machines (SVMs), deep learning neural networks (DLNNs), and Bayesian networks, facilitate a range of tasks, including classification, regression analysis, feature selection, and anomaly identification (Aditya Shastry and Sanjay, 2019).

This chapter explores how the incorporation of AI into plant sciences constitutes a conceptual shift, providing unparalleled abilities in species recognition, breeding, phenotyping, disease management, and sustainable agriculture, while highlighting its diverse applications, advantages, and prospects (Gupta *et al.*, 2024).

2.2 Overview of Multi-omics (Genomics, Transcriptomics, Proteomics, and Metabolomics)

The 'omics' field provides a holistic and integrative framework for exploring different biological systems. When unified under the concept of 'multi-omics,' they enable us to study the intricate interactions among various layers of biological information (Cembrowska-Lech *et al.*, 2023). The term 'multi-omics' refers to the comprehensive integration of several 'omic' fields, each specializing in the study of a specific biological system. This encompasses fields such as genomics, proteomics, transcriptomics, metabolomics, and more (Chandran *et al.*, 2020; Walker *et al.*, 2022). Each omics layer provides a distinct insight into the functional elements of a biological system. Nonetheless, examining these layers in isolation overlooks the comprehensive understanding of how these elements interact and shape the complete phenotype. This is where multi-omics offers a significant advancement (Jamil *et al.*, 2020; Li *et al.*, 2021a). Multi-omics methods seek to combine data from different omics fields to unravel the intricate relations and regulatory systems driving an organism's observable traits or phenotypes. In plant phenotyping, multi-omics offers a holistic understanding of the dynamic relationships between genetic composition, environmental factors, and plant phenotypic traits (Crandall *et al.*, 2020; Li *et al.*, 2022).

Genomics is the study of an organism's complete genome, encompassing its entire DNA and all associated genes. It focuses on exploring the genomes for their structure, function, and evolution. Through genomics, researchers can investigate complex genetic characteristics and learn how multiple genes contribute to them. In plant phenotyping, genomics offers valuable information about the genetic basis of different plant characteristics and supports the advancement in the field of marker-assisted selection methods (Lin and Lane, 2017; Yang *et al.*, 2021). Whole-genome sequencing (WGS) offers a detailed and in-depth analysis of the plant genome, enabling the discovery of previously uncharacterized genes and regulatory factors. In contrast, genotyping-by-sequencing is an economically viable approach for detecting small insertions and deletions (INDELs) and single-nucleotide polymorphisms (SNPs). This technique is especially useful for marker-assisted breeding, genetic mapping, and population genetics research (Morganti *et al.*, 2019; Hui *et al.*, 2020; Zhang *et al.*, 2020a). SNPs and other various genetic alterations form the foundation of genetic variability and can significantly influence desired characteristics in plant science, including fruit color, size, flavor, and different pest and disease resistance. For instance, an SNP within a specific gene might alter the function of a protein, resulting in a phenotypic change in the plant (Bayes-Genis *et al.*, 2020; Indelicato and Boesch, 2021; Florensa *et al.*, 2022). Advances in genomics, driven by next-generation sequencing methods, have greatly expanded opportunities to unravel the complex genetic make-up of plants and accelerate their genetic enhancement (Kim *et al.*, 2021; Rong *et al.*, 2021; Wu *et al.*, 2023).

Transcriptomics, often utilizing RNA sequencing (RNA-seq), examines the transcriptome, which is the complete set of RNA transcripts generated by the genome under particular conditions or within particular cells. This approach provides valuable insights into gene expression dynamics, helping researchers determine which genes are active or inactive throughout various developmental stages or in response to environmental factors. Such findings reveal how genetic data is converted into functional outcomes and can highlight the genes that drive particular plant characteristics (Song *et al.*, 2019). For instance, in horticultural studies, RNA-seq can be employed to elucidate the transcriptomic shifts underlying fruit ripening, while in broader agricultural research, it can reveal how gene expression changes during insect and pest attacks (Chen *et al.*, 2019; Luecken and Theis, 2019; Zappia and Theis, 2021). Transcriptomics generates high-resolution data that offer not only an outline of gene activity at a given moment but also insights into the dynamic depiction of gene expression. This understanding can uncover critical metabolic and regulatory processes in plants, providing valuable knowledge to enhance agricultural practices and crop improvement methods. As a vital link between the genome and phenotype, transcriptomics plays a crucial role in identifying the functional factors of the genome and their applications in plant science (Ahmed *et al.*, 2022; Bai *et al.*, 2022; Lee *et al.*, 2023).

Proteomics examines the full complement of proteins produced by a genome, encompassing their interactions, localization, alterations, and activities. By analyzing protein compositions, functions, and networks, proteomics sheds light on the cellular processes that drive different plant traits. It can identify post-translational alterations, map protein–protein interactions, and assess how environmental conditions affect protein behavior (Cembrowska-Lech *et al.*, 2023). As proteins are the main executors of cellular functions, proteomics complements genomics and transcriptomics by offering a more controlled connection to cellular activity and phenotype. In plant science, proteomics can identify critical proteins that govern fundamental processes like respiration, photosynthesis, stress responses, and signaling cascades. For example, comparing the proteomes of disease-resistant and susceptible cultivars may uncover proteins that underlie resistance. Additionally, mapping of protein–protein interactions illuminates the intricate networks controlling plant development and environmental adaptation. By dissecting the protein interactions involved in fruit ripening, researchers can devise approaches to improve fruit quality and extend shelf life (Wang *et al.*, 2021; Ghag *et al.*, 2022; Mondal *et al.*, 2022).

Metabolomics is the systematic analysis of the distinct chemical signatures left by cellular activities, specifically, the small-molecule metabolic compounds. It offers insights into a plant's physiological state and how it responds to external stimuli. By evaluating the metabolomes of various plants or of the same plant under varying situations, scientists can recognize shifts in metabolic pathways that may drive particular characteristics (White *et al.*, 2017; Pandita *et al.*, 2021).

When combined, these omics fields offer an end-to-end portrait of a biological system, from the genetic blueprint through its transcriptome, proteome, and metabolome. Employing a multi-omics strategy to integrate these layers exposes the intricate networks and interactions that give rise to a plant's observable traits, delivering a truly holistic, molecular-level understanding of plant phenotypes. Additionally, this approach makes it possible to pinpoint molecular markers tied to favorable characteristics, which can then be harnessed for plant breeding and genetic advancement in agriculture.

2.2.1 Application of multi-omics in plant science

Integrating multi-omics datasets offers a potent means to deepen our insights into plant biology and to dramatically speed up advances in agricultural research and breeding initiatives. Below, we present the key ways in which multi-omics integration benefits agriculture.

Holistic overview of biological systems

One of the foremost benefits of multi-omics integration is the all-encompassing view it offers of biological systems. By uniting genomics, transcriptomics, proteomics, and metabolomics, scientists can examine numerous layers of biological findings in parallel. This strategy illuminates how genetic variabilities affect gene expression, protein synthesis, and metabolite abundance, and how these changes translate into observable plant characteristics (Liu *et al.*, 2022; Wan *et al.*, 2022).

Reveal intricate networks and regulatory pathways

By integrating multi-omics datasets, scientists can unveil the intricate networks and control pathways driving plant phenotypes. For example, linking genomic information with transcriptomic, proteomic, or metabolomic profiles enables the identification of how DNA sequence alterations influence gene expression, protein synthesis, and metabolite concentrations. Such insights shed light on the mechanisms by which genetic variants give rise to observable traits (Fan *et al.*, 2022; Jaramillo-Botero *et al.*, 2022; Chang *et al.*, 2023).

Improve prediction power

Combining multi-omics data strengthens the predictive capabilities of models forecasting plant characteristics. By drawing on multiple omics fields, these models capture the interaction among diverse biological practices, resulting in more improved predictions (Hu *et al.*, 2021).

Facilitate precision breeding

Integrating multi-omics data supports precision breeding by pinpointing molecular markers linked to particular plant characteristics across diverse biological layers, enabling breeders to make more targeted selections and develop superior plant varieties (Jiang *et al.*, 2022; Roychowdhury *et al.*, 2023).

Enhance insights into interactions between plant and environment

By integrating multi-omics datasets, researchers can deepen their insights into plant–environment interactions. This approach reveals how environmental stimuli can shape gene expression dynamics, protein synthesis, and metabolic pathways, ultimately influencing plant developmental processes and stress (biotic and abiotic) responses (Shen *et al.*, 2023; Zhang *et al.*, 2023).

Facilitate disease identification and management

By offering an all-encompassing perspective on plant biological systems, multi-omics research supports the detection and control of plant pests and diseases. For instance, it pinpoints molecular markers linked to the genes resistant to disease, steers the breeding of resistant cultivars, and shapes effective control methods (Zapletalova *et al.*, 2023).

2.2.2 Exploring the role of multi-omics in specific molecular pathways in plants

Leveraging multi-omics alongside AI in plant science allows for detailed investigation and comprehension of the intricate molecular pathways that govern plant growth, disease resistance, and responses to stress.

Plant growth and development

The intricate network of genes and their interactions governs plant growth and development. Employing multi-omics approaches lets us delve deeper into these molecular processes and identify the principal contributors.

GENOMIC STUDIES. Genomics delivers a complete portrait of a plant's genetic blueprint, revealing essential genes that drive growth and development. For example, members of the auxin signaling pathway are vital for cell elongation, and organ formation can be recognized and their DNA sequences examined. Genetic variants, such as SNPs or INDELs in these genes, can then be attached to observable morphological differences, enriching our knowledge about plant form and development (Igiehon *et al.*, 2019; Jaiswal *et al.*, 2020; Li *et al.*, 2023).

TRANSCRIPTOMIC STUDIES. A transcriptomic study is done based on genomics by analyzing gene expression dynamics, revealing the time and place of specific genes' activation or inactivation throughout the life cycle of a plant. This adds a temporal and spatial dimension to our knowledge about development. For instance, RNA-seq can track changes in auxin pathway gene expression across various growth stages or in response to environmental cues (Hu *et al.*, 2020; Kashyap *et al.*, 2020).

METABOLOMIC STUDIES. A metabolomics study is conducted based on genetic and transcriptomic analyses by examining the metabolic shifts that occur during plant growth and development. It enables the detection and quantification of countless metabolites, uncovering which biochemical processes are active at different developmental stages. For example, metabolomic profiling can track fluctuations in auxin and related compounds throughout development, offering concrete indicators of physiological changes (Kang *et al.*, 2019; Raza, 2022).

Disease resistance pathways

Plants have developed numerous, intricate resistance pathways to defend against a wide array of pests and diseases. These systems encompass a multitude of genes, proteins, and metabolic compounds, and multi-omics strategies offer an essential toolkit for dissecting these processes at the molecular level.

GENOMIC STUDIES. Resistance (R) genes play a central role in plant defense pathways. Genomic analysis of variations (for example, SNPs and INDELs) within these genes can reveal a plant's inherent capacity to withstand particular diseases (Cai *et al.*, 2020; Yu *et al.*, 2020).

TRANSCRIPTOMIC STUDIES. Transcriptomic analyses, particularly RNA-seq, enable us to track R-gene activation in response to pathogen damage. By quantifying R-gene transcript levels after exposure to various pathogens, we can map the timing and strength of resistance pathway induction and identify which pathogens trigger those defenses (Sun *et al.*, 2019; Zhao *et al.*, 2022).

PROTEOMIC STUDIES. Proteomic analyses offer insights into the post-transcriptional and post-translational alterations that R proteins experience during pathogen attack. These alterations can modulate an R protein's function and activity. By recognizing both the altered proteins and their specific alterations, proteomics reveals the mechanisms through which R proteins mediate disease resistance (Wu *et al.*, 2019; Nawaz *et al.*, 2020)

METABOLOMIC STUDIES. When challenged by pathogens, plants produce an array of defensive metabolites. Metabolomics allows for the detection and quantification of these compounds, such as phytoalexins produced in response to microbial invasion. By profiling these metabolites, researchers capture an overview of the metabolic state of the plant under attack, enriching our knowledge about the biochemical dimensions of plant defense (Sharma *et al.*, 2021; Ncube *et al.*, 2022)

Stress-response pathways

Stationary by nature, plants encounter a wide range of environmental stresses such as salinity, drought, and extreme temperatures, and decoding their molecular responses is key to bolstering crop resilience. Multi-omics strategies offer a thorough toolkit for understanding these complex stress-response pathways:

GENOMIC STUDIES. Genomics enables the discovery of genes involved in stress reactions. For example, researchers have identified numerous genes linked to drought, salinity, and temperature tolerance across different plant species. Many of these genes encode transcription factors that regulate other stress-responsive genes. Examining sequence and structural differences in these genes, one can predict the capacity of the plant to endure various environmental challenges (Kopriva *et al.*, 2019; Sahoo *et al.*, 2020).

TRANSCRIPTOMIC STUDIES. Transcriptomic analysis can monitor how stress-responsive genes are regulated under various situations. For example, RNA-seq can detect the up- or down-regulation of particular genes when plants face abiotic stresses such as drought, salinity, or temperature, offering a dynamic portrait of transcriptome alterations in response to environmental challenges (Xu *et al.*, 2019; Nefissi Ouertani *et al.*, 2021)

PROTEOMIC STUDIES. Proteomics augments genomic and transcriptomic analyses by revealing stress-responsive proteins. For example, specific proteins may be upregulated under stress to shield plant cells from harm. Through proteomic profiling, these proteins can be identified and their abundance tracked across various stress conditions, offering a clear picture of the plant's protein-level response to environmental challenges (Wang *et al.*, 2022).

METABOLOMIC STUDIES. Metabolomics enriches our understanding by characterizing metabolic shifts under stress. In response to adverse conditions, plants accumulate specific metabolites to play protective roles, such as osmolytes during drought and salinity or heat-shock proteins under temperature stress. Identifying these compounds uncovers stress-induced biochemical adaptations, delivering a complete picture of the plant's metabolic defense strategy (You *et al.*, 2019; Muhammad Aslam *et al.*, 2022).

2.3 Application of Artificial Intelligence (AI) in Plant Science

Plant growth is affected by many factors, including soil conditions, insect and pest attacks, changing climatic conditions, etc. The yield of the plant depends on these complex factors. To maximize yield, there is a necessity for large data analysis using artificial intelligence (AI) techniques for bridging farmers' knowledge gaps (Eli-Chukwu, 2019). AI is advantageous in terms of its accuracy, flexibility, high performance, and cost-effectiveness (Fig. 2.1). AI is a field of science that deals with machines or computers that perform activities like human beings. Under the umbrella of AI, deep learning (DL) and machine learning (ML) act as subsets that help teach the machines and train ML models. AI applications in plant science involving various fields are listed below.

2.3.1 AI application in crop management

Crop management starts from the sowing of seeds until the harvesting, storage, and market distribution. Knowing the crop sowing time and input requirements can increase the yield. Precision agriculture is based on the crop and soil input requirements to reduce the wastage of input, protect the soil quality, and optimize the profit (Padhiary *et al.*, 2025). Changing environmental conditions also affect plant growth and development. The prediction of drought incidence is important for minimizing losses to farmers due to changing weather patterns. Therefore, understanding the weather patterns will support farmers in decision making that will

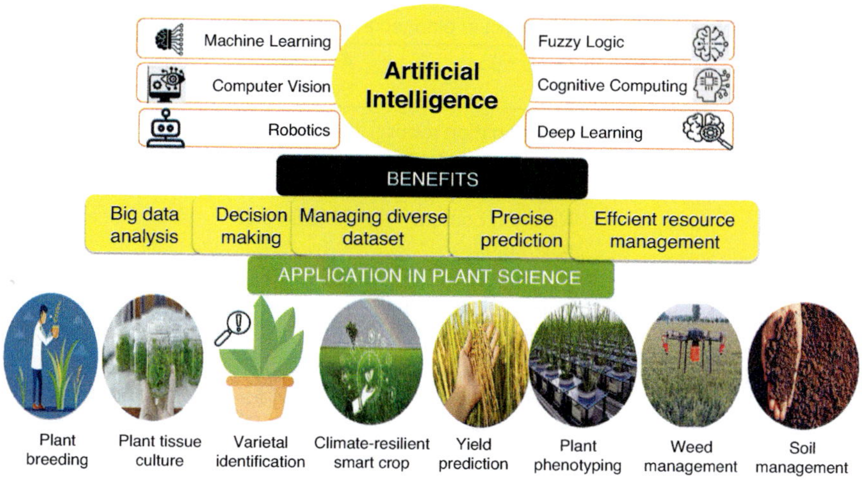

Fig. 2.1. Benefits and applications of artificial intelligence (AI) in plant science (Author's own image)

add to the high-quality crop yield. Some of the AI techniques for crop management are described in Table 2.1, namely, ANN, PROLOG, ROBOTICS, FUZZY Cognitive Map, ANN along with Fuzzy logic, etc. The Artificial Neural Network (ANN) was used for predicting crop yield and detecting crop nutrition disorder (Song and He, 2005; Dahikar and Rode, 2014). PROLOG was used for location-specific selection of useful farm tools and removal of less-used farm tools (Lal *et al.*, 1992). ROBOTICS was used for harvesting crops and resulted in an 80% success rate (van Henten *et al.*, 2002). FUZZY Cognitive Map was used to predict cotton yield and helped in decision management (Papageorgiou *et al.*, 2011). ANN, along with Fuzzy logic, helped to reduce the prevalence of insects (Yang *et al.*, 2003). The detection of nutrient status and disease incidence is another area where AI is applicable. Convolutional neural network (CNN) was used for predicting the deficiency of magnesium and incidence of powdery mildew (Dissanayake *et al.*, 2023).

2.3.2 AI application in plant breeding

AI algorithms are used in the processing of massive datasets that help farmers in decisions related to the specific needs of their crops. Murmu *et al.* (2024) used AI along with the omics techniques that help in understanding the complex gene regulatory pathways. AI applications help in analyzing big datasets, extracting valuable conclusions, decision making, accurate data management, and precise prediction. AI-driven techniques like ML, DL and robotic techniques are involved in predicting superior genotypes based on the phenotypic outcomes by analyzing the large datasets. Random forest (RFs) are ML models that work on supervised algorithms and are used for demonic data analysis (Montesinos López *et al.*, 2022). Support vector machines (SVMs) and ML models are used for varietal classification based on genetic markers analysis (Gutierrez *et al.*, 2015). Gradient boosting algorithms (GBAs) are used for classification tasks and regression analysis and are based on ML algorithms (Yan *et al.*, 2021). DL is classified as a subset of ML and is used for analyzing multidimensional and complex information. Artificial neural networks (ANNs), convolutional neural networks (CNNs), and recurrent neural networks (RNNs) are some of the most commonly used tools in deep learning. CNNs can analyze images or sequences, plant images that help to predict disease conditions or any growth anomalies (Karahan and Nabiyev, 2021). RNNs are used to analyze sequential data depending on time, for example changes in growth patterns over a time period under changing climatic conditions (Mohmed *et al.*,

Table 2.1. Application of artificial intelligence (AI) in plant science (Author's own table)

AI-based tool	Crop	Application	Accuracy/ Success rate	Reference
ROBOTICS	Tomato	As pollination robot system for hormone treatment	69.6%	Yuan *et al.*, 2016
ANN	Rice	Nitrogen content estimation	RMSE = 0.32	Afandi *et al.*, 2016
FUZZY Cognitive Map	Cotton	Decision management and prediction of cotton yield	67–75%	Papageorgiou *et al.*, 2011
ANN and Fuzzy logic	Corn	Weed identification and precision herbicide application	52–66% for corn, 76–85% for weed	Yang *et al.*, 2003
CNN	Bell pepper	Predicting the deficiency of magnesium and the incidence of powdery mildew	97%	Dissanayake *et al.*, 2023
RNNs	Lettuce	Analyzing changes in growth patterns under varying environmental conditions	Minimum MSE (0.18, 0.02 and 1.32 for modeling shoot fresh wt, shoot dry wt and leaf area)	Mohmed *et al.*, 2023
ANNs	Capsicum	Identification of plant species based on morphological traits	100%	Ferreira *et al.*, 2017
GRNN	*Cannabis sativa* L.	Optimization of *in vitro* shoot growth based on changing light and carbohydrate concentration	Prediction accuracy was higher compared with MLP and ANFIS	Pepe *et al.*, 2021
Machine vision and ANN methods	Potato	Identification of varieties	100%	Azizi *et al.*, 2016
Automatic Rice Variety Identification System	Rice	Variety Identification	93%	Komal *et al.*, 2022
Computer-assisted image analysis and ANNs	Wheat	Varietal identification	84–94%	Fumia *et al.*, 2023
CNN and Inception-ResNet	Tomato	Yield prediction	91%	Rahnemoonfar and Sheppard, 2017
AMULET	*Arabidopsis thaliana*	Next-generation high throughput phenotyping	IoU score 0.9948	Zdrazil *et al.*, 2025

Continued

Table 2.1. Continued

AI-based tool	Crop	Application	Accuracy/ Success rate	Reference
Naïve Bayes algorithm	Cotton, blackgram, oilseeds, pigeonpea	Weed management	89.29%	Dasgupta *et al.*, 2020
ANN	Rice	Effect of nutrient on grain yield	R^2 0.997	Shankar *et al.*, 2022

ANN: artificial neural network; CNN: convolutional neural network; RNN: regression neural networks; GRNN: generalized regression neural metworks; AMULET: adaptable and multi-task machine learning for endless phenotyping trait prediction; RMSE: root mean square error; MLP: multi-layer perceptron; ANFIS: adaptive neuro-fuzzy inference system; IoU: Intersection over Union.

2023). ANNs are based on the complex nature of animal and human neural networks that help in the identification of plant species based on morphological traits (Ferreira *et al.*, 2017). In this way, the selection of superior crop varieties can be achieved by using ML and DL tools. The traditional breeding method is based on the selection of improved traits like yield, shelf-life, taste, nutritional quality, and resistance to biotic and abiotic stresses. AI applications help in the selection of these traits that can be used in breeding programs.

The traditional breeding strategies depend on the breeder's experience, accuracy, and efficiency. Manual phenotyping can collect superficial observations, but fails to capture complex interactions at subcellular levels, which affects the overall phenotyping process (Diaz-Garcia *et al.*, 2016). Traditional disease identification methods rely on time-consuming visual inspections during early stages. Furthermore, classic genomic prediction approaches incur problems such as low prediction accuracy and having more markers than observations (Barbosa *et al.*, 2021). Therefore, AI algorithms can help analyze omics data and phenotypic trait relationships, disease detection, and biotic and abiotic stress resilience for crop improvement.

2.3.3 AI application in plant cell and tissue culture

In vitro micropropagation of the plant depends on complex processes that involve many factors that need to be optimized for different plant species. Micropropagation of drug-type cannabis involves the production of high-quality secondary metabolites from the cultured biomass. The optimization of culture media components, such as plant growth regulators and minerals, is useful in this area; however, it is a time-consuming process. Therefore, high-throughput AI techniques are used in the selection and optimization of culture conditions to maximize productivity. Generalized regression neural networks (GRNNs) are used in modelling and predicting the *in vitro* culture conditions (Malabadi *et al.*, 2023). Optimization of culture conditions, namely relative humidity, temperature, light duration, and agar concentration, was predicted for the differentiation rate of melon *in vitro* conditions using a genetic algorithm and ANN that resulted in high accuracy for prediction (Zhang *et al.*, 2020b). Predicting and optimizing cell growth, secondary metabolite production, embryo classification, shoot length, shoot proliferation, hairy root culture, etc., are the areas where the AI models and algorithms have been used (Hesami and Jones, 2020)

2.3.4 AI application for genomic selection/varietal identification

Varietal identification is important for breeding and crop improvement. The identification of potato breeds used five deep learning approaches, namely ResNet50, Inception-v3, Mobile-Net, VGG16, and a customized CNN. These models used visual characteristics unique to each breed, namely, texture, color, shape, size, and skin pattern. The deep learning models were trained on images of different potato breeds. Among the examined models, the customized CNN model outperformed all the other models with an accuracy of 94.84% (Rahman *et al.*, 2024). Azizi *et al.* (2016) developed a method for identifying potato varieties by using machine vision and ANN methods. A total of ten potato varieties and 72 characters, including texture, color, and morphological features, were selected for analysis. The combined method of machine vision and ANN methods had acceptable accuracy for the identification of potato varieties. The Automatic Rice Variety Identification System was developed based on the image dataset containing 9692 images of rice varieties cultivated in the Punjab (India) area. This prototype of the identification system was tested using k-NN, a k-nearest neighbor classifier that showed 93% accuracy (Komal *et al.*, 2022). Computer-assisted image analysis and ANNs can be used for varietal identification of wheat with a classification accuracy of 84–94% for separate varieties (Dubey *et al.*, 2006). The metabolomics data were combined with a machine learning approach to characterize and classify ten new races of *Capsicum annuum* cv. *Jalapeno*. A total of 48 metabolites were identified with NMR metabolomics profiling. The random forest model was used to predict the diversity and similarity of the experimental and established races by using variables, namely yield, length, width, and organic acid and xylose content (Ramírez-Meraz *et al.*, 2020).

2.3.5 AI application for climate-resilient smart crop breeding

Identification of resilient genotypes against different environmental stresses could be beneficial for climate-resilient smart crop breeding. In this area, AI-assisted germplasm selection using genomics and phenomics data could be beneficial. Recently, the selection of heat-tolerant capsicum genotypes was achieved by using k-mean clustering analysis for anomaly analysis combined with a random forest model to classify the heat-tolerant genotypes (Fumia *et al.*, 2023). Light, humidity, temperature, and CO_2 are all components of the environment that affect plant growth and development. Unfavorable environmental conditions cause stress symptoms in the plant and these symptoms can be detected in the early growth stages of the plant. Image-based study and the use of ML models could be beneficial in the detection of stress symptoms in the early growth stages of plants. Islam *et al.* (2024) used SVM and image processing for the classification of the early-stage stress system in pepper plants that could be used for real-time stress monitoring and optimizing environmental conditions by the growers. For the detection of stress conditions experienced by the plants in the field, the combination of deep learning models along with the hyperspectral imaging system helped in the detection of drought-tolerant cultivars (Lapajne *et al.*, 2024). Machine learning models, namely random forest, Ada Boost, Extra Tree, and Extreme Gradient Boosting, were used for the classification of potato varieties based on variability for drought stress tolerance. Parameters in observations included stem mass, maturity, leaves mass, assimilation area, yield per plant, etc. (Boguszewska-Mańkowska *et al.*, 2022). The productivity of crops is influenced by the incidence of leaf diseases. Crop losses can be minimized by controlling their spread. CNNs have been used for the detection of plants by using potential images. Pasalkar *et al.* (2023) used an image-based CNN model for the detection of early blight and late blight disease in potatoes, which achieved 97.4% accuracy for leaf disease identification and can be used for the early detection of diseases and management.

2.3.6 AI application in yield prediction and monitoring productivity

Yield is a complex attribute that is influenced by both genetics and the environment. ML in modern crop breeding helps in the identification of traits and precise phenotyping for yield prediction. ANNs were used to predict the tomato yield in greenhouse cultivation and determine the most important inputs influencing tomato production (Belouz *et al.*, 2022). Here, the ANN model was compared with the multiple linear regression (MLR) technique, and the results showed that ANN predictions were more accurate. Furthermore, the sensitivity analysis resulted in potassium, nitrogen, farmyard manure, fungicides, and electricity as the most significant inputs for greenhouse tomato cultivation. An ML approach was used to predict the genetic control of yield attributes in F_2 to F_5 generations of interspecific hybrids of tomato. ANNs and MLR identified three traits, namely plant height, test weight of seeds, and fruit weight, as the most important traits for fruit yield that can be used in the identification of superior tomato segregates (Acharya *et al.*, 2024). An augmented reality-based mobile application that used region-based convolutional neural network (RCNN) was implemented for counting, grading, and yield prediction. This mobile application obtained images of the pre-harvest stage of a tomato farm. The model was trained and tested using 2083 images for the classification of tomatoes (Balaji Prabhu *et al.*, 2024). Yield prediction based on manual counting of fruits and flowers is time-consuming and difficult to apply to large fields. A deep learning technique based on convolutional neural networks and Inception (ResNet) was trained using 24,000 images and tested on 2400 images. The results showed 91% and 93% accuracy on real and synthetic images, respectively (Rahnemoonfar and Sheppard, 2017). The tomato yield in greenhouse cultivation was compared in two conditions, namely AI-supported and human-operated conditions. The strategies for optimizing lighting facilities were a better strategy than optimizing temperature and CO_2 (Hemming *et al.*, 2020). Simulation-based breeding models using genomic information and phenotypic data were used for the improvement of the yield and flavor of tomato (Yamamoto *et al.*, 2016). The prediction of yield was done using five different ML methods in eggplant, namely ANN, random forest (RF), k-nearest neighbor (k-NN), support vector machine (SVM), and adaptive boosting (AB) (Taşan *et al.*, 2022). The yield and quality model was developed using ANN, which showed that phosphorus concentration was the more important factor, followed by Zn and available Fe in soil, toward the yield prediction. The prediction of crop potential based on the nutritional composition of potato leaf tissue was performed using machine learning approaches, namely k-nearest neighbor, support vector machine algorithms and random forest. The ionome data were used to correlate with the high- and low-yield classes (Coulibali *et al.*, 2020). The selection of appropriate seed cultivars is important for the breeding program; however, small seed sizes and visual similarities make this task difficult. The application of CNN models to classify pepper seeds of different cultivars will be helpful in cultivar differentiation (Sabanci *et al.*, 2022).

2.3.7 AI application in next-generation plant phenotyping

Plant phenotype is a complex interplay of genes and the environment. Plant phenotyping emerged as a high-throughput technique for precision breeding when combined with AI features. It enabled efficient data collection and analysis. It is used for data management, model development, genotype–phenotype–environment interaction studies, etc. The AI-based phenotyping was used for root yield prediction in cassava, identifying the growth stage in wheat and maize, plant segmentation, segmentation of crop and weeds in oilseeds, productivity prediction, and disease resistance prediction, etc. (Brichet *et al.*, 2017; Sadeghi-Tehran *et al.*, 2017; Mahlein *et al.*, 2019; Abdalla *et al.*, 2019; Mochida *et al.*, 2019; Selvaraj *et al.*, 2020; Jung *et al.*, 2021). Adaptable and Multi-task Machine Learning for Endless Phenotyping Trait Prediction is a new high-throughput phenotyping approach that combines phenotyping and machine learning for predicting physiological and morphological traits in *Arabidopsis thaliana*. The results can be utilized to predict plant health status, stress responses, growth, and new biomarkers (Zdrazil *et al.*, 2025).

2.3.8 AI application in weed management for maintaining plant health

Weeds compete with the plant for nutrients, water, and other resources in the field. They cause up to 50% loss in the yield of corn and beans in the field, 48% in wheat, up to 55% in soybean, and 50–75% in sesame (Johansen *et al.*, 2020; Gadade and Kirange, 2021; Tan *et al.*, 2021; Bupi *et al.*, 2023; Chowdhury *et al.*, 2023; de Oliveira Dias *et al.*, 2023). Therefore, weed management is an important aspect of the better performance of the plants in the field. Different AI techniques used in weed management include inverse weed optimization (IWO), ROBOTICS, learning vector quantization, support vector machines (SVMs), digital image analysis (DIA), unmanned aerial vehicles (UAVs), etc. The detection of weeds and unwanted plants is possible through the application of drones and deep learning techniques. Naïve Bayes algorithm was used for the detection of weeds, which resulted in 89.29% accuracy that helped in weed management and reduced the harmful effects on crops due to weeds (Dasgupta *et al.*, 2020). Smart sprayers were developed by using machine learning and AI, which can easily distinguish between target and non-target and result in precise application for weed management. This resulted in fast, precise and reduced agrochemicals spraying, which could reduce costs, chances of crop damage, and herbicide residue problems (Partel *et al.*, 2019). Unmanned vehicles (UMVs) and machine learning (ML) were used for weed management that can contribute to sustainable, efficient, and economic agricultural practices (Ehrampoosh *et al.*, 2025).

2.3.9 AI application in soil management for maintaining plant health

Soil management is an important parameter for determining plant health. Knowledge related to the soil types and conditions, control of soil pollutants, use of organic fertilizers, control of soil-borne pathogens, alternative tillage operations, etc., is an essential parameter for soil management. There are different AI techniques involved in soil management, namely artificial neural network (ANN), Soil Risk Characterization Decision Support System (SRC-DSS), Management-Oriented Modeling (MOM), etc. MOM was used to lower nitrate leaching conditions and maximize production (Li and Yost, 2000). SRC-DSS was used for risk-based classification of soil (López *et al.*, 2008). ANN was utilized to estimate soil-based enzyme activity, moisture content, temperature, texture, structure, and soil nutrient status after erosion (Levine *et al.*, 1996; Behrens *et al.*, 2005; Elshorbagy and Parasuraman, 2008; Zhao *et al.*, 2009; Bilgili, 2011; Tajik *et al.*, 2012). AI is used in soil science for modeling soil processes, soil phenotyping, precision and efficient soil management (Minasny and McBratney, 2025).

2.4 An Integrated Approach of AI and Multi-omics in Predicting Plant Traits

The integration of artificial intelligence (AI) with multi-omics data promises a breakthrough approach to understanding and forecasting complex plant features (Fig. 2.2). This growing discipline of AI, with integration of multi-omics, helps researchers to unravel the intricate biological networks driving agricultural key features (Table 2.2). There are several sub-components under AI and multi-omics domains mentioned below, which support the scientific community working in the area of plant biology and its associated domain to decipher more insights in the world of plant biotechnology (Subramanian *et al.*, 2020; Wieme *et al.*, 2022). In multi-omics domain, genomics covers the elucidation of structure-based genomics and variations in DNA sequences; transcriptomics elucidates the patterns of gene expression in various contexts; proteomics elucidates the quantity of proteins and post-translational changes; metabolomics elucidates the biochemical processes and metabolites of small molecules; and phenomics elucidates about high-throughput assessments of phenotype; while on the other hand, in cases of AI and ML domain, deep learning (DL) helps in elucidation of pattern recognition architectures such as CNNs, RNNs, and Transformers); ensemble method provides information on forecasts from several models; dimensionality reduction method assists in handling of high-dimensional data in omics; and lastly integration of all the AI helps to decipher the selection of model choices and uncover the hidden biological information (Ravisankar and Mathew, 2022; Walker *et al.*, 2022).

2.5 Challenges in the Application of the AI-Based Multi-omics Approach

AI applications have been acknowledged as a significant advancement in the area of plant

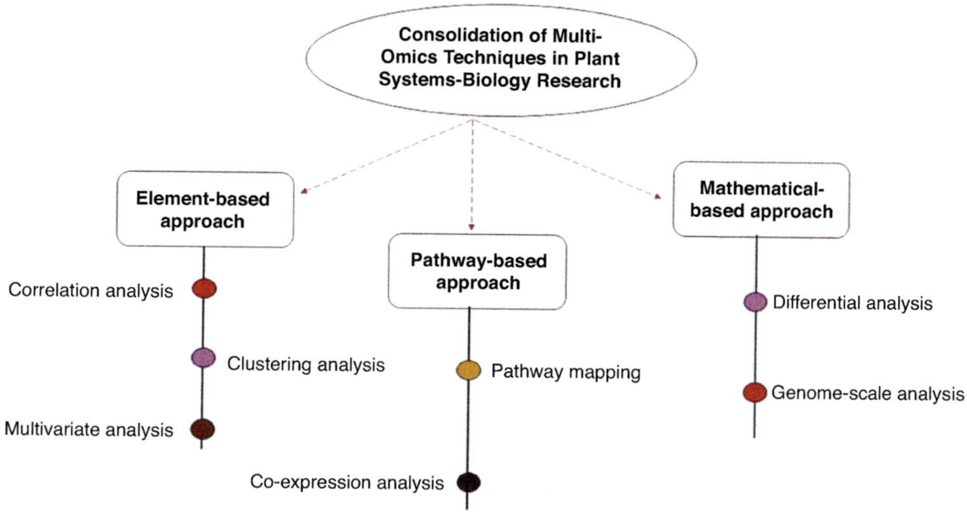

Fig. 2.2. Multi-omics approaches in trends for investigating various research aspects in plant systems biology (Modified from Jamil *et al.*, 2020).

science. However, the effectiveness of these applications is dependent on their accuracy and response time to generate data. Users will adopt such a system with minimum effort and maximum accuracy. Therefore, these two factors determine the use of AI in agriculture. For the performance of a task related to the prediction of plant health status, weed status, cultivar identification, soil management, phenotype, response to environmental stressors, etc., large input data are required to be analyzed by AI systems. A large dataset refers to annotated dataset generated from the interaction of genotypic, phenotypic, and environmental factors (Sun *et al.*, 2024). The high data cost for the farmers living in rural areas can be a limitation, whereas government schemes can be formulated to enable these services for the farmers. The requirement of advanced infrastructure, high-performance computing devices and high computational cost may also pose limitations to the application of AI in remote areas (Jeon *et al.*, 2023). Despite the many advantages of AI, several challenges hinder its potential; quality data and its availability are the most important among them.

Another challenge is the interpretation of model predictions. Lack of understanding and transparency results in limited practical application of AI tools (Graziani *et al.*, 2023). Interpretable and accessible AI-based models are helpful in this area. FAIR (Findable, Accessible, Intelligent, and Reproducible) solutions are available for those lacking computational understanding to aid in disease prediction and biomarker development (Arefolov *et al.*, 2021; Psomopoulos, 2024). SLIDE (Significant Latent Factor Interaction Discovery and Exploration) is an interpretable machine learning-based technique that overcomes some of the challenges related to the interpretability, transparency, and explainability (Rahimikollu *et al.*, 2024). An interdisciplinary approach is required among the data scientists, agronomists, and geneticists to understand the underlying mechanism in the biological system so that the model can be used in the new datasets (Piorkowski *et al.*, 2021). Enhanced model interpretability can be achieved through automated machine learning (AutoML) and explainable AI techniques that will help in increasing usability and trust in the AI-based prediction models (Malounas *et al.*, 2024). The huge and heterogeneous nature of multi-omics data requires expertise to implement and interpret the AI approaches. Omics data may contain missing data, sparse data and outliers, which may require expertise in analyzing such data by using AI tools. AI algorithms require a transparent and explainable method to be applied to complex multi-omics data.

Table 2.2. Correlation of conventional and modern phenotyping techniques (Author's own table)

Sl. No.	Method	Details	Benefits	Constraints	Reference
1.	Conventional phenotyping	Plant characteristics like height, bloom color, fruit size, and disease signs are measured manually and visually.	1. Simple and economic. 2. No need for high-skilled training or instruments.	1. Requires a lot of work and time. 2. Usually helps in collection of surface-level characteristics. 3. Subjective, with the possibility of mistakes and inconsistent results.	Diaz-Garcia *et al.*, 2016; Colaço *et al.*, 2018; Li *et al.*, 2021b
2.	Multi-omics-based phenotyping	This covers genomics, transcriptomics, proteomics, and metabolomics.	1. Offers a thorough comprehension of molecular biology aspects. 2. Provides details about intricate characteristics and mechanisms.	1. Requires specific training and equipment. 2. Because multi-omics data is so large and complicated, data analysis can be challenging.	Subramanian *et al.*, 2020; Zhang *et al.*, 2022; Cao *et al.*, 2022
3.	Artificial intelligence and machine learning-based phenotyping	This category includes analyzing and interpreting data in the form of graphs, pictures, and spectra using machine learning techniques.	1. Able to manage substantial amounts of intricate data. 2. Offers unbiased and reliable analyses. 3. Reveals links and patterns that people cannot see.	1. Requires a significant computing knowledge. 2. Choosing and interpreting models might be difficult. 3. Characteristics of some ML algorithms may cause problems with trust and transparency.	Rathore *et al.*, 2021; Wieme *et al.*, 2022; Ravisankar and Mathew, 2022; Walker *et al.*, 2022

Working in collaboration with other organizations and disciplines may help in the translation of the AI algorithm. The common challenge in AI-mediated multi-omics data analysis is related to the generalization step. The AI model works well with the data used for training, but it becomes difficult to predict new data (Ahmed *et al.*, 2024).

2.6 Conclusion

Multi-omics and artificial intelligence constitute two cutting-edge technologies poised to transform the field of plant science. Their convergence promises unparalleled opportunities to understand the intricate workings of plant biological structure and the creation of highly precise, projecting models of plant systems. AI-driven plant phenotyping enables the extraction of critical traits by pinpointing the features most closely linked to desirable characteristics, thereby accelerating the development of superior crop varieties. When these AI techniques are integrated with multi-omics datasets, they form a comprehensive framework that captures the full complexity of plant biology and yields highly accurate phenotype predictions, addressing the shortcomings of traditional methods that struggle to encompass such biological intricacy.

The rise of genomics, transcriptomics, proteomics, and metabolomics, which are together known as multi-omics, offers a deeply integrative perspective on plant biology. When paired with AI's analytical prowess, this combined approach becomes a potent framework for handling the sheer scale and intricacy of multi-omics datasets, delivering actionable insights and predictive models that could transform agricultural science. Yet, deploying such an AI-powered, multi-omics pipeline brings its own set of hurdles, spanning technical issues like data harmonization and computational resource demands, as well as non-technical concerns such as data governance, ethical considerations, and cross-disciplinary collaboration. Despite significant progress, substantial technical hurdles persist, from the capture and harmonization of diverse datasets to the selection, deployment, and interpretability of AI models. Equally critical are the ethical, legal, and educational frameworks that must underpin these efforts. Looking forward, however, the fusion of AI with multi-omics promises transformative advances: precision agriculture driven by real-time analytics; predictive breeding strategies; more effective management of diseases and environmental stressors; sustainable agricultural practices; deeper understanding of plant ecology and biodiversity; and the scaling-up of cutting-edge agricultural research.

References

Abdalla, A., Cen, H., Wan, L., Rashid, R., Weng, H. *et al.* (2019) Fine-tuning convolutional neural network with transfer learning for semantic segmentation of ground-level oilseed rape images in a field with high weed pressure. *Computers and Electronics in Agriculture* 167, 105091.

Acharya, B., Kumar, P.P., Hazra, S., Dutta, S., Saha, S. *et al.* (2024) Genetic control of important yield attributing characters predicted through machine learning in segregating generations of interspecific crosses of tomato (*Solanum lycopersicum* L.). *Acta Physiologiae Plantarum* 46(8), 78.

Aditya Shastry, K. and Sanjay, H.A. (2019) Data analysis and prediction using big data analytics in agriculture. In: *Internet of Things and Analytics for Agriculture*, Vol. 2. Springer, Singapore, pp. 201–224.

Afandi, S.D., Herdiyeni, Y., Prasetyo, L.B., Hasbi, W., Arai, K. *et al.* (2016) Nitrogen content estimation of rice crop based on near infrared (NIR) reflectance using artificial neural network (ANN). *Procedia Environmental Sciences* 33, 63–69.

Ahmed, U., Xie, Q., Shi, X. and Zheng, B. (2022) Development of reference genes for horticultural plants. *Critical Reviews in Plant Sciences* 41(3), 190–208.

Ahmed, Z., Wan, S., Zhang, F. and Zhong, W. (2024) Artificial intelligence for omics data analysis. *BMC Methods* 1(1), 4.

Arefolov, A., Adam, L., Brown, S., Budovskaya, Y., Chen, C. *et al.* (2021) Implementation of the FAIR data principles for exploratory biomarker data from clinical trials. *Data Intelligence* 3(4), 631–662.

Azizi, A., Abbaspour-Gilandeh, Y., Nooshyar, M. and Afkari-Sayah, A. (2016) Identifying potato varieties using machine vision and artificial neural networks. *International Journal of Food Properties* 19(3), 618–635.

Bai, Y., Liu, H., Lyu, H., Su, L., Xiong, J. *et al.* (2022) Development of a single-cell atlas for woodland strawberry (*Fragaria vesca*) leaves during early *Botrytis cinerea* infection using single-cell RNA-seq. *Horticulture Research* 9, uhab055.

Balaji Prabhu, B.V., Shashank, R., Shreyas, B. and Narsipura, O.S.J. (2024) ARIA: Augmented reality and artificial intelligence enabled mobile application for yield and grade prediction of tomato crops. *Procedia Computer Science* 235, 2693–2702.

Barbosa, I.D.P., da Silva, M.J., da Costa, W.G., de Castro Sant'Anna, I., Nascimento, M. *et al.* (2021) Genome-enabled prediction through machine learning methods considering different levels of trait complexity. *Crop Science* 61(3), 1890–1902.

Bayes-Genis, A., Liu, P.P., Lanfear, D.E., de Boer, R.A., González, A. *et al.* (2020) Omics phenotyping in heart failure: The next frontier. *European Heart Journal* 41(36), 3477–3484.

Behrens, T., Förster, H., Scholten, T., Steinrücken, U., Spies, E.D. *et al.* (2005) Digital soil mapping using artificial neural networks. *Journal of Plant Nutrition and Soil Science* 168(1), 21–33.

Belouz, K., Nourani, A., Zereg, S. and Bencheikh, A. (2022) Prediction of greenhouse tomato yield using artificial neural networks combined with sensitivity analysis. *Scientia Horticulturae* 293, 110666.

Bilgili, M. (2011) The use of artificial neural networks for forecasting the monthly mean soil temperatures in Adana, Turkey. *Turkish Journal of Agriculture and Forestry* 35(1), 83–93.

Boguszewska-Mańkowska, D., Ruszczak, B. and Zarzyńska, K. (2022) Classification of potato varieties drought stress tolerance using supervised learning. *Applied Sciences* 12(4), 1939.

Brichet, N., Fournier, C., Turc, O., Strauss, O., Artzet, S. *et al.* (2017) A robot-assisted imaging pipeline for tracking the growths of maize ear and silks in a high-throughput phenotyping platform. *Plant Methods* 13, 1–12.

Bupi, N., Sangaraju, V.K., Phan, L.T., Lal, A., Vo, T.T.B. *et al.* (2023) An effective integrated machine learning framework for identifying severity of tomato yellow leaf curl virus and their experimental validation. *Research* 6, 0016.

Cai, Y., Cai, X., Wang, Q., Wang, P., Zhang, Y. *et al.* (2020) Genome sequencing of the Australian wild diploid species *Gossypium australe* highlights disease resistance and delayed gland morphogenesis. *Plant Biotechnology Journal* 18(3), 814–828.

Cao, P., Zhao, Y., Wu, F., Xin, D., Liu, C. *et al.* (2022) Multi-omics techniques for soybean molecular breeding. *International Journal of Molecular Sciences* 23(9), 4994.

Cembrowska-Lech, D., Krzemińska, A., Miller, T., Nowakowska, A., Adamski, C. *et al.* (2023) An integrated multi-omics and artificial intelligence framework for advance plant phenotyping in horticulture. *Biology* 12(10), 1298.

Chandran, H., Meena, M. and Sharma, K. (2020) Microbial biodiversity and bioremediation assessment through omics approaches. *Frontiers in Environmental Chemistry* 1, 570326.

Chang, Y., Gong, W., Xu, J., Gong, H., Song, Q. *et al.* (2023) Integration of semi-*in vivo* assays and multi-omics data reveals the effect of galloylated catechins on self-pollen tube inhibition in *Camellia oleifera*. *Horticulture Research* 10(1), uhac248.

Chen, G., Ning, B. and Shi, T. (2019) Single-cell RNA-seq technologies and related computational data analysis. *Frontiers in Genetics* 10, 317.

Chowdhury, R.H., Eti, F.S., Ahmed, R., Gupta, S.D., Jhan, P.K. *et al.* (2023) Drought-responsive genes in tomato: Meta-analysis of gene expression using machine learning. *Scientific Reports* 13(1), 19374.

Colaço, A.F., Molin, J.P., Rosell-Polo, J.R. and Escolà, A. (2018) Application of light detection and ranging and ultrasonic sensors to high-throughput phenotyping and precision horticulture: Current status and challenges. *Horticulture Research* 5, 35.

Coulibali, Z., Cambouris, A.N. and Parent, S.É. (2020) Cultivar-specific nutritional status of potato (*Solanum tuberosum* L.) crops. *PLoS ONE* 15(3), e0230458.

Crandall, S.G., Gold, K.M., Jiménez-Gasco, M.D.M., Filgueiras, C.C. and Willett, D.S. (2020) A multi-omics approach to solving problems in plant disease ecology. *PLoS ONE* 15(9), e0237975.

Dahikar, S.S. and Rode, S.V. (2014) Agricultural crop yield prediction using artificial neural network approach. *International Journal of Innovative Research in Electrical, Electronics, Instrumentation and Control Engineering* 2(1), 683–686.

Dasgupta, I., Saha, J., Venkatasubbu, P. and Ramasubramanian, P. (2020) AI crop predictor and weed detector using wireless technologies: A smart application for farmers. *Arabian Journal for Science and Engineering* 45, 11115–11127.

de Oliveira Dias, F., Magalhães Valente, D.S., Oliveira, C.T., Dariva, F.D., Copati, M.G.F. *et al.* (2023) Remote sensing and machine learning techniques for high throughput phenotyping of late blight-resistant tomato plants in open field trials. *International Journal of Remote Sensing* 44(6), 1900–1921.

Diaz-Garcia, L., Covarrubias-Pazaran, G., Schlautman, B. and Zalapa, J. (2016) GiNA, an efficient and high-throughput software for horticultural phenotyping. *PloS ONE* 11(8), e0160439.

Dissanayake, A., Rajapaksha, I., Gunarathna, R., Jayasinghe, S., De Silva, H. *et al.* (2023) Detection of diseases and nutrition in bell pepper. In: *2023 5th International Conference on Advancements in Computing (ICAC)*, IEEE, pp. 286–291.

Dubey, B.P., Bhagwat, S.G., Shouche, S.P. and Sainis, J.K. (2006) Potential of artificial neural networks in varietal identification using morphometry of wheat grains. *Biosystems Engineering* 95(1), 61–67.

Ehrampoosh, A., Hettiarachchi, P., Koirala, A., Hassan, J., Islam, N. *et al.* (2025) Intelligent weed management using aerial image processing and precision herbicide spraying: An overview. *Crop Protection* 194, 107206.

Eli-Chukwu, N.C. (2019) Applications of artificial intelligence in agriculture: A review. *Engineering, Technology & Applied Science Research* 9(4), 4377–4383.

Elshorbagy, A. and Parasuraman, K. (2008) On the relevance of using artificial neural networks for estimating soil moisture content. *Journal of Hydrology* 362(1–2), 1–18.

Fan, Z., Tieman, D.M., Knapp, S.J., Zerbe, P., Famula, R. *et al.* (2022) A multi-omics framework reveals strawberry flavor genes and their regulatory elements. *New Phytologist* 236(3), 1089–1107.

Ferreira, M.G., Azevedo, A.M., Siman, L.I., da Silva, G.H., Carneiro, C.D.S. *et al.* (2017) Automation in accession classification of Brazilian Capsicum germplasm through artificial neural networks. *Scientia Agricola* 74(3), 203–207.

Florensa, A.F., Kaas, R.S., Clausen, P.T.L.C., Aytan-Aktug, D. and Aarestrup, F.M. (2022) ResFinder – an open online resource for identification of antimicrobial resistance genes in next-generation sequencing data and prediction of phenotypes from genotypes. *Microbial Genomics* 8(1), 000748.

Fumia, N., Kantar, M., Lin, Y.P., Schafleitner, R., Lefebvre, V. *et al.* (2023) Exploration of high-throughput data for heat tolerance selection in capsicum annuum. *The Plant Phenome Journal* 6(1), e20071.

Gadade, H.D. and Kirange, D.K. (2021) Machine learning based identification of tomato leaf diseases at various stages of development. In: *2021 5th International Conference on Computing Methodologies and Communication (ICCMC)*, IEEE, pp. 814–819.

Garske, B., Bau, A. and Ekardt, F. (2021) Digitalization and AI in European agriculture: A strategy for achieving climate and biodiversity targets. *Sustainability* 13(9), 4652.

Ghag, S.B., Ganapathi, T.R., Jain, S.M. and Penna, S. (2022) Omics technologies and breeding of horticultural crops. In: Rout, G.R. and Peter, K.V. (eds) *Omics in Horticultural Crops*. Academic Press, pp. 75–90.

Graziani, M., Dutkiewicz, L., Calvaresi, D., Amorim, J.P., Yordanova, K. *et al.* (2023) A global taxonomy of interpretable AI: Unifying the terminology for the technical and social sciences. *Artificial Intelligence Review* 56(4), 3473–3504.

Gupta, D.K., Pagani, A., Zamboni, P. and Singh, A.K. (2024) AI-powered revolution in plant sciences: Advancements, applications, and challenges for sustainable agriculture and food security. *Exploration of Foods and Foodomics* 2(5), 443–459.

Gutierrez, S., Tardaguila, J., Fernandez-Novales, J. and Diago, M.P. (2015) Support vector machine and artificial neural network models for the classification of grapevine varieties using a portable NIR spectrophotometer. *PloS ONE* 10(11), e0143197.

Hemming, S., Zwart, F.D., Elings, A., Petropoulou, A. and Righini, I. (2020) Cherry tomato production in intelligent greenhouses—sensors and AI for control of climate, irrigation, crop yield, and quality. *Sensors* 20(22), 6430.

Hesami, M. and Jones, A.M.P. (2020) Application of artificial intelligence models and optimization algorithms in plant cell and tissue culture. *Applied Microbiology and Biotechnology* 104(22), 9449–9485.

Hu, H., Campbell, M.T., Yeats, T.H., Zheng, X., Runcie, D.E. *et al.* (2021) Multi-omics prediction of oat agronomic and seed nutritional traits across environments and in distantly related populations. *Theoretical and Applied Genetics* 134, 4043–4054.

Hu, J., Liu, Y., Tang, X., Rao, H., Ren, C. *et al.* (2020) Transcriptome profiling of the flowering transition in saffron (*Crocus sativus* L.). *Scientific Reports* 10(1), 9680.

Hui, Y., Zhang, Y., Wang, K., Pan, C., Chen, H. *et al.* (2020) Goat DNMT3B: An indel mutation detection, association analysis with litter size and mRNA expression in gonads. *Theriogenology* 147, 108–115.

Igiehon, N.O., Babalola, O.O. and Aremu, B.R. (2019) Genomic insights into plant growth promoting rhizobia capable of enhancing soybean germination under drought stress. *BMC Microbiology* 19, 1–22.

Indelicato, E. and Boesch, S. (2021) From genotype to phenotype: Expanding the clinical spectrum of CACNA1A variants in the era of next generation sequencing. *Frontiers in Neurology* 12, 639994.

Islam, S., Samsuzzaman, Reza, M.N., Lee, K.H., Ahmed, S. *et al.* (2024) Image processing and support vector machine (SVM) for classifying environmental stress symptoms of pepper seedlings grown in a plant factory. *Agronomy* 14(9), 2043.

Jaiswal, A.K., Alkan, N., Elad, Y., Sela, N., Graber, E.R. *et al.* (2020) Molecular insights into biochar-mediated plant growth promotion and systemic resistance in tomato against Fusarium crown and root rot disease. *Scientific Reports* 10(1), 13934.

Jamil, I.N., Remali, J., Azizan, K.A., Nor Muhammad, N.A., Arita, M. *et al.* (2020) Systematic multi-omics integration (MOI) approach in plant systems biology. *Frontiers in Plant Science* 11, 944.

Jaramillo-Botero, A., Colorado, J., Quimbaya, M., Rebolledo, M.C., Lorieux, M. *et al.* (2022) The óMICAS alliance, an international research program on multi-omics for crop breeding optimization. *Frontiers in Plant Science* 13, 992663.

Javaid, M., Haleem, A., Singh, R.P. and Suman, R. (2022) Enhancing smart farming through the applications of Agriculture 4.0 technologies. *International Journal of Intelligent Networks* 3, 150–164.

Javaid, M., Haleem, A., Khan, I.H. and Suman, R. (2023) Understanding the potential applications of artificial intelligence in agriculture sector. *Advanced Agrochem* 2(1), 15–30.

Jeon, D., Kang, Y., Lee, S., Choi, S., Sung, Y. *et al.* (2023) Digitalizing breeding in plants: A new trend of next-generation breeding based on genomic prediction. *Frontiers in Plant Science* 14, 1092584.

Jiang, X., Zhang, W., Fernie, A.R. and Wen, W. (2022) Combining novel technologies with interdisciplinary basic research to enhance horticultural crops. *The Plant Journal* 109(1), 35–46.

Johansen, K., Morton, M.J., Malbeteau, Y., Aragon, B., Al-Mashharawi, S. *et al.* (2020) Predicting biomass and yield in a tomato phenotyping experiment using UAV imagery and random forest. *Frontiers in Artificial Intelligence* 3, 28.

Jung, J., Maeda, M., Chang, A., Bhandari, M., Ashapure, A. *et al.* (2021) The potential of remote sensing and artificial intelligence as tools to improve the resilience of agriculture production systems. *Current Opinion in Biotechnology* 70, 15–22.

Kang, Z., Babar, M.A., Khan, N., Guo, J., Khan, J. *et al.* (2019) Comparative metabolomic profiling in the roots and leaves in contrasting genotypes reveals complex mechanisms involved in post-anthesis drought tolerance in wheat. *PLoS ONE* 14(3), e0213502.

Karahan, T. and Nabiyev, V. (2021) Plant identification with convolutional neural networks and transfer learning. *Pamukkale Üniversitesi Mühendislik Bilimleri Dergisi* 27(5), 638–645.

Kashyap, S.P., Prasanna, H.C., Kumari, N., Mishra, P. and Singh, B. (2020) Understanding salt tolerance mechanism using transcriptome profiling and de novo assembly of wild tomato *Solanum chilense*. *Scientific Reports* 10(1), 15835.

Khiarak, J.N., Valizadeh-kamran, R., Heydariyan, A. and Damghani, N. (2018) Big data analysis in plant science and machine learning tool applications in genomics and proteomics. *International Journal of Computational and Experimental Science and Engineering* 4(2), 23–31.

Kim, S., Jo, N., Gil, J., Koo, S.C., Um, Y. *et al.* (2021) Development of genome-wide simple sequence repeat markers in *Codonopsis lanceolata* using next-generation sequencing. *Horticulture, Environment, and Biotechnology* 62, 985–993.

Komal, Sethi, G.K. and Bawa, R.K. (2022) A prototype of automatic rice variety identification system using artificial intelligence techniques. In: *AIP Conference Proceedings*, AIP Publishing LLC (Vol. 2455, No. 1. p. 040004).

Kopriva, S., Malagoli, M. and Takahashi, H. (2019) Sulfur nutrition: Impacts on plant development, metabolism, and stress responses. *Journal of Experimental Botany* 70(16), 4069–4073.

Lal, H., Jones, J.W., Peart, R.M. and Shoup, W.D. (1992) FARMSYS—a whole-farm machinery management decision support system. *Agricultural Systems* 38(3), 257–273.

Lapajne, J., Vojnović, A., Vončina, A. and Žibrat, U. (2024) Enhancing water-deficient potato plant identification: Assessing realistic performance of attention-based deep neural networks and hyperspectral imaging for agricultural applications. *Plants* 13(14), 1918.

Lee, J.S., Jang, H.K., Kang, W.H. and Yeom, S.I. (2023) Transcriptome analysis of pepper–*Phytophthora infestans* interaction based on a pipeline of a simplified and effective RNA-seq analysis (PoRAS). *Horticultural Science and Technology* 41, 100–110.

Levine, E.R., Kimes, D.S. and Sigillito, V.G. (1996) Classifying soil structure using neural networks. *Ecological Modelling* 92(1), 101–108.

Li, M. and Yost, R.S. (2000) Management-oriented modeling: Optimizing nitrogen management with Artificial Intelligence. *Agricultural Systems* 65(1), 1–27.

Li, J., Wang, Y. and Suh, J.H. (2022) Multi-omics approach in tea polyphenol research regarding tea plant growth, development and tea processing: Current technologies and perspectives. *Food Science and Human Wellness* 11(3), 524–536.

Li, P., Xiao, L., Du, Q., Quan, M., Song, Y. *et al.* (2023) Genomic insights into selection for heterozygous alleles and woody traits in *Populus tomentosa*. *Plant Biotechnology Journal* 21(10), 2002–2018.

Li, Y., Ma, L., Wu, D. and Chen, G. (2021a) Advances in bulk and single-cell multi-omics approaches for systems biology and precision medicine. *Briefings in Bioinformatics* 22(5), bbab024.

Li, Y., Wu, X., Xu, W., Sun, Y., Wang, Y. *et al.* (2021b) High-throughput physiology-based stress response phenotyping: Advantages, applications and prospective in horticultural plants. *Horticultural Plant Journal* 7(3), 181–187.

Lin, E. and Lane, H.Y. (2017) Machine learning and systems genomics approaches for multi-omics data. *Biomarker Research* 5, 1–6.

Liu, C., Tai, Y., Luo, J., Wu, Y., Zhao, X. *et al.* (2022) Integrated multi-omics analysis provides insights into genome evolution and phosphorus deficiency adaptation in pigeonpea (*Cajanus cajan*). *Horticulture Research* 9, uhac107.

López, E.M., García, M., Schuhmacher, M. and Domingo, J.L. (2008) A fuzzy expert system for soil characterization. *Environment International* 34(7), 950–958.

Luecken, M.D. and Theis, F.J. (2019) Current best practices in single-cell RNA-seq analysis: A tutorial. *Molecular Systems Biology* 15(6), e8746.

Mahlein, A.K., Kuska, M.T., Thomas, S., Wahabzada, M., Behmann, J. *et al.* (2019) Quantitative and qualitative phenotyping of disease resistance of crops by hyperspectral sensors: Seamless interlocking of phytopathology, sensors, and machine learning is needed. *Current Opinion in Plant Biology* 50, 156–162.

Malabadi, R.B., Nethravathi, T., Kolkar, K.P., Chalannavar, R.K., Mudigoudra, B.S. *et al.* (2023) Applications of Artificial Intelligence (AI) in cannabis industries: *In vitro* plant tissue culture. *International Journal of Research and Innovation in Applied Science* 8(VII), 21–40.

Malounas, I., Paliouras, G., Nikolopoulos, D., Liakopoulos, G., Bresta, P. *et al.* (2024) Early detection of broccoli drought acclimation/stress in agricultural environments utilizing proximal hyperspectral imaging and automl. *Smart Agricultural Technology* 8, 100463.

Minasny, B. and McBratney, A.B. (2025) Machine learning and artificial intelligence applications in soil science. *European Journal of Soil Science* 76(2), e70093.

Mochida, K., Koda, S., Inoue, K., Hirayama, T., Tanaka, S. *et al.* (2019) Computer vision-based phenotyping for improvement of plant productivity: A machine learning perspective. *GigaScience* 8(1), giy153.

Mohmed, G., Heynes, X., Naser, A., Sun, W., Hardy, K. *et al.* (2023) Modelling daily plant growth response to environmental conditions in Chinese solar greenhouse using Bayesian neural network. *Scientific Reports* 13(1), 4379. DOI: 10.1038/s41598-023-30846-y.

Mondal, S., Song, H., Zhang, L. and Cao, Y. (2022) Multi-omics and computational biology in horticultural plants: From genotype to phenotype. *Frontiers in Plant Science* 13, 1073266.

Montesinos López, O.A., Montesinos López, A. and Crossa, J. (2022) Random forest for genomic prediction. In: *Multivariate Statistical Machine Learning Methods for Genomic Prediction*. Springer International Publishing, Cham, pp. 633–681.

Morganti, S., Tarantino, P., Ferraro, E., D'Amico, P., Duso, B.A. *et al.* (2019) Next generation sequencing (NGS): A revolutionary technology in pharmacogenomics and personalized medicine in cancer. *Advances in Experimental Medicine and Biology* 1168, 9–30.

Muhammad Aslam, M., Waseem, M., Jakada, B.H., Okal, E.J., Lei, Z. *et al.* (2022) Mechanisms of abscisic acid-mediated drought stress responses in plants. *International Journal of Molecular Sciences* 23(3), 1084.

Murmu, S., Sinha, D., Chaurasia, H., Sharma, S., Das, R. *et al.* (2024) A review of artificial intelligence-assisted omics techniques in plant defense: Current trends and future directions. *Frontiers in Plant Science* 15, 1292054.

Nawaz, G., Usman, B., Peng, H., Zhao, N., Yuan, R. *et al.* (2020) Knockout of Pi21 by CRISPR/Cas9 and itraq-based proteomic analysis of mutants revealed new insights into *M. oryzae* resistance in elite rice line. *Genes* 11(7), 735.

Ncube, E., Mohale, K. and Nogemane, N. (2022) Metabolomics as a prospective tool for soybean (Glycine max) crop improvement. *Current Issues in Molecular Biology* 44(9), 4181–4196.

Nefissi Ouertani, R., Arasappan, D., Abid, G., Ben Chikha, M., Jardak, R. *et al.* (2021) Transcriptomic analysis of salt-stress-responsive genes in barley roots and leaves. *International Journal of Molecular Sciences* 22(15), 8155.

Padhiary, M., Hoque, A., Prasad, G., Kumar, K. and Sahu, B. (2025) Precision agriculture and AI-driven resource optimization for sustainable land and resource management. In: Ruiz-Vanoye, J.A. and Díaz-Parra, O. (eds) *Smart Water Technology for Sustainable Management in Modern Cities*. IGI Global Scientific Publishing, New York, London, Beijing, pp. 197–232.

Pandita, D., Pandita, A., Wani, S.H., Abdelmohsen, S.A., Alyousef, H.A. *et al.* (2021) Crosstalk of multi-omics platforms with plants of therapeutic importance. *Cells* 10(6), 1296.

Papageorgiou, E.I., Markinos, A.T. and Gemtos, T.A. (2011) Fuzzy cognitive map based approach for predicting yield in cotton crop production as a basis for decision support system in precision agriculture application. *Applied Soft Computing* 11(4), 3643–3657.

Partel, V., Kakarla, S.C. and Ampatzidis, Y. (2019) Development and evaluation of a low-cost and smart technology for precision weed management utilizing artificial intelligence. *Computers and Electronics in Agriculture* 157, 339–350.

Pasalkar, J., Gorde, G., More, C., Memane, S. and Gaikwad, V. (2023) Potato leaf disease detection using machine learning. *Current Agriculture Research Journal* 11(3).

Pepe, M., Hesami, M., Small, F. and Jones, A.M.P. (2021) Comparative analysis of machine learning and evolutionary optimization algorithms for precision micropropagation of *Cannabis sativa*: Prediction and validation of *in vitro* shoot growth and development based on the optimization of light and carbohydrate sources. *Frontiers in Plant Science* 12, 757869.

Piorkowski, D., Park, S., Wang, A.Y., Wang, D., Muller, M. *et al.* (2021) How AI developers overcome communication challenges in a multidisciplinary team: A case study. *Proceedings of the ACM on Human-Computer Interaction* 5(CSCW1), 1–25.

Psomopoulos, F. (2024) FAIR research software as the catalyst for trustworthy AI in life sciences. In: *5th Belgrade Bioinformatics Conference*, Institute of Molecular Genetics and Genetic Engineering, Belgrade, p. 18.

Rahimikollu, J., Xiao, H., Rosengart, A., Rosen, A.B., Tabib, T. *et al.* (2024) SLIDE: Significant latent factor interaction discovery and exploration across biological domains. *Nature Methods* 21(5), 835–845.

Rahman, M.A., Khan, A.A., Hasan, M.M., Rahman, M.S. and Habib, M.T. (2024) Deep learning modeling for potato breed recognition. *IEEE Transactions on AgriFood Electronics* 2, 419–427.

Rahnemoonfar, M. and Sheppard, C. (2017) Deep count: Fruit counting based on deep simulated learning. *Sensors* 17(4), 905.

Ramírez-Meraz, M., Méndez-Aguilar, R., Hidalgo-Martínez, D., Villa-Ruano, N., Zepeda-Vallejo, L.G. *et al.* (2020) Experimental races of *Capsicum annuum* cv. Jalapeno: Chemical characterization and classification by 1H NMR/machine learning. *Food Research International* 138, 109763.

Rathore, M.M., Shah, S.A., Shukla, D., Bentafat, E. and Bakiras, S. (2021) The role of AI, machine learning, and big data in digital twinning: A systematic literature review, challenges, and opportunities. *IEEE Access* 9, 32030–32052.

Ravisankar, V. and Mathew, D. (2022) Multiomics bioinformatics approaches in horticultural crops. In: *Omics in Horticultural Crops*. Academic Press, pp. 27–54.

Raza, A. (2022) Metabolomics: A systems biology approach for enhancing heat stress tolerance in plants. *Plant Cell Reports* 41(3), 741–763.

Rong, T., Chunchun, Z., Wei, G., Yuchen, G., Fei, X. *et al.* (2021) Proteomic insights into protostane triterpene biosynthesis regulatory mechanism after MeJA treatment in *Alisma orientale* (Sam.) Juz. *Biochimica et Biophysica Acta (BBA) – Proteins and Proteomics* 1869(8), 140671.

Roper, J.M., Garcia, J.F. and Tsutsui, H. (2021) Emerging technologies for monitoring plant health *in vivo*. *ACS Omega* 6(8), 5101–5107.

Roychowdhury, R., Das, S.P., Gupta, A., Parihar, P., Chandrasekhar, K. *et al.* (2023) Multi-omics pipeline and omics-integration approach to decipher plant's abiotic stress tolerance responses. *Genes* 14(6), 1281.

Rudin, C., Chen, C., Chen, Z., Huang, H., Semenova, L. *et al.* (2022) Interpretable machine learning: Fundamental principles and 10 grand challenges. *Statistic Surveys* 16, 1–85.

Ryan, M., Isakhanyan, G. and Tekinerdogan, B. (2023) An interdisciplinary approach to artificial intelligence in agriculture. *NJAS: Impact in Agricultural and Life Sciences* 95(1), 2168568.

Sabanci, K., Aslan, M.F., Ropelewska, E. and Unlersen, M.F. (2022) A convolutional neural network-based comparative study for pepper seed classification: Analysis of selected deep features with support vector machine. *Journal of Food Process Engineering* 45(6), e13955.

Sadeghi-Tehran, P., Sabermanesh, K., Virlet, N. and Hawkesford, M.J. (2017) Automated method to determine two critical growth stages of wheat: Heading and flowering. *Frontiers in Plant Science* 8, 252.

Sahoo, J.P., Behera, L., Sharma, S.S., Praveena, J., Nayak, S.K. *et al.* (2020) Omics studies and systems biology perspective towards abiotic stress response in plants. *American Journal of Plant Sciences* 11(12), 2172.

Selvaraj, M.G., Valderrama, M., Guzman, D., Valencia, M., Ruiz, H. *et al.* (2020) Machine learning for high-throughput field phenotyping and image processing provides insight into the association of above and below-ground traits in cassava (*Manihot esculenta* Crantz). *Plant Methods* 16, 1–19.

Shah, V. and Konda, S.R. (2021) Neural networks and explainable AI: Bridging the gap between models and interpretability. *International Journal of Computer Science and Technology* 5(2), 163–176.

Shankar, T., Malik, G.C., Banerjee, M., Dutta, S., Praharaj, S. *et al.* (2022) Prediction of the effect of nutrients on plant parameters of rice by artificial neural network. *Agronomy* 12(9), 2123.

Sharma, V., Gupta, P., Kagolla, P., Hangargi, B., Veershetty, A. *et al.* (2021) Metabolomics intervention towards better understanding of plant traits. *Cells* 10(2), 346.

Shen, S., Zhan, C., Yang, C., Fernie, A.R. and Luo, J. (2023) Metabolomics-centered mining of plant metabolic diversity and function: Past decade and future perspectives. *Molecular Plant* 16(1), 43–63.

Shoaib, M., Shah, B., Ei-Sappagh, S., Ali, A., Ullah, A. *et al.* (2023) An advanced deep learning models-based plant disease detection: A review of recent research. *Frontiers in Plant Science* 14, 1158933.

Singh, A.V., Rosenkranz, D., Ansari, M.H.D., Singh, R., Kanase, A. *et al.* (2020) Artificial intelligence and machine learning empower advanced biomedical material design to toxicity prediction. *Advanced Intelligent Systems* 2(12), 2000084.

Singh, A.V., Chandrasekar, V., Janapareddy, P., Mathews, D.E., Laux, P. *et al.* (2021) Emerging application of nanorobotics and artificial intelligence to cross the BBB: Advances in design, controlled maneuvering, and targeting of the barriers. *ACS Chemical Neuroscience* 12(11), 1835–1853.

Song, H. and He, Y. (2005) Crop nutrition diagnosis expert system based on artificial neural networks. In: *Third International Conference on Information Technology and Applications (ICITA'05)*, IEEE, pp. 357–362.

Song, Y., Xu, X., Wang, W., Tian, T., Zhu, Z. *et al.* (2019) Single cell transcriptomics: Moving towards multi-omics. *Analyst* 144(10), 3172–3189.

Subramanian, I., Verma, S., Kumar, S., Jere, A. and Anamika, K. (2020) Multi-omics data integration, interpretation, and its application. *Bioinformatics and Biology Insights* 14, 1177932219899051.

Sun, J., Zhang, J., Fang, H., Peng, L., Wei, S. *et al.* (2019) Comparative transcriptome analysis reveals resistance-related genes and pathways in *Musa acuminata* banana 'Guijiao 9' in response to Fusarium wilt. *Plant Physiology and Biochemistry* 141, 83–94.

Sun, L., Lai, M., Ghouri, F., Nawaz, M.A., Ali, F. *et al.* (2024) Modern plant breeding techniques in crop improvement and genetic diversity: From molecular markers and gene editing to artificial intelligence—A critical review. *Plants* 13(19), 2676.

Tajik, S., Ayoubi, S. and Nourbakhsh, F. (2012) Prediction of soil enzymes activity by digital terrain analysis: Comparing artificial neural network and multiple linear regression models. *Environmental Engineering Science* 29(8), 798–806.

Talaviya, T., Shah, D., Patel, N., Yagnik, H. and Shah, M. (2020) Implementation of artificial intelligence in agriculture for optimisation of irrigation and application of pesticides and herbicides. *Artificial Intelligence in Agriculture* 4, 58–73.

Tan, L., Lu, J. and Jiang, H. (2021) Tomato leaf diseases classification based on leaf images: A comparison between classical machine learning and deep learning methods. *AgriEngineering* 3(3), 542–558.

Taşan, S., Cemek, B., Taşan, M. and Cantürk, A. (2022) Estimation of eggplant yield with machine learning methods using spectral vegetation indices. *Computers and Electronics in Agriculture* 202, 107367.

van Henten, E.J., Hemming, J., van Tuijl, B.A.J., Kornet, J.G., Meuleman, J. *et al.* (2002) An autonomous robot for harvesting cucumbers in greenhouses. *Autonomous Robots* 13(3), 241–258.

Walker, K.A., Basisty, N., Wilson, D.M. and Ferrucci, L. (2022) Connecting aging biology and inflammation in the omics era. *The Journal of Clinical Investigation* 132, 14.

Wan, H., Zhang, X., Wang, P., Qiu, H., Guo, Y. *et al.* (2022) Integrated multi-omics analysis of developing 'Newhall' orange and its glossy mutant provide insights into citrus fragrance formation. *Horticultural Plant Journal* 8(4), 435–449.

White, III. R.A., Rivas-Ubach, A., Borkum, M.I., Köberl, M., Bilbao, A. *et al.* (2017) The state of rhizospheric science in the era of multi-omics: A practical guide to omics technologies. *Rhizosphere* 3, 212–221.

Wang, Y., Guo, L., Zhao, X., Zhao, Y., Hao, Z. *et al.* (2021) Advances in mechanisms and omics pertaining to fruit cracking in horticultural plants. *Agronomy* 11(6), 1045.

Wang, X., Geng, X., Yang, L., Chen, Y., Zhao, Z. *et al.* (2022) Total and mitochondrial transcriptomic and proteomic insights into regulation of bioenergetic processes for shoot fast-growth initiation in moso bamboo. *Cells* 11(7), 1240.

Wieme, J., Mollazade, K., Malounas, I., Zude-Sasse, M., Zhao, M. *et al.* (2022) Application of hyperspectral imaging systems and artificial intelligence for quality assessment of fruit, vegetables and mushrooms: A review. *Biosystems Engineering* 222, 156–176.

Wu, X., Yan, J., Wu, Y., Zhang, H., Mo, S. *et al.* (2019) Proteomic analysis by iTRAQ-PRM provides integrated insight into mechanisms of resistance in pepper to *Bemisia tabaci* (Gennadius). *BMC Plant Biology* 19, 1–19.

Wu, L., Marcelis, L.F., Kong, F. and Zhu, Y. (2023) Flowering time control in agricultural and horticultural crops. *Frontiers in Plant Science* 14, 1116197.

Xu, S., Wang, J., Wang, H., Bao, Y., Li, Y. *et al.* (2019) Molecular characterization of carbendazim resistance of Fusarium species complex that causes sugarcane pokkah boeng disease. *BMC Genomics* 20, 1–14.

Yamamoto, E., Matsunaga, H., Onogi, A., Kajiya-Kanegae, H., Minamikawa, M. *et al.* (2016) A simulation-based breeding design that uses whole-genome prediction in tomato. *Scientific Reports* 6(1), 19454.

Yan, J., Xu, Y., Cheng, Q., Jiang, S., Wang, Q. *et al.* (2021) LightGBM: Accelerated genomically designed crop breeding through ensemble learning. *Genome Biology* 22, 1–24.

Yang, C.C., Prasher, S.O., Landry, J.A. and Ramaswamy, H.S. (2003) Development of a herbicide application map using artificial neural networks and fuzzy logic. *Agricultural Systems* 76(2), 561–574.

Yang, Y., Saand, M.A., Huang, L., Abdelaal, W.B., Zhang, J. *et al.* (2021) Applications of multi-omics technologies for crop improvement. *Frontiers in Plant Science* 12, 563953.

Yoosefzadeh Najafabadi, M., Hesami, M. and Eskandari, M. (2023) Machine learning-assisted approaches in modernized plant breeding programs. *Genes* 14(4), 777.

You, J., Zhang, Y., Liu, A., Li, D., Wang, X. *et al.* (2019) Transcriptomic and metabolomic profiling of drought-tolerant and susceptible sesame genotypes in response to drought stress. *BMC Plant Biology* 19, 1–16.

Yuan, T., Zhang, S., Sheng, X., Wang, D., Gong, Y. *et al.* (2016) An autonomous pollination robot for hormone treatment of tomato flower in greenhouse. In: *2016 3rd International Conference on Systems and Informatics (ICSAI)*, IEEE, pp. 108–113.

Yu, J., Li, Y., Qin, Z., Guo, S., Li, Y. *et al.* (2020) Plant chloroplast stress response: Insights from thiol redox proteomics. *Antioxidants & Redox Signaling* 33(1), 35–57.

Zapletalova, M., Rancurel, C., Industri, B., Bardin, M., Le Brigand, K. *et al.* (2023) BABA-induced pathogen resistance: A multi-omics analysis of the tomato response reveals a hyper-receptive status involving ethylene. *Horticulture Research* 10(6), uhad068.

Zappia, L. and Theis, F.J. (2021) Over 1000 tools reveal trends in the single-cell RNA-seq analysis landscape. *Genome Biology* 22, 1–18.

Zdrazil, J., Kong, L., Klimeš, P., Jasso-Robles, F.I., Saiz-Fernández, I. *et al.* (2025) Next-generation high-throughput phenotyping with trait prediction through adaptable multi-task computational intelligence. *Computers and Electronics in Agriculture* 235, 110390.

Zhang, J., Yang, J., Zhang, L., Luo, J., Zhao, H. *et al.* (2020a) A new SNP genotyping technology target SNP-seq and its application in genetic analysis of cucumber varieties. *Scientific Reports* 10(1), 5623.

Zhang, Q., Deng, D., Dai, W., Li, J. and Jin, X. (2020b) Optimization of culture conditions for differentiation of melon based on artificial neural network and genetic algorithm. *Scientific Reports* 10(1), 3524.

Zhang, R., Zhang, C., Yu, C., Dong, J. and Hu, J. (2022) Integration of multi-omics technologies for crop improvement: Status and prospects. *Frontiers in Bioinformatics* 2, 1027457.

Zhang, M., Han, Y., Li, D., Xu, S. and Huang, Y. (2023) Smart horticulture as an emerging interdisciplinary field combining novel solutions: Past development, current challenges, and future perspectives. *Horticultural Plant Journal* 10(6), 1257–1273.

Zhao, Z., Chow, T.L., Rees, H.W., Yang, Q., Xing, Z. *et al.* (2009) Predict soil texture distributions using an artificial neural network model. *Computers and Electronics in Agriculture* 65(1), 36–48.

Zhao, L., He, F., Li, B., Gu, X., Zhang, X. *et al.* (2022) Transcriptomic analysis of the mechanisms involved in enhanced antagonistic efficacy of *Meyerozyma guilliermondii* by methyl jasmonate and disease resistance of postharvest apples. *Lwt* 160, 113323.

3 AI-Based Multiple Omics Data Integration for Plant Genomic Prediction

Indrajit Kalita[1*], Sourav Bhattacharjee[2], Bhaben Tanti[1] and Sharmistha Sarma Kalita[1,3]

[1]Department of Botany, Gauhati University, Guwahati, Assam, India; [2]Department of Botany, Pandit Deendayal Upadhyaya Adarsha Mahavidyalaya Tulungia, Bongaigaon, Assam, India; [3]Department of Botany, Salbari College, Salbari, Baksa, Assam, India

Abstract

The integration of artificial intelligence (AI) with multiple omics data is revolutionizing plant genomic prediction by enhancing our understanding of genetic, molecular, and phenotypic traits. In this chapter, we examined the potential of combining genomics, transcriptomics, proteomics, and metabolomics through advanced AI methodologies like machine learning and deep learning. By utilizing comprehensive datasets from these omics technologies, researchers can reveal complex interactions and regulatory mechanisms that influence plant responses to environmental stresses. The integration of multi-omics data provides a holistic perspective on plant biology, facilitating the identification of molecular markers linked to desirable traits and improving predictive models for crop breeding. Additionally, this chapter addresses challenges in data integration and interpretation while proposing innovative solutions. The implications of AI-based multi-omics integration extend to precision agriculture and sustainable crop management, promising advancements in developing resilient plant varieties as data-driven approaches reshape agricultural practices.

Keywords: Artificial intelligence, Deep learning, Machine learning, Molecular markers, Multiple omics, Plant genomic prediction

3.1 Introduction and Historical Background: Concept of AI and ML in Plant Genomic Prediction

The term artificial intelligence was first coined by John McCarthy in the year 1956, and it came into the limelight under a concept that gained exposure through two centuries. Prehistoric traces of artificial intelligence were based on ancient systems of independent machines or 'automations' that were conceptualized by ancient thinkers and philosophers. One such ancient 'automation system' can be traced back to a 'mechanical automatic pigeon' created by a co-worker of the ancient Greek philosopher Plato (Debnath *et al.*, 2025).

Artificial intelligence is rooted in a unique blend of computer science-based algorithms and mathematics. The concept of AI began with the innovative thinking of Alan Turing regarding the possibility of intellectual thinking and decision making by machines, leading to the

*Corresponding author: kalitaindrajit7@gmail.com

 Plant Genomic Prediction: AI-Enhanced Models for Plant Genomics and Crop Improvement (ed. Jen-Tsung Chen)
DOI: 10.1079/9781836991328.0003

mathematical model of 'learning machines' put forward by Turing in 1950 (Shanker, 1995). The concept of AI in medicine was first designated as 'augmented intelligence' for a precise healthcare framework (Ashby, 1956).

This revolution actually began with the most popular invention of the modern computer by Charles Babbage in 1882. This invention laid the foundation stone of 'learning machines', a concept that was later taken up by various eminent workers for development and innovation. The field gained momentum and efficacy through decades of development by various eminent workers from time to time leading to its potential application in revolutionizing diverse areas of research such as healthcare, medicine, industrial efficiency, finance, etc. by moving towards technology-based automation and formulation of algorithm-based computerized programs that normally require human professional intervention (Murmu *et al.*, 2024). One of the most recent fields of study under enhancement due to increased application of AI is plant genomic prediction, in which advancement and revolution are necessary to meet the rising demands of the increasing population. An increase in population affects food security, blended with climate change, which necessitates the need for precision breeding for striking a balance between production and demand of economically important food and medicinal plants (Feng *et al.*, 2024). This need for modern intervention and shift from classical or conventional practices is also felt primarily due to the rapid urbanization, industrialization, and over-exploitation of natural resources that go hand in hand with the massive rise in world population. Such factors, blended with global warming, or climate change, lead to less available agricultural land and more chances of crop loss, low yield, and a decrease in production quality. Classical breeding and agricultural practices are also massively affected owing to biotic, abiotic, and xenobiotic stresses rooted in various natural or anthropogenic causes that are again directly or indirectly linked to climate change and population expansion (Feng *et al.*, 2024). A unique blend of artificial intelligence (AI), machine learning (ML), and bioinformatics has ushered in a transformative era of discovery for increased effectiveness and exponential progress, avoiding the over-exploitation of natural resources to save time and workload.

The solution of New Age problems lies within nature itself, which needs to be retrieved and blended with modern technologies to arrive at distinct innovative practices that can lead to sustainable progress in precision breeding and modern agriculture through plant genomic prediction. Integration of AI-based models can aid in the precise identification of plants with the ability to synthesize bioactive molecules that have desirable therapeutic as well as nutritional values.

AI-based models can also allow us to elucidate potentially resistant plants against biotic, abiotic, and xenobiotic stress conditions (Xu *et al.*, 2022). The sole objective behind the application of predictive AI-based models in breeding and agriculture is to arrive at a sustainable, accessible, and robust pathway in which the phenotypic qualities of a potential lead plant-breeding candidate or the starting plant material can be estimated with precision, even before plantation by the workers. This can lead to efficient forecasting of phenotypic results in the case of breeding experiments and application of accurate propagules in the field with potential for desirable characteristic expressions (Feng *et al.*, 2024).

3.2 AI-Based Understanding of Genetic and Molecular Traits

AI simulation algorithms screen the specific genetic data in order to harness the potential hidden within it. This can lead to the suitable application of the genetic data in order to move towards a comprehensive data-driven method of genomic selection for precision agriculture and a gradual shift from traditional or classical plant breeding approaches (Jubair and Domaratzki, 2023). Genomic selection has thus been conveniently defined as a comprehensive marker-assisted selection (MAS)-based methodology for selection of the most eligible lead breeding candidates with suitable potential phenotypic qualities. AI algorithms can access the entire genome of specific candidates in order to suitably scan for the presence of some specific molecular markers with high polymorphism and whole-genome distribution to predict the possible future phenotypic expression of the plant, making a marked shift from visual observation-based predictions and paving the way for more resilient and sustainable pre-breeding strategies leading to improved agricultural practices (Shook *et al.*, 2021). AI and ML can thus extract meaningful information from high-throughput sequencing results, datasets on growth and yield characteristics, weather spectrum-based variability data, datasets on soil health and pest incidence to perform algorithm-based automated phenotype selection that integrates all the fed datasets to accurately select the most desirable lead breeding candidate as mother plant for further future plant improvement programs that can lead to effective enhancement in overall production efficiency, sustainability, and quality (Wójcik-Gront *et al.*, 2024). The AI-based models scan whole genomic data to accurately identify specific genes responsible for the expression of proteins actively involved in the biochemical pathway of tolerance and resistance against various biotic, abiotic, and xenobiotic stress conditions. Such processes have been applied successfully for AI-driven emergence of drought-stress and heat-stress tolerant maize varieties and salt-stress tolerant rice varieties, leading the drive towards minimization of the gap between sustainable agricultural production and demand (Voss-Fels *et al.*, 2019). Comprehensive application of the genetic information stored within various worldwide sequence storage repositories can be subjected to AI-based downstream processing against a constructed reference genome to map and identify useful allelic variations and mutations to streamline the pre-breeding strategies for better agricultural outputs. More than 1700 genebanks throughout the world store genetic information under germplasm accessions that represent the vast plant genetic diversity existing worldwide. This genetic information stored germplasm accession-wise includes species and various infraspecific categories, which contain a vast potential for application in genotyping and further application in harnessing genomic selection through AI-based modelling methodologies (Farooq *et al.*, 2024). The reference genomes in repositories can be used for detecting intra-specific genetic variations that can be the result of a prolonged exposure to a specific environment, leading to gene diversification and allelic variations for better adaptation. Such variations are indicative of plant genetic diversity that remains hidden in nature but can be efficiently used as mother plants for better yield and productivity against specific environments (Farooq *et al.*, 2024). The broader goal behind such AI-based innovative concepts is to effectively utilize the wild populations of plants and integrate the beneficial genetic variations hidden within them into our cultivated plants in order to combat various challenges triggered by population explosion, anthropogenic interventions, and climate change.

3.3 AI-Based Understanding of Phenotypic Traits

The integration of phenomics with AI requires comprehensive data acquisition that is based on real-time monitoring and capturing of various phenotypic data (Table 3.1). This can be various quantifiable data related to growth monitoring, yield parameters, response to weather pattern changes, reproductive fitness, senescence features with timings, and also interactions with biotic components such as insects and pests on a real-time basis. The locations for tracking or monitoring can be either fixed or automated to move across the field or study area (Sangjan *et al.*, 2025). This is done in order to minimize the imaging limitations owing to various environmental factors such as weather variations and light intensity. Patterns and gradient spectrums are extracted from the acquired big datasets using various AI and ML-based tools such as random forest (RF)and support vector machines (SVMs). SVMs work to identify a specific decision boundary that can effectively demarcate the linear or nonlinear acquired phenotypic dataset into distinct classes, even when the datasets are not such that they can be linearly categorized into classes. This geometric bifurcation using a distinct hyperplane surface can be instrumental in working out various correlation and regression functions to extract valuable information for data-driven decision making during plant breeding approaches for the next crop.

Selection of desirable mother plants can be done so that they can be propagated for further desirable agronomic outputs (Novielli *et al.*, 2024). Random forest (RF) further specifies a resilient AI-based strategy that deals with

Table 3.1. Comparative overview of AI tools and omics layers (Author's own table)

Omics layer	Key data type	AI tools supplied	Purpose in prediction
Genomics	DNA sequences, SNPs, CREs	DeepBind, SVM, CNN	Identify genetic markers, and map genes
Transcriptomics	RNA expression profiles	DNN, random forest	Identify stress response, gene regulation
Proteomics	Protein modifications and interactions	ML models, PCA	Decode regulatory mechanisms, stress proteins
Metabolomics	Metabolic pathways and intermediates	Clustering, CNN	Understand the biochemical pathway's impact on phenotype
Phenomics	Imaging, growth m	YOLO, RF, CNN	Real-time monitoring, trait quantification

decision trees based on the Classification and Regression Tree (CART) method (Gordon *et al.*, 1984) of processing trees based on binary decision problems. Random forest assembles various computer-generated decision trees into a single unit in order to accurately predict a single selection of lead crop breeding candidate based on a combination of decision trees that can be taken for further crop improvement (Yasrab *et al.*, 2019). Random forest is specialized for the selection of the most suitable features from a big dataset, which makes it a highly reliable and popular AI-based algorithm in precision breeding. Analysis of complex spectrums of phenotypic data generated from real-time monitoring of potential lead breeding candidates can require the application of more robust and resilient AI-based algorithms such as deep neural networks (DNNs) or convolutional neural networks (CNNs), derivatives of DNN that are highly specialized to extract and process complex spectrums of visual datasets to extract specific phenotypic marker features from the data and lead towards accurate predictions for sustainable agricultural practices (Monaco *et al.*, 2021).

3.4 AI-Based Integration of Multi-omics – Necessity, Processing, and Challenges

Genomic data can be a suitable resource for gaining a deep-rooted understanding into plant stress biology involving various biotic, abiotic, and xenobiotic stress conditions with the associated biochemical and metabolic response pathways that are specific to various potential lead breeding candidates (Farooq *et al.*, 2024). Deep neural networks (DNNs) or convolutional neural networks (CNNs) are also used to analyze DNA/RNA sequence data in order to scan and efficiently identify molecular markers such as repetitive sequences, single-nucleotide polymorphisms (SNPs), and regulatory regions in order to successfully establish useful relationships between genomic (DNA) sequences and phenotypic expressions for accurate pre-breeding predictions for data-driven decision making during agricultural practices. AI-based algorithms use binary functions to scan and access the presence or absence of some specific genetic markers throughout the sequence, such as SNPs. Such numerical notations are also used to identify homozygosity or heterozygosity based on positions in the number line. The value of −1 toward the left side of 0 indicates a homozygous recessive combination, whereas +1 toward the right side of 0 represents a homozygous dominant combination, respectively, where 0 (zero) represents heterozygosity (Jubair and Domaratzki, 2023). The specific marker throughout a genome is treated as an informative marker only when it crosses a threshold value in uniformity of position across the genome of the lead breeding candidate under study (Ma *et al.*, 2025). Plant genomes are categorized as complex genomes as compared with animal genomes, due to the prominent presence of repetitive elements, transposable elements, structural and numerical chromosomal aberrations contributing toward a highly complex genome organization that is difficult to decipher (Ruperao *et al.*, 2025). AI-based approaches are used in conjunction with next-generation sequencing technology in order to proceed with seamless interpretation and analysis of the sequencing results for precise identification of genetic features that are suitable to be utilized in plant breeding for improvement in the yield potential of various economically, medicinally, and industrially important plants.

AI and ML-based models such as DeepBind are aimed at systematically scanning the whole genome sequences in lead breeding candidate plants and identifying specific regulatory elements, one of which is the *cis*-regulatory elements (CREs) that are present as regions of gene regulation (non-coding DNA) which are specifically recognized by regulatory proteins known as transcription factors (TFs) involved in the transcription process. These regulatory elements control the gene expression in genes present in neighboring positions. The CREs, also known as *cis*-regulatory modules (CRMs), are short sequences that are the binding sites for transcription factors, and thus such sequences that include silencers, promoters, and enhancers control the gene expression mechanism and intensity (Maderazo *et al.*, 2022). AI and ML-based identification of such regulatory regions is essential for precision breeding towards sustainable agriculture because such regions are the key pathways for elucidating gene functions and molding the gene expression in anticipation of desirable agronomic and positive stress-response characteristics in suitable lead breeding candidate crops (Peleke *et al.*, 2024). Such an identification has been successfully conducted for improvement in economically important food plants like *Zea mays* (Peleke *et al.*, 2024). DNNs or CNNs are applied to decipher genetic variation with potential for better phenotypic traits. Such approaches can go a long way toward a futuristic plan of devising cultivars that have the best possible phenotypic expression suitable for climate resilience and stress tolerance with suitable production quality (Maraveas, 2024).

Genomic data alone remain inadequate for suitable predictive modeling in a robust manner. This is solely due to the integrated nature of biological systems, especially for organisms with complex genomes like plants (Fig. 3.1). The specificities of transcription and post-transcriptional processing can be understood from transcriptomic analysis, while the intricacies of translation and post-translational modification can be read through proteomics and epigenomic analysis. AI-based algorithms and tools can move us toward a deep dive into a detailed understanding of the multiple-omics data that has significantly changed the understanding of biological systems and provides a deep insight into various intricate specifications and inter-relationships in molecular, biochemical, and metabolic pathways that are ultimately responsible for expression of desirable phenotype (Xiao *et al.*, 2022) (Fig. 3.1). The gradual replacement and substitution of conventional statistical methodologies for analysis of solitary-omics data have shifted towards AI-based analysis of comprehensive, sophisticated multi-omics data for higher resilience, accuracy, and efficiency. Multi-omics integration is necessary for understanding the multidimensional comprehensive insights into the contributions of genomics, transcriptomics, proteomics, metabolomics, and epigenomics for the expression of the ultimate phenotype. Single-omics data analysis remains insufficient in deducing the minute intricacies in phenotypic expression, because complex biological systems are a result of the augmentation and amalgamation of various omics in an integrated manner (Fumia *et al.*, 2023). This is especially significant because a multidimensional integrated approach can suitably allow us to efficiently identify any potential off-target effects owing to untimely, unintended interactions in the biological system. Such delicate identifications are necessary for a formidable, robust approach in the dissemination of decision making during the final stages of predictive modeling in precision breeding processes (Crossa *et al.*, 2025).

AI-based data processing in predictive modeling depends on the training or foundational data that augments the machine learning algorithm for appropriate processing and prediction. The model is fed with both genotypic data (genetic marker data) and phenotypic (on-site observational) data for training or acclimatization. The genotype data processing

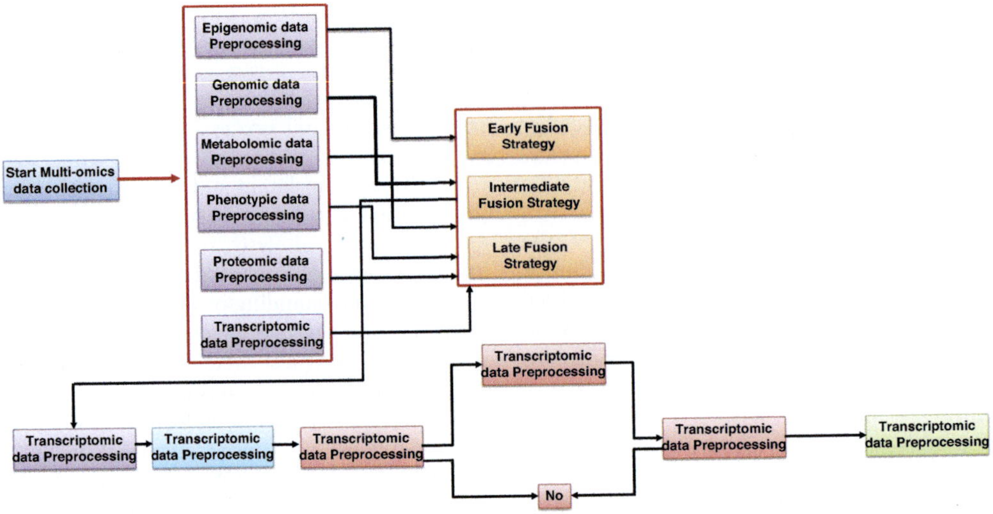

Fig. 3.1. Workflow diagram illustrating the strategies for integrating multi-omics data (genomic, metabolomic, phenotypic, proteomic, and transcriptomic) using early, intermediate, and late fusion approaches. The diagram also outlines the pre-processing steps for each omics type and explores downstream analysis pathways such as drug–target interaction prediction, disease gene prediction, cancer subtyping, and patient stratification (Author's own image).

and phenotype data processing are done during the training phase, taking the foundational data as the reference point for augmentation and normalization. The training phase makes the algorithm normalized with or adjusted with observational data (foundational data). At this point, a crucial step of the training phase is executed, where the data is subjected to efficient whole or segmented processing, error normalization, and optimization (iteration stage). This is followed by acclimatization of the model, in which the model is subjected to validation data in order to systematically acclimatize the algorithm-based model to get fine-tuned with each new dataset and not get used to a specific set of training data. The input of genotypic data is subjected to the model for it to process the prediction of phenotype. The estimated difference in actual and predicted phenotype is analyzed, followed by the running of successive epochs or the data processing and scanning cycles. This is terminated on arriving at the most successful epoch with no further scope for augmentation, improvement, and normalization (Runcie and Cheng, 2019). At this point, the AI-based DL or ML model gets completely acclimatized and augmented with the specific climatic condition or circumstance-based observational data pattern as taught in the training phase. During the predictive state, the observation portion is left to be done by the algorithm, and it is exclusively subjected to the genotypic data without observational data or phenotype that is to be deduced or predicted by the model. This stage marks a critical point in predictive modeling where the phenotypic traits or characteristic features are to be mapped or elucidated based on the input genetic markers (Tong and Nikoloski, 2021).

The prediction is followed by categorization or positioning of the predicted genotypes. The algorithm-based models are subjected to the genotype data and are then equipped for ranking of the predicted phenotypes based on various standard ranking criteria that can be various agronomic parameters, such as yield, production quality, and biotic, abiotic, or xenobiotic stress resistance (Zubair, 2024). The ranking of the predicted phenotypes is done by the model, and then, based on the ranking position, the best genotypes are taken for further field trials, as their predicted phenotypes have been selected based on some pre-defined criteria. Such a sequential methodology for analysis of multi-omics data can result in accurate predictions of gene–environment and gene–disease interactions. Such sets of big comprehensive data having large volume and variability can bring various crucial non-linear correlations that remain unnoticed in single-omics analysis processes with much narrower datasets.

Technical challenges in data processing and data integration during multi-omics data with real-time phenomics data, such as field trials, environmental monitoring, pest interactions, soil health, etc., during an integrational screening, are emerging as one of the significant causes of concern in the modern era. The challenges arise solely due to the variations in prerequisites during experimentation, experimental condition intricacies, data extraction methodologies, size, margin of error, properties, nature, and format of data that represent various omics of biological systems to be integrated to perform a comprehensive multi-omics screening of biological data (Harrison and Caccamo, 2022). Such technical specifications also lack proper management and compatibility in the case of real-time phenomics data. All such large datasets require augmentation and standardization at multiple levels for efficient integration. Highly variable data can be challenging for marking of important features, causing the need for dimensionality minimization functions like principal component analysis (PCA) that reduces the dimensionality (correlated variables) but preserves the actual core variability of the dataset, leading to a pinpoint approach causing efficient normalization of the data (Wang *et al.*, 2023). Big datasets during integration of multi-omics can be efficiently managed in terms of statistical analysis by using dynamic AI and ML-based tools such as random forest (RF) and support vector machines (SVMs) to identify specific patterns within the dataset as a prerequisite for predictive modeling. In addition to sophistication in analysis, multi-omics data can have technical specifications as per compatibility in format for execution, retrieval, storage, and analysis. This enables compatible computing systems with technical intricacy to efficiently handle the dataset without system glitches or failures. This mandates New Age computing methodologies like parallelism in computing (parallel computing) for bifurcating the necessary processing operations into small components for simultaneous parallel processing all at once in an augmented manner (Navarro *et al.*, 2014). Combinational computing approaches can be suitably efficient for pre-processing and processing the reads through a cluster of high-performance computing devices with smooth compatibility with the datasets. Compatibility mismatches and additional demands for computational resources can be retrieved on a real-time basis over the web through sophisticated cloud-computing techniques (Grzesik *et al.*, 2021). Interpretations of complex data can be augmented by AI-based visualization algorithms such as DNN or CNN that are specialized for working in visualization problems, processing of reads in pathway analysis models to suitably correspond them with respective biological pathways. Such intricate technical reforms can work suitably in elucidation of biological systems through efficient analysis of multi-omics data for application in predictive modelling and genomic selection-based precision breeding with sufficient transparency, authenticity, and reproducibility.

3.5 Integration of Environmental Dynamism with AI

Environmental dependence of agricultural practices makes the system of predictive modeling much more intricate and complex. The roots can be traced back to the classical concepts of genetics, such as polygenic inheritance or quantitative inheritance, in which there is no exclusive governance of genetic factors on the phenotypic expression; rather, the phenotypic output is a result of the combined effect of genetic and environmental factors. Such traits or characteristic features are not under the influence of a single gene and are thus polygenic, representing a spectrum of phenotypes in the F2 generation that are a representation of the combined effects of genes and environment on phenotype (Crouch and Bodmer, 2020). Polygenic traits or metric traits are quantitative in nature and have suitable agronomic and economic importance in agriculture. Variation in environmental conditions demands better acclimatization and resilience for better adaptation and stress tolerance among the selected lead breeding candidates (Zhang and Belsky, 2020). Conventional selection mechanisms repeatedly fail to address the site-specific non-uniformity in intra-cellular, extra-cellular, as well as the surrounding environment of the lead breeding candidate crop. AI-based algorithms such as random forest (RF), which specifies a resilient strategy dealing with decision trees based on the CART method, have been proven to be a strong alternative in processing and prediction of phenotypic attributes that are polygenic in nature (Crossa *et al.*, 2017). Environmental influence is significant for various characteristic features such as yield, disease resistance, flowering incidence, fruiting, growth parameters, bioactive and therapeutic component concentrations, etc. (Cuevas *et al.*, 2020). Incidence of comprehensive big data with sufficient variability from multi-omics gives a competitive edge on possible environment acclimatization, but in order to ensure robustness in the predictive modelling, multi-dimensional data from trials in different agro-climatic zones can be used to acclimatize the AI-based model to accurately predict the phenotypic expression in variable environmental conditions (multi-environment trial system). Models having the capacity to integrate both the environmental and genotypic influences for predictive uniqueness will go a long way as a successful model in precision breeding for sustainable agricultural improvement (Xu *et al.*, 2023). Environmental factors include topography, soil health, weather patterns, weather extremes, photoperiod, humidity, pest incidence, and other stress-inducing factors. Consideration of such factors makes the AI-based models more realistic and robust for prolonged reliability. The gene-based predictive model (G) and environment-based predictive model (E) are integrated to deduce a gene–environment interplay (G × E) model. The gene-based predictive model (G) solely depends on the single-omics or multi-omics-based genotypic data, whereas the environment-based predictive model (E) depends on the multi-dimensional (N-dimensional) environmental factors as specified according to the requirements of the specific agro-climatic zone (Feng *et al.*, 2024). Furthermore, the marker characteristic features sorted from the two different models (G and E) are processed in a combinational approach in order to chalk out features for feeding into a newly devised gene–environment interplay (G × E) model based on element-wise multiplication of fundamental data structures (tensors) or direct inclusion operations for yielding convergent dimensions. Such features are then augmented into the G × E model for navigation through the training, testing, and validation stages before harvesting the final output of the model. The final output of such a G × E model provides a suitable phenotypic prediction for a specific lead candidate crop that is then subjected to ranking and selection for further precision breeding programs under the specified environmental conditions ensuring reliability, resilience, and robustness in the whole process of predictive modelling while navigating through site-specific environmental dynamism (Wang *et al.*, 2021).

3.6 AI-Based Multi-omics Techniques in Plant Defense Systems

Land plants face various categories of environmental disturbances and imbalances that affect their various aspects of growth and development. These environmental changes or imbalances occur in and around the ecosystem where the plant grows and completes its life cycle. Biotic and abiotic factors cause significant losses in annual yield and production quality of various food and economically important crop plants all over the world. The abiotic factors of the environment are perhaps the most concerning forces that cause a huge number of losses to the world's agricultural produce each and every year. These forces may appear singly or can also be in various combinations, badly affecting the growth and developmental processes of the plant (Wani *et al.*, 2016; Kumar, 2020; Ma *et al.*, 2020). The abiotic factors that affect the growth and development are various physical and chemical forces, such as temperature imbalances and extremes, salinity imbalances, exposure to harmful ultraviolet radiations, soil water-content extremes, soil pH imbalances, heavy metal and metalloid levels beyond tolerance, etc. (Suzuki *et al.*, 2014; Sulmon *et al.*, 2015; Wani *et al.*, 2016; He *et al.*, 2018; Wang *et al.*, 2018). The ever-increasing climate change, which is directly or indirectly due to human interventions and environmental pollution, just adds to the intensity of impact that these forces have on plant growth and development (He *et al.*, 2018).

In addition to abiotic factors, plants also face threats from biotic factors as they are exposed to primary inoculum. Pathogens adopt various strategies for successful infection and colonization that lead to the appearance of visible symptoms on the plant body. Such symptoms imply pathogen colonization and disease cycle initiation. Plants have their own lines of defense mechanisms to combat pathogens and disrupt the disease cycle. Such lines include morphological, anatomical, biochemical, cellular, and even cytoplasmic barriers against pathogen infestation and colonization (Murmu *et al.*, 2024). Pests infest the plant parts and cause significant damage, alongside their role as vectors that disseminate pathogens such as viruses into the plant system. Such factors have significant impacts on food security, biodiversity, ecosystem homeostasis, and public health.

Plants have internal defense mechanisms for protection against such biotic stressors at various levels within their system. Such mechanisms and responses have specific molecular intricacies that are blended into complex biological responses encompassing contributions from genomics, transcriptomics, proteomics, metabolomics, and phenomics, involving gene expression in response to stressors, RNA formation, and protein synthesis with specific post-translational modification and targeting leading to a specific alteration in phenotype. Such intricate mechanisms must be comprehensively understood for a holistic approach to multi-omics integration for sustainable agricultural productivity. AI-assisted approaches, such as deep learning models, support vector machines and random forest, work to synchronize the complex datasets obtained during abiotic stress response or plant–pathogen interaction situations, leading to systemization of the data to identify various patterns and hidden markers to include new aspects in predictive modeling for sustainable crop management (Wu and Xie, 2025).

AI-based algorithms such as SVMs have been documented as successfully identifying genes responsible for tolerance and defense responses against biotic and abiotic stressors in plants such as *Arabidopsis thaliana* (Liang *et al.*, 2023). In addition, identification of vulnerable genes through deciphering transcriptomic data where expression is altered in response to pathogen infestation has also been successfully done using AI-based approaches such as neural network models. It has also been used to decipher *cis*-regulatory elements of non-coding DNA that are responsible for transcriptional upregulation or downregulation of specific genes responsible for stress resistance in plant cells. AI-assisted deep learning algorithms can also dive deeper into the understanding of the differential expression of genes in response to stress that becomes evident from integrated multi-omics analysis of genomic and transcriptomic data (Vakilian, 2020). Such algorithms have also been used to identify the non-coding RNAs, such as long non-coding RNAs that perform a regulatory function of spearheading the epigenetic modifications in gene expressions guided by post-translational histone modifications in response to biotic or abiotic stressors. Long non-coding (lnc) RNAs alter the chromatin coiling states in order to reinstate gene expression in response to stress conditions (Pradhan *et al.*, 2023). Support vector machines and random forest approaches have been applied for accurately deciphering complex interactions involving pathogen-associated molecular patterns (PAMPs) and effector molecules secreted by the pathogens for successful infection. Such mechanisms involve protein–protein interactions between plants and pathogens that underscore intimate biochemical regulatory pathways controlling pathogen identification, disease tolerance, and early recovery by plants. AI-based visual identification and pattern recognition tools such as You Only Look Once (YOLO) version 3, devised by Joseph Redmon, work on real-time monitoring and processing of images to detect multiple hidden objects with suitable annotations, rapidity, and accuracy. Such tools have reshaped real-time phenomics with sufficiently fast and accurate on-site monitoring of crop plants during field trials for the detection of minute traces of symptoms caused as a result of successful pathogen infection or pest infestation, or abiotic stressors (Wen *et al.*, 2023). Stripe rust of wheat, a fungal disease caused by the prevalent fungus *Puccinia striiformis*, and leaf spot disease of tomato caused by the fungus *Septoria lycopersici* were efficiently detected in wheat and tomato plants, respectively, at initial stages through accurate real-time AI-based video processing and image detection systems (Liu *et al.*, 2020; Cardellicchio *et al.*, 2023). The application of CNN algorithms in image processing and real-time detection of phenomics posed a challenge in careful scanning of every specific image grid location by systematically applying altered frame sizes to accurately identify and characterize a hidden object for the detection of pest infestations, as well as pathogen infection symptoms on crop plants. This limitation was augmented by using a further suitable region proposal network (RPN) approach, in which there is a selective searching mechanism by selecting regions of interest for suitable scanning and detection. This makes the real-time monitoring of phenomics data much more reliable and rapid (Yuan *et al.*, 2023). AI-based tools such as ASRpro, DeepAProt, ASLncR, and PLncPRO are widely applied for real-time capturing and processing of phenomics data (Pradhan *et al.*, 2023; Kundu *et al.*, 2024). Such approaches have also been applied to detect changes in metabolic pathway intricacies due to the release of some effectors by the pathogens to initiate the disease cycle inside plant cells. Such effector molecules inside plant cells interact biologically with various enzymes of metabolic pathways to cause disruption in significant steps of the pathway and a change in the overall product of the biochemical pathway. Such changes inside plant cells are detected using AI-based algorithms for valuable insights on host–parasite interaction, defense mechanisms, disease cycle, and pathogen dissemination (Yuan *et al.*, 2023).

In addition to biotic and abiotic stressors, AI-based approaches have also become instrumental in early detection and study of xenobiotic stress conditions involving heavy metals and metalloids. The elements that have been listed as deleterious or toxic heavy metals and metalloids include the heavy metals Cr, Ni, Cu, Zn, Hg, Pb and Cd, and the metalloid arsenic (As) (Raffa *et al.*, 2021). Among these, the metalloid As is one of the most toxic elements, and may be present in soil, soil water, water and sediments (Chung *et al.*, 2014; Adeloju *et al.*, 2021). Arsenic has been known as a toxic element since ancient times, as early as 300 BC. It is regarded as one of the most lethal metalloids on Earth. It is a component of the Lithosphere and is naturally found in bound and blended form in approximately 200 minerals (Sharma, 2013). There have been lots of reports and research on severe contamination of soil water due to As. It has already been reported by various workers that the xenobiotic stress not only alters the physiological and biochemical state of the plants but also changes signaling pathways within them (Rodziewicz *et al.*, 2013; Zhu, 2016; Waqas *et al.*, 2019). The phytotoxic effects of heavy metal stress, such as arsenic, include degeneration of chlorophyll and oxidative stress (Ahmad *et al.*, 2025). Cell death can be caused by arsenic, as As(III) causes loss of enzyme and protein functions as it reacts with the sulfahydryl groups of proteins. As As(V) is analogous to phosphate, it can act as a toxic substitute for phosphate in various important molecules, like ATP in plant cells (Quaghebeur and Rengel, 2003). It has been repeatedly reported that heavy metals such as As are one of the most lethal elements for plants, not only because of their negative impacts on plant growth and development but also because they have no beneficial impacts on plant growth and functioning (Chibuike and Obiora, 2014; Gupta and Khan, 2015). AI-assisted algorithms such as eXtreme Gradient Boosting (XGboost), support vector machines, remote sensing imagery algorithms, etc., can effectively detect the heavy metal contamination state of soil, water, and plants using spectral imagery methods to elucidate areas having elevated concentrations of heavy metals and metalloids (Hu *et al.*, 2020; Chen *et al.*, 2024; Tavallaie Nejad *et al.*, 2025). The spectrum of images from continuous and comprehensive monitoring can be analyzed to find minor traces of heavy metal contamination symptoms on plant morphology and development, and to immediately opt for suitable remedial measures.

3.7 Conclusion and Prospects

AI-based approaches in predictive modeling lead to a comprehensive, significant shift from conventional hybridization and classical plant breeding approaches that have far less reliability and accuracy. They aim to finally find the missing link between genotype and phenotype for better-informed decision making in modern agriculture. Such transformative approaches guide us smoothly for the selection of appropriate phenotypes to provide a futuristic, feasible, resilient, and sustainable solution to the ever-increasing disproportionation in agricultural output corresponding to the rise in world population. It leads to efficient breeding cycles with reliable phenotypic outcomes that have better acclimatization with the surrounding environment, with superior and desirable traits. Integration of multi-omics data for a far greater clarity-based understanding of complex biological pathways in plants poses significant challenges in data management and efficient integration that need to be efficiently managed for a holistic, data-driven approach in precision breeding. Future research can focus on streamlining data management and the various intricate norms that can address data integrity and the reliable application of AI-based models in agricultural practices. The issues pertaining to unknown off-target effects arising from the application of AI-based models on consumers, ecosystem health, and homeostasis of the environment can be suitably addressed for better reliability. The implementation of AI-based big data analytics in multi-omics for sustainable agriculture will heavily depend on technical skill and expertise for the extraction of foundational data. Technical skill and expertise must be intricate and reliable in every step from data extraction, model designing, to training, testing, augmentation, and optimization phases. Feasibility issues in real-time monitoring for comprehensive extraction of phenomics–environment data need to be addressed in the future so as to maintain a uniform system for suitable on-site monitoring of biotic and abiotic factors specific to the field. This also needs to be addressed for real-time monitoring of phenotypic parameters, including growth, development, stress tolerance, and environmental interactions. Ethical issues on accountability, integrity, and safety must be well-defined at an appropriate level for exponential progress throughout the world. Large-scale implementation of AI-based multi-omics data integration for predictive modeling in precision breeding should lead us towards a positive reshaping of agricultural practices while maintaining the scientific accountability, data integrity, environmental homeostasis, and human safety for a sustainable future. This can be a resilient solution toward the growing concern on global food security challenges, sustainable resource utilization and efficient waste management.

References

Adeloju, S.B., Khan, S. and Patti, A.F. (2021) Arsenic contamination of groundwater and its implications for drinking water quality and human health in under-developed countries and remote communities—A review. *Applied Sciences* 11(4). DOI: 10.3390/app11041926.

Ahmad, J., Baig, M.A., Ali, A.A., Beg, M.A., Al-Huqail, A.A. *et al.* (2025) Arsenic-induced phytotoxicity in *Trigonella foenum-graecum* and its regulation by thiol metabolism and ROS quenching enzymes. *Journal of Hazardous Materials Letters* 6, 100153. DOI: 10.1016/j.hazl.2025.100153.

Ashby, W. (1956) An introduction to cybernetics. *Journal of Mental Science* 104(435), 590–592. DOI: 10.1192/bjp.104.435.590-b.

Cardellicchio, A., Solimani, F., Dimauro, G., Petrozza, A., Summerer, S. *et al.* (2023) Detection of tomato plant phenotyping traits using YOLOv5-based single stage detectors. *Computers and Electronics in Agriculture* 207, 107757. DOI: 10.1016/j.compag.2023.107757.

Chen, L., He, P., Zhang, H., Peng, W., Qiu, J. *et al.* (2024) Applications of machine learning tools for biological treatment of organic wastes: Perspectives and challenges. *Circular Economy* 3(2), 100088. DOI: 10.1016/j.cec.2024.100088.

Chibuike, G.U. and Obiora, S.C. (2014) Heavy metal polluted soils: Effect on plants and bioremediation methods. *Applied and Environmental Soil Science* 2014, 1–12. DOI: 10.1155/2014/752708.

Chung, J., Yu, S. and Hong, Y. (2014) Environmental source of arsenic exposure. *Journal of Preventive Medicine and Public Health* 47(5), 253–257. DOI: 10.3961/jpmph.14.036.

Crossa, J., Pérez-Rodríguez, P., Cuevas, J., Montesinos-López, O., Jarquín, D. *et al.* (2017) Genomic selection in plant breeding: Methods, models, and perspectives. *Trends in Plant Science* 22(11), 961–975. DOI: 10.1016/j.tplants.2017.08.011.

Crossa, J., Montesinos-Lopez, O.A., Costa-Neto, G., Vitale, P., Martini, J.W. *et al.* (2025) Machine learning algorithms translate big data into predictive breeding accuracy. *Trends in Plant Science* 30(2), 167–184. DOI: 10.1016/j.tplants.2024.09.011.

Crouch, D.J. and Bodmer, W.F. (2020) Polygenic inheritance, GWAS, polygenic risk scores, and the search for functional variants. *Proceedings of the National Academy of Sciences* 117(32), 18924–18933. DOI: 10.1073/pnas.2005634117.

Cuevas, A.G., Mann, F.D., Williams, D.R. and Krueger, R.F. (2020) Discrimination and anxiety: Using multiple polygenic scores to control for genetic liability. *Proceedings of the National Academy of Sciences* 118(1). DOI: 10.1073/pnas.2017224118.

Debnath, S., Devi, K. and Majumder, S. (2025) Artificial intelligence (AI) and data-driven plant breeding. In: Khan, Z. (ed.) *Plant Breeding Technology: Future Trends and Challenges*. CABI, Wallingford, UK, pp. 126–137. DOI: 10.1079/9781800626638.0006.

Farooq, M.A., Gao, S., Hassan, M.A., Huang, Z., Rasheed, A. *et al.* (2024) Artificial intelligence in plant breeding. *Trends in Genetics* 40(10), 891–908. DOI: 10.1016/j.tig.2024.07.001.

Feng, W., Gao, P. and Wang, X. (2024) AI breeder: Genomic predictions for crop breeding. *New Crops* 1, 100010. DOI: 10.1016/j.ncrops.2023.12.005.

Fumia, N., Nair, R., Lin, Y., Lee, C., Chen, H. *et al.* (2023) Leveraging genomics and phenomics to accelerate improvement in mungbean: A case study in how to go from GWAS to selection. *The Plant Phenome Journal* 6(1), 88–102. DOI: 10.1002/ppj2.20088.

Gordon, A.D., Breiman, L., Friedman, J.H., Olshen, R.A. and Stone, C.J. (1984) Classification and regression trees. *Biometrics* 40(3), 874. DOI: 10.2307/2530946.

Grzesik, P., Augustyn, D.R., Wyciślik, Ł. and Mrozek, D. (2021) Serverless computing in omics data analysis and integration. *Briefings in Bioinformatics* 23(1), bbab349. DOI: 10.1093/bib/bbab349.

Gupta, M. and Khan, E. (2015) Mechanism of arsenic toxicity and tolerance in plants: Role of silicon and signalling molecules. In: Tripathi, B.N. and Müller, M. (eds) *Stress Responses in Plants*. Springer Nature, pp. 143–157. DOI: 10.1007/978-3-319-13368-3_6.

Harrison, R.J. and Caccamo, M. (2022) Managing data in breeding, selection and in practice: A hundred year problem that requires a rapid solution. In: Williamson, H.F. and Leonelli, S. (eds) *Towards Responsible Plant Data Linkage: Data Challenges for Agricultural Research and Development*. Springer Nature, pp. 37–64. DOI: 10.1007/978-3-031-13276-6_3.

He, M., He, C. and Ding, N. (2018) Abiotic stresses: General defenses of land plants and chances for engineering multistress tolerance. *Frontiers in Plant Science* 9, 71–88. DOI: 10.3389/fpls.2018.01771.

Hu, B., Xue, J., Zhou, Y., Shao, S., Fu, Z. *et al.* (2020) Modelling bioaccumulation of heavy metals in soil-crop ecosystems and identifying its controlling factors using machine learning. *Environmental Pollution* 262, 114308. DOI: 10.1016/j.envpol.2020.114308.

Jubair, S. and Domaratzki, M. (2023) Crop genomic selection with deep learning and environmental data: A survey. *Frontiers in Artificial Intelligence* 5, 95–109. DOI: 10.3389/frai.2022.1040295.

Kumar, S. (2020) Abiotic stresses and their effects on plant growth, yield and nutritional quality of agricultural produce. *International Journal of Food Science and Agriculture* 4(4), 367–378. DOI: 10.26855/ijfsa.2020.12.002.

Kundu, S., Saini, D.K., Meena, R.K., Bahuguna, R.N. and Jagadish, S.V. (2024) High-throughput phenotyping and AI technologies for deciphering crop resilience to heat stress. *Plant Physiology Reports* 29(4), 699–715. DOI: 10.1007/s40502-024-00821-4.

Liang, M., Tang, Q., Zhong, J. and Ai, Y. (2023) Machine learning empowered multi-stress level electromechanical phenotyping for high-dimensional single cell analysis. *Biosensors and Bioelectronics* 225, 115086. DOI: 10.1016/j.bios.2023.115086.

Liu, G., Nouaze, J.C., Touko Mbouembe, P.L. and Kim, J.H. (2020) YOLO-tomato: A robust algorithm for tomato detection based on YOLOv3. *Sensors* 20(7), 2145. DOI: 10.3390/s20072145.

Ma, J., Cheng, Z. and Cao, Y. (2025) Artificial intelligence-assisted breeding for plant disease resistance. *International Journal of Molecular Sciences* 26(11), 5324. DOI: 10.3390/ijms26115324.

Ma, Y., Dias, M.C. and Freitas, H. (2020) Drought and salinity stress responses and microbe-induced tolerance in plants. *Frontiers in Plant Science* 11, 591911. DOI: 10.3389/fpls.2020.591911.

Maderazo, D., Flegg, J.A., Algama, M., Ramialison, M. and Keith, J. (2022) Detection and identification of *cis*-regulatory elements using change-point and classification algorithms. *BMC Genomics* 23(1). DOI: 10.1186/s12864-021-08190-0.

Maraveas, C. (2024) Image analysis artificial intelligence technologies for plant phenotyping: Current state of the art. *Agri Engineering* 6(3), 3375–3407. DOI: 10.3390/agriengineering6030193.

Monaco, S., Greco, S., Farasin, A., Colomba, L., Apiletti, D. *et al.* (2021) Attention to fires: Multi-channel deep learning models for wildfire severity prediction. *Applied Sciences* 11(22), 11060. DOI: 10.3390/app112211060.

Murmu, S., Sinha, D., Chaurasia, H., Sharma, S., Das, R. *et al.* (2024) A review of artificial intelligence-assisted omics techniques in plant defense: Current trends and future directions. *Frontiers in Plant Science* 15, 54–67. DOI: 10.3389/fpls.2024.1292054.

Navarro, C.A., Hitschfeld-Kahler, N. and Mateu, L. (2014) A survey on parallel computing and its applications in data-parallel problems using GPU architectures. *Communications in Computational Physics* 15(2), 285–329. DOI: 10.4208/cicp.110113.010813a.

Novielli, P., Romano, D., Pavan, S., Losciale, P., Stellacci, A.M. *et al.* (2024) Explainable artificial intelligence for genotype-to-phenotype prediction in plant breeding: A case study with a dataset from an almond germplasm collection. *Frontiers in Plant Science* 15, 29–47. DOI: 10.3389/fpls.2024.1434229.

Peleke, F.F., Zumkeller, S.M., Gültas, M., Schmitt, A. and Szymański, J. (2024) Deep learning the *cis*-regulatory code for gene expression in selected model plants. *Nature Communications* 15(1), 56–69. DOI: 10.1038/s41467-024-47744-0.

Pradhan, U.K., Meher, P.K., Naha, S., Rao, A.R. and Gupta, A. (2023) ASLncR: A novel computational tool for prediction of abiotic stress-responsive long non-coding RNAs in plants. *Functional & Integrative Genomics* 23(2), 51–67. DOI: 10.1007/s10142-023-01040-0.

Quaghebeur, M. and Rengel, Z. (2003) The distribution of arsenate and arsenite in shoots and roots of *Holcus lanatus* is influenced by arsenic tolerance and arsenate and phosphate supply. *Plant Physiology* 132(3), 1600–1609. DOI: 10.1104/pp.103.021741.

Raffa, C.M., Chiampo, F. and Shanthakumar, S. (2021) Remediation of metal/metalloid-polluted soils: A short review. *Applied Sciences* 11(9), 4134. DOI: 10.3390/app11094134.

Rodziewicz, P., Swarcewicz, B., Chmielewska, K., Wojakowska, A. and Stobiecki, M. (2013) Influence of abiotic stresses on plant proteome and metabolome changes. *Acta Physiologiae Plantarum* 36(1), 1–19. DOI: 10.1007/s11738-013-1402-y.

Runcie, D. and Cheng, H. (2019) Pitfalls and remedies for cross validation with multi-trait genomic prediction methods. *G3 Genes Genomes Genetics* 9(11), 3727–3741. DOI: 10.1534/g3.119.400598.

Ruperao, P., Rangan, P., Shah, T., Sharma, V., Rathore, A. *et al.* (2025) Developing pangenomes for large and complex plant genomes and their representation formats. *Journal of Advanced Research* 52–67. DOI: 10.1016/j.jare.2025.01.052.

Sangjan, W., Kick, D.R. and Washburn, J.D. (2025) Improving plant breeding through AI-supported data integration. *Theoretical and Applied Genetics* 138(6), 2–19. DOI: 10.1007/s00122-025-04910-2.

Shanker, S. (1995) Turing and the origins of AI. *Philosophia Mathematica* 3(1), 52–85. DOI: 10.1093/philmat/3.1.52.

Sharma, I. (2013) Arsenic stress in plants: An inside story. In: Hakeem, K.R., Ahmad, P. and Ozturk, M. (eds) *Crop Improvement: New Approaches and Modern Techniques*. Springer, New York, pp. 379–400. DOI: 10.1007/978-1-4614-7028-1_13.

Shook, J., Wu, L., Gangopadhyay, T., Ganapathysubramanian, B., Sarkar, S. *et al.* (2021) Integrating genotype and weather variables for soybean yield prediction using deep learning. *Preprint bioRXiv* 3315, 61–78. DOI: 10.1101/331561.

Sulmon, C., Van Baaren, J., Cabello-Hurtado, F., Gouesbet, G., Hennion, F. *et al.* (2015) Abiotic stressors and stress responses: What commonalities appear between species across biological organization levels? *Environmental Pollution* 202, 66–77. DOI: 10.1016/j.envpol.2015.03.013.

Suzuki, N., Rivero, R.M., Shulaev, V., Blumwald, E. and Mittler, R. (2014) Abiotic and biotic stress combinations. *New Phytologist* 203(1), 32–43. DOI: 10.1111/nph.12797.

Tavallaie Nejad, A., Vila, M.C., Paneiro, G. and Baptista, J.S. (2025) A systematic review of machine learning algorithms for soil pollutant detection using satellite imagery. *Remote Sensing* 17(7), 1207. DOI: 10.20944/preprints202503.0068.v1.

Tong, H. and Nikoloski, Z. (2021) Machine learning approaches for crop improvement: Leveraging phenotypic and genotypic big data. *Journal of Plant Physiology* 257, 153354. DOI: 10.1016/j.jplph.2020.153354.

Vakilian, K.A. (2020) Machine learning improves our knowledge about miRNA functions towards plant abiotic stresses. *Scientific Reports* 10(1), 6–19. DOI: 10.1038/s41598-020-59981-6.

Voss-Fels, K.P., Stahl, A., Wittkop, B., Lichthardt, C., Nagler, S. *et al.* (2019) Breeding improves wheat productivity under contrasting agrochemical input levels. *Nature Plants* 5(7), 706–714. DOI: 10.1038/s41477-019-0445-5.

Wang, L., Bai, P., Yuan, X., Chen, H., Wang, S. *et al*. (2018) Genetic diversity assessment of a set of introduced mung bean accessions (*Vigna radiata* L.). *The Crop Journal* 6(2), 207–213. DOI: 10.1016/j.cj.2017.08.004.

Wang, H., Ye, M., Fu, Y., Dong, A., Zhang, M. *et al*. (2021) Modeling genome-wide by environment interactions through omnigenic interactome networks. *Cell Reports* 35(6), 109114. DOI: 10.1016/j.celrep.2021.109114.

Wang, W., Guo, W., Le, L., Yu, J., Wu, Y. *et al*. (2023) Integration of high-throughput phenotyping, GWAS, and predictive models reveals the genetic architecture of plant height in maize. *Molecular Plant* 16(2), 354–373. DOI: 10.1016/j.molp.2022.11.016.

Wani, S.H., Kumar, V., Shriram, V. and Sah, S.K. (2016) Phytohormones and their metabolic engineering for abiotic stress tolerance in crop plants. *The Crop Journal* 4(3), 162–176. DOI: 10.1016/j.cj.2016.01.010.

Waqas, M.A., Kaya, C., Riaz, A., Farooq, M., Nawaz, I. *et al*. (2019) Potential mechanisms of abiotic stress tolerance in crop plants induced by thiourea. *Frontiers in Plant Science* 10, 36–51. DOI: 10.3389/fpls.2019.01336.

Wen, T., Li, J., Wang, Q., Gao, Y., Hao, G. *et al*. (2023) Thermal imaging: The digital eye facilitates high-throughput phenotyping traits of plant growth and stress responses. *Science of The Total Environment* 899, 165626. DOI: 10.1016/j.scitotenv.2023.165626.

Wójcik-Gront, E., Zieniuk, B. and Pawełkowicz, M. (2024) Harnessing AI-powered genomic research for sustainable crop improvement. *Agriculture* 14(12), 2299. DOI: 10.3390/agriculture14122299.

Wu, Y. and Xie, L. (2025) AI-driven multi-omics integration for multi-scale predictive modeling of genotype-environment-phenotype relationships. *Computational and Structural Biotechnology Journal* 27, 265–277. DOI: 10.1016/j.csbj.2024.12.030.

Xiao, Q., Bai, X., Zhang, C. and He, Y. (2022) Advanced high-throughput plant phenotyping techniques for genome-wide association studies: A review. *Journal of Advanced Research* 35, 215–230. DOI: 10.1016/j.jare.2021.05.002.

Xu, C., Ganesh, S.K. and Zhou, X. (2023) MtPGS: Leverage multiple correlated traits for accurate polygenic score construction. *The American Journal of Human Genetics* 110(10), 1673–1689. DOI: 10.1016/j.ajhg.2023.08.016.

Xu, Y., Zhang, X., Li, H., Zheng, H., Zhang, J. *et al*. (2022) Smart breeding driven by big data, artificial intelligence, and integrated genomic-enviromic prediction. *Molecular Plant* 15(11), 1664–1695. DOI: 10.1016/j.molp.2022.09.001.

Yasrab, R., Atkinson, J.A., Wells, D.M., French, A.P., Pridmore, T.P. *et al*. (2019) RootNav 2.0: Deep learning for automatic navigation of complex plant root architectures. *Giga Science* 8(11), 23–36. DOI: 10.1093/gigascience/giz123.

Yuan, H., Song, M., Liu, Y., Xie, Q., Cao, W. *et al*. (2023) Field phenotyping monitoring systems for high-throughput: A survey of enabling technologies, equipment, and research challenges. *Agronomy* 13(11), 2832. DOI: 10.3390/agronomy13112832.

Zhang, X. and Belsky, J. (2020) Three phases of gene × environment interaction research: Theoretical assumptions underlying gene selection. *Development and Psychopathology* 34(1), 295–306. DOI: 10.1017/s0954579420000966.

Zhu, J. (2016) Abiotic stress signaling and responses in plants. *Cell* 167(2), 313–324. DOI: 10.1016/j.cell.2016.08.029.

Zubair, M. (2024) Clinical applications of artificial intelligence in identification and management of bacterial infection: Systematic review and meta-analysis. *Saudi Journal of Biological Sciences* 31(3), 103934. DOI: 10.1016/j.sjbs.2024.103934.

4 AI-Based Genotype-to-Phenotype Prediction in Crop Breeding

Dhivyapriya Dharmaraj[1]*, Kalirajan Murugasridevi[2], Manonmani Karthikeyan[1] and Murali Shankar Perumal[3]
[1]Department of Genetics and Plant Breeding, Amrita School of Agricultural Sciences, Amrita Vishwa Vidyapeetham, Coimbatore; [2]Department of Agricultural Entomology, Amrita School of Agricultural Sciences, Amrita Vishwa Vidyapeetham, Coimbatore; [3]Department of Plant Pathology, Pushkaram College of Agricultural Sciences, Tamil Nadu Agricultural University, Pudukottai

Abstract
Advancements in genomics revolutionized crop breeding, and there has been significant knowledge gained in understanding the genotype and phenotypic interaction. Artificial intelligence (AI) is being included in crop breeding, marking a paradigm shift toward data-driven agricultural techniques. The AI-empowered attempt aims to close this gap by combining sophisticated machine learning models to improve genotype-to-phenotype interpretability and predictive accuracy, which will increase their dependability and facilitate better breeding choices. In crop breeding, machine learning (ML) is considered to be a computational method for identifying predictive patterns in data that has enormous potential to transform genotype-to-phenotype predictions. Furthermore, explainable artificial intelligence (XAI) algorithms were created as a result of recent developments in machine learning with the goal of clarifying the inner workings of machine learning models. Tree-based machine learning models (Random Forests, AdaBoost, and Gradient Boosting) can also be used for genomic prediction using germplasm data. This chapter will discuss the various AI tools involved in genomic prediction and their present-day applications, obstacles, and prospective potential to achieve precision farming.

Keywords: Artificial intelligence, Crop breeding, Genomic prediction, Precision agriculture, Sustainable agriculture

4.1 Introduction

With the world's population growing and the effects of climate change-related weather events intensifying, increasing food output is still a difficult task. The advances in selection intensity, additive genetic variation, and the inherited genetic pattern are required to achieve genetic gain, which is a measure of enhanced crop productivity over time. The use of particular technologies, i.e. speed-breeding, phenomics, and crop genomics, will accelerate the rate of genetic gain. Traditional linear regression models may not be able to fully capture the complex interactions between genotypes and phenotypes, despite being useful tools in genetic research. A substantial gap was identified, and, in order to bridge it, AI-empowered tools are

*Corresponding author: d_dhivyapriya@cb.amrita.edu

 Plant Genomic Prediction: AI-Enhanced Models for Plant Genomics and Crop Improvement (ed. Jen-Tsung Chen)
DOI: 10.1079/9781836991328.0004

necessary. Large datasets are made to increase the effectiveness and accuracy. The potential of AI, especially in genotype and phenotypic prediction, will lead to a deeper knowledge of basic plant physiology and the genetic mechanisms behind it. Even with great advancements, it is still difficult to predict phenotypes from high-dimensional genomic data, especially when determining the main genetic variables affecting these predictions. These models frequently make the assumption that biological interactions are linear, which may not always be the case. Furthermore, they might have trouble handling high-dimensional genomic data, which could result in problems like multi-collinearity and overfitting that lower predicted accuracy and dependability. The advancement in omics technologies has created large multidimensional datasets to explore and address the problems. The traditional methods will not employ these accurate imaging data, and the possibility of linking a genotype to a phenotype has been less biased since the insights are from high-throughput sequencing.

The fascinating plant science can be produced in an intelligent way by AI technologies. Phenotypes can be predicted by identifying nonlinear correlations in high-dimensional data. Deep learning (DL) models such as DeepAT and ResDeepGS have demonstrated greater performance in phenotypic prediction. Climate-resilient breeding populations can be developed in a short span by utilizing machine learning (ML) algorithms, which can be utilized in pre-breeding methodologies, tightly linked markers and genic traits, and finally selecting the best cultivars. By locating genomic areas linked to desired qualities like yield and stress resistance, artificial intelligence (AI) increases the precision of genomic selection. Additionally, the dynamic processes that lead to genetic variability, which is essential for plant adaptation to shifting climates, can be preserved and restored by these algorithms. In addition, AI is used to monitor and control plant stressors through the integration of advanced sensing technologies, including RGB, multi-spectral, and thermal images. AI solves issues with high-throughput phenotyping and gene functional analysis and creates new breeding opportunities. Combining high-throughput phenotyping (HTP) data with AI and machine learning algorithms has opened up new avenues for crop improvement by offering precise and effective methods for trait selection, which in turn has decreased the time and expense needed for variety development. Along with phenotypic prediction to support rapid breeding, AI-powered genomic research and big data can increase the efficacy, specificity, and safety of sophisticated crop enhancement techniques like genome editing and genomic selection. In conclusion, the integration of AI and machine learning algorithms with high-throughput phenotyping data is being used to overcome the difficulties in genotype-to-phenotype prediction in crop breeding.

4.2 Challenges in the Prediction of Genotype in Crops

The key factors in predicting the genotype and the genotype–environment (G × E) interaction, especially in unfavorable environments, are highly challenging in crops (Crossa *et al.*, 2017). Maize, wheat, rice, sorghum, soybean, and barley have higher G × E interaction than other crops when the environmental condition is unfavorable. Maize grows in diverse environments, responsive to nitrogen application, but severe drought or low nitrogen in the soil cannot be managed by the crop. The International Maize and Wheat Improvement Center (CIMMYT) highly focuses on the prediction of genotype under these unfavorable conditions. Epistatic interactions prevail in the crops, so the genome cannot be predicted. Soybean is sensitive to photoperiod and drought. The institutes are working on the use of G × E analysis to modify soybean lines to specific microclimates. The polygenic traits are complex, and they can be predicted with little accuracy (Spindel *et al.*, 2015). The low mapping resolution due to single-nucleotide polymorphism (SNP) chips can reduce the power and accuracy of predicting the genotypes. The functional annotation of the genome in crops is limited (Varshney *et al.*, 2014). The genomic prediction models like deep learning and Bayesian methods require large computing power (Montesinos-López *et al.*, 2018).

4.3 Genotypic and Phenotypic Datasets

Biological big data will be generated by means of whole-genome sequencing, where the number of single-nucleotide variations will be identified across the genome. A commonly used SNP encoding method in ML/DL is one-hot encoding, where a binary system is employed for phenotype prediction. For the four nucleotides, four columns are created, with presence marked as '1' and absence marked as '0' (Khaki and Wang, 2019). During marker trait association, there is an imbalance between the number of samples and features from phenotypic and genotypic data. It can be addressed by reducing the SNPs based on minor allele frequency. Apart from SNPs, other genomic variations like methylation data can also be used for phenotypic prediction (Champigny *et al.*, 2020).

Phenotypic data can be collected manually or by using high-throughput phenotyping facilities. These traits can be quantitative or qualitative, and the external appearance of an organism is influenced by genotype and environmental factors (Grinberg *et al.*, 2020). For ordinal or qualitative data, one-hot encoding should be used to score the presence or absence of specific traits. It is particularly used for the identification of symptoms of diseases, pests, and nutrient deficiencies. All phenotypic data will be subjected to pre-processing to handle the missing values, remove outliers, and normalize data.

4.3.1 Structured datasets

The major challenge in AI-assisted phenotype prediction is the lack of structured datasets with sufficient data and variability. Even though large volumes of phenotypic data are generated from various research projects, it is not readily accessible for broad use. This is because of missing values in the datasets, or the data is stored in separate repositories with limited access to the scientific community (Lobet, 2017; Bayer and Edwards, 2020). Hence, researchers rely on their own data for integrating with AI tools. Going forward, it is crucial to adopt data-driven hypotheses for experimental research. In order to maximize the utilization of datasets, FAIR principles (easily findable, highly accessible, interoperable, and reusable) need to be adhered to (Wilkinson *et al.*, 2016).

The second concern is a lack of standard description, including field experimental layout, management practices, and environmental factors. This creates a gap between the predicted and actual phenotype of a plant. In order to increase the utilization, researchers should follow the guidelines provided by the Minimum Information About a Plant Phenotyping Experiment project (MIAPPE) (Papoutsoglou *et al.*, 2020). As per the guide, it is necessary to maintain metadata with a standard format for the datasets generated from the research projects (Mayernik and Acker, 2018). The AgBioData consortium, which was initiated in 2015, aims to consolidate the genomic and breeding database and tools to improve data interoperability (Harper *et al.*, 2018). Additionally, a genotype–phenotype working group was formed to address the data-related challenges by integrating large-scale genotype and phenotype datasets. Deng *et al.* (2023) proposed several recommendations to enhance the reuse of data generated from scientific research projects. These include implementing standard data collection protocol, developing a centralized data-sharing platform, complete data annotation, quality check and promoting data integration, encouraging collaborative efforts among researchers, and enforcing legal policies to support data sharing.

4.3.2 Databases

GGB stands for genomics, genetics and breeding databases. Several GGB databases are available to provide published datasets for reuse and interoperability. Some of the databases are the AgBioData consortium, CottonGen, GrainGenes, Gramene, MaizeGDB, Soybase, Legume Information System, PeanutBase, Triticeae ToolBox, etc. (Deng *et al.*, 2023). Phenomics databases such as GnpIS, PGP Repository, Cartograplant, Plant-crops PathoPlant, Indian crop Phenome DB, etc., are also available for the utilization of phenotype and disease data for future research.

Table 4.1. Genotype to phenotype prediction for key agronomic traits in crop plants.

S. No.	Trait	Crop	AI model	References
1	Yield	Hybrid rice	AI_HIB	Sabouri and Sajadi, 2022
		Wheat	DeepAT	Li *et al.*, 2024
		Wheat	PheGeMIL	Togninalli *et al.*, 2023
		Tomato	ANN combined with sensitivity analysis	Belouz *et al.*, 2022
2	Multiple agronomic traits	Rice	DeepCCR	Ma *et al.*, 2024
		Cotton, maize, rice and tomato	SoyDNGB	Gao *et al.*, 2023
		Rice	LGT	Dong *et al.*, 2025
		Wheat	CropQuant-Air	Chen *et al.*, 2023
		Maize	AutoGP	Wu *et al.*, 2025
		Flax	Ridge Regression	Hoque *et al.*, 2024
3	Growth rate	Rice	ANN, GEP	Liu *et al.*, 2021
4	Leaf diseases	Chilli	LGENetB4CA	Van *et al.*, 2025
5	Northern leaf blight	Maize	Mark R-CNN	Stewart *et al.*, 2019

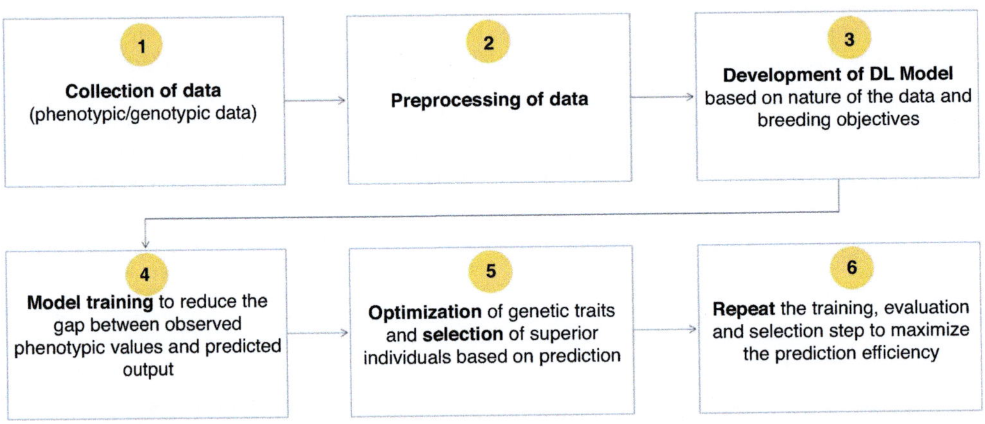

Fig. 4.1. Steps involved in designing AI-assisted crop breeding.

4.4 Methods for Genotype-to-Phenotype Prediction

Genomic selection is a method that predicts the phenotype of the individual using genetic marker data obtained through whole-genome sequencing (WGS) (Table 4.1). Phenotype prediction is easy when the environmental factors are under control. The commonly used linear model in genomic selection is genomic best linear unbiased prediction (GBLUP) (Meuwissen *et al.*, 2001). Some other linear models are Bayesian least absolute shrinkage and selection operator (LASSO), and ridge regression BLUP (RR-BLUP). However, the major challenges of these linear models are handling the large volume of marker data and difficulty in capturing the complex marker–trait interactions . In order to overcome this, nonlinear-based methods like machine learning (ML) and deep learning (DL) models have been identified for phenotype prediction (Montesinos-López *et al.*, 2021) (Fig. 4.1). These approaches offer several advantages, including handling high-dimensional genome data, integrating multimodal data, and providing more accurate predictions (Cuevas *et al.*, 2016). When compared with the traditional linear models, nonlinear models like random forest (RF), convolutional neural networks (CNNs), and feedforward deep neural networks will perform better in phenotype prediction (Spindel *et al.*, 2015; Ma *et al.*, 2018; Sandhu *et al.*, 2021).

Ensemble models combine multiple ML approaches, often performing better than individual ML approaches (Banerjee *et al.*, 2020). A comparison between a deep learning model and GBLUP by Montesinos-López *et al.* (2018) revealed that when genotype-by-environment interaction is included, GBLUP performed better than the DL model. Additionally, GBLUP allows multi-trait prediction. Conversely, another study on octoploid strawberry and tetraploid blueberry found that the DL model performed better than linear and Bayesian models when epistatic interactions were considered (Zingaretti *et al.*, 2020).

Cropformer is a deep learning platform that combines CNN with many self-attention mechanisms to increase the prediction accuracy. This helps for identification of genes associated with the particular trait and also phenotype prediction (Wang *et al.*, 2025). Its prediction efficiency is better for more than 20 traits of five major crops: maize, rice, wheat, foxtail millet, and tomato.

In addition to the above mentioned, genotype-to-phenotype prediction is also achieved by utilizing classical models such as support vector machine (SVM) and random forest (RF); neural network models such as multi-layer perceptron (MLP), CNN, deep neural network for genome prediction (DNNGP), deep belief networks (DBNs); and specialized models such as Soybean deep neural genome prediction (SoyDNGP) and DeepCCR (CCN plus bidirectional long short-term memory) (Tong and Nikoloski, 2021; Gao *et al.*, 2023; Wang *et al.*, 2023; Ma *et al.*, 2024).

4.5 Development and Evaluation of a Field-Based High-Throughput Phenotyping Platform for Biotic and Abiotic Stress Assessment

The real-world challenge to develop resilient crop varieties is mainly due to climate change and increasing environmental stress factors. Although molecular breeding methods have achieved advancements, phenotyping of varieties remained until now a great burden (Fiorani and Schurr, 2013). Conventional phenotypic techniques remain laborious and descriptive, need more expertise, and lack sustainability. High-throughput phenotyping (HTP), particularly under field conditions, mitigates these constraints by facilitating quick, harmless, and multi-temporal evaluation of plant reactions to various environmental challenges (Araus and Cairns, 2014). Field-based HTP solutions amalgamate imaging systems (such as RGB, thermal, and hyperspectral), proximal sensors, robots, and analytical tools to gather dynamical plant attribute datasets. This framework is essential for comprehending genotype × environment × management (G×E×M) relationships in practical settings (Tardieu *et al.*, 2017).

4.6 Platform Design and Technological Components

4.6.1 Structural configurations

Field-based high-throughput phenotyping tools are crucial for assessing plant responses to environmental factors under field and semi-field situations. There are three primary categories of these platforms: (i) ground-based mobile units; (ii) aerial systems; and (iii) fixed installations. Each variety is optimal for specific locations, plants, and characteristics.

Ground-based mobile units

Ground-based HTP devices include tractor-mounted systems, gantry-mounted tracks, and automated buggies. These carry a number of sensors, i.e. thermal imagers, spectral sensors, LiDAR systems, and RGB cameras. In Germany,

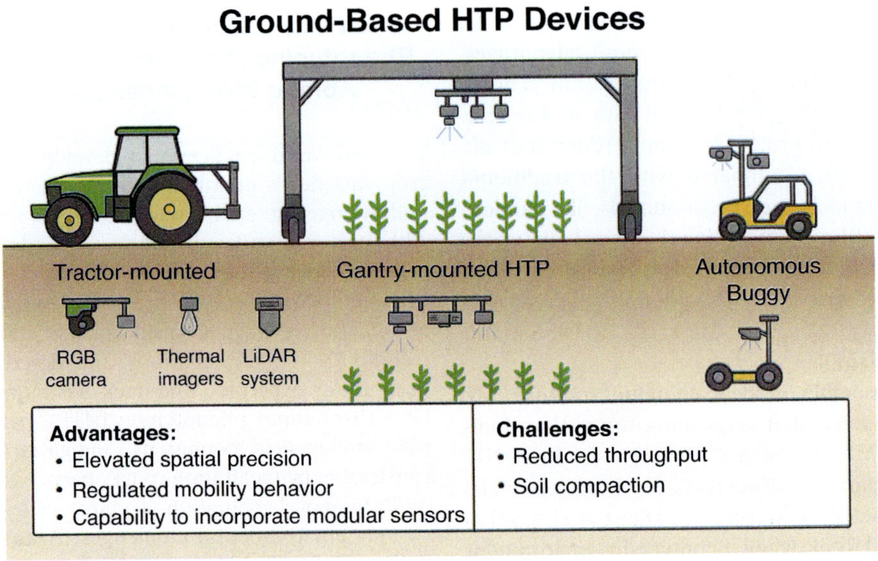

Fig. 4.2. Types of ground-based mobile units, their advantages and challenges (Author's own image)

Table 4.2. Sensors utilized for trait capture and stress indication (Author's own table)

Sensor type	Trait captured	Stress indicator
RGB cameras	Canopy cover, lesion area, growth rate	Disease spread, pest damage
Multispectral sensors	NDVI, PRI, RECI, chlorophyll indices	Nutrient stress, drought, and early disease
Hyperspectral sensors	Reflectance signatures across 400–2500 nm	Physiological disorders, salt toxicity
Thermal infrared	Canopy temperature	Stomatal closure, drought/heat stress
LiDAR	Plant height, biomass, and architecture	Growth retardation due to any stress
Soil/environmental	Soil moisture, EC, temperature	Contextual abiotic stress drivers

at Leibniz Institute of Plant Genetics and Crop Plant Research, a gantry system is installed which enables high-precision photographs and covers acres of fields (Virlet *et al.*, 2017). The advantages and challenges of these mobile platforms are presented in Fig. 4.2 (Pauli *et al.*, 2016).

Aerial systems

Drones, also known as unmanned aerial vehicles (UAVs), are becoming more and more popular for agricultural phenotyping because they are fast, flexible, and can cover a lot of ground. RGB, multispectral, hyperspectral, and thermal cameras can be attached to UAVs to take pictures of the canopy level across whole areas in just a few minutes. These tools are especially good at keeping an eye on NDVI (Normalized Difference Vegetation Index), the height of plants, and canopy coverage. Thermal imaging is used to find stress caused by drought (Table 4.2). and making orthomosaic field maps for analyzing traits on a large scale (Zaman-Allah *et al.*, 2015). Recent improvements in lightweight sensors, autonomous navigation, and image stitching software have made UAV-based phenotyping more accurate and possible (Madec *et al.*, 2017). But there are still problems with the rules, how long the energy lasts, and how much they depend on weather conditions like wind and cloud cover.

Fixed installations

Fixed phenotyping platforms like stationary gantries and rainout shelters are good for applying semi-controlled plant stress. These kinds of structures usually cover huge areas and use linear rail systems or suspended rigs to move sensors over the crops. The PhénoField® system in France is a great example of a framework that uses multi-sensor imaging and portable rainout shelters to study how wheat and barley respond to drought in conditions that are similar to those in the real world (Cabrera-Bosquet *et al.*, 2016). One benefit of stationary installations is that they allow the control of exactly how much water and fertilizer is used and keep an eye on the surroundings throughout the growing period of time. Attributes are more inclined to be transmitted and heritable since the environment does not vary greatly. Such structures help evaluate remote sensing equipment and research the effects of stress within plants. Nevertheless, they remain expensive and have less scalability.

4.6.2 Data processing pipeline

The data processing pipeline is an important part of HTP systems because it turns raw sensor data into phenotypic traits that are biologically useful. There are four main steps in this process: pre-processing, feature extraction, analysis, and visualization.

Pre-processing

Pre-processing makes sure that data from sensors are consistent and can be trusted for later analysis. This step includes filtering out noise to get rid of artifacts caused by things like ambient light, wind moving plants, or sensors that are not stable. Spectral calibration fixes wavelength drift and makes sure that sensor outputs match up with known reference standards. Georectification, also called spatial referencing, uses GPS and ground control points to line up images with real-world coordinates. This is especially important for UAV- and gantry-based imaging systems.

Feature extraction

This step is to figure out the vegetation indices and structural traits that show how healthy a plant is, what its structure is like, and how stressed it is. NDVI, PRI, and RECI (Red-Edge Chlorophyll Index) are some of the most common indices. RGB, LiDAR, or stereo image data can be used to get structural traits like plant height, canopy volume, and leaf area index. If trying to find stress, this step might also involve finding disease lesions, chlorosis, or really hot spots.

Analysis

Machine learning (ML) models are used to figure out what the features mean and to group or regress traits linked to biotic and abiotic stress. Random forests, support vector machines, and convolutional neural networks are examples of supervised learning algorithms that are often used to find patterns in datasets with more than one dimension. These models can tell the difference between different types of stress, guess the values of traits (like biomass and stomatal conductance), and guess how different genotypes will react to certain situations in the environment.

Visualization

The last step is to make visual outputs that will help with understanding and making decisions. Heat maps and chlorophyll distribution maps show how stressed something is and where it is in space. Stress progression maps or time-lapse visualizations can show how things change over time. Three-dimensional crop models made from LiDAR or stereo vision data can help us understand how plants are built and how well they can handle stress.

Table 4.3. Example biotic stress detection using HTP (Author's own table)

Crop	Pest/disease	Sensor	Trait measured	References
Tomato	*Alternaria solani*	RGB + CNN	Lesion count, color change	Mohanty *et al.*, 2016
Rice	*Xanthomonas oryzae*	Hyperspectral	Chlorosis, red edge shift	Mahlein *et al.*, 2013
Maize	*Spodoptera frugiperda*	Thermal, RGB	Heat spots, chewing damage	Wang *et al.*, 2018
Wheat	*Puccinia striiformis* (Rust)	Multispectral, RGB	Stripe pattern, necrosis	Bauriegel *et al.*, 2011
Potato	*Phytophthora infestans*	Hyperspectral	Water content, spectral indices	Wahabzada *et al.*, 2015
Soybean	*Heterodera glycines* (Nematode)	Chlorophyll Fluorescence	Root zone stress, photosynthetic efficiency	Zhang *et al.*, 2020
Cotton	*Bemisia tabaci* (Whitefly)	UAV-RGB, Thermal	Leaf wilting, abnormal temperature patterns	Sankaran *et al.*, 2015
Grapevine	*Erysiphe necator* (Powdery mildew)	Multispectral, RGB	Fungal patches, canopy color changes	Zarco-Tejada *et al.*, 2015
Banana	*Fusarium oxysporum* (Panama disease)	Hyperspectral	Vascular wilt, leaf chlorosis	Lu *et al.*, 2020
Sugarcane	*Melanaphis sacchari* (Aphid)	Thermal, NDVI drone	Canopy temperature, vigor index drop	Raza *et al.*, 2021

4.6.3 Biotic stress phenotyping

Insect pests, plant pathogens (fungi, bacteria, viruses), and parasitic weeds are all examples of biotic stresses that pose a serious threat to the productivity of crops around the world. Conventional methods of biotic stress measurement, such as quantitative disease assessment and visual scoring, are laborious, subjective, and not suitable for large-scale application. These issues can be overcome with the aid of field-based HTP platforms. These platforms help in early recognition, warning, assessment of pests and diseases, and surveillance on a large scale under real field situations (Mahlein, 2016; Murugasridevi *et al.*, 2024, 2025a). Table 4.3 represents the deployment of HTP technology for biotic stress detection.

Early and non-invasive detection

HTP systems encompassing sensors mainly pertaining to spectral imaging, such as hyperspectral, thermal, and multispectral cameras, easily discover the minute level changes in the crop before and after pest and disease infestation. Specifically, the moisture level of the leaves, morphology, anatomy of the leaves, and pigment concentration may differ in visible and near-IR regions. These signs specify that there is an attack by pest or disease (Zarco-Tejada *et al.*, 2012). This technique ensures easy, earlier detection and helps to fasten the appropriate management tactics.

Disease detection

Mahlein *et al.* (2013) denoted that before the appearance of any clear symptoms, disintegration of pigments in leaves and cell death can be detected with changes in wavelength by utilizing imagery based on multispectral. Paddy leaves affected by *Xanthomonas oryzae* show refraction of light in 705–740 nm, which is the red edge area. By integrating RGB imaging and CNN models, *Alternaria* infestation in tomato leaves can be detected with 90% accuracy (Mohanty *et al.*, 2016). Generally, HTP platforms deploy ML algorithms and advanced visual analysis to diagnose and differentiate different pathogens and assess their intensity using high-resolution images. However, compared with conventional assessment, these advanced tools present accurate, reliable, and easily replicable values under complex situations. This feature is highly suitable for genetic scoring in plant breeding studies that concentrate on breeding disease-resistant crops (Mahlein *et al.*, 2013).

High-throughput monitoring of spatial and temporal dynamics

HTP platforms are highly precise for monitoring the stress felt by crops under different spatial and temporal conditions. Drones play a crucial role in detecting the diseases present in the whole field with enabled sensors. Simko *et al.* (2017) highlighted that disease spread, insect vectors, causal organisms of diseases, host plants, and conducive environment have to be investigated.

Pest damage quantification

Defoliation by fall armyworm, *Spodoptera frugiperda*, can be easily detected in the leaves with thermal imagery technique; as larvae feed and move, there will be a generation of heat which can be detected with sensors. Also, these injury indices can be assessed using RGB algorithms and monitoring of larval infestation over a period of time. Data acquired via HTP technologies can be efficiently employed in crop management systems and decision-support tools to improve real-time monitoring and precision interventions. Thermal imaging technologies allow for the detection of specific temperature anomalies associated with pest activity (Murugasridevi *et al.*, 2025b), thereby enabling the targeted delivery of pesticides. This focused strategy enhances pest management effectiveness while minimizing environmental impact and lowering input costs, fostering sustainable agriculture practices.

4.6.4 Abiotic stress

Phenotyping for abiotic stress using HTP systems is increasingly significant in contemporary plant research, owing to its ability to non-invasively and regularly assess the responses of crops in actual field circumstances. Stresses, including drought, heat, salinity, and waterlogging, affect several physiological and morphological characteristics, many of which can be measured and analyzed by thermal, spectral, and structural imaging techniques. These platforms offer scalable, objective, and temporally accurate information that can be used in agricultural improvement initiatives. Table 4.4 represents the ML/DL models employed for detecting and phenotyping abiotic stress.

Drought and heat stress

Drought and heat constitute the most prevalent abiotic threats to the world food supply. HTP platforms are capable of identifying earlier physiological reactions to such stimuli. Canopy temperature serves as a dependable indicator of stomatal conductance during drought stress, reflecting transpiration efficiency (TE) and water utilization (WU). Increased crown temperatures frequently indicate stomatal closure resulting from shortages of water (Jones *et al.*, 2009). Vegetation indices, including normalized difference vegetation index (NDVI) and PRI, diminish due to the breakdown of chlorophyll caused by dryness. The 'stay-green' trait, assessed via RGB imaging and SPAD indices, offers insight into the thermal robustness of genotypes under extreme temperatures. This characteristic has proven especially advantageous in cereals such as sorghum and maize (Borrell *et al.*, 2000).

Salinity and submergence stress

Salinity affects ionic and osmotic homeostasis in plants, leading to leaf senescence and stunted growth. Hyperspectral imaging enables early detection by capturing subtle changes in leaf reflectance related to salt-induced damage. Indices like MCARI (Modified Chlorophyll Absorption Ratio Index) and TCARI (Transformed Chlorophyll Absorption in Reflectance Index) are often used for this purpose. Submergence stress, relevant to lowland rice systems, can be monitored using UAVs that detect changes in canopy structure and regrowth post-flooding. These approaches offer a non-destructive means of screening for flood recovery capacity (Zheng *et al.*, 2020).

Table 4.4. ML/DL models in abiotic stress detection and phenotyping (Author's own table)

Stress type	Trait	Sensor used	Example crop	References
Drought	Canopy temperature, NDVI	Thermal, multispectral	Wheat, maize	Jones *et al.*, 2009
Heat	Stay-green index	RGB, SPAD, NDVI	Sorghum	Borrell *et al.*, 2000
Salinity	Leaf reflectance, EC mapping	Hyperspectral, soil sensor	Rice	Zheng *et al.*, 2020
Cold/frost	Chlorophyll fluorescence, leaf temperature	Thermal imaging, fluorometer	Barley, potato	Vishal *et al.*, 2021
Waterlogging	Oxygen concentration, leaf wilting	Soil O_2 sensor, RGB	Maize, soybean	Setter and Waters, 2003
Nutrient deficiency (e.g. N, P)	SPAD, PRI, leaf color	Chlorophyll meter, multispectral	Rice, maize	Pinter *et al.*, 2003
Light stress (shading)	Photosynthetic efficiency, leaf angle	Fluorometer, LiDAR	Soybean, tomato	Murchie and Lawson, 2013
Heavy metal stress	Leaf spectral signature, chlorosis	Hyperspectral sensor	Spinach, rice	Yu *et al.*, 2019
Wind stress	Stem bending, lodging index	RGB, 3D Imaging	Wheat, barley	Berry *et al.*, 2003
UV radiation	Anthocyanin content, leaf morphology	Fluorescence sensor, RGB	Lettuce, tomato	Robson *et al.*, 2005

4.7 Machine Learning in Trait Prediction

The use of AI and machine learning in HTP systems has revolutionized trait predictions by facilitating the evaluation of extensive, intricate, and dimensional datasets. Such algorithms are exceptionally effective in identifying nuanced phenotypic differences linked to biotic and abiotic pressures that may be missed by conventional approaches. By analyzing associations and trends in multi-sensor data, machine learning models enable the early identification of stress signs and symptoms, accurately characterize morphological features, and enhance the forecasting efficacy of decision making in crop breeding initiatives.

4.7.1 Model applications

Numerous algorithms for supervised learning have been effectively utilized to forecast stress-related characteristics in various crops and environments. CNNs demonstrate exceptional efficacy in image classification tasks, including the detection of leaf disease symptoms in crops such as tomatoes, achieving accuracies of 90% (Mohanty *et al.*, 2016). When we compare RF with SVM, RF consists of decision trees that have more potential to categorize drought to 92% precision using the environmental dataset. Whereas, SVM manages small datasets and achieves 88% accuracy (Rahaman *et al.*, 2018; Singh *et al.*, 2021). ML models utilized for stress trait prediction are presented in Table 4.5.

4.8 Future Thrust

We can establish AI-integrated technologies in the field of omics, i.e. genomics, transcriptomics, proteomics, metabolomics, with the help of phenotypic data. The standardization of data has to be uniform across the breeding programs across the world. AI models can be developed for all sorts of crops, including minor crops, and the environment. The multi-tasking and

Table 4.5. Machine learning models for stress trait prediction (Author's own table)

Model	Application	Accuracy (%)	Reference
CNN	Leaf disease classification (tomato)	>90	Mohanty *et al.*, 2016
Random forest (RF)	Drought stress classification	85–92	Singh *et al.*, 2021
SVM	Genotype-stress interaction modeling	88	Rahaman *et al.*, 2018
LSTM	Time-series prediction of temperature stress	90–94	Malhotra *et al.*, 2020
Decision Tree	Soil salinity level classification	80–85	Kumar *et al.*, 2019
KNN	Drought tolerance classification in maize genotypes	82	Zhao *et al.*, 2020
ANN	Nitrogen deficiency detection in wheat using SPAD data	87	Sharma *et al.*, 2021
Deep CNN + LSTM	Heat stress temporal prediction	91	Wang *et al.*, 2022
XGBoost	Multi-stress classification in crops (heat, drought)	89–93	Joshi *et al.*, 2023
AutoML (TPOT)	Optimizing stress phenotyping pipelines	92	Patel *et al.*, 2022
GAN (Generative Adversarial Network)	Synthetic stress image generation for training	NA (used for augmentation)	Ahmed *et al.*, 2020
Logistic Regression	Classifying moderate versus severe stress from indices	78	Banerjee *et al.*, 2021

graph neural networks would actively have engaged in bringing out the genetic interactions, pleiotropy, and G × E interactions. AI can replace or make faster the process of selecting the superior genotypes when compared with the traditional selection models. The genomic selection can be accelerated through predictive analytics to reduce the breeding cycles. It is highly useful in the field of *in silico* breeding and trait arrangement. Edge-AI sensors, satellites, and drones can be used to characterize crops in real time. For high-throughput trait analysis in the field, we can combine AI with computer vision and hyperspectral imagery. The phenotypic responses under various circumstances of extreme environments can be studied with the aid of AI-based digital crop modeling. Before the employment of gene editing technology like CRISPR/Cas9, AI could be used to study the phenotypic effects of crop plants. The studies on the multi-trait optimization models will be useful for gene editing targets. The AI models can be transparent in predicting the genes, pathways, and tools, i.e. SHAP, LIME, or integrated gradients, which should be built into genotype-to-phenotype models. Based on the male and female crosses, the structure of the crossing pattern can be derived.

4.9 Conclusion

Artificial intelligence takes a paradigm transformation in crop breeding programs by the genotype-to-phenotype prediction, making it so accurate, faster, and more reliable. Plant breeders can decode the complex genetic architecture, which is available, and it can forecast performance in different environmental conditions. AI can be integrated with deep learning machines and multi-omics analytics, which further favors the development of climate-resilient and high-yielding crops. Although we know about the AI potential, the attention must be toward an interdisciplinary collaboration between plant scientists, data scientists, and policymakers that will be critical to turning this vision into reality. Finally, AI-driven genotype-to-phenotype prediction models will revolutionize crop breeding programs by ensuring global food security, sustainability, and resilience in an era of climate uncertainty.

References

Ahmed, S., Hasan, M.M. and Rahman, M.A. (2020) Generation of synthetic plant disease images using generative adversarial networks. *Computers and Electronics in Agriculture* 175, 105528.

Araus, J.L. and Cairns, J.E. (2014) Field high-throughput phenotyping: The new crop breeding frontier. *Trends in Plant Science* 19(1), 52–61.

Banerjee, A., Singh, B. and Gupta, S. (2021) Logistic regression model for abiotic stress classification in rice using vegetation indices. *Agricultural Research* 10(1), 45–53.

Banerjee, R., Marathi, B. and Singh, M. (2020) Efficient genomic selection using ensemble learning and ensemble feature reduction. *Journal of Crop Science and Biotechnology* 23(4), 311–323.

Bauriegel, E., Giebel, A. and Herppich, W.B. (2011) Hyperspectral and chlorophyll fluorescence imaging for early detection of plant diseases, with special reference to Fusarium spec Infections on Wheat. *Agronomy* 1(3), 191–207.

Bayer, P.E. and Edwards, D. (2020) Machine learning in agriculture: From silos to marketplaces. *Plant Biotechnology Journal* 19(4), 648.

Belouz, K., Nourani, A., Zereg, S. and Bencheikh, A. (2022) Prediction of greenhouse tomato yield using artificial neural networks combined with sensitivity analysis. *Scientia Horticulturae* 293, 110666.

Berry, P.M., Sterling, M., Baker, C.J., Spink, J.H. and Sparkes, D.L. (2003) A calibrated model of wheat lodging compared with field measurements. *Agronomy Journal* 95(5), 1032–1040.

Borrell, A.K., Hammer, G.L. and Henzell, R.G. (2000) Does maintaining green leaf area in sorghum improve yield under drought? I. Leaf Growth and Senescence. *Crop Science* 40(4), 1026–1037.

Cabrera-Bosquet, L., Fournier, C., Brichet, N., Welcker, C. and Tardieu, F. (2016) Phenomics of maize leaves in the field: Relationships between leaf growth parameters and water deficit. *Plant Physiology* 172(2), 749–763.

Champigny, M.J., Unda, F., Skyba, O., Soolanayakanahally, R.Y., Mansfield, S.D. *et al.* (2020) Learning from methylomes: Epigenomic correlates of *Populus balsamifera* traits based on deep learning models of natural DNA methylation. *Plant Biotechnology Journal* 18(6), 1361–1375.

Chen, J., Zhou, J., Li, Q., Li, H., Xia, Y. *et al.* (2023) CropQuant-air: An AI-powered system to enable phenotypic analysis of yield-and performance-related traits using wheat canopy imagery collected by low-cost drones. *Frontiers in Plant Science* 14, 1219983.

Crossa, J., Pérez-Rodríguez, P., Cuevas, J., Montesinos-López, O., Jarquín, D. *et al.* (2017) Genomic selection in plant breeding: Methods, models, and perspectives. *Trends in Plant Science* 22(11), 961–975. DOI: 10.1016/j.tplants.2017.08.011.

Cuevas, J., Crossa, J., Soberanis, V., Pérez-Elizalde, S., Pérez-Rodríguez, P. *et al.* (2016) Genomic prediction of genotype × environment interaction kernel regression models. *The Plant Genome* 9(3).

Deng, C.H., Naithani, S., Kumari, S., Cobo-Simón, I., Quezada-Rodríguez, E.H. *et al.* (2023) Genotype and phenotype data standardization, utilization and integration in the big data era for agricultural sciences. *Database* 2023, baad088.

Dong, Y., Chen, S. and Zhang, J. (2025) Learnable group transform: Enhancing genotype-to-phenotype prediction for rice breeding with small, structured datasets. *Preprint arXiv* 2503.11180.

Fiorani, F. and Schurr, U. (2013) Future scenarios for plant phenotyping. *Annual Review of Plant Biology* 64, 267–291.

Gao, P., Zhao, H., Luo, Z., Lin, Y., Feng, W. *et al.* (2023) SoyDNGP: A web-accessible deep learning framework for genomic prediction in soybean breeding. *Briefings in Bioinformatics* 24(6), bbad349.

Grinberg, N.F., Orhobor, O.I. and King, R.D. (2020) An evaluation of machine-learning for predicting phenotype: Studies in yeast, rice, and wheat. *Machine Learning* 109(2), 251–277.

Harper, L., Campbell, J., Cannon, E.K., Jung, S., Poelchau, M. *et al.* (2018) AgBioData consortium recommendations for sustainable genomics and genetics databases for agriculture. *Database* 2018, bay088.

Hoque, A., Anderson, J.V. and Rahman, M. (2024) Genomic prediction for agronomic traits in a diverse flax (*Linum usitatissimum* L.) germplasm collection. *Scientific Reports* 14(1), 3196.

Jones, H.G., Serraj, R., Loveys, B.R., Xiong, L., Wheaton, A. *et al.* (2009) Thermal infrared imaging of crop canopies for the remote diagnosis and quantification of plant responses to water stress in the field. *Functional Plant Biology* 36(11), 978–989.

Joshi, R., Mishra, D. and Singh, A. (2023) Comparative assessment of machine learning models for abiotic stress classification in crops. *Computers and Electronics in Agriculture* 206, 107578.

Khaki, S. and Wang, L. (2019) Crop yield prediction using deep neural networks. *Frontiers in Plant Science* 10, 621.

Kumar, M., Dutta, S. and Rathore, A. (2019) Decision tree-based model for salinity stress classification using soil EC data. *Journal of the Indian Society of Remote Sensing* 47(3), 489–496.

Li, J., He, Z., Zhou, G., Yan, S. and Zhang, J. (2024) DeepAT: A deep learning wheat phenotype prediction model based on genotype data. *Agronomy* 14(12), 2756.

Liu, L.W., Ma, X., Wang, Y.M., Lu, C.T. and Lin, W.S. (2021) Using artificial intelligence algorithms to predict rice (*Oryza sativa* L.) growth rate for precision agriculture. *Computers and Electronics in Agriculture* 187, 106286.

Lobet, G. (2017) Image analysis in plant sciences: Publish then perish. *Trends in Plant Science* 22(7), 559–566.

Lu, Y., Zhang, J., Liu, Y. and Wang, L. (2020) Early detection of Fusarium wilt in banana using hyperspectral imaging and machine learning. *Remote Sensing* 12(3), 546.

Ma, W., Qiu, Z., Song, J., Li, J., Cheng, Q. *et al.* (2018) A deep convolutional neural network approach for predicting phenotypes from genotypes. *Planta* 248, 1307–1318.

Ma, X., Wang, H., Wu, S., Han, B., Cui, D. *et al.* (2024) DeepCCR: Large-scale genomics-based deep learning method for improving rice breeding. *Plant Biotechnology Journal* 22(10), 2691.

Madec, S., Baret, F., de Solan, B., Thomas, S., Dutartre, D. *et al.* (2017) High-throughput phenotyping of plant height: Comparing unmanned aerial vehicles and ground lidar estimates. *Frontiers in Plant Science* 8, 2002.

Mahlein, A.K. (2016) Plant disease detection by imaging sensors – Parallels and specific demands for precision agriculture and plant phenotyping. *Plant Disease* 100(2), 241–251.

Mahlein, A.K., Rumpf, T., Welke, P., Dehne, H.W., Plümer, L. *et al.* (2013) Development of spectral indices for detecting and identifying plant diseases. *Remote Sensing of Environment* 128, 21–30.

Malhotra, R., Kumar, V. and Singh, P. (2020) LSTM model for forecasting crop temperature stress using time series data. *IEEE Access* 8, 173547–173556.

Mayernik, M.S. and Acker, A. (2018) Tracing the traces: The critical role of metadata within networked communications. *Journal of the Association for Information Science and Technology* 69(1), 177–180.

Meuwissen, T.H., Hayes, B.J. and Goddard, M. (2001) Prediction of total genetic value using genome-wide dense marker maps. *Genetics* 157(4), 1819–1829.

Mohanty, S.P., Hughes, D.P. and Salathé, M. (2016) Using deep learning for image-based plant disease detection. *Frontiers in Plant Science* 7, 1419.

Montesinos-López, O.A., Montesinos-López, A., Hernandez-Suarez, C.M., Barrón-López, J.A. and Crossa, J. (2021) Deep-learning power and perspectives for genomic selection. *The Plant Genome* 14(3), e20122.

Montesinos-López, A., Montesinos-López, O.A., Gianola, D., Crossa, J. and Hernández-Suárez, C.M. (2018) Multi-environment genomic prediction of plant traits using deep learners with dense architecture. *G3: Genes, Genomes, Genetics* 8(12), 3813–3828.

Murchie, E.H. and Lawson, T. (2013) Chlorophyll fluorescence analysis: A guide to good practice and understanding some new applications. *Journal of Experimental Botany* 64(13), 3983–3998.

Murugasridevi, K., Mageshen, V.R., Aswitha, K. and Saranya, M. (2024) AI-based regulation of water supply and pest management in farming. In: Chouhan, S.S., Saxena, A., Singh, U.P. and Jain, S. (eds) *Artificial Intelligence Techniques in Smart Agriculture*. Springer Nature, Singapore, pp. 195–215.

Murugasridevi, K., Mageshen, V.R. and Ramya, A. (2025a) The role of robotics in sustainable agriculture and waste management. In: Azrour, M., Hossain, I. and Mahmudul Haque, A.K.M. (eds) *Addressing Environmental Challenges With AI, Robotics, and Augmented Reality*. IGI Global Scientific Publishing, Hershey, Pennsylvania, pp. 47–82.

Murugasridevi, K., Dhivya, V., Kirubhadharshini, L., Mageshen, V.R. and Sowjanya, S. (2025b) Role of remote sensing and geographic information system (GIS) in integrated pest management (IPM). In:

Singh, S., Dahiya, N., Mohammed, M., Alzharihi, A. and Dutt, V. (eds) *Hyperspectral Remote Sensing for Sustainable Agriculture*. Bentham Science, Singapore.

Papoutsoglou, E.A., Faria, D., Arend, D., Arnaud, E., Athanasiadis, I.N. *et al*. (2020) Enabling reusability of plant phenomic datasets with MIAPPE 1.1. *New Phytologist* 227(1), 260–273.

Patel, R., Mehta, R. and Soni, V. (2022) Automated machine learning for crop stress trait prediction: A TPOT-based pipeline. *Computers and Electronics in Agriculture* 193, 106687.

Pauli, D., Chapman, S.C., Bart, R., Topp, C.N., Lawrence-Dill, C.J. *et al*. (2016) The quest for understanding phenotypic variation via integrated approaches in the field environment. *Plant Physiology* 172(2), 622–634.

Pinter, P.J., Hatfield, J.L., Schepers, J.S., Barnes, E.M., Moran, M.S. *et al*. (2003) Remote sensing for crop management. *Photogrammetric Engineering & Remote Sensing* 69(6), 647–664.

Rahaman, M.M., Chen, D., Gillani, Z., Klukas, C. and Chen, M. (2018) Advanced phenotyping and phenotype data analysis for the study of plant growth and development. *Frontiers in Plant Science* 9, 617.

Raza, A., Razzaq, A., Mehmood, S. and Abbas, S. (2021) Remote sensing applications for sugarcane aphid detection using UAV-based NDVI and thermal imaging. *Precision Agriculture* 22(4), 1093–1110.

Robson, T.M., Klem, K., Urban, O. and Jansen, M.A. (2005) Re-interpreting plant morphological responses to UV-b radiation. *Plant, Cell & Environment* 28(6), 529–541.

Sabouri, H. and Sajadi, S.J. (2022) Predicting hybrid rice performance using AIHIB model based on artificial intelligence. *Scientific Reports* 12(1), 9709.

Sandhu, K.S., Lozada, D.N., Zhang, Z., Pumphrey, M.O. and Carter, A.H. (2021) Deep learning for predicting complex traits in spring wheat breeding program. *Frontiers in Plant Science* 11, 613325.

Sankaran, S., Mishra, A., Ehsani, R. and Davis, C. (2015) A review of advanced techniques for detecting plant diseases. *Computers and Electronics in Agriculture* 72(1), 1–13.

Setter, T.L. and Waters, I. (2003) Review of prospects for germplasm improvement for waterlogging tolerance in wheat, barley and oats. *Plant and Soil* 253(1), 1–34.

Sharma, R., Thakur, A. and Singh, D. (2021) Artificial neural networks for nitrogen deficiency detection in wheat using chlorophyll measurements. *Journal of Applied Remote Sensing* 15(4), 046503.

Simko, I., Jimenez-Berni, J.A. and Sirault, X.R. (2017) Phenomic approaches to improve plant breeding efficiency. *Functional Plant Biology* 44(1), 3–6.

Singh, D., Meena, R.K. and Yadav, S. (2021) Random forest-based drought stress classification in wheat using remote sensing data. *Agricultural Water Management* 248, 106770.

Spindel, J., Begum, H., Akdemir, D., Virk, P., Collard, B. *et al*. (2015) Genomic selection and association mapping in rice (*Oryza sativa*): Effect of trait genetic architecture, training population composition, marker number and statistical model on accuracy of rice genomic selection in elite, tropical rice breeding lines. *PLOS Genetics* 11(6), e1005350.

Stewart, E.L., Wiesner-Hanks, T., Kaczmar, N., DeChant, C., Wu, H. *et al*. (2019) Quantitative phenotyping of northern leaf blight in UAV images using deep learning. *Remote Sensing* 11(19), 2209.

Tardieu, F., Cabrera-Bosquet, L., Pridmore, T. and Bennett, M. (2017) Plant phenomics, from sensors to knowledge. *Current Biology* 27(15), R770–R783.

Togninalli, M., Wang, X., Kucera, T., Shrestha, S., Juliana, P. *et al*. (2023) Multi-modal deep learning improves grain yield prediction in wheat breeding by fusing genomics and phenomics. *Bioinformatics* 39(6), btad336.

Tong, H. and Nikoloski, Z. (2021) Machine learning approaches for crop improvement: Leveraging phenotypic and genotypic big data. *Journal of Plant Physiology* 257, 153354.

Van, H.T., Van Vu, G., Tuan, T.T., Vo, B. and Chung, Y.S. (2025) LGENetB4CA: A novel deep learning approach for chili germplasm differentiation and leaf disease classification. *Computers and Electronics in Agriculture* 233, 110149.

Varshney, R.K., Thudi, M., Nayak, S.N., Gaur, P.M., Kashiwagi, J. *et al*. (2014) Genetic dissection of drought tolerance in chickpea (*Cicer arietinum* L.). *Theoretical and Applied Genetics* 127(2), 445–462. DOI: 10.1007/s00122-013-2230-6.

Virlet, N., Sabermanesh, K., Sadeghi-Tehran, P. and Hawkesford, M.J. (2017) Field scanalyzer: An automated robotic field phenotyping platform for detailed crop monitoring. *Functional Plant Biology* 44(1), 143–153.

Vishal, R., Singh, S. and Tripathi, R. (2021) Early detection of frost injury using chlorophyll fluorescence in potato. *International Journal of Remote Sensing* 42(2), 504–519.

Wahabzada, M., Mahlein, A.K., Bauckhage, C., Steiner, U., Oerke, E.C. *et al.* (2015) Plant phenotyping using probabilistic topic models: Uncovering the hyperspectral language of plants. *Scientific Reports* 6, 22482.

Wang, H., Zhang, C. and Liu, X. (2018) Detection of insect feeding damage on maize leaves using thermal imaging and machine learning. *Computers and Electronics in Agriculture* 154, 482–489.

Wang, H., Yan, S., Wang, W., Chen, Y., Hong, J. *et al.* (2025) Cropformer: An interpretable deep learning framework for crop genomic prediction. *Plant Communications* 6(3).

Wang, J., Hu, J. and Zhang, X. (2022) Deep learning model integrating CNN and LSTM for heat stress prediction in crops. *IEEE Transactions on Geoscience and Remote Sensing* 60, 1–11.

Wang, K., Abid, M.A., Rasheed, A., Crossa, J., Hearne, S. *et al.* (2023) DNNGP, a deep neural network-based method for genomic prediction using multi-omics data in plants. *Molecular Plant* 16(1), 279–293.

Wilkinson, M.D., Dumontier, M., Aalbersberg, I.J., Appleton, G., Axton, M. *et al.* (2016) The FAIR guiding principles for scientific data management and stewardship. *Scientific Data* 3(1), 1–9.

Wu, H., Han, R., Zhao, L., Liu, M., Chen, H. *et al.* (2025) AutoGP: An intelligent breeding platform for enhancing maize genomic selection. *Plant Communications* 6(4).

Yu, H., Huang, S. and Zhang, Y. (2019) Hyperspectral imaging for mapping heavy metal stress in spinach leaves. *Sensors* 19(20), 4487.

Zaman-Allah, M., Vergara, O., Araus, J.L. and Cairns, J.E. (2015) Unmanned aerial platform-based multi-spectral imaging for field phenotyping of maize. *Plant Methods* 11, 35.

Zarco-Tejada, P.J., González-Dugo, V. and Berni, J.A.J. (2012) Fluorescence, temperature and narrow-band indices acquired from a UAV platform for water stress detection using a micro-hyperspectral imager and a thermal camera. *Remote Sensing of Environment* 117, 322–337.

Zarco-Tejada, P.J., Hubbard, N. and Loudjani, P. (2015) Precision agriculture: An opportunity for EU farmers. *European Parliament, Policy Department B: Structural and Cohesion Policies.*

Zhang, M., Qin, Z., Liu, X. and Ustin, S.L. (2020) Detection of soybean cyst nematode (*Heterodera glycines*) stress using remote sensing techniques. *Remote Sensing* 12(9), 1423.

Zhao, Y., Chen, D. and Li, X. (2020) Classification of drought-tolerant maize genotypes using KNN and remote sensing-derived indices. *Agricultural Systems* 184, 102913.

Zheng, H., Guo, H. and Zhang, Y. (2020) Monitoring rice salinity stress responses using hyperspectral imagery and physiological parameters. *Agricultural Water Management* 231, 105992.

Zingaretti, L.M., Gezan, S.A., Ferrão, L.F.V., Osorio, L.F., Monfort, A. *et al.* (2020) Exploring deep learning for complex trait genomic prediction in polyploid outcrossing species. *Frontiers in Plant Science* 11, 25.

5 Data-Driven Genomic Design for Next-Generation Crop Breeding

Rahele Ghanbari Moheb Seraj[1] **and Vavar Vafaee**[1,2*]
[1]*Department of Horticultural Science and Engineering, Faculty of Agriculture, University of Kurdistan, Sanandaj, Iran;* [2]*Medicinal Plants Breeding, Development Research Institute, University of Kurdistan, Sanandaj, Iran*

Abstract
Plant breeding is undergoing a profound transformation, shifting from a phenotype-based discipline to a data-driven science. This chapter explores the paradigm of genomic design as the cornerstone of this revolution. We detail how the integration of high-throughput genotyping (e.g., SNP arrays, sequencing) and phenotyping (e.g., UAVs, IoT sensors) generates massive, multidimensional datasets. The chapter provides a foundational overview of genomic prediction models, from traditional statistical methods like GBLUP and Bayesian approaches to advanced machine learning and deep learning algorithms that capture complex, nonlinear genotype–phenotype–environment interactions. We emphasize the critical role of integrating multi-omics data (genomics, transcriptomics, proteomics, metabolomics) and environmental information to build predictive models for complex traits such as yield, stress resilience, and nutritional quality. Furthermore, we examine precise breeding strategies, including genomic selection (GS), marker-assisted selection (MAS), and genome editing (CRISPR/Cas9), which leverage these predictions to accelerate the development of superior cultivars. The chapter also reviews essential computational tools, cloud-based platforms, and bioinformatics pipelines that manage and analyze this big data. Finally, we discuss the prevailing challenges – data standardization, ethical considerations, and model interpretability – and outline future directions, including the synergistic potential of synthetic biology and digital twins. This work serves as a comprehensive guide for leveraging data-driven genomic design to enhance the precision, efficiency, and sustainability of next-generation crop breeding programs, ultimately contributing to global food security.

Keywords: Data-driven breeding, Genomic prediction, Genomic selection, Multi-omics integration, Precision breeding

5.1 Introduction

5.1.1 An overview of genomic design in crop breeding

Since the domestication of plants around 10,000 years ago, plant breeding has played a fundamental role in ensuring food security and advancing human civilization. From the mid-20th century, traditional pre-genomics breeding methods, relying on phenotype-based selection and field trials, successfully produced modern cultivars that dramatically increased the yield of many staple crops (Fedoroff, 2010).

*Corresponding author: y.vafaee@uok.ac.ir

 Plant Genomic Prediction: AI-Enhanced Models for Plant Genomics and Crop Improvement (ed. Jen-Tsung Chen)
DOI: 10.1079/9781836991328.0005

However, the complexity of quantitative traits – often governed by multiple genes and strongly influenced by environmental conditions – has always posed a major challenge for these approaches. The advent of genomics marked a turning point in plant breeding, enabling the direct study of entire genomes and the precise identification of genotype–phenotype relationships. This revolution transformed breeding from a trial-and-error process into a data-driven and predictable approach (Varshney *et al.*, 2021a, b).

Genomic design, as one of the cutting-edge approaches in plant breeding, leverages an array of advanced technologies and genomic resources to select and develop desirable cultivars. Within this framework, molecular markers play a pivotal role by revealing the positions of genes or genomic regions associated with important traits. These markers fall into two main categories: classical markers, including morphological, cytological, and biochemical characteristics; and DNA markers, which can be identified through various methods such as RFLP, RAPD, AFLP, SSR, and especially single-nucleotide polymorphisms (SNPs). As the most common variations in plant genomes, SNPs are widely used in modern breeding due to their high accuracy, relatively low cost, and compatibility with modern technologies such as genotyping-by-sequencing (GBS). Furthermore, newer techniques such as ILP, DArT, PARMS, and iSNAP have enabled more precise trait selection and a more comprehensive analysis of genetic diversity (Myles *et al.*, 2009; Hasan *et al.*, 2021). A detailed categorization of classical and DNA-based molecular markers, including their advantages, disadvantages, and applications, is provided in Table 5.1.

Alongside these tools, innovative approaches such as genome-wide association studies (GWAS), marker-assisted selection (MAS), and genomic selection (GS) have empowered breeders to more accurately predict genotype performance and select desirable traits with greater confidence. Moreover, the emergence of gene-editing technologies – particularly the CRISPR/Cas9 system – has brought another revolution to genomic design. This technology not only facilitates precise and targeted modifications to gene sequences but also significantly reduces the time and cost required to develop new cultivars, opening new avenues for enhancing stress resistance, increasing yield, and improving crop quality (Riaz *et al.*, 2021).

Genomic resources, including reference genomes, genetic and protein databases, transcriptomes, and gene expression profiles, provide a rich framework for identifying genes influencing desirable traits, discovering QTLs, and detecting overlooked alleles with small effects. Integrating these resources with predictive models and bioinformatics analyses has transformed plant breeding from a time-consuming process into a targeted, rapid, and evidence-based system (Perez-de-Castro *et al.*, 2012).

Genomic design not only increases the precision and efficiency of plant breeding but also paves the way for developing cultivars better adapted to the challenges of climate change, disease pressures, and the growing nutritional needs of societies. With the continuous advancement of omics technologies, the development of artificial intelligence algorithms, and the expanding accessibility of genomic databases, this approach is expected to become the backbone of global breeding programs in the decades to come (Kumar *et al.*, 2024).

5.1.2 The importance of data-driven approaches

Data-driven approaches in plant breeding, as one of the transformative pillars of modern agriculture, play an indispensable role in optimizing the processes of cultivar selection and development. In today's world – faced with challenges such as climate change, dwindling natural resources, and the growing need for increased food production – the utilization of large-scale data and advanced analytical technologies offers an efficient, evidence-based pathway for improving crop performance. The use of big data analytics, data mining, and predictive statistical models such as genomic selection enables breeders to accurately identify and select complex traits, including disease resistance, drought tolerance, and resource-use efficiency. This approach not only accelerates the identification of superior genotypes but also improves the accuracy of parent selection for targeted crosses (Xu *et al.*, 2017).

Table 5.1. Categorization of molecular markers in plant breeding, including classical and DNA-based types, with examples, advantages, and applications (Author's own table).

Marker type	Category	Examples	Advantages	Disadvantages	Applications in plant breeding	References
Classical markers	Morphological	Plant height, flower color, and seed shape	Easy to observe, no need for lab equipment	Influenced by the environment, limited polymorphism	Initial trait selection, hybrid identification	Fedoroff, 2010
Classical markers	Cytological	Chromosome banding, karyotype variations	Useful for structural genomics	Low resolution, labor-intensive	Studying chromosomal aberrations, aneuploidy detection	Dhar and Koul, 2024
Classical markers	Biochemical	Isozymes, protein profiles	Co-dominant, no DNA extraction needed	Limited number, tissue-specific	Genetic diversity assessment, cultivar identification	Hasan *et al*., 2021
DNA markers	Restriction-based	RFLP (restriction fragment length polymorphism)	Co-dominant, highly reproducible	Radioactive labeling, time-consuming, high cost	Linkage mapping, QTL analysis	Myles *et al*., 2009
DNA markers	PCR-based	RAPD (random amplified polymorphic DNA)	Quick, low cost, no prior sequence knowledge	Dominant, low reproducibility	Genetic diversity, fingerprinting	Varshney *et al*., 2021a
DNA markers	PCR-based	AFLP (amplified fragment length polymorphism)	High polymorphism, no sequence info needed	Dominant, complex banding patterns	Genome mapping, diversity studies	Hasan *et al*., 2021
DNA markers	PCR-based	SSR (simple sequence repeats)	Co-dominant, highly polymorphic, reproducible	Requires sequence knowledge	Marker-assisted selection, genotyping	Riaz *et al*., 2021
DNA markers	Sequence-based	SNPs (single-nucleotide polymorphisms)	Abundant, co-dominant, high-throughput	Biallelic, requires sequencing	GWAS, genomic selection, high-density mapping	Kumar *et al*., 2024
DNA markers	Sequence-based	ILP (intron length polymorphism)	Co-dominant, gene-specific	Limited to intronic regions	Functional genomics, trait association	Badoni *et al*., 2016

DNA markers	Array-based	DArT (diversity arrays technology)	High-throughput, cost-effective	Dominant markers	Genetic mapping, diversity analysis	Ballesta *et al.*, 2020
DNA markers	Sequence-based	PARMS (penta-primer amplification refractory mutation system)	Multiplexing capability, accurate	Requires assay design	SNP genotyping in breeding populations	Sinha *et al.*, 2023
DNA markers	Sequence-based	iSNAP (InDel SNP Amplification Panel)	Detects InDels and SNPs, with high specificity	Assay development needed	Precision breeding, variant detection	Sinha *et al.*, 2023

With the rapid expansion of high-throughput sequencing and phenotyping technologies, massive volumes of genomic, phenotypic, and environmental data are being generated – data that can only be effectively analyzed and leveraged through advanced information technologies and robust data management infrastructures. Integrated databases, cloud computing, and distributed storage systems facilitate the collection, organization, sharing, and efficient analysis of such datasets. In parallel, the use of real-time monitoring tools, including sensors, Internet of Things (IoT) technologies, and remote sensing, enables continuous and precise assessment of plant responses to environmental conditions in real-world settings (Xiao *et al.*, 2022).

The integration of artificial intelligence (AI) and machine learning algorithms has further opened new horizons in plant breeding. By analyzing complex patterns in genomic and phenotypic data – particularly for traits controlled by multiple genes – these algorithms deliver highly accurate predictive models that enhance efficiency, reduce experimental costs, and improve selection accuracy. Furthermore, deep learning enables the automated analysis of image-based phenotyping data, accelerating and refining the evaluation of traits such as plant health, growth, and yield (Kamilaris and Prenafeta-Boldú, 2018; van Dijk *et al.*, 2021).

Ultimately, the convergence of data mining, information technology, and AI in plant breeding has driven the shift from traditional practices to intelligent, precision-based approaches – approaches capable of effectively addressing emerging global challenges such as food security, resource sustainability, and climate change, while providing the foundation for producing resilient, high-yielding cultivars tailored to specific climatic and environmental conditions.

5.1.3 Key challenges in traditional plant breeding methods

Traditional plant breeding methods, such as mass selection, pure line selection, and hybridization, have historically played a significant role in improving desirable traits in crop plants. However, due to their empirical and time-consuming nature, these methods face multiple limitations and challenges that have reduced their efficiency, especially in the context of modern agriculture with its fast-paced and complex demands (Bhuiyan *et al.*, 2025). This section outlines the most critical challenges of traditional approaches:

Time-intensive breeding process

One of the primary challenges in traditional methods is the requirement for multiple generations to stabilize desirable traits. Since breeding is based on phenotypic selection and successive crossings, achieving breeding goals can take several years. This delay hampers the ability to respond quickly to urgent agricultural needs in the face of environmental changes or biological threats.

Low accuracy in selecting polygenic traits

Traditional methods often rely on visible traits and are less effective in identifying and selecting complex, quantitative traits such as yield, nutritional quality, or resistance to drought and diseases. These traits are usually controlled by multiple genes with small effects, and their expression is also influenced by environmental interactions. Consequently, accurate selection based solely on phenotype is difficult.

Susceptibility to environmental influences

In traditional breeding, phenotypic evaluation is strongly affected by environmental factors such as temperature, humidity, and soil quality. This makes it difficult to distinguish between genetic and environmental effects on traits, reducing selection accuracy and increasing the risk of errors in identifying superior genotypes.

Limited use of genetic diversity

Historically, breeders have primarily utilized the diversity present in local crop populations or available cultivars. The absence of precise tools to identify and exploit the hidden genetic diversity in plant resources has meant that many beneficial genes remain inaccessible to breeders or are lost during breeding processes.

Table 5.2. Comparative overview of challenges in traditional plant breeding methods versus advantages of genomic breeding, highlighting time, accuracy, environmental susceptibility, genetic diversity, predictability, and trait visibility (Author's own table).

Aspect	Challenges in Traditional Breeding	Advantages in Genomic Breeding	References
Time-intensive process	Requires multiple generations (years) for trait stabilization due to phenotypic selection	Accelerates cycles (months) using molecular markers and predictive models	Bhuiyan *et al.*, 2025
Accuracy in selecting polygenic traits	Low for complex traits controlled by multiple genes; relies on visible phenotypes	High precision via GWAS, MAS, and GS capturing small-effect genes	Xu *et al.*, 2017
Susceptibility to environmental influences	Phenotypes affected by G × E interactions, hard to distinguish from genetic effects	Accounts for G × E through multi-omics integration and enviromic prediction	Xu *et al.*, 2017
Limited use of genetic diversity	Relies on local populations; hidden diversity is often lost	Exploits global diversity via pan-genomes, WGS, and epigenetic data	Varshney *et al.*, 2021a
Lack of genetic predictability	Trial-and-error, chance-based success	Data-driven predictions with statistical models (e.g. Bayesian, ML)	Crossa *et al.*, 2017
Limitations in breeding for non-visible traits	Overlooks traits like stress resistance, not observable early	Enables selection for non-visible traits using high-throughput phenotyping and genotyping	Xiao *et al.*, 2022

Lack of genetic predictability

In the absence of genomic information and molecular markers, selections made through traditional breeding have limited predictive power. In other words, success in improving desirable traits depends heavily on chance and trial-and-error, making it difficult to reproduce results under different conditions.

Limitations in breeding for non-visible traits

Many important traits, such as the concentration of medicinal compounds, photosynthetic potential, or physiological stress resistance, are not directly observable during early growth stages. As a result, in traditional breeding, which relies heavily on the observation of visible traits, these characteristics are often overlooked or excluded from selection.

A comparative summary of the key challenges in traditional breeding versus the advantages offered by genomic breeding is presented in Table 5.2.

Although traditional plant breeding has laid the foundation for the successes of modern agriculture, its fundamental challenges, such as slow progress, low precision, and dependence on environmental factors, highlight the growing need to integrate these methods with modern technologies such as genomic breeding, data mining, and machine learning (Bhuiyan *et al.*, 2025).

5.1.4 Objectives and scope

The primary objective of this chapter is to provide a comprehensive and practical exposition of data-driven genomic design as an innovative and pioneering approach for breeding the next generation of agricultural crops. By focusing on scientific foundations and emerging technologies, this chapter aims to establish a clear framework for the optimal utilization of genomic, phenotypic, and environmental data, enabling the application of advanced genomic selection methods, complex trait engineering, and precise genome modification to substantially enhance crop performance and sustainability.

The scope of this chapter encompasses a wide range of interrelated topics, including the principles and models of genomic selection, diverse data resources and strategies for integrating multi-omics datasets, the application of machine learning and artificial intelligence in genomic prediction and breeding, and the computational tools and pipelines employed in this field. It also addresses precise breeding strategies such as genome editing and marker-assisted selection, while reviewing successful case studies that illustrate the practical application of these technologies in crop improvement.

In addition, this chapter highlights the pivotal role of remote sensing technologies and the Internet of Things (IoT) in accurately monitoring plant growth and responses to environmental conditions, thereby improving the capacity for data-informed breeding decisions. Finally, special emphasis is placed on environmental sustainability, resource-use efficiency, and the enhancement of the nutritional value of crops, outlining practical and scientific strategies for developing climate-resilient and smart agriculture. This chapter presents data-driven genomic design as the backbone of modern plant breeding and delineates clear pathways for developing the next generation of crops with high performance, environmental sustainability, and improved nutritional quality.

5.2 Foundations of Genomic Prediction and Selection

5.2.1 Concepts of genomic selection (GS)

Genomic selection (GS) is an advanced and innovative approach in plant and animal breeding designed to predict the genetic merit of individuals through extensive use of genomic data (Meuwissen *et al.*, 2001). Unlike marker-assisted selection (MAS), which focuses only on a limited number of markers associated with major QTLs, GS utilizes all available markers across the genome, even when the exact location or effect of QTLs is unknown (Hickey *et al.*, 2017). This feature makes GS particularly suitable for complex quantitative traits controlled by numerous small-effect alleles distributed throughout the genome (Bhat *et al.*, 2015).

In GS, a training population consisting of individuals with both phenotypic and genotypic data is first established. Statistical models are then trained using this dataset to capture the relationship between genotype and phenotype. The trained model is subsequently applied to predict the genetic merit of individuals without phenotypic data in the validation or test population (Norman *et al.*, 2018). This process eliminates the need for lengthy and costly phenotypic evaluations in later breeding stages, thereby significantly shortening and improving the efficiency of breeding cycles (Bhat *et al.*, 2016). The schematic workflow of the genomic selection process, from training population to prediction, is illustrated in Fig. 5.1.

MAS is inherently limited in its genomic coverage and its focus on large-effect QTLs, making it less effective for improving complex traits, such as yield, fruit quality, and stress tolerance, that are regulated by many small-effect genes (Bernardo, 2008; Xu and Crouch, 2008). In contrast, GS, by employing thousands of genomic markers spanning the entire genome, can comprehensively model the cumulative effects of genes, thereby preventing the loss of genetic variance contributed by small-effect loci (Meuwissen *et al.*, 2001). A direct comparison of the genomic coverage, efficiency, and accuracy of MAS and GS is provided in Table 5.3.

A variety of statistical models are employed in GS, including parametric methods such as ridge regression (RR) and LASSO, Bayesian models, semi-parametric models such as reproducing kernel Hilbert space (RKHS), as well as machine learning algorithms including artificial neural networks (ANN), support vector machines (SVM), and random forests (RF) (Gianola *et al.*, 2006; Long *et al.*, 2011; Holliday *et al.*, 2012). Each model operates under different assumptions regarding the distribution of marker effects, and the optimal choice depends on factors like crop type, population structure, environment, and genomic data characteristics (Budhlakoti *et al.*, 2020).

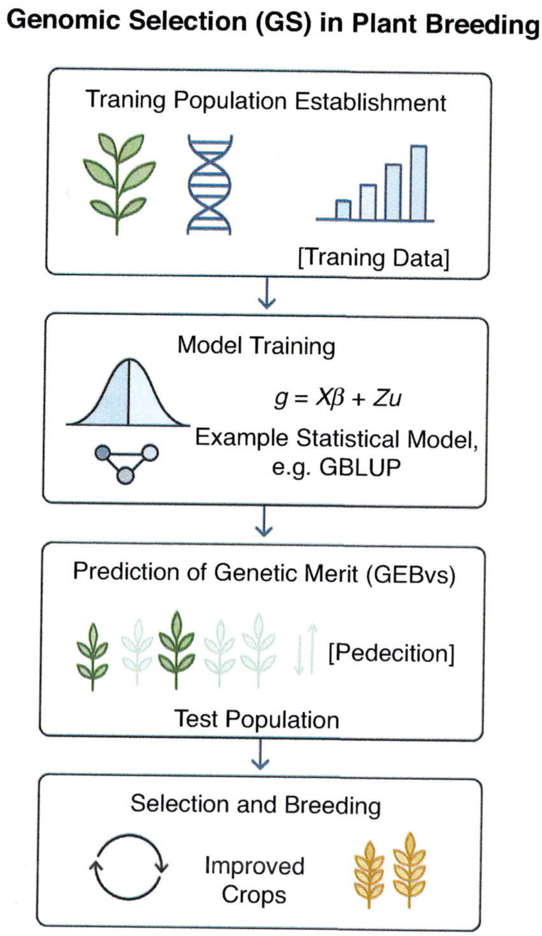

Fig. 5.1. Schematic workflow of genomic selection (GS), depicting the steps from training population establishment to prediction of genetic merit in test populations (Author's own image).

Studies have shown that prediction accuracy in GS is significantly higher than in MAS. For instance, genomic prediction accuracy has been reported to be three times greater in maize and twice as high in wheat (Heffner *et al.*, 2009). Consequently, GS is recognized as a key tool for improving complex traits and serves as an effective alternative to phenotypic selection and MAS (Crossa *et al.*, 2017).

The advancement of next-generation sequencing (NGS) technologies and the reduction in genotyping costs have facilitated the widespread use of high-resolution genomic data, greatly enhancing the applicability of GS (Gorjanc *et al.*, 2015). Ultimately, by integrating phenotypic and genotypic data into predictive models and applying these models in the early stages of breeding, GS accelerates, refines, and reduces the costs of breeding, securing a prominent place in modern crop improvement programs.

5.2.2 Advances in quantitative genetics for crop performance improvement

Quantitative genetics, as a cornerstone of modern plant breeding, plays a crucial role in enhancing crop performance and productivity. Recent advances in this field,

Table 5.3. Comparative analysis of marker-assisted selection (MAS) versus genomic selection (GS), including genomic coverage, suitability for traits, efficiency, and prediction accuracy (Author's own table).

Aspect	Marker-assisted selection (MAS)	Genomic selection (GS)	References
Genomic coverage	Limited to a few markers/ QTLs	Genome-wide, thousands of markers	Hickey *et al.*, 2017
Suitability for traits	Effective for major QTLs/ simple traits	Ideal for complex/ polygenic traits	Bhat *et al.*, 2015
Efficiency	Requires QTL mapping, slower for large populations	Faster, no need for prior QTL identification	Bernardo, 2008
Prediction accuracy	Lower for small-effect genes (e.g. 1–2 × in wheat/ maize)	Higher overall (e.g. 2–3 × in maize/wheat)	Heffner *et al.*, 2009
Cost and resources	Moderate, targeted genotyping	High initially, but cost-effective with NGS	Gorjanc *et al.*, 2015
Limitations	Misses cumulative small effects	Requires large training populations	Xu and Crouch, 2008

particularly through simulation techniques and mathematical modeling, have enabled the recreation of biological conditions and the early validation of breeding hypotheses. These approaches bridge the gap between theory and practice, offering effective solutions for breeding programs by optimizing genetic gain and resource utilization.

Research has shown that the choice of breeding strategy should align with the heritability of the trait, the timing of selection, and the maintenance of genetic diversity. While strategies that maximize trait values can increase performance, they often lead to a reduction in genetic diversity. Therefore, combining different approaches and considering genotype–environment interactions is essential to avoid bias in early selections (Lanzl *et al.*, 2023).

The use of molecular markers can accelerate breeding cycles when precise marker placement, balanced foreground and background selection, and efficient QTL management are ensured. Genomic selection, in particular, improves genetic gain by shortening breeding cycles and enhancing parental selection, especially for traits with low heritability and complex genetic architectures. The success of GS, however, depends on regularly updating training populations (Silva *et al.*, 2021).

In terms of statistical approaches, Bayesian models have demonstrated better performance for traits governed by fewer genes in early breeding cycles, whereas BLUP methods show greater stability for traits controlled by a large number of QTLs. RR-BLUP offers high flexibility under diverse conditions. Increasing the size of breeding populations can also improve genetic gain, provided that clear objectives and sufficient germplasm resources are available (Villiers *et al.*, 2024).

For low-heritability traits, using multi-trait analysis in combination with high-heritability traits enhances selection accuracy. Moreover, low-density genotyping combined with imputation methods provides a cost-effective alternative to high-density genotyping, delivering comparable results. Precise targeting of populations and managing their genetic relationships can further improve the accuracy and efficiency of selection.

Overall, advances in quantitative genetics highlight the need to balance short-term genetic gain with long-term potential. Implementing fast-cycle genomic selection programs, coupled with diverse breeding strategies and measures to preserve rare alleles, can simultaneously improve crop performance and maintain genetic diversity (Vieira *et al.*, 2025).

5.2.3 The role of high-throughput phenotyping and genotyping

Successful implementation of genomic selection (GS) requires the integration of advanced technologies in high-throughput phenotyping and genotyping (HTP and HTG). GS depends on the quality and precision of input data; therefore, the availability of accurate, repeatable phenotypes that represent the target environmental conditions is of paramount importance.

High-throughput phenotyping, using platforms such as unmanned aerial vehicles (UAVs) equipped with RGB, thermal infrared, and multispectral cameras, enables the measurement of key traits such as canopy temperature, leaf area index (LAI), vegetation indices like NDVI, and symptoms of biotic and abiotic stresses, including wilting and chlorosis (Messina and Modica, 2020; Awais *et al.*, 2023). These aerial platforms can monitor spatial heterogeneity and temporal changes in crop performance throughout the growing season. For instance, thermal imaging has proven highly effective in identifying heat- and drought-tolerant genotypes by quantifying reductions in canopy temperature, which serve as indicators of transpiration efficiency (Ashfaq *et al.*, 2022).

Additionally, ground-based sensors such as chlorophyll fluorometers for measuring photosynthetic efficiency and infrared gas analyzers for assessing stomatal conductance and transpiration rate provide detailed insights into the physiological mechanisms underlying stress responses (Wang, 2022; Driever *et al.*, 2023). Advanced systems such as X-ray computed tomography (X-ray CT) and rhizotrons for root phenotyping enable precise evaluation of root traits such as length, branching, and depth, which are critical for water and nutrient uptake (McGrail *et al.*, 2020). These data are collected non-destructively and repeatedly, enabling dynamic phenotyping that tracks plant responses to stresses over time. Such an approach facilitates the identification of genotypes with desirable physiological attributes and links phenotypic traits to underlying molecular mechanisms.

Despite recent advances in digital imaging, machine vision, and phenomics, field-based phenotyping remains costly and time-consuming. To address this, strategies have been proposed to optimize the design of training population subsets (TRS) using statistical algorithms and machine learning methods. These approaches reduce the number of individuals required and optimize the prediction accuracy of genomic selection (Akdemir *et al.*, 2021; Atanda *et al.*, 2021).

In genotyping, technologies such as genotyping-by-sequencing (GBS) and SNP arrays have brought remarkable advances. However, studies have shown that increasing marker density does not always lead to improved prediction accuracy (Hemingway *et al.*, 2021). Therefore, selecting a subset of markers that effectively represents the genetic diversity of the population is crucial. Strategies such as filtering based on minor allele frequency (MAF), removing highly correlated markers (LD pruning), and ensuring uniform marker distribution across the genome are common in optimizing marker density (Kainer *et al.*, 2018; Ballesta *et al.*, 2020). An overview of key high-throughput phenotyping and genotyping technologies and their applications in genomic selection is summarized in Table 5.4.

Integrating high-throughput phenotypic and genotypic data paves the way for advanced analyses such as genome-wide association studies (GWAS), quantitative trait loci (QTL) analysis, and genomic selection, providing breeders with powerful tools to identify desirable genotypes in plant breeding programs.

5.2.4 Statistical and computational models in genomic selection

Genomic selection, as a modern approach in plant breeding, has significantly enhanced the accuracy of predicting genomic estimated breeding values (GEBVs) through the development of advanced statistical and computational models. By incorporating genomic, phenotypic, and more recently multi-omics data, these models can capture complex patterns and interactions at various biological levels. The following sections review the main and emerging models in this field.

Reduced-bias regression (RLR) models

When the number of genetic markers (p) greatly exceeds the number of samples (n), the ordinary

Table 5.4. Overview of high-throughput phenotyping and genotyping technologies, including tools, measured traits, and applications in genomic selection (Author's own table).

Technology type	Tools/Platforms	Measured traits	Applications in genomic selection	References
Phenotyping	UAVs with RGB/thermal/ multispectral cameras	Canopy temperature, NDVI, LAI, stress symptoms	Monitoring spatial/ temporal changes, stress tolerance prediction	Messina and Modica, 2020; Awais *et al.*, 2023
Phenotyping	Ground-based sensors (chlorophyll fluorometers, IR gas analyzers)	Photosynthetic efficiency, stomatal conductance	Physiological stress responses, dynamic phenotyping	Wang, 2022; Driever *et al.*, 2023
Phenotyping	X-ray CT, rhizotrons	Root length, branching, depth	Root architecture for nutrient uptake, non-destructive evaluation	McGrail *et al.*, 2020
Genotyping	Genotyping-by-sequencing (GBS)	SNPs, genetic variations	High-resolution data for GS models	Hemingway *et al.*, 2021
Genotyping	SNP arrays	Marker density, polymorphisms	Optimizing marker subsets, diversity representation	Kainer *et al.*, 2018
Integrated	Machine learning algorithms for TPS optimization	Prediction accuracy enhancement	Reducing population size, improving GS efficiency	Akdemir *et al.*, 2021; Atanda *et al.*, 2021

least squares (OLS) model becomes inefficient, leading to overfitting and unstable coefficient estimates. To address this issue, penalized regression models such as ridge regression (RR) and LASSO have been developed. These models apply an ℓ2 penalty (RR) or ℓ1 penalty (LASSO) to regression coefficients, thereby reducing variance and aiding in the selection of key variables (Szymczak *et al.*, 2009; Friedman *et al.*, 2010).

RR shrinks coefficients toward zero without eliminating them, making it suitable for markers with small and correlated effects. In contrast, LASSO completely removes some coefficients, enabling variable selection, but when markers are correlated, it retains only one (Ogutu *et al.*, 2012). The Elastic Net model, which combines ℓ1 and ℓ2 penalties, addresses the limitations of RR and LASSO and performs better in the presence of correlated markers (Tutz and Ulbricht, 2009).

Mixed linear models (MLM)

Mixed linear models, which account for both fixed and random effects, are among the most widely used approaches in GS. Their foundation is the best linear unbiased prediction (BLUP) method, introduced by Henderson (1975). The genetic relationship matrix in these models can be derived from pedigree information (ABLUP) or genomic data (GBLUP).

GBLUP, by utilizing a genomic relationship matrix constructed from SNP markers, enables the calculation of actual genetic similarities among genotypes without requiring pedigree data, thus offering an advantage over ABLUP (De los Campos *et al.*, 2009). The RR-BLUP variant, based on ridge regression, assumes that marker effects follow a normal distribution with equal variance (Hoerl and Kennard, 1970; Heslot *et al.*, 2012).

Bayesian prediction models

Bayesian methods rely on probability theory and the combination of prior and posterior distributions of parameters, performing well in situations where the number of SNPs far exceeds the number of samples (Gelman *et al.*, 1995; Habier *et al.*, 2011). Various models, including Bayes A, Bayes B, Bayesian LASSO, and Bayesian ridge regression (BRR), have been developed, each with distinct characteristics and the ability to model larger marker effects (Meuwissen *et al.*, 2001; Legarra *et al.*, 2011; De Los Campos *et al.*, 2013).

The main advantage of Bayesian models over GBLUP is their ability to focus on SNPs with large effects, thereby increasing prediction accuracy for complex traits. However, they require high computational power and careful selection of appropriate prior distributions (Hayes *et al.*, 2010).

Machine learning and deep learning models in genomic selection

With the growing availability of high-dimensional multi-omics datasets, more sophisticated predictive models have been developed to analyze nonlinear interactions and hidden patterns across genomic, transcriptomic, proteomic, metabolomic, and phenotypic layers (Mahmood *et al.*, 2022; Zafar *et al.*, 2023). These models aim to predict traits such as yield, stress tolerance, and disease resistance with high accuracy, thereby accelerating breeding cycles.

Among machine learning algorithms, random forests (RFs), extreme gradient boosting (XGBoost), and support vector machines (SVMs) are widely used in plant trait prediction. RF improves prediction stability by constructing ensembles of decision trees on bootstrap samples and can handle multi-omics data without complex parameter tuning (Tembhare *et al.*, 2024; Sánchez *et al.*, 2025). XGBoost optimizes decision trees, efficiently handling heterogeneous and missing data, and is effective in integrating genomic, metabolomic, and environmental datasets. SVM enhances complex classifications by mapping data into higher-dimensional spaces (Wu *et al.*, 2024; Xu *et al.*, 2025).

Deep learning goes a step further by automatically extracting hierarchical representations from multi-omics data, enabling the detection of complex spatial, temporal, and nonlinear dependencies. Convolutional neural networks (CNNs) are applied to genomic sequence data and spatial imaging, with some studies reporting up to a 15% improvement in prediction accuracy for stress-related traits (Liang *et al.*, 2024; Amin *et al.*, 2025). Recurrent neural networks (RNNs) and long short-term memory (LSTM) models are suitable for time-series data such as transcriptomics, while autoencoders facilitate the extraction of key features by reducing the dimensionality of large datasets (Lin, 2024). Moreover, multimodal deep learning enables the simultaneous integration of heterogeneous omics and imaging data (Haji, 2024).

Although these models achieve high accuracy, challenges such as limited interpretability and the need for large training datasets remain. The adoption of modern statistical and computational models has substantially improved the prediction of complex traits and accelerated plant breeding, thereby supporting the advancement of precision agriculture (Montesinos-López *et al.*, 2024). The assumptions, advantages, and applications of the major categories of statistical and computational models used in GS are compared in Table 5.5.

5.3 Data Sources and Integration in Genomic Design

5.3.1 Genomic data: SNPs, whole-genome sequencing, and epigenetics

In plant stress biology, genomic data provide a fundamental resource for identifying the genetic mechanisms underlying adaptation to biotic and abiotic stresses. These datasets encompass various forms of genetic variation recorded at different structural and functional levels of plant genomes. Among them, single-nucleotide polymorphisms (SNPs), as the most common type of genetic variation, play a key role in investigating traits associated with stress tolerance. SNP identification using next-generation sequencing (NGS) technologies enables the implementation of genome-wide association studies (GWAS) and quantitative trait loci (QTLs) mapping, both of which are

Table 5.5. Comparative overview of statistical and computational models in genomic selection, including reduced-bias regression (RLR), mixed linear models (MLM), Bayesian models, and machine/deep learning approaches, with assumptions, advantages, and applications (Author's own table).

Model category	Examples	Assumptions	Advantages	Applications	References
Reduced-bias regression (RLR)	Ridge regression (RR), LASSO, Elastic Net	Penalized coefficients to handle p > n	Reduces overfitting, variable selection	High-dimensional marker data, correlated effects	Szymczak *et al.*, 2009; Friedman *et al.*, 2010
Mixed linear models (MLMs)	GBLUP, rrBLUP	Fixed/random effects, normal distribution	Accounts for relationships, stable for polygenic traits	Pedigree/genomic data integration, breeding value estimation	De los Campos *et al.*, 2009; Heslot *et al.*, 2012
Bayesian models	Bayes A/B, Bayesian LASSO, BRR	Prior/posterior distributions	Handles large effects, uncertainty quantification	Complex traits with few/many QTLs	Habier *et al.*, 2011; De Los Campos *et al.*, 2013
Machine/deep learning	RF, SVM, ANN, CNN, RNN	Nonlinear patterns, ensembles	Captures interactions, high accuracy for multi-omics	Trait prediction in nonlinear data, image-based phenotyping	Mahmood *et al.*, 2022; Montesinos-López *et al.*, 2024

highly effective in pinpointing genomic regions linked to stress response traits (Silaiyiman *et al.*, 2025).

Whole-genome sequencing provides researchers with high-resolution nucleotide-level genetic information, offering a comprehensive understanding of the structure, function, and diversity of crop genomes. However, relying on a single reference genome cannot fully capture the genetic diversity within a species. In this context, the pan-genome approach has been developed, encompassing a combination of genomes from multiple accessions of the same species. This approach categorizes genes into three main groups: core genes, dispensable (variable) genes, and unique genes. Many genes associated with responses to environmental stresses fall into the variable and unique categories and are absent from conventional reference genomes (Whibley *et al.*, 2021).

Beyond sequence variation, epigenetic mechanisms – such as DNA methylation, chromatin remodeling, and the expression of non-coding RNAs – play crucial roles in regulating plant stress responses. These epigenetic changes, without altering DNA sequence, can mediate temporary or stable modifications in gene expression under stress conditions, and in some cases, can be transmitted to subsequent generations. Techniques such as bisulfite sequencing for DNA methylation analysis and ChIP-seq for profiling protein–DNA interactions can reveal deeper regulatory layers of stress responses (Amir *et al.*, 2020; Sherman and Salzberg, 2020).

Ultimately, integrating genomic (sequence and variation) data with epigenetic information provides a powerful framework for identifying key genes and molecular pathways involved in stress tolerance. This integrated approach enables plant breeders to utilize more precise genetic markers and advanced predictive models in breeding programs aimed at climate adaptation (Tyagi *et al.*, 2024; Sarawad *et al.*, 2025). The types of genomic data, their associated techniques, and their role in understanding stress tolerance are categorized in Table 5.6.

5.3.2 Phenotypic data: high-resolution imaging and field trials

Phenotyping, powered by advanced imaging technologies and precision sensors, plays a pivotal role in analyzing phenotypic traits associated with stress tolerance in plants.

Table 5.6. Categorization of genomic data types, including SNPs, whole-genome sequencing, and epigenetic mechanisms, with techniques, applications, and roles in stress tolerance (Author's own table).

Data type	Techniques	Applications	Role in stress tolerance	References
SNPs	NGS, GWAS, QTL mapping	Variant detection, association studies	Identifying tolerance-linked regions	Silaiyiman *et al.*, 2025
Whole-genome sequencing	NGS, Pan-Genome Assembly	Comprehensive genome analysis	Capturing core/ variable genes for adaptation	Whibley *et al.*, 2021; Sarawad *et al.*, 2025
Epigenetics (DNA methylation)	Bisulfite sequencing, MeDIP	Methylation profiling	Regulating gene expression under stress	Sherman and Salzberg, 2020
Epigenetics (chromatin remodeling)	ChIP-seq, ATAC-seq	Protein–DNA interaction mapping	Stable modifications for heritable resistance	Amir *et al.*, 2020
Epigenetics (non-coding RNAs)	RNA-seq, Small RNA sequencing	Expression analysis	Mediating stress responses, transgenerational effects	Tyagi *et al.*, 2024

High-throughput phenotyping (HTP) systems employ a wide array of tools and equipment, including unmanned aerial vehicles (UAVs), multispectral and thermal cameras, fluorescence imaging, and dedicated root phenotyping chambers, to record plant physiological and morphological characteristics non-destructively and with high precision and resolution. Parameters measured by these systems include canopy temperature, leaf water content, stomatal conductance, chlorophyll fluorescence, leaf area, and root system architecture (Zhang and Zhang, 2018; Angidi *et al.*, 2025).

These technologies enable precise and continuous monitoring of plant responses to environmental stresses under controlled laboratory conditions as well as in field trials. Combining phenotypic data with molecular and genomic datasets provides deeper insights into the genotype–phenotype relationship and enhances the accuracy of predictive models in plant breeding. Furthermore, employing these technologies in field trials allows for the assessment of actual plant performance in natural and variable climatic conditions – an essential factor in breeding stress-resilient cultivars (Li *et al.*, 2021).

5.3.3 Environmental data: Climate, soil, and management practices

In predictive plant breeding, integrating high-resolution environmental data with phenomic and omics information opens new horizons for developing precise and context-specific models that align with real growth conditions. Environmental datasets include climatic factors such as temperature, precipitation, soil moisture, solar radiation, and wind; physical and chemical soil properties; and information on management practices such as irrigation, fertilization, and planting dates (Xu *et al.*, 2022). Satellite-based remote sensing provides accurate spatiotemporal data on climatic variables at regional scales. At the field level, Internet of Things (IoT) sensors, installed at multiple locations, continuously record microclimatic data, including real-time temperature, relative humidity, light intensity, and wind speed (Ali *et al.*, 2015; Levizzani and Cattani, 2019).

Integrating these datasets with genomic, transcriptomic, metabolomic, and phenotypic information facilitates the development of environmentally informed selection indices and predictive models tailored to actual growth conditions. For example, Singh *et al.* (2024) demonstrated that combining hourly temperature and humidity data with multi-omics datasets in wheat improved predictions of heat stress tolerance and aided in the selection of genotypes adapted to specific climatic zones. Such integrated approaches underpin dynamic breeding programs capable of adapting to temporal and spatial variability in climate and soil, playing a key role in the forward-looking development of crop improvement strategies (Singh *et al.*, 2024).

However, practical implementation of these approaches faces several challenges, including the lack of standardization in data formats, the heterogeneity and fragmentation of data sources, and the need for algorithms capable of simultaneously modeling genotype–phenotype–environment interactions. Overcoming these challenges will pave the way for precise, climate-oriented breeding strategies that ensure maximum genetic gains under variable environmental conditions (Deng *et al.*, 2023).

5.3.4 Multi-omics data integration: Transcriptomics, proteomics, metabolomics

The integration of multi-omics data, including transcriptomics, proteomics, and metabolomics, has gained increasing importance as a comprehensive approach for systems biology analysis and phenotypic trait prediction. These distinct data layers each represent different dimensions of biological expression and regulation, and their combination can provide a deeper understanding of molecular mechanisms, particularly in response to biotic and abiotic stresses (Azodi *et al.*, 2019; Wang *et al.*, 2023).

Specifically, transcriptomics data reveal gene expression under various conditions and serve as an intermediary between the genome and cellular function. Proteomics reflects the expression levels and post-translational modifications of proteins, which are vital for proper cellular function. Meanwhile, metabolomics provides information on the end products of metabolic pathways and the physiological state of the organism (Zingaretti *et al.*, 2020; Costa-Neto *et al.*, 2022).

Multi-omics data integration strategies are generally categorized into horizontal, vertical, and diagonal conceptual models. In horizontal integration, homogeneous data types (e.g. multiple transcriptomics datasets) from different conditions or populations are combined to increase analytical power and robustness (Ulfenborg, 2019; Vahabi and Michailidis, 2022; Hayes *et al.*, 2024). Vertical integration analyzes different omics layers (genome, transcriptome, proteome, metabolome) from a single sample and reveals hierarchical structures of biological regulation (Ullah *et al.*, 2022). Diagonal integration involves the use of multi-omics data from related but non-identical samples connected by shared traits or pathways. These conceptual models guide experimental design and computational analysis pipelines, with method selection dependent on research objectives and available data (Akdemir *et al.*, 2020; Santiago-Rodriguez and Hollister, 2021). The conceptual models (horizontal, vertical, diagonal) for multi-omics data integration are depicted in Fig. 5.2.

To implement this complex integration, various tools have been developed. The mixOmics package in R, employing multivariate statistical methods such as canonical correlation analysis (CCA) and Partial Least Squares, enables the integrated analysis of heterogeneous data (Rohart *et al.*, 2017; Misra *et al.*, 2019). This tool has assisted in simultaneous transcriptome and metabolome analysis under cold stress in rice, identifying molecular markers related to cold tolerance. Bayesian factor-based models like MOFA + can handle incomplete data and detect shared or specific variations across omics types (Calado, 2020; John Martin *et al.*, 2024).

Similarly, network-based tools like iOmicsPASS integrate multi-omics features with phenotypic data to prioritize functional clusters and biomarkers. Weighted Gene Co-expression Network Analysis (WGCNA), by generating co-expression modules, facilitates the analysis of trait-associated modules across multiple omics layers. The DIABLO algorithm, a subset of mixOmics, is used for targeted integration in biomarker discovery and classification (Xue, 2020; Roychowdhury *et al.*, 2023).

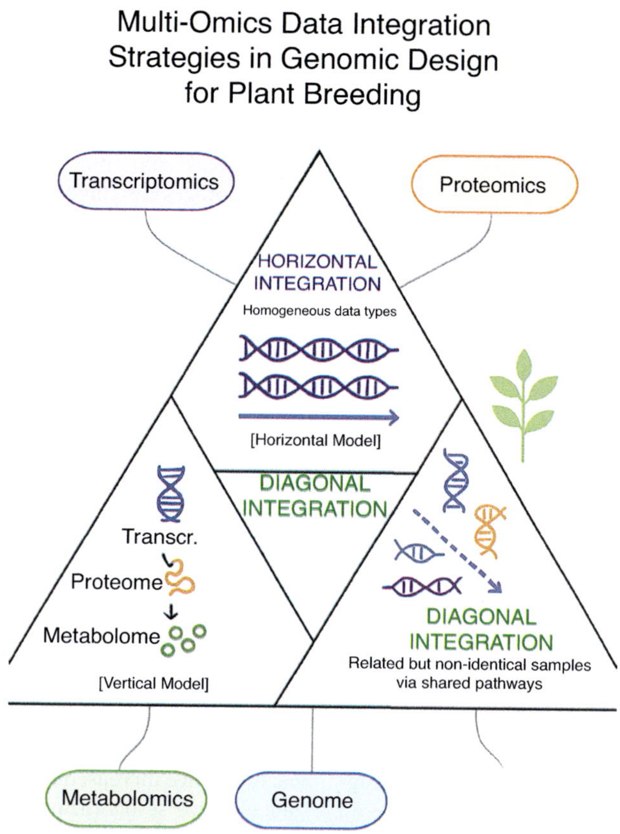

Fig. 5.2. Conceptual workflow of multi-omics data integration strategies, illustrating horizontal, vertical, and diagonal models for combining transcriptomics, proteomics, and metabolomics data (Author's own image).

Using machine learning and deep learning algorithms, such as convolutional neural networks and hybrid models, allows modeling of nonlinear and complex relationships between omics layers and phenotypes. Recent studies have demonstrated that integrating transcriptome, proteome, and metabolome data into a comprehensive model not only improves prediction accuracy but also identifies key biological components of the traits studied (Liang *et al.*, 2023).

Applications of such integrative approaches in plant breeding and precision medicine enable more accurate identification of molecular pathways involved in stress resistance, product quality, and biomarker development. These integrated approaches, combined with advanced computational tools, open new horizons in systems biology and its practical applications (Singh *et al.*, 2016; Pei *et al.*, 2017).

5.4 Machine Learning and Artificial Intelligence in Genomic Selection

5.4.1 Application of machine learning for genomic prediction

In recent years, machine learning (ML) has emerged as an effective and novel tool for improving the accuracy of genomic selection (GS) prediction and has received extensive attention. ML employs statistical techniques to enable systems to learn from data without explicit programming and to build models capable of developing algorithms that learn from existing data and make predictions on new data. These approaches provide better prediction accuracy compared with traditional GS methods (Yoosefzadeh-Najafabadi *et al.*, 2022).

Traditional statistical GS models are mainly parametric and based on the assumption of a linear relationship between genetic markers and phenotypes, whereas ML offers highly flexible non-parametric models capable of adapting to complex and nonlinear relationships between data and outcomes. This is especially important given the increasing volume of genomic data, the inherent complexity of genes, unique organism features, and the influence of environmental variables (Greener *et al.*, 2022). Indeed, ML can extract hidden patterns and unknown structures that parametric models are unable to detect (Gianola, 2013).

Since statistical analysis of quantitative traits in plants under complex conditions, such as pleiotropy, gene-by-gene interactions, and gene-by-environment interactions, is challenging, traditional methods may fail to capture all marker effects and face issues like the 'large p, small n' problem (more variables than samples), causing over-parameterization. In contrast, ML methods improve prediction accuracy by learning from repeated experiences and processing large datasets (Gianola *et al.*, 2006).

Machine learning is broadly categorized into supervised and unsupervised learning. Supervised learning is tasked with predicting desired phenotypic values based on input data, whereas unsupervised learning focuses on discovering clusters and relationships among data without a specified output. Many ML models applied in GS are supervised, including algorithms such as Bayesian networks, decision trees, naïve Bayes, and nearest neighbors, implemented as random forest (RF), artificial neural networks (ANNs), and support vector machines (SVMs) (González-Camacho *et al.*, 2018).

Random forest is considered a powerful method for analyzing complex individual traits using dense genetic markers. RF has good predictive power and can measure the importance of each marker. This method does not require inheritance assumptions, accounts for nonlinear and non-additive effects, operates very efficiently with a large number of variables, and is applicable for both regression and classification models (González-Recio *et al.*, 2010).

Support vector machines also have particular advantages for classification and regression problems and excel in identifying subtle and complex patterns in large datasets. This algorithm can shape the decision boundary through kernel functions to improve prediction accuracy and can model nonlinear relationships between phenotype and genotype (Noble, 2006).

Artificial neural networks, especially those enabling deep learning, are recognized as powerful models for enhancing GS performance. ANNs can identify complex patterns in data as universal approximators and accurately estimate factors affecting genetic interactions such as epistasis and dominance without the need for phenotypic distribution assumptions (Gianola *et al.*, 2011; Rosado *et al.*, 2020). Key characteristics, advantages, and applications of popular supervised ML algorithms in genomic prediction are outlined in Table 5.7.

Moreover, the applications of machine learning have increased recently with the emergence of diverse algorithms and the development of hybrid methods. For example, algorithms such as SVM, RF, and multi-layer perceptron (MLP) have shown good performance in genomic data of various species, including pigs, maize, and soybean. Using hyperspectral reflectance data combined with RF has led to accurate prediction of soybean genotype performance, and important within-season traits for optimized selection have been identified (Yoosefzadeh-Najafabadi *et al.*, 2021).

Although considerable evidence shows that ML models outperform classical models in some cases, their performance depends on the data type, genetic structure of the population, phenotypic features, and model settings. Therefore, effective use of these methods requires a deep understanding of algorithms, appropriate feature selection, parameter tuning, and intelligent integration with traditional models (Parmley *et al.*, 2019).

Additionally, unsupervised learning algorithms play an important role in discovering hidden structures in genomic data, clustering

Table 5.7. Overview of supervised machine learning algorithms in genomic prediction, including random forest (RF), support vector machines (SVMs), and artificial neural networks (ANNs), with characteristics, advantages, and applications (Author's own table).

Algorithm	Characteristics	Advantages	Applications	References
Random forest (RF)	Ensemble of decision trees, bootstrap sampling	Handles nonlinearity, measures feature importance, robust to overfitting	Complex trait analysis, dense marker prediction	González-Recio *et al.*, 2010
Support vector machines (SVMs)	Kernel-based, decision boundaries in high dimensions	Effective for subtle patterns, nonlinear relationships	Classification/ regression in large datasets	Noble, 2006
Artificial neural networks (ANNs)	Layered neurons, deep learning variants	Captures complex interactions, universal approximators	Epistasis/ dominance estimation, multi-omics integration	Gianola *et al.*, 2011; Rosado *et al.*, 2020

genotypes, and dimensionality reduction, providing a foundation for the development of intelligent and hybrid methods in future genomic prediction (Griffel *et al.*, 2018).

5.4.2 Application of deep learning in trait prediction

Deep learning, a branch of machine learning, has recently emerged as a novel tool in genetic trait prediction and genomic selection (GS). Although research on deep learning applications in this field is still limited, significant studies have compared the performance of these methods with conventional statistical models (Montesinos-López *et al.*, 2018; Montesinos-López *et al.*, 2019).

In a study by Montesinos-López *et al.* (2018), deep learning models and GBLUP models with densely connected network architectures were compared on nine genomic datasets (including three maize and six wheat datasets). The results showed that when genotype × environment (G × E) interaction was not considered, deep learning performed well in six out of nine datasets. However, when this interaction was included, the GBLUP model performed better in eight out of nine datasets (Montesinos-López *et al.*, 2018). Another study (Montesinos-López *et al.*, 2019) indicated that machine learning methods such as multi-layer perceptron (MLP) and support vector machines (SVMs) are computationally more efficient than Bayesian models and have competitive capabilities.

Artificial neural networks (ANNs) form the basis of deep learning and are constructed by mimicking the structure of human brain neurons. When three or more ANN layers are stacked sequentially, a deep neural network (DNN) is formed (LeCun *et al.*, 2015). Various types of deep learning networks are applied in genetic trait prediction, the most important of which include multilayer perceptron (MLP), convolutional neural networks (CNNs), and recurrent neural networks (RNNs) (Montesinos-López *et al.*, 2021).

The multi-layer perceptron (MLP) has a simple structure that transfers data unidirectionally from the input layer to hidden layers and then to the output. This model is suitable for general prediction problems but may suffer from overfitting during training, leading to reduced accuracy on real data (Abdollahi-Arpanahi *et al.*, 2020).

Convolutional neural networks (CNNs), primarily used in image recognition, reduce the input size and share parameters, thus decreasing the number of trainable parameters, which enhances efficiency and reduces training time. This architecture involves convolution operations, nonlinear transformations, and pooling that preserve important input information (Pook *et al.*, 2020).

Recurrent neural networks (RNNs) contain feedback loops that allow them to incorporate previous information when processing current data. These networks are suitable for sequential and temporal data but require more computational resources due to their high computational complexity (Montesinos-López *et al.*, 2021).

Although deep learning applications in genomic selection are still in early stages, these methods have advantages, including direct processing of raw data without pre-processing, the ability to model genetic diversity and non-linear effects, and a reduction in the number of parameters needed through parameter sharing. However, challenges such as overfitting and the need for precise model parameter tuning remain obstacles to widespread adoption, for which Bayesian frameworks have been proposed as solutions (Neal, 2012).

Studies indicate that deep learning can achieve acceptable performance in genetic trait prediction, but in many cases, its performance is still comparable to or slightly lower than classical statistical models. Therefore, further experiments, more diverse data, including various omics, climatic data, and breeder experiences, and optimal topology designs for deep learning are needed in the future (Jeon *et al.*, 2023).

5.4.3 Feature selection and dimensionality reduction in Big Data

In the era of big and complex data, dealing with high-dimensional data is one of the greatest challenges in the fields of machine learning and data mining. Such data typically includes a large number of features, many of which are often irrelevant, redundant, or contain duplicate information. The presence of these features leads to reduced accuracy of machine learning models, increased computational costs and training time, as well as problems such as overfitting. To address these challenges, two main methods – feature selection and dimensionality reduction – are widely used, each designed to reduce the number of features while preserving important information in the data (Bolón-Canedo *et al.*, 2016).

The process of feature selection involves identifying and removing irrelevant and redundant features from the dataset so that the learning model can analyze the data with higher accuracy and speed. This process is sometimes also referred to as variable selection or attribute selection. Feature selection methods are generally divided into two broad categories: subset selection methods and feature ranking methods. In subset selection, all or part of different combinations of features are evaluated using search algorithms such as forward selection or backward elimination, and the best subset is selected. Due to the large number of possible combinations and high computational cost, this approach typically has high complexity. Conversely, feature ranking methods evaluate each feature independently using criteria such as information gain, symmetric uncertainty, or gain ratio, and select the best features based on their scores. This method is faster and less costly but may be less effective in eliminating redundant features (Rouhi and Nezamabadi-Pour, 2020).

From a general approach perspective, feature selection is categorized into four main types: Wrapper, Embedded, Filter, and Hybrid methods. In the Wrapper method, feature selection is performed with the help of a learning model, and each feature subset is evaluated using a classification algorithm, which increases accuracy but also computational cost. In the Embedded method, feature selection occurs simultaneously with model training, where important features are identified during training. This method is model-dependent and has a computational cost between the Wrapper and Filter methods. The Filter method selects features independently of the learning model based on statistical criteria, increasing generalizability and reducing computational complexity, making it well-suited for high-dimensional data. Finally, the Hybrid method combines the previous two approaches, aiming to retain the advantages of both (Manthena *et al.*, 2022).

Conversely, dimensionality reduction is a method that reduces the number of features by creating new and compact combinations of existing features, such that important information and data structure are preserved. Besides reducing data volume, this method can improve the efficiency of machine learning models and prevent complexity and overfitting. Various dimensionality reduction methods exist, which can be divided into linear and nonlinear techniques. One of the most widely used linear methods is principal component analysis (PCA), which extracts the main and low-dimensional features of the data through linear combinations of original features. Nonlinear methods such as kernel PCA, autoencoders, and deep learning techniques can also preserve more complex data structures in a low-dimensional space. Advantages of dimensionality reduction include facilitating data visualization, reducing computational costs, and eliminating indirect noise; however, disadvantages include difficulty in interpreting new features and the potential loss of some information (Gnana *et al.*, 2016).

Feature selection and dimensionality reduction are two complementary approaches for managing big and high-dimensional data. Intelligent selection and appropriate combination of these methods based on the nature of the data and analysis goals can significantly improve the performance of machine learning models and enable more accurate and faster knowledge extraction. A comparative analysis of the main feature selection methods, including their computational costs and suitability for genomic data, is provided in Table 5.8.

5.4.4 Case studies: Plant improvement using artificial intelligence

One of the studies on plant improvement using artificial intelligence was conducted by Yoosefzadeh-Najafabadi *et al.* (2025). This study examined the vital role of plant breeding in addressing challenges related to food insecurity and global hunger – issues expected to intensify in the coming years. The study emphasized the importance of a deep understanding of related fields such as phenomics and genomics and introduced five key areas as strategies to accelerate crop improvement: germplasm identification, genome assembly, genomic selection, gene function identification, and gene editing. Over the past decade, the integration of machine learning algorithms in this field to merge diverse omics and biological data and develop accurate predictive models has attracted considerable attention. The study also reviewed the advantages and challenges of using machine learning in genomic selection, including higher prediction accuracy, improved breeding efficiency, data biases, and implementation barriers. Successful applications of artificial intelligence in enhancing crop resistance and productivity through multi-omics data integration and addressing data biases were also highlighted. Ultimately, this case study demonstrated the transformative potential of artificial intelligence in genomic plant breeding and identified future research gaps and opportunities.

5.5 Computational Tools and Pipelines for Genomic Design

5.5.1 Bioinformatics platforms for genomic data processing

Processing genomic data in plant sciences has become a central domain in biotechnology and bioinformatics, mainly due to next-generation sequencing (NGS) technologies that can produce vast volumes of data rapidly and accurately. This massive amount of genomic data, including diverse formats such as FASTQ (raw sequencing reads), BAM (aligned sequences), and VCF (variant call files), requires advanced and integrated bioinformatics platforms capable of performing preprocessing, analysis, and interpretation steps reliably and precisely.

In the pre-processing stage, quality control of data using tools like FastQC and sequencing error correction via software such as Trimmomatic and Cutadapt are essential to minimize errors arising from laboratory and sequencing procedures and to standardize input data for downstream analyses.

Among popular platforms in this area, Galaxy stands out with its web-based graphical user interface that enables users without extensive programming expertise to conduct various genomic data analyses, including reference-guided or *de novo* genome assembly, gene annotation, and variant detection. Galaxy's support for diverse tools and customizable workflows has made it widely adopted in many plant genomic projects.

The Bioconductor project, developed within the R programming environment, plays a key role in genomic data analysis and beyond, due to its wide variety of packages and adaptability to complex analyses such as transcriptomics, gene

Table 5.8. Comparative analysis of feature selection methods (Wrapper, Embedded, Filter, and Hybrid), including approaches, advantages, computational costs, and suitability for genomic data (Author's own table).

Method	Approaches	Advantages	Computational cost	Suitability for genomic data	References
Wrapper	Subset evaluation with ML model (e.g. Forward/Backward Selection)	High accuracy, model-specific	High (evaluates subsets)	Small datasets, precise selection	Bolón-Canedo *et al.*, 2016
Embedded	Integrated in model training (e.g. LASSO penalties)	Balances accuracy and efficiency	Medium	High-dimensional, correlated markers	Manthena *et al.*, 2022
Filter	Statistical criteria (e.g. Information Gain, Chi-Square)	Fast, model-independent	Low	Initial screening in large genomic sets	Rouhi and Nezamabadi-Pour, 2020
Hybrid	Combines Filter + Wrapper/Embedded	Retains the strengths of both	Medium–high	Comprehensive for omics data integration	Manthena *et al.*, 2022

expression pattern identification, gene network analysis, and multi-omics data integration.

Tools like the Genome Analysis Toolkit (GATK) are specifically designed for accurate identification of genetic variants such as SNPs and InDels. With advanced error correction algorithms and data filtering, GATK enables high-quality detection of genetic variations that are crucial for understanding genetic diversity and plant trait improvement.

Concurrently, with the explosive growth of genomic data, cloud computing technologies such as Amazon Web Services and Google Cloud Platform have provided scalable, flexible solutions for storage, management, and parallel analysis of data. These technologies allow execution of computationally intensive workflows without the need for complex local hardware infrastructure.

Moreover, machine learning and AI algorithms have been increasingly integrated into bioinformatics platforms to uncover complex and hidden patterns in genomic data, predict gene functions and metabolic pathways, and optimize annotation processes. For instance, classification and clustering methods, deep neural networks, and gene prediction models in multilayered omics analyses significantly enhance the accuracy and speed of biological knowledge extraction. Together, these tools and technologies form the foundation of modern genomic research in plant sciences, enabling advanced investigations, development of improved plant varieties, and a deeper understanding of biological processes (Popescu *et al.*, 2016; Demidchik *et al.*, 2020). A summary of key bioinformatics platforms and tools for genomic data processing is presented in Table 5.9.

5.5.2 Cloud computing and big data analytics in genomic breeding

The advent of NGS technologies has led to the generation of enormous volumes of genomic data, posing significant challenges in data storage, processing, and analysis. For example, the human genome contains approximately 3 billion base pairs, corresponding to about 100 gigabytes of raw data. Estimates indicate that by the end of 2011 the global sequencing capacity reached approximately 13 quadrillion bases annually, with this trend accelerating rapidly. This enormous data scale necessitates novel, reliable, and efficient storage and analysis methods to enable researchers to gain scientific insights swiftly and cost-effectively (Driscoll *et al.*, 2013).

In bioinformatics workflows, after sequencing data acquisition, extensive computational analyses are performed to process data, validate experimental results, and derive scientific conclusions. Therefore, beyond investments in

Table 5.9. Overview of bioinformatics platforms and tools for genomic data processing, including pre-processing, analysis, and integration tools, with descriptions and applications in plant sciences (Author's own table).

Platform/Tool	Description	Applications in Plant Sciences	References
Galaxy	Web-based GUI for workflows, supports assembly/annotation	Variant detection, de novo assembly in crop genomes	Popescu *et al.*, 2016
Bioconductor	R-based packages for omics analysis	Transcriptomics, gene networks, multi-omics integration	Popescu *et al.*, 2016
GATK	Variant calling toolkit with error correction	SNP/InDel identification in plant diversity studies	Demidchik *et al.*, 2020
FastQC	Quality control for sequencing data	Pre-processing raw reads from NGS	Liu *et al.*, 2014
Trimmomatic/Cutadapt	Adapter trimming and error correction	Standardizing data for downstream analysis	Driscoll *et al.*, 2013
AWS/Google Cloud	Cloud storage and parallel processing	Handling large-scale plant genomic datasets	Krampis *et al.*, 2012

sequencing equipment, substantial investments in computational hardware, expert bioinformatics teams, and data analysis software are essential. However, existing tools often face limitations when handling large-scale data, and software installation and configuration pose significant challenges and time consumption for medical and biomedical researchers (Krampis *et al.*, 2012).

Cloud computing offers flexible, scalable, and pay-as-you-go computational and storage resources, presenting new opportunities to overcome these challenges. Cloud platforms can automatically configure computing clusters and allocate resources dynamically based on project requirements. For instance, integrating scientific workflow systems like Galaxy with cloud cluster management tools such as Globus Provision enables rapid, automated, and scalable deployment of analytical environments (Thain *et al.*, 2005; Liu *et al.*, 2012).

A key challenge in big data analysis is the secure and efficient transfer of massive datasets. Services like Globus Transfer facilitate high-performance data movement while ensuring data integrity across geographically distributed sites. Additionally, integration of job queue management systems like HTCondor with Galaxy facilitates parallel and distributed execution of computational tasks, improving resource utilization and efficiency (Foster, 2011).

Furthermore, developing user-tailored specialized tools and incorporating semantic validation mechanisms for designing and executing analytical workflows are innovations that reduce potential errors and accelerate analytical processes. Cloud computing, combined with integrated workflow systems and advanced data transfer tools, provides a comprehensive solution for managing and analyzing large-scale genomic big data, lowering technical barriers and enabling large-scale genomic breeding projects for researchers (Liu *et al.*, 2014).

5.5.3 Open-source tools and databases for genomic prediction

Significant advances in DNA sequencing technologies, especially next-generation and third-generation sequencing, alongside the development of open-source tools and databases, have fundamentally transformed the field of genomic prediction and targeted plant breeding. These technologies, by increasing the capacity for generating genomic data and enabling rapid and accurate analysis, have accelerated the development of improved varieties and substantially reduced the costs associated with breeding programs.

One of the key pillars of genomic advances is the development of open-source software for assembly and mapping of genomic data. Tools such as Roche 454 GSassembler, Celera Assembler, and Mira are utilized for *de novo* genome assembly, while after constructing a reference genome, software like Bowtie, BWA, and TopHat are employed for data mapping, and tools such as SAMtools and GigaBayes are used for identifying single-nucleotide polymorphisms (SNPs). This suite of software, together with extensive open-access databases such as GenBank, EMBL, DDBJ, Ensembl, UCSC Genome Browser, and dbSNP, provides researchers and breeders with comprehensive access, storage, and analytical capabilities for genomic data.

Furthermore, species-specific specialized databases such as SGN, Phytozome, Gramene, and CropNet, by offering targeted genomic, proteomic, and metabolic data, play an effective role in facilitating breeding programs and research. Protein databases, including UniProt, Swiss-Prot, Gene Ontology, Protein Data Bank, InterPro, KEGG, Pfam, and STRING, significantly aid in the analysis of protein function and interactions (Medina *et al.*, 2010).

On the other hand, gene expression analysis is a key tool in understanding plant molecular processes and selecting genetic targets. Methods such as qRT-PCR and microarrays are used for quantitative and high-throughput gene expression measurements, and open-source software and databases like Babelomics, Bioconductor, MeV, Genevestigator, GEO, and ArrayExpress, provide extensive functionalities for analysis and access to expression data (Wang *et al.*, 2019).

In the domain of genomic prediction, statistical and analytical approaches such as quantitative trait loci (QTL) mapping and genome-wide association studies (GWAS) hold a prominent position. Open-source software such as QTL Cartographer, QTL IciMapping, R/qtl, and MetaQTL offer efficient tools for identifying genetic regions linked to quantitative traits and performing meta-analyses (84 and 105). Advanced statistical GWAS models, leveraging open-source software and mixed linear models, have enabled more precise identification of genes and factors influencing complex traits in diverse populations (Jannink *et al.*, 2001; Kaler *et al.*, 2020).

Additionally, novel genome editing technologies such as the CRISPR/Cas system, zinc finger nucleases (ZFNs), and TALENs provide the ability to introduce targeted and rapid modifications in DNA sequences, serving as powerful tools for genetic improvement of economically important traits. The development of diverse CRISPR systems, including Cas9, Cas12, and Cas13, has significantly expanded the scope of applications of this technology in genetic engineering (Smithies *et al.*, 1985; Jinek *et al.*, 2012).

Overall, the integration of this extensive range of advanced and open-source tools and databases has established a robust platform for rapid, precise, and cost-effective breeding and genomic prediction. These infrastructures ensure easy access to data and analytical results and contribute to the continuous improvement of economically important traits in agricultural products, ultimately leading to sustainable agriculture and enhanced global food security.

5.5.4 Integrated pipelines for genomic and phenotypic data

The integration of genomic and phenotypic data is one of the fundamental challenges in modern plant breeding studies, aiming to achieve a deeper and more precise understanding of the genotype–phenotype relationship. In this regard, advanced and integrated analytical pipelines, utilizing cutting-edge technologies such as artificial intelligence, play a key role in processing and analyzing vast volumes of molecular and agronomic trait data (van Dijk *et al.*, 2021).

Genomic data, including whole-genome sequencing, single-nucleotide polymorphisms (SNPs), and regulatory information, are available at very high volumes and complexity. On the other hand, phenotypic data consist of a broad set of morphological, physiological, and environmental traits recorded under diverse environmental conditions. The main challenge in integrating these data is synchronizing and merging information with different temporal, spatial, and technological scales (Wang *et al.*, 2020).

Integration pipelines, aided by artificial intelligence algorithms (especially deep neural networks and machine learning) are capable of extracting hidden features and complex relationships between genomic and phenotypic data. These technologies, by simultaneously analyzing multifunctional data, enable the identification of genetic patterns influencing agronomic traits that were inaccessible with traditional analytical methods (Altman *et al.*, 2021).

A key advantage of using integrated pipelines is the ability to precisely predict plant performance under varying environmental conditions. These predictions, supported by deep learning models that extract hidden data layers and their interrelations, significantly enhance the quality and efficiency of breeding programs. Furthermore, these methods play an important role in accelerating the identification of target genes and *cis*-regulatory elements (CREs), which can lead to structural and functional improvements in agricultural products (Kwon *et al.*, 2020).

From a practical perspective, data integration pipelines provide efficient management and analysis of Big Data. This includes empowering breeders to exploit diverse archives of genetic and phenotypic data that are increasingly available. Additionally, these pipelines have the capacity to analyze interactions of environmental factors such as climate change, biotic and abiotic stresses, and meteorological parameters, which are critical for improving crop resistance and performance (Razzaq *et al.*, 2021).

Ultimately, the development and application of integrated genomic and phenotypic data pipelines will transform modern breeding and enhance sustainable agricultural productivity. These pipelines, as powerful tools, pave the way for novel discoveries in plant molecular biology and provide practical solutions for improving the quality and quantity of agricultural products (Divya and Kumar, 2024).

5.6 Precise Breeding Strategies Based on Genomic Data

5.6.1 Genome-assisted breeding compared with traditional methods

Since the domestication of plants about 10,000 years ago, traditional breeding has been able to develop various crops and cultivars using phenotypic trait-based methods, which have significantly increased agricultural productivity and sustainability (Fedoroff, 2010). These methods, especially until the mid-20th century, played a crucial role in boosting the production of major crops, but had limitations such as time-consumption, low precision, and challenges in improving complex polygenic traits. With advances in genomics, advanced tools for whole-genome analysis of plants and direct analysis of the genotype–phenotype relationship became available to breeders, marking a major transformation in plant breeding processes (Varshney *et al.*, 2021a). The use of molecular markers such as SNPs, which have much higher precision and efficiency than classical markers, has enabled the accurate identification of genomic regions associated with key traits. Moreover, novel techniques like genome-wide association studies (GWAS), marker-assisted selection (MAS), and genomic selection (GS) allow breeders to select desirable traits in a targeted manner and predict breeding outcomes with greater accuracy (Hasan *et al.*, 2021; Riaz *et al.*, 2021). The CRISPR/Cas9 genome editing technology, as an innovative method, enables precise and rapid gene modification and has accelerated the breeding process compared with traditional methods (Perez-de-Castro *et al.*, 2012). Overall, genomic approaches, especially in breeding complex traits influenced by multiple genes and environmental factors, have been much more effective and efficient than traditional methods. Genome-assisted breeding has replaced or complemented traditional methods and significantly improved the accuracy, speed, and effectiveness of breeding programs.

5.6.2 Genome editing (CRISPR, TALEN) for crop improvement

In recent decades, genome editing technology has emerged as one of the most revolutionary biotechnology tools for improving agronomic traits in crop plants. This technology allows precise and targeted modifications of DNA sequences, enabling the alteration of specific genes associated with important agronomic traits including yield, disease resistance, and nutritional quality (Karlson *et al.*, 2021). Among various genome editing tools, the CRISPR/Cas and TALEN systems have attracted more attention than others due to their efficiency, precision, and broad applicability (Shan *et al.*, 2013).

The CRISPR/Cas system, originating from bacterial defense mechanisms against phages, has rapidly become a popular tool in agricultural biotechnology because of its ease of design and use (Jinek *et al.*, 2012). In this system, the Cas9 enzyme is guided by a single guide RNA (sgRNA) to a specific genomic region where it creates a double-strand break in the DNA. The cell then repairs the break using natural pathways such as non-homologous end joining (NHEJ) or homology-directed repair (HDR), which forms the basis for generating mutations or precise gene replacements (Illouz-Eliaz *et al.*, 2020). One major advantage of CRISPR is that only the RNA guide sequence needs to be changed to target different sites, without the need to redesign the protein, making it simpler and more cost-effective compared with TALEN or ZFN. Furthermore, the development of various Cas enzymes such as Cas12, Cas13, and Cas14 has increased the precision, flexibility, and range of applications of this technology (Van der Oost and Patinios, 2023).

In contrast, the TALEN system relies on DNA-binding proteins derived from the bacterium *Xanthomonas*. These proteins consist of modular repeats, each recognizing a specific nucleotide, which together recognize the target DNA sequence (Bae *et al.*, 2019). After two TALENs bind to opposite sides of the target sequence, the FokI nuclease domains dimerize and induce a double-strand break in the DNA. Although TALEN design is more time-consuming and complex than CRISPR, its high specificity and lower off-target cutting rate keep it in use for some specific applications (Jiang and Doudna, 2017).

The application of these technologies in crop improvement has yielded remarkable achievements. For instance, editing genes related to cell growth can increase grain yield or biomass. Also, by knocking out or modifying genes sensitive to biotic and abiotic stresses, plants resistant to diseases, drought, salinity, or high temperatures have been developed. Metabolic pathway modifications have allowed enhancement of beneficial nutritional compounds such as vitamins and essential fatty acids or reduction of anti-nutritional and allergenic compounds. Moreover, the development of herbicide- and pest-resistant crops has reduced chemical input use and improved environmental sustainability in agricultural systems (Jinek *et al.*, 2012).

Despite these successes, challenges remain in expanding the use of these technologies. Concerns exist regarding off-target effects, horizontal gene transfer, and societal acceptance of edited products. Furthermore, regulatory frameworks for the approval and commercialization of genome-edited crops are still incomplete in many countries (Li *et al.*, 2023). However, extensive efforts are underway to optimize delivery systems, improve targeting accuracy, develop new enzymes, and establish clear regulations to facilitate broader adoption of these technologies (Yang *et al.*, 2022).

In summary, genome editing tools like CRISPR and TALEN have revolutionized plant breeding and promise a new era of smart and sustainable agriculture. These technologies enable precise modification of specific genes without introducing foreign DNA, opening new horizons for developing resilient, high-yielding, high-quality crops adapted to climate change, which can play a vital role in ensuring global food security.

5.6.3 Marker-assisted selection (MAS) and genomic selection (GS)

Marker-assisted selection (MAS) and genomic selection (GS) are both molecular information-based breeding strategies aimed at increasing selection accuracy and reducing breeding time. MAS primarily relies on the identification of markers that are directly associated with major quantitative trait loci (QTLs) controlling the traits of interest, and it is typically applied to traits with high heritability governed by a few major genes (Kumar *et al.*, 2024). This method is efficient when major-effect QTLs have been identified and precise markers have been developed for them. In contrast, GS does not require prior identification of QTLs; instead, it uses genome-wide markers to model the relationship between genotype and phenotype and is more effective for quantitative and complex traits controlled by numerous QTLs with small effects (Badoni *et al.*, 2016). Indeed, GS can simultaneously incorporate the cumulative small effects of multiple genes into the model, whereas MAS focuses solely on major genes. This characteristic allows GS to outperform MAS in species with complex genetic architectures such as polyploids (e.g. blueberry). Ultimately, MAS and GS are complementary approaches; however, genomic selection, by enabling rapid, accurate, and multidimensional selection, has opened new horizons in plant and animal breeding (de Bem Oliveira *et al.*, 2020).

With advances in biotechnology and genomics, novel methods have been developed to improve efficiency and speed in breeding programs, among which MAS and GS are recognized as two effective molecular information-based approaches. Both methods share the common goal of enhancing selection accuracy and reducing breeding duration, but they differ fundamentally in implementation and in the types of traits for which they are best suited (Sinha *et al.*, 2023).

MAS mainly depends on identifying markers that are directly linked to major QTLs controlling the target traits and is generally suitable for highly heritable traits governed by a limited number of major genes. This method achieves maximum efficiency when major QTLs have been previously identified and accurate markers have been designed for them. By focusing on major genes, MAS plays a significant role in gene pyramiding, reducing the need for extensive phenotyping, and enabling early selection of superior individuals (Budhlakoti *et al.*, 2022).

Conversely, GS utilizes genome-wide marker data without the necessity of QTL identification to model the genotype–phenotype relationship. This feature makes GS a powerful tool for quantitative and complex traits influenced by numerous QTLs with small effects. Unlike MAS, which concentrates on major genes, GS can incorporate the combined small effects of many genes simultaneously in the model. Therefore, in species with complex genetic structures such as polyploids (e.g. blueberry), genomic selection often outperforms MAS (Heffner *et al.*, 2009).

The GS process involves three main steps: establishing a training population with both genotypic and phenotypic data; developing a genomic estimated breeding value (GEBV) prediction model; and predicting GEBVs for new individuals based solely on genotypic data. By leveraging diverse statistical models and machine learning algorithms, GS enables precise and multidimensional selection of complex traits at early plant growth stages (Mueller and Van Eenennaam, 2022).

In conclusion, it is important to emphasize that MAS and GS are complementary, and, depending on trait type, available resources, and breeding program objectives, both methods can be used synergistically. Nonetheless, genomic selection, by providing rapid, accurate, and multidimensional selection capabilities, has unveiled new prospects in plant and animal breeding (Kriaridou *et al.*, 2020).

5.6.4 Case studies: Successful applications in plant breeding

In recent years, technological advancements in data mining and genomics have led to the emergence of a new generation of plant breeding approaches. One notable example is a study on genomic selection in blueberry (*Vaccinium corymbosum* L.), which utilized a combination of data-driven and genetic methods to optimize marker design and training population composition. This study, leveraging a historical dataset of over 3000 genotypes, demonstrated that targeted selection of genetic markers and training populations can enhance the accuracy of quantitative trait prediction and, in the long term, improve genetic gain. Besides optimal marker selection, optimization algorithms were employed to allocate phenotyping resources, thereby improving the predictive performance of models. The results indicated that the use of data-driven methods not only improves the prediction of genetic value but also serves as a strategic tool for breeding decision-making under resource-limited conditions. This case study illustrates that the informed integration of genomic data, machine learning algorithms, and statistical principles can facilitate the development of novel plant breeding approaches and pave the way for producing more resilient, higher-yielding, and climate-adapted cultivars (Adunola *et al.*, 2024).

Another prominent example of this trend is the 3D-breeding approach, which combines genomics, farmers' indigenous knowledge, and environmental analyses within a decentralized, data-driven system. A successful case study in durum wheat (*Triticum durum* Desf.) from the Ethiopian highlands showed that this decentralized method could double the accuracy of performance prediction. By leveraging extensive field trials across more than 1000 farmer-managed farms, this approach identified genotypes with high local adaptation that exhibited superior and stable performance under variable environmental conditions. The results not only validate the capabilities of 3D-breeding in exploiting genetic and environmental diversity but also present an innovative solution to complement and enhance traditional plant breeding programs. This approach can especially secure food production in challenging growing regions and improve the livelihoods of millions of smallholder farmers (De Sousa *et al.*, 2021).

The study by Razzaq *et al.* (2021) represents a distinguished example of applying advanced technologies in plant breeding to address the challenges of food security and climate change. The authors comprehensively reviewed a set of advanced breeding approaches that have been successfully employed in the genetic improvement of key crop species. Among the most significant technologies is genomic editing based on extensive pangenomic diversity. Pangenomes, by providing a comprehensive representation of intraspecific genetic variation, enable the identification and recovery of lost genes. This approach has been applied in species such as wheat, rice, and maize, resulting in enhanced performance and resistance. Additionally, advanced gene editing systems such as CRISPR/Cas and its novel variants, including Prime Editing and Base Editing, have facilitated precise and targeted modification of genes involved in resistance to biotic and abiotic stresses. The use of various Cas enzymes, including Cas9, Cas12, Cas13, and Cas14, has further expanded the efficiency and scope of this technology in improving many crop species. Razzaq *et al.* also emphasized the critical role of high-throughput phenotyping, artificial intelligence (AI), and machine learning (ML) in future agricultural automation. The integration of these tools with the 'Speed Breeding' approach has significantly accelerated the process of identifying effective genes and the development of superior genotypes. The authors concluded that by combining gene editing technologies, pangenomics, AI, and speed breeding, a multidisciplinary and next-generation platform for producing climate-resilient crops can be established. This case study stands as a prime example of the success of innovative plant breeding strategies in response to the global food security crisis.

This integrative, next-generation breeding strategy is conceptually summarized in Fig. 5.3.

5.7 Challenges and Future Directions

5.7.1 Data quality, biases, and standardization issues

One of the fundamental obstacles in data-driven breeding is the quality and diversity of data obtained from various sources. Genomic, phenotypic, and environmental data are often collected in different formats, structures, and scales, leading to heterogeneity and difficulties in their integration. The lack of common standards for data formats, metadata, and processing methods severely limits the efficiency and reusability of data (Hickey *et al.*, 2017).

Moreover, data may contain biases resulting from non-random selection of genotypes, specific environmental conditions, or the use of particular equipment and platforms in phenotyping and sequencing. These biases can lead to inaccurate or non-generalizable results. For example, phenotyping with advanced systems in specialized experimental farms may not be generalizable to the real conditions of smallholder farms (Xu *et al.*, 2017).

Other challenges include the shortage of computational and bioinformatics infrastructure in some regions or small breeding institutions. Additionally, skill gaps in data science and bioinformatics among plant breeders hinder the full utilization of advanced technologies. Addressing these challenges requires the development of data standardization frameworks, investment in infrastructure and training, and strengthening inter-institutional collaborations (Ågren *et al.*, 2017; Ewa *et al.*, 2022). The primary data-related challenges in modern breeding and proposed solutions to address them are summarized in Table 5.10.

5.7.2 Ethical and regulatory considerations in genomic breeding

Data-driven plant breeding, particularly using technologies such as gene editing (CRISPR/Cas9), raises significant ethical and regulatory questions. One of the primary concerns is genetic data privacy. Genomic information of crop plants, especially in proprietary breeding programs, is often regarded as sensitive assets and may be vulnerable to misuse without appropriate data governance frameworks (Ewens, 2012).

Intellectual property rights are also critical in this context. Newly bred varieties, especially genetically modified or gene-edited plants, are often subject to exclusive patents that may restrict access for small-scale farmers and breeders, creating inequalities in benefiting from scientific advancements. This issue is particularly pronounced in developing countries (Esquinas-Alcázar, 2005).

In the field of gene editing, concerns such as off-target effects, stability of genetic changes, and potential environmental consequences (e.g. gene flow to wild species or reduction in biodiversity) have led some countries to impose stringent regulations on the use of these technologies. Such regulatory approaches may cause delays or even bans on the deployment of advanced tools in certain regions (Metje-Sprink *et al.*, 2019).

To ensure responsible use of novel breeding technologies, the development of harmonized legal and ethical frameworks at the international

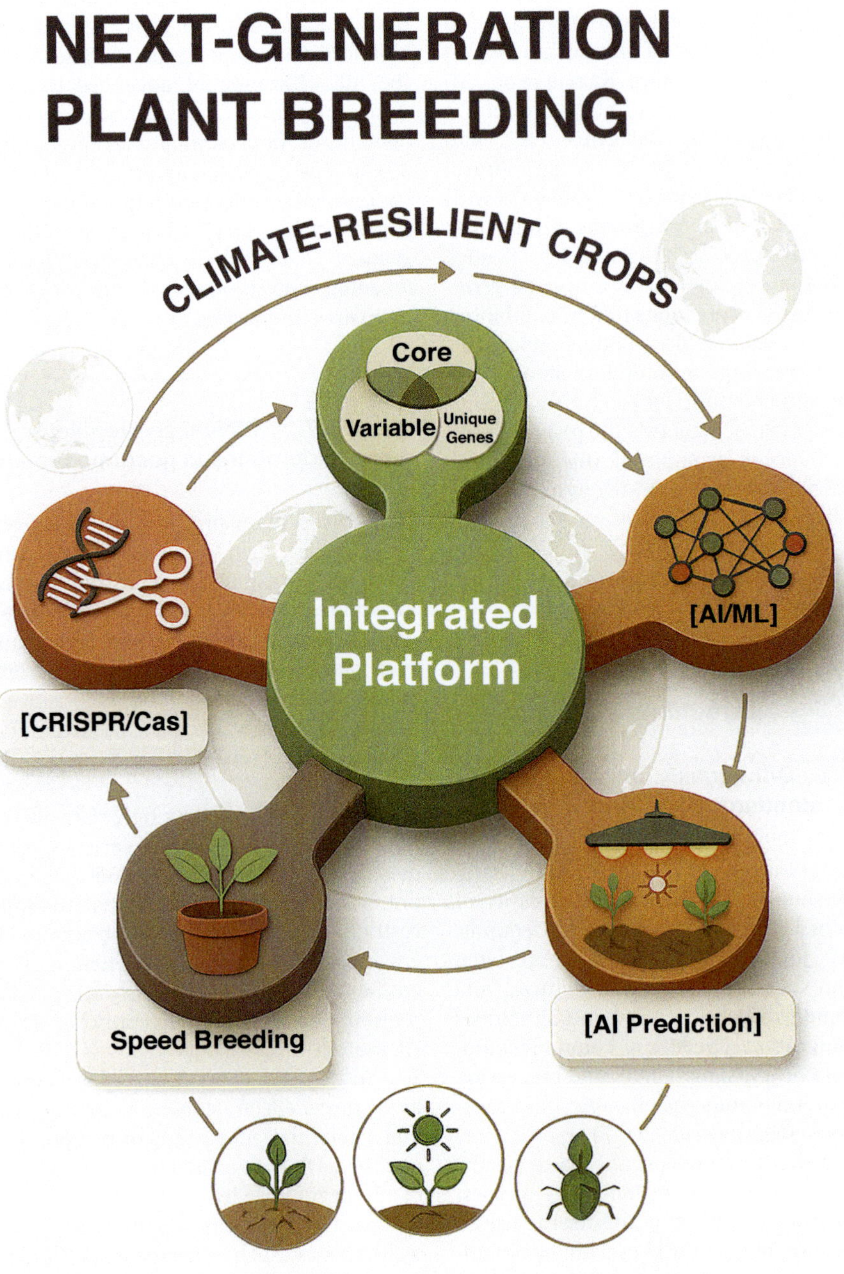

Fig. 5.3. Conceptual framework of next-generation breeding strategies, integrating gene editing, pangenomics, AI, and speed breeding for climate-resilient crop development (Author's own image).

Table 5.10. Key challenges in data-driven breeding, including data quality issues, biases, standardization problems, and proposed solutions (Author's own table).

Challenge	Description	Proposed solutions	References
Data quality	Heterogeneity in formats, incomplete datasets	Implement quality control pipelines, use standardized protocols	Hickey *et al.*, 2017
Biases	Non-random genotype selection, environmental skew	Diverse training populations, bias-correction algorithms	Xu *et al.*, 2017
Standardization	Lack of common metadata, data fragmentation	Develop frameworks like FAIR principles, integrated databases	Wilkinson *et al.*, 2016; Deng *et al.*, 2023
Infrastructure	Limited computing in small institutions	Cloud-based platforms, collaborations	Driscoll *et al.*, 2013
Skill Gaps	Lack of bioinformatics expertise	Training programs, interdisciplinary teams	Ewa *et al.*, 2022

level is essential – frameworks that preserve scientific innovation while upholding transparency, fairness, and public interest.

5.7.3 Next-generation technologies: Synthetic biology and digital twins

The growing global population and climate change present serious and complex challenges to food security. While the human population has increased from approximately one billion to over eight billion in the past two centuries and is projected to reach 11 billion by the end of this century, agricultural production faces multiple stresses, including abiotic factors such as drought, extreme climate variability, soil acidification, and nutrient deficiencies, as well as biotic factors such as pests, diseases, and fungal agents. To achieve the United Nations Sustainable Development Goal 2 (Zero Hunger), agricultural productivity must increase by at least 70%, necessitating the development of crop plants with high adaptability and resistance to variable environmental conditions (Roser *et al.*, 2013; Kumar *et al.*, 2022).

In this regard, two novel and complementary technologies – synthetic biology and digital twins – play a very important role in the future of agriculture.

Synthetic biology, leveraging advanced genetic engineering technologies, including precise gene-editing tools such as CRISPR/Cas9, enables targeted and rapid modification of plant genomes. This technology aids in producing plants with higher resistance to environmental stresses such as drought, salinity, and cold, improving the quality and quantity of yields, and reducing harmful traits like allergens. Beyond direct genome editing, synthetic biology allows the design and synthesis of new metabolic pathways in plants, leading to the production of pharmaceuticals, biopolymers, and bio-based valuable compounds. Additionally, diagnostic systems based on this technology can rapidly and accurately detect diseases and biotic stresses, facilitating optimal resource management for farmers. These technologies, through precise genetic trait regulation, provide a new pathway for sustainable agriculture development and food security enhancement (Gabr *et al.*, 2020; Ehyaei Rad, 2025).

On the other hand, digital twins, as data-driven simulation models based on accurate and up-to-date environmental and genetic information, enable the prediction and analysis of plant population behaviors under various climatic and biological conditions. Utilizing occurrence data of wild relatives of crops (CWR) and cumulative modeling techniques, digital twins can map suitable habitats for these species and identify correlations between plant adaptation traits and environmental factors. This technology, by automating data flows, implementing advanced models, uncertainty analysis, and timely alerts, assists breeders, policymakers, and conservation scientists in faster and more precise identification and utilization of valuable genetic resources. Furthermore, the development of ecogeographic land characterization (ELC) maps within this system helps identify gaps in germplasm collections and plan more targeted collecting efforts (Chala *et al.*, 2024; Melesse, 2025).

The combination of these two technologies (synthetic biology and digital twins) creates unprecedented capabilities for intelligent management of genetic resources and addressing complex environmental and food security challenges. As next-generation technologies, they will play a key role in achieving sustainable development and enhancing agricultural resilience.

5.7.4 Future perspectives on AI-based genomic design

Genomic design, as a central pillar of modern plant breeding, is on the verge of a fundamental transformation accelerated by the convergence of artificial intelligence (AI), data science, and molecular biology technologies. With increasing access to genomic, phenotypic, transcriptomic, and environmental data, the application of machine learning (ML) and deep learning (DL) algorithms in the analysis of these multifaceted datasets has paved a novel path for the intelligent design of genotypes with desirable traits (Kamilaris and Prenafeta-Boldú, 2018).

AI can identify hidden patterns and non-linear relationships among genes, traits, and environmental conditions, providing accurate predictions of phenotypic performance. This capability enables breeders to utilize predictive models for designing targeted and efficient genomes instead of relying solely on time-consuming field experiments. In particular, models based on deep neural networks (such as CNNs and RNNs) have demonstrated remarkable performance in predicting traits such as stress tolerance, grain yield, and nutritional quality (Sundmaeker *et al.*, 2022).

In the future, genomic design is expected to evolve into an interactive process based on reinforcement learning, where intelligent models simulate various genetic combination scenarios and propose optimal breeding strategies considering biological and climatic constraints. Furthermore, the use of generative models like generative adversarial networks (GANs) for virtual generation of phenotypic data or designing innovative genetic sequences is expanding (Ronald, 2023).

Integrating AI with genomic design not only accelerates the breeding process but also facilitates the development of 'customized' cultivars tailored to the climatic, agronomic, and nutritional needs of diverse regions. This is particularly strategic in designing crops resistant to climate change, optimizing resource use, and producing nutritionally enriched foods (Benos *et al.*, 2021).

However, fully realizing the potential of AI-based genomic design requires overcoming challenges such as data quality and diversity, data standardization, model interpretability, and ensuring that predictive outcomes align with biological reality. In this regard, interdisciplinary collaboration among plant scientists, data miners, biotechnologists, and software engineers is key to developing accurate, generalizable, and practical algorithms (Yoosefzadeh-Najafabadi *et al.*, 2023).

The future outlook of genomic design, through integrating AI algorithms with technologies like genome editing (CRISPR/Cas), remote sensing, the Internet of Things, and blockchain, promises the creation of an intelligent plant breeding ecosystem; a system in which the design, evaluation, modification, and deployment of new cultivars are conducted dynamically, precisely, and data-driven. Such an approach can play a crucial role in achieving food security, sustainable agriculture, and smart production goals (Farooq *et al.*, 2024).

5.8 Conclusion

5.8.1 Summary of key findings

Technological advancements in data-driven genomic design have opened new horizons in breeding the next generation of agricultural crops. The key findings of this chapter emphasize the central role of data-centric technologies in enhancing the accuracy, efficiency, and sustainability of breeding programs.

Precise genetic modification based on data

The integration of novel genome editing methods such as CRISPR/Cas9 with advanced data analytics has enabled precise and targeted genome modifications. These technologies, supported by bioinformatics models, facilitate the selection of optimal gene insertion sites, design of gRNAs, and prediction of off-target effects, thereby improving the safety and efficacy of genetic transformation (Metje-Sprink *et al.*, 2019; Gao, 2021).

Enhancement of mass selection efficiency

High-throughput phenotyping and machine learning models have made it possible to simultaneously evaluate thousands of genotypes, accelerating the selection of desirable cultivars while preserving genetic diversity. This approach allows for precise identification of traits influencing performance, quality, and climatic adaptability (Ronald, 2023).

Application of data-driven modeling in breeding

Multi-layer data analysis, including genomic, phenotypic, and environmental information, has facilitated the identification of genetic markers linked to desirable traits. This enables breeders to perform more accurate selections based on predicted genotype performance under diverse conditions (Xu *et al.*, 2017).

5.8.2 Implications for agricultural crop breeding programs

Data-driven genomic design holds great potential to transform agricultural breeding programs; however, its effective implementation requires attention to technical, social, regulatory, and environmental implications.

From a technical and infrastructural perspective, applying these approaches necessitates advanced infrastructures, precise phenotyping systems, Big Data processing equipment, and specialized personnel skilled in bioinformatics, data science, and quantitative genetics. In many regions, the lack of such resources may pose barriers to widespread adoption of these technologies (Ewens, 2012).

Regarding regulations and social acceptance, there are significant differences in rules governing genetically modified organisms and gene-edited products across countries. These disparities, combined with societal concerns about biosafety, bioethics, and data privacy, may challenge the commercialization and public acceptance of crops developed using these technologies (Esquinas-Alcázar, 2005).

From an environmental standpoint, the use of advanced genomic design technologies may have unforeseen impacts on biodiversity, including gene flow to wild populations or effects on natural ecosystems. These consequences highlight the need for comprehensive management strategies, environmental monitoring systems, and coherent legal frameworks to ensure responsible and sustainable use of these technologies (Vrbničanin *et al.*, 2017).

Addressing these implications and developing aligned policies is essential for the effective and safe utilization of genomic design in breeding future crop generations.

5.8.3 Future directions for research and development

The transformative progress of data-driven technologies in plant breeding has opened new avenues for future research and development in agriculture. The adoption of information technology (IT), artificial intelligence (AI), machine learning (ML), the Internet of Things (IoT), and blockchain, beyond simple tools, has become a foundational pillar in designing forward-looking modern breeding systems. These technologies provide the basis for precise, adaptable, and sustainable systems capable of rapidly responding to biological and climatic challenges.

Future research and development should focus on interdisciplinary integration of data science, systems biology, biotechnology, and agricultural sciences. The use of advanced machine learning algorithms for predicting complex traits, dynamic modeling of gene–environment interactions, and multi-omics data analysis will be key areas. High-throughput phenotyping systems, real-time on-field sensing via IoT, and AI-driven predictive analytics can guide breeding processes on much shorter time scales and with higher precision.

Moreover, developing data-driven infrastructures such as standardized databases, blockchain networks for secure data sharing, and AI-based decision support platforms will be the cornerstone of global collaboration in future plant breeding. These frameworks enable shared utilization of genomic, phenotypic, and environmental data, promoting transparency and synergy among research centers, universities, and policymaking institutions, thereby enhancing productivity and sustainability in agricultural production systems.

Going forward, research and development must pursue multidimensional objectives: increasing yield, improving nutritional quality, enhancing resource use efficiency (water, soil, fertilizer), and resistance to biotic and abiotic stresses. All these goals should be framed within sustainable development and aligned with the United Nations Sustainable Development Goals (SDGs). Thus, novel breeding strategies should be designed with a systemic approach, based on large-scale data analytics, and consider economic, environmental, and social indicators.

Ultimately, the future pathway in plant breeding will not merely lead to the production of new cultivars but will result in the creation of a sustainable, resilient, and intelligent agricultural system; a system capable of addressing complex food security and climate change challenges through science, technology, and global collaboration.

References

Abdollahi-Arpanahi, R., Gianola, D. and Peñagaricano, F. (2020) Deep learning versus parametric and ensemble methods for genomic prediction of complex phenotypes. *Genetics Selection Evolution* 52, 12.

Adunola, P., Ferrão, L.F.V., Benevenuto, J., Azevedo, C.F. and Munoz, P.R. (2024) Genomic selection optimization in blueberry: Data-driven methods for marker and training population design. *The Plant Genome* 17(3), e20488.

Ågren, J., Oakley, C.G., Lundemo, S. and Schemske, D.W. (2017) Adaptive divergence in flowering time among natural populations of *Arabidopsis thaliana*: Estimates of selection and QTL mapping. *Evolution* 71(3), 550–564.

Akdemir, D., Knox, R. and Isidro y Sánchez, J. (2020) Combining partially overlapping multi-omics data in databases using relationship matrices. *Frontiers in Plant Science* 11, 947.

Akdemir, D., Rio, S. and Isidro y Sánchez, J. (2021) TrainSel: An R package for selection of training populations. *Frontiers in Genetics* 12, 655287.

Ali, I., Greifeneder, F., Stamenkovic, J., Neumann, M. and Notarnicola, C. (2015) Review of machine learning approaches for biomass and soil moisture retrievals from remote sensing data. *Remote Sensing* 7(12), 16398–16421.

Altman, A., Fan, L. and Foyer, C. (2021) Past and future milestones of plant breeding. *Trends in Plant Science* 26(6), 530–538.

Amin, A., Zaman, W., Rahman, S.U., Rahman, L.U. and Park, S. (2025) Harnessing multi-omics and predictive modeling for climate-resilient crop breeding: From genomes to fields. *Genes* 16(7), 809.

Amir, R., Maqsood, W., Munir, F., Fatima, N., Siddiqa, A. *et al.* (2020) Pan-genomics of plant pathogens and its applications. In: Barh, D., Soares, S., Tiwari, S. and Azevedo, V. (eds) *Pan-Genomics: Applications, Challenges, and Future Prospects*. Academic Press, pp. 121–145.

Angidi, S., Madankar, K., Tehseen, M.M. and Bhatla, A. (2025) Advanced high-throughput phenotyping techniques for managing abiotic stress in agricultural crops—a comprehensive review. *Crops* 5(2), 8.

Ashfaq, W., Brodie, G., Fuentes, S. and Gupta, D. (2022) Infrared thermal imaging and morpho-physiological indices used for wheat genotypes screening under drought and heat stress. *Plants* 11(23), 3269.

Atanda, S.A., Olsen, M., Burgueño, J., Crossa, J., Dzidzienyo, D. *et al.* (2021) Maximizing efficiency of genomic selection in CIMMYT's tropical maize breeding program. *Theoretical and Applied Genetics* 134(1), 279–294.

Awais, M., Li, W., Cheema, M.J.M., Zaman, Q.U., Shaheen, A. *et al.* (2023) UAV-based remote sensing in plant stress imaging using high-resolution thermal sensor for digital agriculture practices: A meta-review. *International Journal of Environmental Science and Technology* 20(1), 1135–1152.

Azodi, C.B., Bolger, E., McCarren, A., Roantree, M., De Los Campos, G. *et al.* (2019) Benchmarking parametric and machine learning models for genomic prediction of complex traits. *G3: Genes, Genomes, Genetics* 9(11), 3691–3702.

Badoni, S., Das, S., Sayal, Y.K., Gopalakrishnan, S., Singh, A.K. *et al.* (2016) Genome-wide generation and use of informative intron-spanning and intron-length polymorphism markers for high-throughput genetic analysis in rice. *Scientific Reports* 6, 23765.

Bae, T., Hur, J.W., Kim, D. and Hur, J.K. (2019) Recent trends in CRISPR-Cas system: Genome, epigenome, and transcriptome editing and CRISPR delivery systems. *Genes & Genomics* 41(8), 871–877.

Ballesta, P., Bush, D., Silva, F.F. and Mora, F. (2020) Genomic predictions using low-density SNP markers, pedigree and GWAS information: A case study with the non-model species *Eucalyptus cladocalyx*. *Plants* 9(1), 99.

Benos, L., Tagarakis, A.C., Dolias, G., Berruto, R., Kateris, D. *et al.* (2021) Machine learning in agriculture: A comprehensive updated review. *Sensors* 21(11), 3758.

Bernardo, R. (2008) Molecular markers and selection for complex traits in plants: Learning from the last 20 years. *Crop Science* 48(5), 1649–1664.

Bhat, J.A., Salgotra, R.K. and Dar, M.Y. (2015) Phenomics: A challenge for crop improvement in genomic era. *Molecular Plant Breeding* 6(22), 1–14.

Bhat, J.A., Ali, S., Salgotra, R.K., Mir, Z.A., Dutta, S. *et al.* (2016) Genomic selection in the era of next generation sequencing for complex traits in plant breeding. *Frontiers in Genetics* 7, 221.

Bhuiyan, M.M.R., Noman, I.R., Aziz, M.M., Rahaman, M.M., Islam, M.R. *et al.* (2025) Transformation of plant breeding using data analytics and information technology: Innovations, applications, and prospective directions. *Frontiers in Bioscience-Elite* 17(1), 27936.

Bolón-Canedo, V., Sánchez-Maroño, N. and Alonso-Betanzos, A. (2016) Feature selection for high-dimensional data. *Progress in Artificial Intelligence* 5(2), 65–75.

Budhlakoti, N., Rai, A. and Mishra, D.C. (2020) Statistical approach for improving genomic prediction accuracy through efficient diagnostic measure of influential observation. *Scientific Reports* 10, 8408.

Budhlakoti, N., Kushwaha, A.K., Rai, A., Chaturvedi, K.K., Kumar, A. *et al.* (2022) Genomic selection: A tool for accelerating the efficiency of molecular breeding for development of climate-resilient crops. *Frontiers in Genetics* 13, 832153.

Calado, R.A. (2020) *Statistical Methods for the Integrative Analysis of Single-Cell Multi-Omics Data*. PhD thesis, University of Cambridge, UK.

Chala, D., Kusch, E., Weiland, C., Andrew, C., Grieb, J. *et al.* (2024) Prototype biodiversity digital twin: Crop wild relatives genetic resources for food security. *Research Ideas and Outcomes* 10, e125192.

Costa-Neto, G., Crespo-Herrera, L., Fradgley, N., Gardner, K., Bentley, A.R. *et al.* (2022) Envirome-wide associations enhance multi-year genome-based prediction of historical wheat breeding data. *G3: Genes, Genomes, Genetics* 12(12), jkac313.

Crossa, J., Pérez-Rodríguez, P., Cuevas, J., Montesinos-López, O., Jarquín, D. *et al.* (2017) Genomic selection in plant breeding: Methods, models, and perspectives. *Trends in Plant Science* 22(11), 961–975.

de Bem Oliveira, I., Amadeu, R.R., Ferrão, L.F.V. and Muñoz, P.R. (2020) Optimizing whole-genomic prediction for autotetraploid blueberry breeding. *Heredity* 125(6), 437–448.

De los Campos, G., Gianola, D. and Rosa, G.J. (2009) Reproducing kernel Hilbert spaces regression: A general framework for genetic evaluation. *Journal of Animal Science* 87(6), 1883–1887.

De Sousa, K., Etten, J., Poland, J., Fadda, C., Jannink, J.L. *et al.* (2021) Data-driven decentralized breeding increases prediction accuracy in a challenging crop production environment. *Communications Biology* 4, 944.

De Los Campos, G., Vazquez, A.I., Fernando, R., Klimentidis, Y.C. and Sorensen, D. (2013) Prediction of complex human traits using the genomic best linear unbiased predictor. *PLOS Genetics* 9(7), e1003608.

Demidchik, V.V., Shashko, A.Y., Bandarenka, U.Y., Smolikova, G.N., Przhevalskaya, D.A. *et al.* (2020) Plant phenomics: Fundamental bases, software and hardware platforms, and machine learning. *Russian Journal of Plant Physiology* 67(3), 397–412.

Deng, C.H., Naithani, S., Kumari, S., Cobo-Simón, I., Quezada-Rodríguez, E.H. *et al.* (2023) Genotype and phenotype data standardization, utilization and integration in the big data era for agricultural sciences. *Database* 2023, baad088.

Dhar, M.K. and Koul, A.K. (2024) Plant cytogenetics in the era of genome editing. *Nucleus* 67, 595–609.

Divya, D. and Kumar, Y.D. (2024) Utilizing AI to enhance the effectiveness of plant breeding for the development of climate-resilient smart food crops. *Agricultural Biotechnology Journal* 16(3), 171–188.

Driever, S.M., Mossink, L., Ocaña, D.N. and Kaiser, E. (2023) A simple system for phenotyping of plant transpiration and stomatal conductance response to drought. *Plant Science* 329, 111626.

Driscoll, A.O., Daugelaite, J. and Sleator, R.D. (2013) 'Big data', Hadoop and cloud computing in genomics. *Journal of Biomedical Informatics* 46(5), 774–781.

Ehyaei Rad, N. (2025) Synthetic biology in genomics: Redefining genetic engineering and its applications in personalized medicine. *Advanced Therapies Journal* 7(22), 61–69.

Esquinas-Alcázar, J. (2005) Protecting crop genetic diversity for food security: Political, ethical and technical challenges. *Nature Reviews Genetics* 6(12), 946–953.

Ewa, W.G., Agata, T., Milica, P., Anna, B., Dennis, E. *et al.* (2022) Public perception of plant gene technologies worldwide in the light of food security. *GM Crops & Food* 13(1), 218–236.

Ewens, W.J. (2012) James F. Crow and the stochastic theory of population genetics. *Genetics* 190(2), 287–290.

Farooq, M.A., Gao, S., Hassan, M.A., Huang, Z., Rasheed, A. *et al.* (2024) Artificial intelligence in plant breeding. *Trends in Genetics* 40(10), 891–908.

Fedoroff, N.V. (2010) The past, present and future of crop genetic modification. *New Biotechnology* 27(5), 461–465.

Foster, I. (2011) Globus online: Accelerating and democratizing science through cloud-based services. *IEEE Internet Computing* 15(3), 70–73.

Friedman, J.H., Hastie, T. and Tibshirani, R. (2010) Regularization paths for generalized linear models via coordinate descent. *Journal of Statistical Software* 33(1), 1–22.

Gabr, H., El Ghamrawy, M.K., Almaeen, A.H., Abdelhafiz, A.S., Hassan, A.O.S. *et al.* (2020) CRISPR-mediated gene modification of hematopoietic stem cells with beta-thalassemia IVS-1-110 mutation. *Stem Cell Research & Therapy* 11, 390.

Gao, C. (2021) Genome engineering for crop improvement and future agriculture. *Cell* 184(6), 1621–1635.

Gelman, A., Carlin, J.B., Stern, H.S. and Rubin, D.B. (1995) *Bayesian Data Analysis*. Chapman and Hall/CRC, London.

Gianola, D. (2013) Priors in whole-genome regression: The Bayesian alphabet returns. *Genetics* 194(3), 573–596.

Gianola, D., Fernando, R.L. and Stella, A. (2006) Genomic-assisted prediction of genetic value with semiparametric procedures. *Genetics* 173(3), 1761–1776.

Gianola, D., Okut, H., Weigel, K.A. and Rosa, G.J. (2011) Predicting complex quantitative traits with Bayesian neural networks: A case study with jersey cows and wheat. *BMC Genetics* 12, 87.

Gnana, D.A.A., Balamurugan, S.A.A. and Leavline, E.J. (2016) Literature review on feature selection methods for high-dimensional data. *International Journal of Computer Applications* 136(1), 9–17.

González-Camacho, J.M., Ornella, L., Pérez-Rodríguez, P., Gianola, D., Dreisigacker, S. *et al.* (2018) Applications of machine learning methods to genomic selection in breeding wheat for rust resistance. *The Plant Genome* 11(2), 170104.

González-Recio, O., Weigel, K.A., Gianola, D., Naya, H. and Rosa, G.J. (2010) L2-Boosting algorithm applied to high-dimensional problems in genomic selection. *Genetics Research* 92(3), 227–237.

Gorjanc, G., Cleveland, M.A., Houston, R.D. and Hickey, J.M. (2015) Potential of genotyping-by-sequencing for genomic selection in livestock populations. *Genetics Selection Evolution* 47, 12.

Greener, J.G., Kandathil, S.M., Moffat, L. and Jones, D.T. (2022) A guide to machine learning for biologists. *Nature Reviews Molecular Cell Biology* 23(1), 40–55.

Griffel, L., Delparte, D. and Edwards, J. (2018) Using support vector machines classification to differentiate spectral signatures of potato plants infected with potato virus Y. *Computers and Electronics in Agriculture* 153, 318–324.

Habier, D., Fernando, R.L., Kizilkaya, K. and Garrick, D.J. (2011) Extension of the Bayesian alphabet for genomic selection. *BMC Bioinformatics* 12, 186.

Haji, A. (2024) Comparative analysis of autoencoder and PCA for dimensionality reduction in gene expression data. *Bioinformatics and Systems Biology* 8(2), 45–52.

Hasan, N., Choudhary, S., Naaz, N., Sharma, N. and Laskar, R.A. (2021) Recent advancements in molecular marker-assisted selection and applications in plant breeding programmes. *Journal of Genetic Engineering and Biotechnology* 19, 128.

Hayes, B.J., Pryce, J., Chamberlain, A.J., Bowman, P.J. and Goddard, M.E. (2010) Genetic architecture of complex traits and accuracy of genomic prediction: Coat colour, milk-fat percentage, and type in Holstein cattle as contrasting model traits. *PLOS Genetics* 6(9), e1001139.

Hayes, C.N., Nakahara, H., Ono, A., Tsuge, M. and Oka, S. (2024) From omics to multi-omics: A review of advantages and tradeoffs. *Genes* 15(12), 1551.

Heffner, E.L., Sorrells, M.E. and Jannink, J.L. (2009) Genomic selection for crop improvement. *Crop Science* 49(1), 1–12.

Hemingway, J., Schnebly, S.R. and Rajcan, I. (2021) Accuracy of genomic prediction for seed oil concentration in high-oleic soybean populations using a low-density marker panel. *Crop Science* 61(6), 4012–4021.

Henderson, C.R. (1975) Best linear unbiased estimation and prediction under a selection model. *Biometrics* 31(2), 423–447.

Heslot, N., Yang, H.P., Sorrells, M.E. and Jannink, J.L. (2012) Genomic selection in plant breeding: A comparison of models. *Crop Science* 52(1), 146–160.

Hickey, J.M., Chiurugwi, T., Mackay, I. and Powell, W. (2017) Genomic prediction unifies animal and plant breeding programs to form platforms for biological discovery. *Nature Genetics* 49(9), 1297–1303.

Hoerl, A.E. and Kennard, R.W. (1970) Ridge regression: Applications to nonorthogonal problems. *Technometrics* 12(1), 69–82.

Holliday, J.A., Wang, T. and Aitken, S. (2012) Predicting adaptive phenotypes from multilocus genotypes in Sitka spruce (*Picea sitchensis*) using random forest. *G3: Genes, Genomes, Genetics* 2(9), 1085–1093.

Illouz-Eliaz, N., Nissan, I., Nir, I., Ramon, U., Shohat, H. *et al.* (2020) Mutations in the tomato gibberellin receptors suppress xylem proliferation and reduce water loss under water-deficit conditions. *Journal of Experimental Botany* 71(12), 3603–3612.

Jannink, J.L., Bink, M.C. and Jansen, R.C. (2001) Using complex plant pedigrees to map valuable genes. *Trends in Plant Science* 6(8), 337–342.

Jeon, D., Kang, Y., Lee, S., Choi, S., Sung, Y. *et al.* (2023) Digitalizing breeding in plants: A new trend of next-generation breeding based on genomic prediction. *Frontiers in Plant Science* 14, 1092584.

Jiang, F. and Doudna, J.A. (2017) CRISPR–Cas9 structures and mechanisms. *Annual Review of Biophysics* 46, 505–529.

Jinek, M., Chylinski, K., Fonfara, I., Hauer, M., Doudna, J.A. *et al.* (2012) A programmable dual-RNA–guided DNA endonuclease in adaptive bacterial immunity. *Science* 337(6096), 816–821.

John Martin, J.J., Song, Y., Hou, M., Zhou, L., Liu, X. *et al.* (2024) Multi-omics approaches in oil palm research: A comprehensive review of metabolomics, proteomics, and transcriptomics based on low-temperature stress. *International Journal of Molecular Sciences* 25(14), 7695.

Kainer, D., Stone, E.A., Padovan, A., Foley, W.J. and Külheim, C. (2018) Accuracy of genomic prediction for foliar terpene traits in *Eucalyptus polybractea*. *G3: Genes, Genomes, Genetics* 8(8), 2573–2583.

Kaler, A.S., Gillman, J.D., Beissinger, T. and Purcell, L.C. (2020) Comparing different statistical models and multiple testing corrections for association mapping in soybean and maize. *Frontiers in Plant Science* 10, 486047.

Kamilaris, A. and Prenafeta-Boldú, F.X. (2018) Deep learning in agriculture: A survey. *Computers and Electronics in Agriculture* 147, 70–90.

Karlson, C.K.S., Mohd-Noor, S.N., Nolte, N. and Tan, B.C. (2021) CRISPR/dCas9-based systems: Mechanisms and applications in plant sciences. *Plants* 10, 2055.

Krampis, K., Booth, T., Chapman, B., Tiwari, B., Bicak, M. *et al.* (2012) Cloud BioLinux: Pre-configured and on-demand bioinformatics computing for the genomics community. *BMC Bioinformatics* 13, 42.

Kriaridou, C., Tsairidou, S., Houston, R.D. and Robledo, D. (2020) Genomic prediction using low density marker panels in aquaculture: Performance across species, traits, and genotyping platforms. *Frontiers in Genetics* 11, 124.

Kumar, L., Chhogyel, N., Gopalakrishnan, T., Hasan, M.K., Jayasinghe, S.L. *et al.* (2022) Climate change and future of agri-food production. In: Bhat, R. (ed.) *Future Foods: Global Trends, Opportunities, and Sustainability Challenges*. Academic Press, Cambridge, pp. 49–79.

Kumar, R., Das, S.P., Choudhury, B.U., Kumar, A., Prakash, N.R. *et al.* (2024) Advances in genomic tools for plant breeding: Harnessing DNA molecular markers, genomic selection, and genome editing. *Biological Research* 57, 80.

Kwon, M.S., Lee, B.T., Lee, S.Y. and Kim, H.U. (2020) Modeling regulatory networks using machine learning for systems metabolic engineering. *Current Opinion in Biotechnology* 65, 163–170.

Lanzl, T., Melchinger, A.E. and Schön, C. (2023) Influence of the mating design on the additive genetic variance in plant breeding populations. *Theoretical and Applied Genetics* 136, 247.

LeCun, Y., Bengio, Y. and Hinton, G. (2015) Deep learning. *Nature* 521(7553), 436–444.

Legarra, A., Robert-Granié, C., Croiseau, P., Guillaume, F. and Fritz, S. (2011) Improved Lasso for genomic selection. *Genetics Research* 93(1), 77–87.

Levizzani, V. and Cattani, E. (2019) Satellite remote sensing of precipitation and the terrestrial water cycle in a changing climate. *Remote Sensing* 11(19), 2301.

Li, D., Quan, C., Song, Z., Li, X., Yu, G. *et al.* (2021) High-throughput plant phenotyping platform (HT3P) as a novel tool for estimating agronomic traits from the lab to the field. *Frontiers in Bioengineering and Biotechnology* 8, 623705.

Li, Y., Liang, J., Deng, B., Jiang, Y., Zhu, J. *et al.* (2023) Applications and prospects of CRISPR/Cas9-mediated base editing in plant breeding: From genomes to fields. *Current Issues in Molecular Biology* 45(2), 918–935.

Liang, H., Zhou, Y., Lu, Y., Pei, S., Xu, D. *et al.* (2024) Evaluation of soybean drought tolerance using multimodal data from an unmanned aerial vehicle and machine learning. *Remote Sensing* 16, 11.

Liang, M., Cao, S., Deng, T., Du, L., Li, K. *et al.* (2023) MAK: A machine learning framework improved genomic prediction via multi-target ensemble regressor chains and automatic selection of assistant traits. *Briefings in Bioinformatics* 24(2), bbad043.

Lin, C. (2024) *Analysis of Complex Dynamical Systems by Combining Recurrent Neural Networks and Mechanistic Models*. PhD thesis, University of Ottawa, Canada.

Liu, B., Sotomayor, B., Madduri, R., Chard, K. and Foster, I. (2012) Deploying bioinformatics workflows on clouds with galaxy and globus provision. In: *Proceedings of SC Companion: High Performance Computing, Networking Storage and Analysis*, pp. 1087–1095.

Liu, B., Madduri, R.K., Sotomayor, B., Chard, K., Lacinski, L. *et al.* (2014) Cloud-based bioinformatics workflow platform for large-scale next-generation sequencing analyses. *Journal of Biomedical Informatics* 49, 119–133.

Long, N., Gianola, D., Rosa, G.J. and Weigel, K.A. (2011) Application of support vector regression to genome-assisted prediction of quantitative traits. *Theoretical and Applied Genetics* 123(7), 1065–1074.

Mahmood, U., Li, X., Fan, Y., Chang, W., Niu, Y. *et al.* (2022) Multi-omics revolution to promote plant breeding efficiency. *Frontiers in Plant Science* 13, 1062952.

Manthena, V., Jarquín, D., Varshney, R.K., Roorkiwal, M., Dixit, G.P. *et al.* (2022) Evaluating dimensionality reduction for genomic prediction. *Frontiers in Genetics* 13, 958780.

McGrail, R.K., Van Sanford, D.A. and McNear, Jr, D.H. (2020) Trait-based root phenotyping as a necessary tool for crop selection and improvement. *Agronomy* 10(9), 1328.

Medina, I., Carbonell, J., Pulido, L., Madeira, S.C., Goetz, S. *et al.* (2010) Babelomics: An integrative platform for the analysis of transcriptomics, proteomics, and genomic data with advanced functional profiling. *Nucleic Acids Research* 38(Web Server issue), W210–W213.

Melesse, T.Y. (2025) Digital twin-based applications in crop monitoring. *Heliyon* 11(2), e40245.

Messina, G. and Modica, G. (2020) Applications of UAV thermal imagery in precision agriculture: State of the art and future research outlook. *Remote Sensing* 12(9), 1491.

Metje-Sprink, J., Menz, J., Modrzejewski, D. and Sprink, T. (2019) DNA-free genome editing: Past, present and future. *Frontiers in Plant Science* 9, 1957.

Meuwissen, T.H., Hayes, B.J. and Goddard, M. (2001) Prediction of total genetic value using genome-wide dense marker maps. *Genetics* 157(4), 1819–1829.

Misra, B.B., Langefeld, C., Olivier, M. and Cox, L.A. (2019) Integrated omics: Tools, advances and future approaches. *Journal of Molecular Endocrinology* 62, R21–R45.

Montesinos-López, A., Montesinos-López, O.A., Gianola, D., Crossa, J. and Hernández-Suárez, C.M. (2018) Multi-environment genomic prediction of plant traits using deep learners with dense architecture. *G3: Genes, Genomes, Genetics* 8, 3813–3828.

Montesinos-López, O.A., Martín-Vallejo, J., Crossa, J., Gianola, D., Hernández-Suárez, C.M. *et al.* (2019) A benchmarking between deep learning, support vector machine and Bayesian threshold best linear unbiased prediction for predicting ordinal traits in plant breeding. *G3: Genes, Genomes, Genetics* 9, 601–618.

Montesinos-López, O.A., Montesinos-López, A., Pérez-Rodríguez, P., Barrón-López, J.A., Martini, J.W. *et al.* (2021) A review of deep learning applications for genomic selection. *BMC Genomics* 22, 19.
Montesinos-López, O.A., Chavira-Flores, M., Kismiantini, dfd, Crespo-Herrera., L., Saint Piere, C. *et al.* (2024) A review of multimodal deep learning methods for genomic-enabled prediction in plant breeding. *Genetics* 228, iyae161.
Mueller, M.L. and Van Eenennaam, A.L. (2022) Synergistic power of genomic selection, assisted reproductive technologies, and gene editing to drive genetic improvement of cattle. *CABI Agriculture and Bioscience* 3, 13.
Myles, S., Boyko, A.R., Owens, C.L., Brown, P.J., Grassi, F. *et al.* (2009) Genetic structure and domestication history of the grape. *Proceedings of the National Academy of Sciences* 106, 2325–2330.
Neal, R.M. (2012) *Bayesian Learning for Neural Networks*. Springer Science & Business Media, New York.
Noble, W.S. (2006) What is a support vector machine? *Nature Biotechnology* 24, 1565–1567.
Norman, A., Taylor, J., Edwards, J. and Kuchel, H. (2018) Optimising genomic selection in wheat: Effect of marker density, population size and population structure on prediction accuracy. *G3: Genes, Genomes, Genetics* 8, 2889–2899.
Ogutu, J.O., Schulz-Streeck, T. and Piepho, H.P. (2012) Genomic selection using regularized linear regression models: Ridge regression, lasso, elastic net and their extensions. *BMC Proceedings* 6(Suppl 2), S10.
Parmley, K., Nagasubramanian, K., Sarkar, S., Ganapathysubramanian, B. and Singh, A.K. (2019) Development of optimized phenomic predictors for efficient plant breeding decisions using phenomic-assisted selection in soybean. *Plant Phenomics* 2019, 5809404.
Pei, G., Chen, L. and Zhang, W. (2017) WGCNA application to proteomic and metabolomic data analysis. *Methods in Enzymology* 585, 135–158.
Perez-de-Castro, A.M., Vilanova, S., Cañizares, J., Pascual, L., Blanca, M.J. *et al.* (2012) Application of genomic tools in plant breeding. *Current Genomics* 13(3), 179–195.
Pook, T., Freudenthal, J., Korte, A. and Simianer, H. (2020) Using local convolutional neural networks for genomic prediction. *Frontiers in Genetics* 11, 561497.
Popescu, G.V., Noutsos, C. and Popescu, S.C. (2016) Big data in plant science: Resources and data mining tools for plant genomics and proteomics. In: *Data Mining Techniques for the Life Sciences*. Springer, New York, pp. 533–547.
Razzaq, A., Kaur, P., Akhter, N., Wani, S.H. and Saleem, F. (2021) Next-generation breeding strategies for climate-ready crops. *Frontiers in Plant Science* 12, 620420.
Riaz, A., Kanwal, F., Börner, A., Pillen, K., Dai, F. *et al.* (2021) Advances in genomics-based breeding of barley: Molecular tools and genomic databases. *Agronomy* 11, 894.
Rohart, F., Gautier, B., Singh, A. and Lê Cao, K.A. (2017) MixOmics: An R package for omics feature selection and multiple data integration. *PLOS Computational Biology* 13, e1005752.
Ronald, P.C. (2023) Enhancing sustainable development through plant genetics. *Nature Reviews Genetics* 24, 659–660.
Rosado, R.D.S., Cruz, C.D., Barili, L.D., de Souza Carneiro, J.E., Carneiro, P.C.S. *et al.* (2020) Artificial neural networks in the prediction of genetic merit to flowering traits in bean cultivars. *Agriculture* 10, 638.
Roser, M., Ritchie, H., Ortiz-Ospina, E. and Rodés-Guirao, L. (2013) World population growth. *Our World in Data* 412, 543.
Rouhi, A. and Nezamabadi-Pour, H. (2020) Feature selection in high-dimensional data. In: *Optimization, Learning, and Control for Interdependent Complex Networks*. Springer International, Cham, pp. 85–128.
Roychowdhury, R., Das, S.P., Gupta, A., Parihar, P., Chandrasekhar, K. *et al.* (2023) Multi-omics pipeline and omics-integration approach to decipher plant's abiotic stress tolerance responses. *Genes* 14, 1281.
Sánchez, J.C.M., Mesa, H.G.A., Espinosa, A.T., Castilla, S.R. and Lamont, F.G. (2025) Improving wheat yield prediction through variable selection using support vector regression, random forest, and extreme gradient boosting. *Smart Agricultural Technology* 11, 100791.
Santiago-Rodriguez, T.M. and Hollister, E.B. (2021) Multi 'omic data integration: A review of concepts, considerations, and approaches. *Seminars in Perinatology* 45, 151456.
Sarawad, A., Hosagoudar, S. and Parvatikar, P. (2025) Pan-genomics: Insight into the functional genome, applications, advancements, and challenges. *Current Genomics* 26, 2–14.

Shan, Q., Wang, Y., Li, J., Zhang, Y., Chen, K. *et al.* (2013) Targeted genome modification of crop plants using a CRISPR-Cas system. *Nature Biotechnology* 31, 686–688.

Sherman, R.M. and Salzberg, S.L. (2020) Pan-genomics in the human genome era. *Nature Reviews Genetics* 21, 243–254.

Silaiyiman, S., Liu, J., Wu, J., Ouyang, L., Cao, Z. *et al.* (2025) A systematic review of the advances and new insights into copy number variations in plant genomes. *Plants* 14, 1399.

Silva, E.D.B., Xavier, A. and Faria, M.V. (2021) Impact of genomic prediction model, selection intensity, and breeding strategy on the long-term genetic gain and genetic erosion in soybean breeding. *Frontiers in Plant Science* 12, 637133.

Singh, A., Gautier, B., Shannon, C.P., Vacher, M., Rohart, F. *et al.* (2016) DIABLO–An integrative, multi-omics, multivariate method for multi-group classification. *bioRxiv* 067611.

Singh, S., Praveen, A., Dudha, N. and Bhadrecha, P. (2024) Integrating physiological and multi-omics methods to elucidate heat stress tolerance for sustainable rice production. *Physiology and Molecular Biology of Plants* 30, 1185–1208.

Sinha, S., Singh, S., Kumar, M., Singh, R.S., Satyendra. *et al.* (2023) Recent advancements in molecular marker technologies and their applications in crop improvement. In: *Molecular Marker Techniques: A Potential Approach of Crop Improvement*. Springer Nature Singapore, Singapore, pp. 319–337.

Smithies, O., Gregg, R.G., Boggs, S.S., Koralewski, M.A. and Kucherlapati, R.S. (1985) Insertion of DNA sequences into the human chromosomal β-globin locus by homologous recombination. *Nature* 317, 230–234.

Sundmaeker, H., Verdouw, C., Wolfert, S. and Freire, L.P. (2022) Internet of food and farm 2020. In: Vermesan, O. and Friess, P. (eds) *Digitising the Industry Internet of Things Connecting the Physical, Digital and Virtual Worlds*. River Publishers, Delft, Netherlands, pp. 129–151.

Szymczak, S., Biernacka, J.M., Cordell, H.J., González-Recio, O., König, I.R. *et al.* (2009) Machine learning in genome-wide association studies. *Genetic Epidemiology* 33(Suppl 1), S51–S57.

Tembhare, K., Sharma, T., Kasibhatla, S.M., Achalere, A. and Joshi, R. (2024) Multi-ensemble machine learning framework for omics data integration: A case study using breast cancer samples. *Informatics in Medicine Unlocked* 47, 101507.

Thain, D., Tannenbaum, T. and Livny, M. (2005) Distributed computing in practice: The condor experience. *Concurrency and Computation: Practice and Experience* 17, 323–356.

Tutz, G. and Ulbricht, J. (2009) Penalized regression with correlation-based penalty. *Statistics and Computing* 19, 239–253.

Tyagi, A., Mir, Z.A., Almalki, M.A., Deshmukh, R. and Ali, S. (2024) Genomics-assisted breeding: A powerful breeding approach for improving plant growth and stress resilience. *Agronomy* 14, 1128.

Ulfenborg, B. (2019) Vertical and horizontal integration of multi-omics data with miodin. *BMC Bioinformatics* 20, 649.

Ullah, M.A., Abdullah-Zawawi, M.R., Zainal-Abidin, R.A., Sukiran, N.L., Uddin, M.I. *et al.* (2022) A review of integrative omic approaches for understanding rice salt response mechanisms. *Plants* 11, 1430.

Vahabi, N. and Michailidis, G. (2022) Unsupervised multi-omics data integration methods: A comprehensive review. *Frontiers in Genetics* 13, 854752.

Van der Oost, J. and Patinios, C. (2023) The genome editing revolution. *Trends in Biotechnology* 41, 396–409.

van Dijk, A.D.J., Kootstra, G., Kruijer, W. and de Ridder, D. (2021) Machine learning in plant science and plant breeding. *iScience* 24, 101890.

Varshney, R.K., Bohra, A., Yu, J., Graner, A., Zhang, Q. *et al.* (2021a) Designing future crops: Genomics-assisted breeding comes of age. *Trends in Plant Science* 26, 631–649.

Varshney, R.K., Thudi, M. and Kumar, A. (2021b) Advancements in genomics tools for plant breeding and their applications. *Nature Reviews Genetics* 22, 597–610.

Vieira, R.A., Nogueira, A.P.O. and Fritsche-Neto, R. (2025) Optimizing the selection of quantitative traits in plant breeding using simulation. *Frontiers in Plant Science* 16, 1495662.

Villiers, K., Voss-Fels, K.P., Dinglasan, E., Jacobs, B., Hickey, L. *et al.* (2024) Evolutionary computing to assemble standing genetic diversity and achieve long-term genetic gain. *Plant Genome* 17, e20467.

Vrbničanin, S., Božić, D. and Pavlović, D. (2017) Gene flow from herbicide-resistant crops to wild relatives. In: *Herbicide Resistance in Weeds and Crops*. IntechOpen, pp. 307–333.

Wang, N. (2022) *The Potential of UAV-based Sun-induced Chlorophyll Fluorescence in Understanding Crop Photosynthesis*. PhD thesis, Wageningen University and Research, Wageningen, The Netherlands.

Wang, H., Cimen, E., Singh, N. and Buckler, E. (2020) Deep learning for plant genomics and crop improvement. *Current Opinion in Plant Biology* 54, 34–41.

Wang, K., Abid, M.A., Rasheed, A., Crossa, J., Hearne, S. *et al.* (2023) DNNGP, a deep neural network-based method for genomic prediction using multi-omics data in plants. *Molecular Plant* 16, 279–293.

Wang, Z., Lachmann, A. and Ma'ayan, A. (2019) Mining data and metadata from the Gene Expression Omnibus. *Biophysical Reviews* 11(1), 103–110.

Whibley, A., Kelley, J.L. and Narum, S.R. (2021) The changing face of genome assemblies: Guidance on achieving high-quality reference genomes. *Molecular Ecology Resources* 21, 641–652.

Wilkinson, M.D., Dumontier, M., Aalbersberg, I.J., Appleton, G., Axton, M. *et al.* (2016) The FAIR guiding principles for scientific data management and stewardship. *Scientific Data* 3, 160018.

Wu, C., Luo, J. and Xiao, Y. (2024) Multi-omics assists genomic prediction of maize yield with machine learning approaches. *Molecular Breeding* 44, 14.

Xiao, Q., Bai, X., Zhang, C. and He, Y. (2022) Advanced high-throughput plant phenotyping techniques for genome-wide association studies: A review. *Journal of Advanced Research* 35, 215–230.

Xu, Y. and Crouch, J.H. (2008) Marker-assisted selection in plant breeding: From publications to practice. *Crop Science* 48, 391–407.

Xu, Y., Li, P., Zou, C., Lu, Y., Xie, C. *et al.* (2017) Enhancing genetic gain in the era of molecular breeding. *Journal of Experimental Botany* 68, 2641–2666.

Xu, Y., Zhang, X., Li, H., Zheng, H., Zhang, J. *et al.* (2022) Smart breeding driven by big data, artificial intelligence, and integrated genomic-enviromic prediction. *Molecular Plant* 15, 1664–1695.

Xu, Y., Yang, W., Qiu, J., Zhou, K., Yu, G. *et al.* (2025) Metabolic marker-assisted genomic prediction improves hybrid breeding. *Plant Communications* 6, 100733.

Xue, Z. (2020) *Integration of High-throughput Phenotyping and Genomics Data to Explore Arabidopsis Natural Variation*. PhD thesis, Université Paris-Saclay, Paris, France.

Yang, Y., Xu, C., Shen, Z. and Yan, C. (2022) Crop quality improvement through genome editing strategy. *Frontiers in Genome Editing* 3, 819687.

Yoosefzadeh-Najafabadi, M., Earl, H.J., Tulpan, D., Sulik, J. and Eskandari, M. (2021) Application of machine learning algorithms in plant breeding: Predicting yield from hyperspectral reflectance in soybean. *Frontiers in Plant Science* 11, 624273.

Yoosefzadeh-Najafabadi, M., Rajcan, I. and Eskandari, M. (2022) Optimizing genomic selection in soybean: An important improvement in agricultural genomics. *Heliyon* 8, e11809.

Yoosefzadeh-Najafabadi, M., Hesami, M. and Eskandari, M. (2023) Machine learning-assisted approaches in modernized plant breeding programs. *Genes* 14, 777.

Yoosefzadeh-Najafabadi, M., Xavier, A., Eskandari, M. and Hesami, M. (2025) Machine learning after a decade: Is it still a missing keystone in genomic-based plant breeding? *Artificial Intelligence Review* 58, 260.

Zafar, I., Anwar, S., Yousaf, W., Nisa, F.U., Kausar, T. *et al.* (2023) Reviewing methods of deep learning for intelligent healthcare systems in genomics and biomedicine. *Biomedical Signal Processing and Control* 86, 105263.

Zhang, Y. and Zhang, N. (2018) Imaging technologies for plant high-throughput phenotyping: A review. *Frontiers of Agricultural Science and Engineering* 5, 406–419.

Zingaretti, L.M., Gezan, S.A., Ferrão, L.F.V., Osorio, L.F., Monfort, A. *et al.* (2020) Exploring deep learning for complex trait genomic prediction in polyploid outcrossing species. *Frontiers in Plant Science* 11, 25.

6 Plant Growth Models in Genomic Prediction

Mohamed Kouighat[1]*, Abdelaaziz Farhaoui[2,3] and Mariam El Harsal[1]

[1]National Institute of Agricultural Research (INRA), Rabat, Morocco; [2]Equipe de Recherche en Didactique des Sciences et Sciences de l'Education et TICE, Centre Régional des Métiers de l'Education et de la Formation de Fès Meknès, Annexe Meknès, Morocco; [3]Department of Plant Protection, Ecole National of Agriculture Meknes, Meknes, Morocco

Abstract

Modern agriculture faces increasing challenges, with the most important being the selection of appropriate varieties for adaptation to climate change. While varietal selection is a complex process considering the interaction of the genotype with its environment, this complex interaction should be made more practical and relevant for plant breeding. In this context, the integration of plant growth models (PGMs) into genomic prediction (GP) offers an innovative application for improving crop productivity and resilience. These models allow the simulation and understanding of plant growth dynamics based on genetic and environmental factors. Decision Support System for Agricultural Technology Transfer (DSSAT), the Simulation Tool for Cropping Systems (STICS), and World Food Studies (WOFOST) are the common PGMs used in plant breeding. However, the effectiveness of integrating PGMs into genomic selection depends on the quality, diversity, and structure of the data used. This chapter explores the foundations of these opportunities, applications, and key strategies for integrating PGMs into GP, which enables a better approximation of genotype–environment (G×E) interactions.

Keywords: Climate change, Crop resilience, Genomic prediction (GP), Plant growth model (PGM)

6.1 Introduction

The global agricultural sector faces significant challenges, including increasing food needs due to population growth, the adverse effects of climate change, and the need to sustainably manage diminishing natural resources. It is projected that to meet the food needs of an estimated 9.7 billion people by 2050, agricultural production will need to increase by about 70% (Li and Siddique, 2020). This context highlights the urgent need for innovative solutions to improve crop productivity and resilience to environmental challenges. A promising approach to address these challenges lies in the integration of Plant Growth Models (PGMs) and Genomic Prediction (GP) (Messina *et al.*, 2018). PGMs are advanced mathematical models that simulate plant physiological processes and growth under various environmental conditions, providing essential information on potential yields and resource use efficiency. These models have been crucial for understanding the complex interactions between plants and their environment,

*Corresponding author: kouighat84@gmail.com

 Plant Genomic Prediction: AI-Enhanced Models for Plant Genomics and Crop Improvement (ed. Jen-Tsung Chen)
DOI: 10.1079/9781836991328.0006

helping to inform agronomic decisions. In parallel, GP relies on large-scale genomic data to predict phenotypic characteristics from genetic information, facilitating the selection of high-performance genotypes in breeding programs (Xu *et al.*, 2022). Through the use of dense molecular markers, GP allows for more accurate estimation of breeding values, which accelerates the development of crop varieties with improved performance characteristics. The combination of PGM and GP provides a comprehensive framework to improve crop performance by considering both genetic and environmental factors. This integrated approach promotes a deep understanding of genotype–environment interactions (G × E), which are essential in developing varieties that can thrive in diverse and changing environments (Heslot *et al.*, 2014). The inclusion of environmental variables in genomic selection (GS) models has been shown to improve the accuracy of predictions for complex traits (Xu *et al.*, 2022). Recent advances have further refined this integration. A novel reaction norm model for plant growth, combining spline and random forest methods, has been proposed to better understand daily growth responses to environmental factors, providing enhanced predictive capabilities under stress conditions (Toda *et al.*, 2024). The use of machine learning techniques, such as convolutional neural networks, in combination with crop growth models has improved the prediction of integrative traits, such as aboveground biomass accumulation (Larue *et al.*, 2024). However, integrating PGMs with GP presents several challenges. Accurate data acquisition is paramount, requiring high-quality phenotypic and genotypic datasets to calibrate and validate the models effectively (Resende *et al.*, 2012). Furthermore, modeling G × E interactions requires the use of advanced computational methods and tools to manage the complexity of biological systems (Burgueño *et al.*, 2012). The scarcity of quality data and inconsistencies in metadata annotation further complicate the development of robust predictive models (Morota *et al.*, 2022).

This chapter reviews the fundamental principles of PGMs and GP, examining their theoretical frameworks, key parameters, and primary applications. It explores the conceptual underpinnings of their integration, discusses the data requirements and associated challenges, and highlights practical applications in crop breeding programs. Finally, current challenges and future perspectives will be discussed, in particular the potential of artificial intelligence, digital twins, and high-throughput phenotyping platforms to improve the efficiency of this integrated approach, contributing to sustainable and adaptable agriculture.

6.2 Overview of Plant Growth Models

6.2.1 Definition and concept

PGMs are essential tools in agronomy, used to simulate plant development by integrating various physiological and environmental factors. These models allow understanding and predicting plant growth in different contexts, thus providing crucial information to optimize agricultural practices and improve yields. Their role is particularly relevant in the context of precision agriculture, varietal selection, as well as the assessment of climate change impacts (Larue *et al.*, 2024). PGMs are mainly divided into two categories: empirical models and mechanistic models.

Empirical models rely on historical data and statistical correlations and are used to predict plant growth when detailed physiological knowledge is lacking or when rapid estimation is required. These models, such as linear and nonlinear regression models, have been widely used to estimate crop yields based on meteorological variables such as temperature, precipitation, and solar radiation (Wang *et al.*, 2024). With recent advances, machine learning approaches, including deep learning, are now being employed to analyze large amounts of data and improve the accuracy of yield predictions (Liakos *et al.*, 2018). However, while these models are effective for rapid predictions, their main drawback is that they do not take into account the underlying physiological processes. This limits their ability to provide reliable predictions under novel environmental conditions. Mechanistic models, or process-based models, simulate plant growth by considering essential physiological processes such as photosynthesis, respiration, nutrient uptake, and transpiration.

Unlike empirical models, these models integrate biological and physical principles to predict plant development under varying environmental conditions.

6.2.2 Commonly used PGMs

Among the most widely used mechanistic models, DSSAT (Decision Support System for Agrotechnology Transfer) is a major tool that simulates soil–plant–atmosphere interactions. This model has been used in many studies to predict crop yields under various climate scenarios and management practices (Dhakar *et al.*, 2018). Recent improvements have enhanced its ability to assess the impact of climate change on agricultural systems, making it a valuable tool for crop management and the formulation of adapted agricultural policies (Mubeen *et al.*, 2020). Another widely used model is STICS (Scientific, Technical and Interdisciplinary simulator of soil-Crop System functioning; in French: Simulateur mulTIdisciplinaire pour les Cultures Standard; or: Simulation Tool for Cropping Systems), which offers a detailed simulation of soil–crop–atmosphere interactions. Through its application, STICS has provided insights into the effects of soil properties and climate variability on crop performance (Beaudoin *et al.*, 2023). In addition, it has recently been used to study crop resilience to water stress, an area of particular relevance in regions subject to increased drought (Beaudoin *et al.*, 2023). The GreenLab model also represents an important advance in plant functional-structural modeling. This model combines the three-dimensional structure of plants with their physiological processes, allowing for a better understanding of the dynamic interplay between biomass, resource allocation, and growth. GreenLab has been used to optimize crop yields, as well as to study plant responses to environmental stresses (Vos *et al.*, 2010). More recently, the simulation of the effects of biotic and abiotic stresses has been improved, providing a more accurate representation of real-world crop conditions (Patil *et al.*, 2023). The WOFOST (World Food Studies) model is another key mechanistic model that simulates the effects of environmental factors such as temperature, water, and radiation on plant development. This model has been widely used in large-scale agricultural studies and has recently been adjusted to improve its predictions in the face of climate change (Singh *et al.*, 2025). By combining these diverse approaches, PGMs continue to play a critical role in improving the understanding of plant growth under different conditions, while also helping to optimize agricultural practices and predict the impact of environmental changes on crop productivity. These models are constantly being improved to address new challenges, including those related to climate change and sustainable agriculture (Table 6.1).

6.3 Genomic Prediction in Plant Breeding

Genomic prediction (GP) represents a major advance in plant breeding, allowing one to anticipate the genetic potential of an individual based on genome-wide genomic marker data. By exploiting high-throughput genotyping techniques coupled with advanced statistical models, GP optimizes the efficiency of breeding processes and accelerates genetic gain by significantly reducing the duration of breeding cycles (Crossa *et al.*, 2017). Unlike conventional breeding methods that rely mainly on phenotypic evaluations and pedigree information, GP leverages single-nucleotide polymorphism (SNP) data to estimate complex traits. This approach is particularly effective for polygenic traits, which are governed by multiple loci across the genome (Hickey *et al.*, 2024). By reducing the reliance on field trials, which are often limited by environmental and time constraints, GP improves selection accuracy and decision-making in varietal improvement programs (Wang *et al.*, 2018). The application of GP relies on statistical models capable of establishing relationships between genotypes and phenotypes, thus facilitating the prediction of the performance of untested individuals. These models, based on Bayesian approaches or machine learning algorithms, allow better exploitation of genetic data and reduce the uncertainties associated with traditional methods (Hickey *et al.*, 2024). Increasingly integrated into improvement programs, GP plays a key role in adapting

Table 6.1. Commonly used mechanistic models

Model	Type	Applications	Key processes simulated	Recent advances/ updates	References
WOFOST	Mechanistic	Large-scale agricultural studies, crop growth predictions.	Impact of environmental factors (temperature, water, radiation).	Updated for improved prediction accuracy under changing climatic conditions.	(Singh *et al.*, 2025)
DSSAT	Mechanistic	Crop simulation, yield prediction, climate impact assessment.	Soil-plant-atmosphere interactions; Photosynthesis, respiration.	Enhanced capacity for modeling climate change effects on agricultural systems.	(Mubeen *et al.*, 2020)
GreenLab	Functional-Structural	Studying plant architecture, biomass allocation, crop optimization	Biomass allocation, organ development, environmental stimuli	Refinements to simulate biotic and abiotic stresses	(Kang, 2013)
STICS	Mechanistic	Understanding soil properties, climate variability effects on crop yield	Soil-crop-atmosphere interactions; Water stress, nutrient uptake	application to assessing resilience to water stress	(Constantin *et al.*, 2015)

agriculture to climate challenges and the need for sustainable production (Begna *et al.*, 2021).

Genomic selection (GS) is an innovative approach in plant breeding, allowing us to estimate the genetic value of individuals from genomic markers covering the entire genome, without resorting to direct phenotypic evaluations. Unlike marker-assisted selection (MAS), which is restricted to a limited number of major effect loci, GS exploits the entire genetic variability by integrating the combined effects of small-effect loci, playing a key role in complex traits (Larue *et al.*, 2024). This global strategy improves the precision of selection and accelerates genetic progress in varietal improvement programs.

The GS process is based on the establishment of a statistical model, calibrated from a reference population combining genotypic and phenotypic data. Once trained, this model makes it possible to predict the genetic value of untested individuals on the sole basis of their genomic profile (Gaynor *et al.*, 2017). Several statistical methods are used in this framework, ranging from ridge regression best linear unbiased predictor (RR-BLUP) to Bayesian approaches (BayesA, BayesB, and BayesCπ), including machine learning algorithms such as support vector machines (SVMs) and deep neural networks (Montesinos-López *et al.*, 2024). These different approaches vary in their ability to capture the underlying genetic architecture, with Bayesian methods being particularly suited to traits influenced by loci with variable and heterogeneous effects. A major advantage of GS is the significant reduction in the time required for selection cycles. With this approach, selection decisions can be made early in plant development, without the need for exhaustive phenotypic evaluation. This time-saving is particularly advantageous for long-cycle species, such as perennial crops and fruit trees, where varietal improvement by conventional methods is time-consuming and costly (Beyene *et al.*, 2021). In addition, the integration of genotype–environment interactions into predictive GS models allows for the optimization of selection in various agro-climatic contexts, thus improving the adaptability and resilience of new varieties (Crossa *et al.*, 2021).

Recent advances in this field integrate the use of multi-omics to improve the accuracy of GP. The incorporation of this functional information allows for the refinement of genetic estimations and a deeper understanding of the molecular mechanisms involved in agronomic traits (Montesinos-López *et al.*, 2024). With the evolution of sequencing technologies and the increase in computing capacities, the application of GS is expected to continue to expand, opening new perspectives for more efficient and sustainable genetic improvement. The application of multi-omics approaches, encompassing genomics, transcriptomics, proteomics, metabolomics, and epigenomics, has significantly improved the accuracy of GP models in plant breeding (Table 6.2). Unlike conventional GS, which mainly relies on genome-wide SNPs, the integration of multiple levels of biological information provides a better understanding of the molecular mechanisms governing the variability of complex traits (Crossa *et al.*, 2021). This comprehensive approach enhances the efficiency of breeding strategies by refining the estimation of genetic potential and taking into account the regulatory networks involved in the expression of agronomic traits (Yang *et al.*, 2021). The integration of these multi-omics data into GP models thus makes it possible to refine predictions and improve the efficiency of varietal selection strategies by exploiting a deeper understanding of the mechanisms underlying the traits of interest.

Table 6.2. Contribution of different omic layers to the improvement of genomic prediction and plant breeding (Author's own table)

Omics layer	Applications in plant breeding	Key contributions to GP	References
Genomics	Improved accuracy in predicting complex traits in large crops such as wheat and maize.	Provides the individual's genetic map based on SNPs, facilitating GWAS and GS.	Meuwissen *et al.* (2022)
Transcriptomics	Optimizing selection for stress tolerance and disease resistance.	Analyzes gene expression to identify activated biological pathways and their impact on phenotypic traits.	Ramu *et al.* (2016)
Proteomics	Improved prediction of disease resistance-related traits by integrating protein profiles into GP models.	Studies protein interactions and post-translational modifications that influence cellular physiology.	Razzaq *et al.* (2019)
Metabolomics	A better understanding of metabolic responses to abiotic and biotic stresses to improve selection.	Analyzes the metabolites implicated in the response to stress and the development of plants.	Larue *et al.* (2024)
Epigenomics	Integrating genotype-environment interactions to improve crop resilience to climate change.	Studies heritable chromatin modifications, such as DNA methylation and histone modifications, that influence gene expression.	Agarwal *et al.* (2020)
Integration of multi-omics	Acceleration of selection cycles and development of varieties adapted to changing environmental conditions.	Combines different layers of omics data to improve the accuracy of GP models of breeding cycles and develop varieties adapted to changing environmental conditions.	Xu *et al.* (2022)

6.4 Integration of PGMs with GP

The integration of PGMs with GP has become a pivotal focus in predictive plant breeding. This synthesis leverages the physiological realism of PGMs and the statistical power of GP to improve the prediction of complex traits under variable environmental conditions. As climate variability intensifies and genotype–environment interactions become more pronounced, combining these two modeling paradigms offers a promising pathway to develop climate-resilient, high-performing crop varieties. The following sections explore the conceptual basis and modeling strategies that underpin this integrative approach.

6.4.1 Conceptual framework

Integrating PGMs with GP continues to emerge as a critical innovation in modern plant breeding, particularly for resolving G×E interactions and modeling complex, integrative traits. While GP leverages genome-wide markers to predict genetic merit, it traditionally struggles with traits influenced by non-additive effects, epistasis, and strong environmental modulation. Conversely, PGMs simulate phenotypic development via dynamic, mechanistic processes involving environmental and physiological interactions, though they cannot often generalize to novel genotypes without additional data (Onogi, 2022).

To bridge these limitations, researchers have developed integrative approaches, most notably the PGM–WGP (Plant Growth Model–Whole Genome Prediction) strategy (Cooper *et al.*, 2016), which jointly leverages the biological fidelity of PGMs and the statistical power of GP (Fig. 6.1). Two primary strategies have been proposed: GP-assisted PGM, where genomic data are used to estimate genotype-specific parameters (GSPs) for input into PGMs; and PGM-assisted GP, where outputs from PGMs are used to inform or augment genomic prediction models (Onogi, 2022). Both approaches benefit from recent advances in high-throughput phenotyping and multiscale modeling techniques. This integration enables the decomposition of complex phenotypes into more

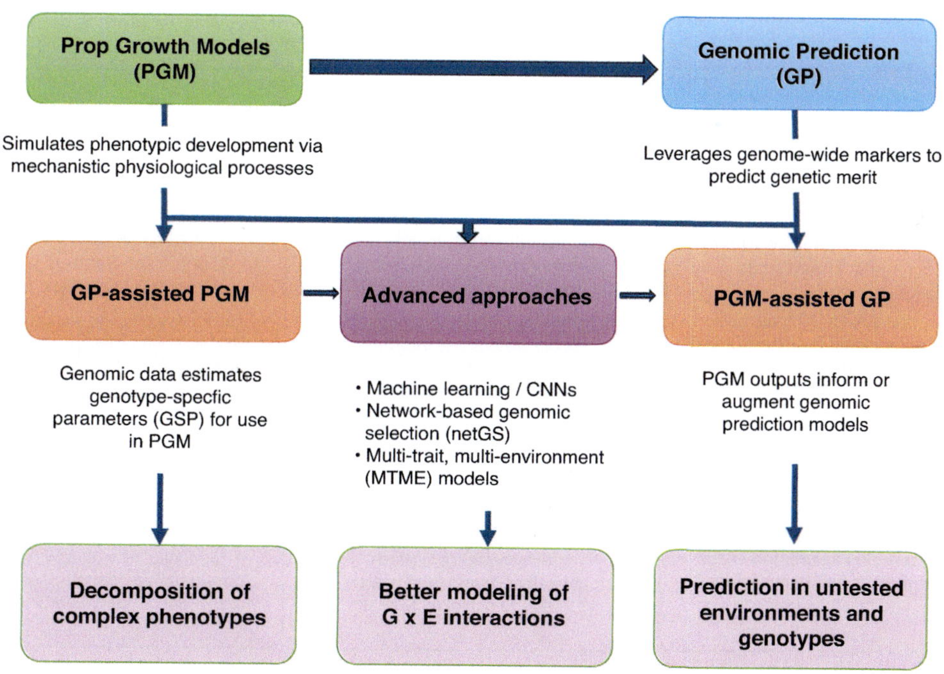

Fig. 6.1. Conceptual framework for integrating plant growth models with genomic prediction (Author's own image)

heritable component traits, better captures G×E interactions, and supports prediction in untested environments or genotypes (Larue *et al.*, 2024).

The limitations of classical GP, such as linearity assumptions and poor performance on non-additive or polygenic traits, have spurred interest in nonlinear machine learning methods, including convolutional neural networks (CNNs), gradient-boosting decision trees (GBDTs), and ensemble models. These tools can model epistatic interactions and high-dimensional feature spaces effectively (Yu *et al.*, 2022; Larue *et al.*, 2024). In parallel, the advent of 'smart breeding' driven by integrated genomic–environment prediction (iGEP) underscores a shift toward predictive breeding grounded in Big Data, AI, and spatiotemporal omics (Xu *et al.*, 2022).

Recent advancements also include the integration of molecular markers into metabolic models, exemplified by the network-based genomic selection (netGS) approach. By incorporating genotype-specific reaction rates from metabolic flux models, netGS significantly improves prediction accuracy for traits like biomass, both within and across environments, compared with classical GS (Tong *et al.*, 2020). This approach enables the modeling of traits closely tied to primary metabolism, offering enhanced biological interpretability and environment-specific predictability.

Furthermore, Toda *et al.* (2020) proposed a two-step model that predicts biomass from intermediate growth-related traits using PGMs, with these traits themselves predicted by GP. This strategy reflects a modular integration of GP and PGMs, leveraging physiological intermediates to improve model performance under varying environmental conditions. Ultimately, PGM–GP integration is not just a fusion of tools but a systemic methodology for predictive phenotyping, supporting ideotype design, adaptive trait targeting, and resource-efficient breeding under climate variability and environmental heterogeneity (Xu *et al.*, 2022).

6.4.2 Modeling approaches

A wide range of modeling strategies (Fig. 6.2) has been developed to operationalize the

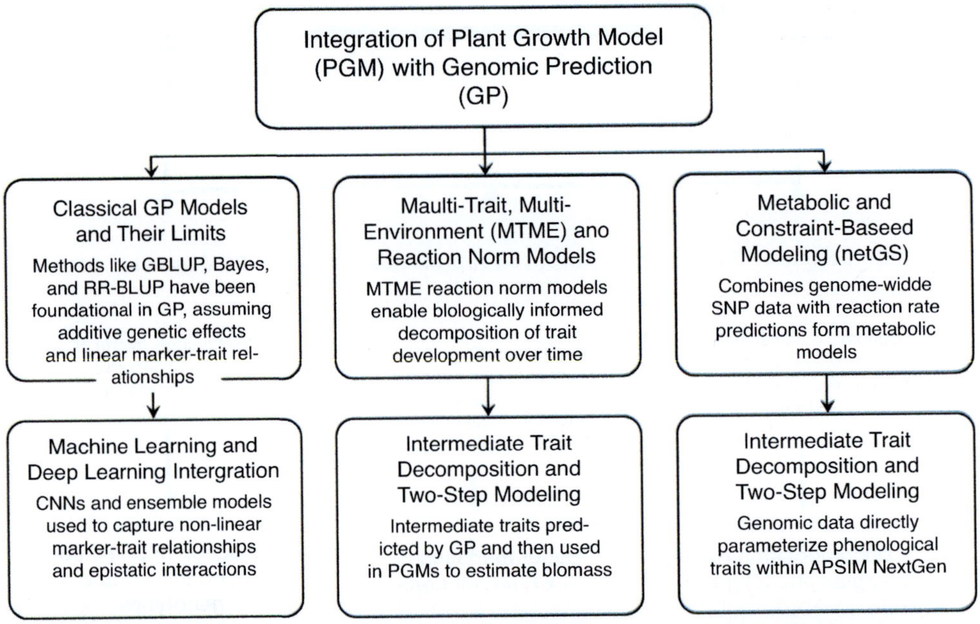

Fig. 6.2. Different modeling approaches to operationalize the integration of PGMs and GP (Author's own image)

Table 6.3. Genomic prediction-assisted plant growth models (GP-assisted PGMs) (Author's own table)

PGMs	Species	Trait	Type of approach	Number of parameters	Reference
Simplified PGM	Maize	Yield	Joint	4	Technow *et al.* (2015)
Simplified PGM with drought stress	Maize	Yield	Joint	5	Cooper *et al.* (2016)
Simplified PGM	Pepper	Yield	Independent	3	Alimi (2016)
Phenology model	Barley	Heading date	Independent	3	Uptmoor *et al.* (2017)
FSPM (MappleT)	Apple	Morphological traits	Independent	4	Migault *et al.* (2016)
Phenology model	Cauliflower	Curding time	Independent	3	Rosen *et al.* (2018)
Simplified PGM with drought stress	Maize	Yield	Joint	4	Messina *et al.* (2018)
Simplified PGM	Rice	Biomass	Independent	4	Toda *et al.* (2020)
Phenology model	Rice	Heading date	Independent	3	Chen *et al.* (2020)

integration of PGMs and GP (Table 6.3), each differing in complexity, interpretability, and predictive performance. Modeling strategies for PGM–GP integration are rapidly evolving. The most effective approaches combine mechanistic plant modeling, advanced machine learning, and multilayered biological data to improve prediction accuracy, address G × E interactions, and accelerate the design of climate-resilient crop ideotypes.

In the case of the GP-assisted PGM strategy, two primary strategies – an independent approach and a joint approach – have been developed to integrate PGMs with GP (Fig. 6.3), each differing in how GSPs are estimated and linked to genetic data.

- For the independent approach, GSPs are first estimated for each genotype using ecophysiological models based on phenotypic and environmental data. These parameters are treated as known quantities and are subsequently used as response variables in QTL mapping or genomic prediction models. This sequential procedure is relatively straightforward and compatible with many established PGMs, such as DSSAT and APSIM. However, it may introduce compounding errors from the first stage in the genetic analysis, and it does not account for uncertainty in parameter estimation. Moreover, because each genotype is modeled separately, there is no borrowing of information across genotypes, which can limit predictive power and robustness (Onogi, 2022).
- The joint approach integrates parameter estimation and genomic prediction within a single hierarchical model. GSPs for all genotypes are estimated simultaneously using both phenotypic data and genomic information, typically through Bayesian frameworks or machine learning algorithms. This integrated modeling enables partial pooling, allowing related genotypes to share information, which improves estimation accuracy and model stability. Joint approaches also better capture the uncertainty of parameter estimates and can provide more robust predictions of phenotype under varying environmental

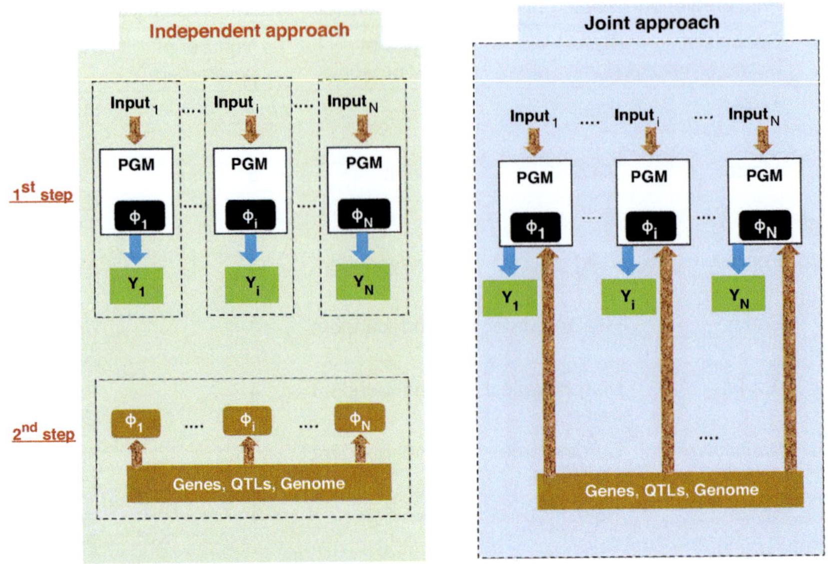

Fig. 6.3. Joint and independent strategies for combining PGMs with GP. In the independent (parallel) strategy, PGMs are calibrated separately for each genotype based on phenotypic data (Y), allowing model parameters (Φ) to be estimated for N genotypes (1st step). These parameters are then subjected to genetic analysis (2nd step) in conjunction with the genotypic data (QTLs, genes, or genomic markers). In contrast, the joint (integrated) strategy simultaneously performs PGM calibration and genetic analysis, directly exploiting genomic information to model interactions between genotypes and phenotypic responses (Author's own image).

conditions. While computationally more demanding, this strategy is conceptually more aligned with the dynamic nature of PGMs and the continuous genetic variation captured by genomic prediction (Poudel *et al.*, 2023).

The integration of machine learning and deep learning algorithms in multi-omics data analysis is a major advance for GP. These approaches overcome the challenges related to the complexity and high dimensionality of multi-omics data. In particular, deep neural networks (DNNs) and ensemble learning methods have demonstrated increased efficiency in predicting complex traits by exploiting the interactions between different omics layers (Zhang *et al.*, 2022). However, several challenges persist, including the standardization of data integration protocols, the management of missing values, and the harmonization of different data types from genomics, transcriptomics, proteomics, and metabolomics. Better structuring of analytical pipelines is needed to maximize the exploitation of omics data in plant breeding programs. With continued advances in sequencing technologies and decreasing data acquisition costs, the use of multi-omics approaches in genomic breeding is expected to increase. This development offers promising prospects for optimizing breeding efficiency, reducing breeding cycles, and developing plant varieties that are more resilient to environmental stresses, thereby contributing to sustainable agriculture (Hina *et al.*, 2024).

6.5 Applications of PGMs in Crop Breeding Programs

In recent years, PGMs have experienced remarkable growth as complementary tools to genomic selection. Their ability to simulate crop growth dynamics based on environmental and genetic factors makes them powerful allies for accelerating varietal improvement.

In crops such as maize, early pioneering work incorporated simplified PGMs to predict yield by combining both genetic data and environmental variables (Table 6.2). Technow *et al.* (2015) used a model with only four parameters, showing that even a relatively simple approach could provide significant added value. Cooper *et al.* (2016) and Messina *et al.* (2018) enhanced these models by taking into account water stress, increasing the number of parameters to five to better capture the effects of drought, an increasingly pressing constraint with climate change. Other crops have also benefited from these approaches. In rice, several teams (Chen *et al.*, 2020) have used phenology models to predict heading date, a key trait for adaptation to different climatic conditions. Similar applications have been observed in horticultural species. Alimi (2016) applied a simplified PGM to pepper to predict yield, while Migault *et al.* (2016) used a functional-structural model (FSPM, via MappleT) to model morphological traits in apple, thus paving the way for a better understanding of plant architecture under different growing conditions.

Beyond classical mechanistic models, PGMs are now associated with advanced genomic selection techniques (Table 6.4). For example, in forest species such as *Cryptomeria japonica* and *Eucalyptus globulus*, several Bayesian methods (BayesA, BayesB, BayesCπ, Bayesian Lasso) and kernel models (RKHS, GBLUP) have been used to predict traits such as tree height, trunk straightness, or disease resistance (Maldonado *et al.*, 2020; Ejima *et al.*, 2023). PGMs are also combining with innovative machine learning approaches to capture dynamic growth. In lettuce (*Lactuca sativa*), different teams have exploited real-time images, chlorophyll fluorescence, and even spatiotemporal sequences to predict fresh biomass or growth dynamics (Nagano *et al.*, 2019; Liu *et al.*, 2022; Hosoda *et al.*, 2024). Rizkiana *et al.* (2021) even used neural networks to predict growth under controlled greenhouse conditions. In broader agricultural regions, the integration of PGMs with hydrological models (such as WEAP-PGM) has made it possible to simulate the impact of climate change on water consumption and crop yields (Yaykiran *et al.*, 2019), demonstrating the value of these tools well beyond the experimental field.

In short, the integration of PGMs into genomic breeding programs illustrates a major shift toward more precise and rapid breeding. By modeling the complex interactions between genetics, environment, and agronomic management, these models allow for better anticipation of the future performance of genotypes, even in rapidly changing climates. They are thus essential tools in the race to develop the agricultural varieties of tomorrow.

6.6 Challenges and Future Perspectives

Integrating GP with crop growth or ecophysiological models represents a powerful step forward for plant breeding, enabling better prediction of complex traits under variable environments. However, the effectiveness of this integration hinges on the quality, diversity, and structure of the data used.

6.6.1 Data acquisition

Successful integration of GP with crop models depends on three principal types of data: genomic, phenotypic, and environmental. These must be detailed, high-quality, and harmonized for meaningful use. Genome-wide single-nucleotide polymorphism (SNP) markers form the foundation of most GP models. Advances in resequencing, pangenomes, and SNP arrays have vastly expanded the genomic resources available for breeding (Danilevicz *et al.*, 2022). Genomic prediction of genotype-specific parameters (GSPs), such as photoperiod sensitivity or radiation use efficiency, requires models that can estimate these values from marker data, often using methods such as GBLUP, BayesB, or machine learning (Poudel *et al.*, 2023).

The increasing availability of high-throughput phenotyping (HTP) technologies has enabled dynamic, non-destructive measurement of traits such as biomass, canopy temperature, or plant height under field and controlled conditions (Danilevicz *et al.*, 2022). Longitudinal and image-based phenotyping is particularly valuable for calibrating plant models that simulate development over time

Table 6.4. Applications of PGMs for predicting traits of interest in different plant species (Author's own table)

Plant species	Objective/ Predicted traits	Model(s) used	Input variables/Data used	Reference
Cryptomeria japonica D	Tree height, wood stiffness, male flower quantity	BayesA, BayesB, BayesCπ, Bayesian Lasso, BRR	Phenotypic + Genotypic data	Ejima *et al.* (2023)
Eucalyptus globulus Labill	Volume, stem straightness, diameter at breast height, branch cluster frequency	BRR, Bayesian LASSO, Bayes B, Bayes C, RKHS, GBLUP, BRNN	Phenotypic + SNP + Markers data	Maldonado *et al.* (2020)
Fraxinus excelsior L.	Resistance to Hymenoscyphus fraxineus	GBLUP, BRR, Bayes B, GBLUP with RR and GAUSS	Phenotypic and Genotypic data	Meger *et al.* (2024)
Hevea brasiliensis	Rubber production, circumference at breast height	RKHS, RRBLUP + variable selection, Bayesian LASSO	Climatic, Genetic, Agronomic data	Cros *et al.*, 2019
Picea glauca	Diameter, volume, acoustic velocity, pungenol concentration in needles	GBLUP	–	Lenz *et al.* (2020)
Pinus sylvestris	Diameter, microfibril angle, modulus of elasticity	GBLUP, BRR, Bayesian LASSO	–	Calleja-Rodriguez *et al.* (2020)
Populus nigra L.	Height, circumference, bud flush, dormancy, rust resistance	GBLUP, Bayesian LASSO	–	Rincent *et al.* (2018)
Pseudotsuga menziesii	Juvenile height	RRBLUP with variable selection, Bayes B	–	Thistlethwaite *et al.* (2019)
Lettuce (*Lactuca sativa*)	Prediction of growth in a plant factory	ANN (Resilient Backpropagation)	Temperature, humidity, EC, light intensity	Rizkiana *et al.* (2021)
Lettuce	Prediction of dynamic fresh weight	Naive Bayesian Network	Canopy images, temperature, radiation, humidity	Liu *et al.* (2022)
Lettuce	Fresh weight via leaf movement	Optical Flow + ML	Images every 20 min (338 plants)	Nagano *et al.* (2019)
Lettuce	Growth via chlorophyll fluorescence and circadian rhythm	Gradient Boosted Regression (GBR)	Chlorophyll fluorescence, day/ night cycles	Hosoda *et al.* (2024)
Wheat	Prediction of organ growth from images	ST-LSTM + MIM (Spatio-Temporal Deep Learning)	Temporal image sequences	Wang *et al.* (2019)

Continued

Table 6.4. Continued

Plant species	Objective/ Predicted traits	Model(s) used	Input variables/Data used	Reference
Agricultural regions	Estimation of water consumption and yield under climate change	WEAP-PGM	Hydrological data (runoff, infiltration, evapotranspiration)	Yaykiran *et al.* (2019)

(Crossa *et al.*, 2021). Moreover, the integration of environmental data (radiation, temperature, and soil moisture) is critical for modeling G × E interactions. Detailed agronomic management data (e.g. sowing date, irrigation, fertilization) are also needed for accurate simulations (Poudel *et al.*, 2023). Recent frameworks propose the use of environmental similarity matrices or latent environmental indices to better represent complex environmental effects (Xu *et al.*, 2022). Furthermore, complex data types such as transcriptomic, metabolomic, and image-derived features (e.g. vegetation indices) are increasingly integrated into predictive models to improve trait resolution and biological realism (Danilevicz *et al.*, 2022). Multi-modal deep learning architectures can leverage these varied data streams to improve prediction accuracy (Xu *et al.*, 2022).

Simulations and data augmentation techniques now play an important role in GP. Approaches like 'mixup' create new training samples by interpolating between known data points, increasing training diversity and improving robustness (Montesinos-López *et al.*, 2024). This is particularly beneficial in settings with limited samples or imbalanced data (Xu *et al.*, 2022).

6.6.2 Data challenges

Despite significant advances in data acquisition technologies and platforms, integrating these datasets into predictive models still poses major challenges. Environmental variability, data imbalance, high dimensionality, and computational limitations can hinder model accuracy and generalizability. Unpredictable weather patterns and inconsistent management practices introduce noise into phenotypic and environmental datasets, complicating model calibration and reducing their applicability across diverse environments. To overcome these issues, models require detailed, high-resolution environmental data and the capacity to extrapolate beyond historical observations (Onogi, 2022; Montesinos-López *et al.*, 2024).

Climate change exacerbates these difficulties by amplifying environmental variability and increasing the frequency of extreme events such as heatwaves and droughts. Consequently, predictive models must be designed to adapt to novel conditions, while data collection systems need to provide more frequent and precise environmental measurements. As emphasized by Onogi (2022), models trained solely on historical data may struggle to remain accurate under future climate scenarios. Jeon *et al.* (2023) highlighted the importance of 'digital breeding' pipelines that integrate real-time environmental monitoring, while genomic vulnerability assessments rely on georeferenced data to trace the impact of climate change on genetic diversity (Hoffmann *et al.*, 2021).

Limited phenotypic and environmental observations, particularly for rare genotypes or marginal conditions, also constrain model training. Underrepresented samples can introduce bias and reduce prediction accuracy. Strategies such as data augmentation and balanced sampling are critical to mitigate these limitations (Xu *et al.*, 2022; Montesinos-López *et al.*, 2024). Moreover, the high dimensionality of genomic datasets, often comprising thousands of markers but relatively few samples, can lead to overfitting. Feature selection techniques, such as GWAS-based filtering, minor allele frequency thresholds, and linkage disequilibrium (LD) pruning, are essential to reduce redundancy, enhance interpretability, and improve computational efficiency (Tomura *et al.*, 2025).

Integrating heterogeneous data types across experiments and platforms is further complicated by inconsistent data formats and missing metadata. Standardization efforts, such as the MIAPPE (Minimum Information About a Plant Phenotyping Experiment) guidelines and interoperable APIs like BrAPI, are essential to ensure data scalability, reproducibility, and integration (Danilevicz *et al.*, 2022; Xu *et al.*, 2022).

Estimating genotype-specific parameters (GSPs) from genomic data remains computationally intensive, given the inherent noise and uncertainty of models. While approaches such as hierarchical Bayesian inference, approximate Bayesian computation, and machine learning offer promising solutions, they often require large, high-quality training datasets and well-informed priors (Poudel *et al.*, 2023).

Successfully integrating genomic, phenomic, and environmental data into predictive frameworks also demands high-performance computing and sophisticated algorithmic tuning. Deep learning and ensemble models (e.g. GBLUP, RKHS, random forests, SVM) offer the flexibility to capture complex relationships, but they require careful optimization and rigorous validation to prevent overfitting (Shahi *et al.*, 2025). Given the heterogeneity of traits and datasets, no single model consistently outperforms others across all scenarios. Instead, ensemble approaches – by aggregating diverse algorithms – often improve robustness and reduce prediction error, following the Diversity Prediction theorem and the No Free Lunch theorem (Lourenço *et al.*, 2024; Tomura *et al.*, 2025).

Despite these advancements, overfitting and limited generalizability remain persistent concerns. Robust validation strategies – including k-fold cross-validation, uncertainty quantification, and independent testing on external datasets – are crucial to ensure predictive stability across diverse genotypes and environmental conditions (Lourenço *et al.*, 2024). Altogether, these challenges underscore the need for robust experimental design, integrated and standardized data pipelines, and scalable computational tools. As genomic prediction continues to evolve, unlocking its full potential will depend on the synergistic integration of omics layers (genomic, phenomic, and enviromic) alongside advanced feature selection, optimized marker densities, and improved model architectures.

6.7 Conclusion

Plant breeding remains a strategic tool for adapting crops to environmental aliases. The integration of plant growth models into genomic prediction marks a turning point in crop improvement strategies. By taking advantage of the accuracy and power of computer models, we can take into account both phenotype and plant physiology during plant breeding. Integrating these models into plant genomic selection allows for more effective prediction of varietal performance in variable and changing environments. Indeed, as mentioned in this chapter, this integration has resulted in the selection of genotypes with improved agronomic traits and drought tolerance. While many challenges remain, particularly in data acquisition and harmonization, rapid advances in artificial intelligence, environmental sensors, and high-throughput phenotyping offer promising prospects. This integrated approach will not only accelerate breeding cycles but also allow for the design of varieties better adapted to future constraints, thus contributing to more sustainable and resilient agriculture.

References

Agarwal, G., Kudapa, H., Ramalingam, A., Choudhary, D., Sinha, P. *et al.* (2020) Epigenetics and epigenomics: Underlying mechanisms, relevance, and implications in crop improvement. *Functional & Integrative Genomics* 20(6), 739–761.

Alimi, N.A. (2016) *Statistical methods for QTL mapping and genomic prediction of multiple traits and environments: Case studies in pepper*. Doctoral dissertation, Wageningen University. Available at: http://edepot.wur.nl/390205

Beaudoin, N., Lecharpentier, P., Ripoche-Wachter, D., Strullu, L., Mary, B. *et al.* (2023) *STICS Soil-Crop Model: Conceptual Framework, Equations and Uses*. Editions Quae. Available at: https://library.oapen.org/handle/20.500.12657/85111 (accessed 7 February 2025).

Begna, T., Yesuf, H., Abdurezake, M. and Eshetu, G. (2021) Genetic mapping in crop plants. *Open Journal of Plant Science* 6(1), 019–026.

Beyene, Y., Gowda, M., Pérez-Rodríguez, P., Olsen, M., Robbins, K.R. *et al.* (2021) Application of genomic selection at the early stage of breeding pipeline in tropical maize. *Frontiers in Plant Science* 12, 685488.

Burgueño, J., de Los Campos, G., Weigel, K. and Crossa, J. (2012) Genomic prediction of breeding values when modeling genotype × environment interaction using pedigree and dense molecular markers. *Crop Science* 52(2), 707–719.

Calleja-Rodriguez, A., Pan, J., Funda, T., Chen, Z., Baison, J. *et al.* (2020) Evaluation of the efficiency of genomic versus pedigree predictions for growth and wood quality traits in scots pine. *BMC Genomics* 21(1), 796. DOI: 10.1186/s12864-020-07188-4.

Chen, L., Onogi, A., Watanabe, M., Mochizuki, T. and Iwata, H. (2020) Prediction of heading date in rice using QTL-based models with environment-specific parameters. *Theoretical and Applied Genetics* 133, 2867–2878.

Constantin, J., Willaume, M., Murgue, C., Lacroix, B. and Therond, O. (2015) The soil-crop models STICS and AqYield predict yield and soil water content for irrigated crops equally well with limited data. *Agricultural and Forest Meteorology* 206, 55–68.

Cooper, M., Technow, F., Messina, C., Gho, C. and Totir, L.R. (2016) Use of crop growth models with whole-genome prediction: Application to a maize multienvironment trial. *Crop Science* 56, 2141–2156. DOI: 10.2135/cropsci2015.02.0125.

Cros, D., Mbo-Nkoulou, L., Bell, J.M., Oum, J., Masson, A. *et al.* (2019) Within-family genomic selection in rubber tree (*Hevea brasiliensis*) increases genetic gain for rubber production. *Industrial Crops and Products* 138, 111464. DOI: 10.1016/j.indcrop.2019.111464.

Crossa, J., Pérez-Rodríguez, P., Cuevas, J., Montesinos-López, O., Jarquín, D. *et al.* (2017) Genomic selection in plant breeding: Methods, models, and perspectives. *Trends in Plant Science* 22(11), 961–975.

Crossa, J., Fritsche-Neto, R., Montesinos-Lopez, O.A., Costa-Neto, G., Dreisigacker, S. *et al.* (2021) The modern plant breeding triangle: Optimizing the use of genomics, phenomics, and enviromics data. *Frontiers in Plant Science* 12, 651480.

Danilevicz, M.F., Tay Fernandez, C.G., Bayer, P.E., Soler, A., Jacobson, A. *et al.* (2022) Data integration and predictive modelling for genomic selection in crops. *Theoretical and Applied Genetics* 135, 2595–2617. DOI: 10.1007/s00122-022-04099-7.

Dhakar, R., Sarath Chandran, M.A., Nagar, S., Visha Kumari, V., Subbarao, A.V.M. *et al.* (2018) Field crop response to water deficit stress: Assessment through crop models. In: Bal, S.K., Mukherjee, J. and Choudhury, B.U. (eds) *Advances in Crop Environment Interaction*. Springer Singapore, Singapore, pp. 287–315. Available at: http://link.springer.com/10.1007/978-981-13-1861-0_11 (accessed 7 February 2025).

Ejima, A., Uchiyama, K., Mori, H. and Tsumura, Y. (2023) Genomic prediction of tree height, wood stiffness, and male flower quantity traits across two generations in selected individuals of *Cryptomeria japonica* D. Don (Japanese Cedar). *Forests* 14(10), 2097. DOI: 10.3390/F14102097.

Gaynor, R.C., Gorjanc, G., Bentley, A.R., Ober, E.S., Howell, P. *et al.* (2017) A two-part strategy for using genomic selection to develop inbred lines. *Crop Science* 57(5), 2372–2386.

Heslot, N., Akdemir, D., Sorrells, M.E. and Jannink, J.-L. (2014) Integrating environmental covariates and crop modeling into the genomic selection framework to predict genotype by environment interactions. *Theoretical and Applied Genetics* 127, 463–480. DOI: 10.1007/s00122-013-2231-5.

Hickey, J., Williams, J.P. and Hector, E.C. (2024) Transfer learning with uncertainty quantification: Random effect calibration of source to target (RECaST). *Journal of Machine Learning Research* 25(338), 1–40.

Hina, A., Abbasi, A., Arshad, M., Imtiaz, S., Shahid, S. *et al.* (2024) Utilization of multi-omics approaches for crop improvement. In: Fiaz, S. and Prakash, C.S. (eds) *OMICs-based Techniques for Global Food Security*. Wiley, Hoboken, New Jersey, pp. 91–121. Available at: https://onlinelibrary.wiley.com/doi/10.1002/9781394209156.ch5 (accessed 7 February 2025).

Hoffmann, A.A., Griffin, P.C., Dillon, S.K., Catullo, R.A., Rane, R.V. *et al.* (2021) A framework for incorporating evolutionary genomics into biodiversity conservation and management. *Molecular Ecology Resources* 21, 859–877.

Hosoda, Y., Tada, T. and Goto, H. (2024) Lettuce fresh weight prediction in a plant factory using plant growth models. *IEEE Access* 12, 97226–97234.

Jeon, M., Crossa, J., Montesinos-López, O.A. and Poland, J. (2023) Digital breeding: Integrating genomic prediction with real-time environmental data in climate-smart crop improvement. *Frontiers in Plant Science* 14, 1092584. DOI: 10.3389/fpls.2023.1092584.

Kang, F. (2013) *Plant growth models and methodologies adapted to their parameterization for the analysis of phenotypes*. PhD Thesis, Ecole Centrale Paris. Available at: https://theses.hal.science/tel-01164965/ (accessed 7 February 2025).

Larue, C., Mier, B., Teran, J.C., Cortés, A.J., Montesinos-López, A. *et al.* (2024) Deep learning models for predictive phenotyping: A framework for integrating phenomics and genomic data. *Nature Plants* 10, 112–128.

Lenz, P.R.N., Nadeau, S., Azaiez, A., Gérardi, S., Deslauriers, M. *et al.* (2020) Genomic prediction for hastening and improving efficiency of forward selection in conifer polycross mating designs: An example from white spruce. *Heredity* 124(4), 562–578. DOI: 10.1038/s41437-019-0290-3.

Li, X. and Siddique, K.H.M. (2020) Future smart food: Harnessing the potential of neglected and underutilized species for Zero Hunger. *Maternal & Child Nutrition* 16(S3), e13008.

Liakos, K.G., Busato, P., Moshou, D., Pearson, S. and Bochtis, D. (2018) Machine learning in agriculture: A review. *Sensors* 18(8), 2674.

Liu, L., Yuan, J., Gong, L., Wang, X. and Liu, X. (2022) Dynamic fresh weight prediction of substrate-cultivated lettuce grown in a solar greenhouse based on phenotypic and environmental data. *Agriculture* 12, 11. DOI: 10.3390/agriculture12111959.

Lourenço, J., Pérez-Rodríguez, P., Akdemir, D. and Crossa, J. (2024) Robust genomic prediction across environments through ensemble learning. *G3: Genes, Genomes, Genetics* 14, jkad231. DOI: 10.1093/g3journal/jkad231.

Maldonado, C., Mora-Poblete, F., Contreras-Soto, R.I., Ahmar, S., Chen, J.T. *et al.* (2020) Genome-wide prediction of complex traits in two outcrossing plant species through Deep Learning and Bayesian Regularized Neural Network. *Frontiers in Plant Science* 11, 593897. DOI: 10.3389/fpls.2020.593897.

Meger, J., Ulaszewski, B., Pałucka, M., Kozioł, C. and Burczyk, J. (2024) Genomic prediction of resistance to *Hymenoscyphus fraxineus* in common ash (*Fraxinus excelsior* L.) populations. *Evolutionary Applications* 17(5), e13694. DOI: 10.1111/eva.13694.

Messina, C.D., Technow, F., Tang, T., Totir, R., Gho, C. *et al.* (2018) Leveraging biological insight and environmental variation to improve phenotypic prediction: Integrating crop growth models with whole-genome prediction. *European Journal of Agronomy* 100, 151–162. DOI: 10.1016/j.eja.2018.01.007.

Meuwissen, T., Hayes, B., MacLeod, I. and Goddard, M. (2022) Identification of genomic variants causing variation in quantitative traits: A review. *Agriculture* 12(10), 1713.

Migault, V., Pallas, B. and Costes, E. (2016) Combining genome-wide information with a functional-structural plant model to simulate one-year-old apple tree architecture. *Frontiers in Plant Science* 7, 2065. DOI: 10.3389/fpls.2016.02065.

Montesinos-López, O.A., Martín-Vallejo, J., Crossa, J., Gianola, D., Hernández-Suárez, C.M. *et al.* (2024) Improving genomic prediction through multimodal deep learning and data augmentation strategies. *BMC Genomics* 25, 544. DOI: 10.1186/s12864-024-10438-4.

Morota, G., Jarquin, D., Campbell, M.T., Campbell, M.T. and Iwata, H. (2022) Statistical methods for the quantitative genetic analysis of high-throughput phenotyping data. In: Lorence, A. and Medina Jimenez, K. (eds) *High-Throughput Plant Phenotyping. Methods in Molecular Biology*. Springer US, New York, pp. 269–296. Available at: https://link.springer.com/10.1007/978-1-0716-2537-8_21 (accessed 6 February 2025).

Mubeen, M., Ahmad, A., Hammad, H.M., Awais, M., Farid, H.U. *et al.* (2020) Evaluating the climate change impact on water use efficiency of cotton-wheat in semi-arid conditions using DSSAT model. *Journal of Water and Climate Change* 11(4), 1661–1675.

Nagano, S., Moriyuki, S., Wakamori, K., Mineno, H. and Fukuda, H. (2019) Leaf-movement-based growth prediction model using optical flow analysis and machine learning in plant factory. *Frontiers in Plant Science* 10, 227. DOI: 10.3389/fpls.2019.00227.

Onogi, A. (2022) Integration of crop growth models and genomic prediction. In: Pérez-Enciso, M., Akdemir, R. and Crossa, J. (eds) *Complex Trait Prediction*. Springer, New York, pp. 359–396. DOI: 10.1007/978-1-0716-2205-6_13.

Patil, S.M., Henke, M., Chandramouli, M. and Jagarlapudi, A. (2023) Role of virtual plants in digital agriculture. In: Chaudhary, S., Biradar, C.M. and Divakaran, S. (eds) *Digital Ecosystem for Innovation in*

Agriculture. Studies in Big Data, Springer Nature Singapore, Singapore, pp. 157–182. Available at: https://link.springer.com/10.1007/978-981-99-0577-5_8 (accessed 7 February 2025).

Poudel, B., Lehermeier, C. and Technow, F. (2023) Integrating Bayesian parameter estimation with crop models and genomic prediction for genotype-by-environment interaction. *Theoretical and Applied Genetics* 136, 45. DOI: 10.1007/s00122-022-04261-1.

Ramu, V.S., Paramanantham, A., Ramegowda, V., Mohan-Raju, B., Udayakumar, M. *et al.* (2016) Transcriptome analysis of sunflower genotypes with contrasting oxidative stress tolerance reveals individual-and combined-biotic and abiotic stress tolerance mechanisms. *PLOS ONE* 11(6), e0157522.

Razzaq, A., Sadia, B., Raza, A., Khalid Hameed, M. and Saleem, F. (2019) Metabolomics: A way forward for crop improvement. *Metabolites* 9(12), 303.

Resende, M.F.R., Muñoz, P., Acosta, J.J., Peter, G.F., Davis, J.M. *et al.* (2012) Accelerating the domestication of trees using genomic selection: Accuracy of prediction models across ages and environments. *New Phytologist* 193(3), 617–624.

Rincent, R., Charpentier, J.P., Faivre-Rampant, P., Paux, E., Le Gouis, J. *et al.* (2018) Phenomic selection is a low-cost and high-throughput method based on indirect predictions: Proof of concept on wheat and poplar. *G3: Genes, Genomes, Genetics* 8(12), 3961–3972. DOI: 10.1534/g3.118.200760.

Rizkiana, A., Nugroho, A.P., Salma, N.M., Afif, S., Masithoh, R.E. *et al.* (2021) Plant growth prediction model for lettuce (*Lactuca sativa*.) in plant factories using artificial neural network. In: *IOP Conference Series: Earth and Environmental Science*, IOP Publishing, p. 012027 (Vol. 733, No. 1). DOI: 10.1088/1755-1315/733/1/012027.

Rosen, C.J., Thompson, M. and Gupta, S.C. (2018) Integration of modeling and genomics for predicting cauliflower curding time. *HortScience* 53, 11–17.

Shahi, D., Montesinos-López, O.A. and Crossa, J. (2025) High-performance computing strategies for large-scale genomic prediction using deep ensemble methods. *G3: Genes, Genomes, Genetics* 15, jkab037.

Singh, J., Singh, S.P., Kaur Kingra, P., Biswas, B., Kaur, V. *et al.* (2025) Comparative evaluation of CERES-maize, WOFOST-maize, and ensemble of models for predicting maize phenology, growth, and grain yield. *Communications in Soil Science and Plant Analysis* 56(9), 1356–1380.

Technow, F., Messina, C.D., Totir, L.R. and Cooper, M. (2015) Integrating crop growth models with whole genome prediction through approximate Bayesian computation. *PLOS ONE* 10, e0130855. DOI: 10.1371/journal.pone.0130855.

Thistlethwaite, F.R., Ratcliffe, B., Klápště, J., Porth, I., Chen, C. *et al.* (2019) Genomic selection of juvenile height across a single-generational gap in douglas-fir. *Heredity* 122(6), 848–863. DOI: 10.1038/s41437-018-0172-0.

Toda, M., Watanabe, M. and Onogi, A. (2020) A modular approach to integrating genomic Prediction with crop modeling: predicting intermediate traits for biomass estimation. *Plant and Cell Physiology* 61, 1787–1799. DOI: 10.1093/pcp/pcaa111.

Toda, Y., Sasaki, G., Ohmori, Y., Yamasaki, Y., Takahashi, H. *et al.* (2024) Reaction norm for genomic prediction of plant growth: Modeling drought stress response in soybean. *Theoretical and Applied Genetics* 137(4), 77.

Tomura, S., Wilkinson, M.J., Powell, O. and Cooper, M. (2025) Ensemble analysis with interpretable genomic prediction (EasiGP): Computational tool for interpreting ensembles of genomic prediction models. *The Plant Genome* 18(4), e70138. DOI: 10.1002/tpg2.70138.

Tong, H., Nikoloski, Z. and Kliebenstein, D.J. (2020) Network-based genomic selection: Predicting metabolic traits from genome-wide SNPs via constraint-based modeling. *Plant Cell* 32, 729–745. DOI: 10.1105/tpc.19.00690.

Uptmoor, R., Wenzel, W. and Wurschum, T. (2017) A model-based strategy for QTL analysis of heading date in barley under multi-environment field trials. *Euphytica* 213, 106.

Vos, J., Evers, J.B., Buck-Sorlin, G.H., Andrieu, B., Chelle, M. *et al.* (2010) Functional–structural plant modelling: A new versatile tool in crop science. *Journal of Experimental Botany* 61(8), 2101–2115.

Wang, E., Brown, H.E., Rebetzke, G.J., Zhao, Z., Zheng, B. *et al.* (2019) Improving process-based crop models to better capture genotype × environment × management interactions. *Journal of Experimental Botany* 70, 2389–2401.

Wang, X., Xu, Y., Hu, Z. and Xu, C. (2018) Genomic selection methods for crop improvement: Current status and prospects. *The Crop Journal* 6(4), 330–340.

Wang, Y., Zhang, Q., Yu, F., Zhang, N., Zhang, X. *et al*. (2024) Progress in research on deep learning-based crop yield prediction. *Agronomy* 14(10), 2264. DOI: 10.3390/agronomy14102264.

Xu, Y., Zhang, X., Li, H., Zheng, H., Zhang, J. *et al*. (2022) Smart breeding driven by big data, artificial intelligence, and integrated genomic-enviromic prediction. *Molecular Plant* 15(11), 1664–1695. DOI: 10.1016/j.molp.2022.09.001.

Yang, Y., Saand, M.A., Huang, L., Abdelaal, W.B., Zhang, J. *et al*. (2021) Applications of multi-omics technologies for crop improvement. *Frontiers in Plant Science* 12, 563953.

Yaykiran, S., Cuceloglu, G. and Ekdal, A. (2019) Estimation of water budget components of the Sakarya River Basin by using the WEAP-PGM model. *Water* 11(2), 271. DOI: 10.3390/w11020271.

Yu, J., Pressoir, G., Briggs, W.H., Vroh Bi, I., Yamasaki, M. *et al*. (2022) A unified mixed-model method for association mapping that accounts for multiple levels of relatedness. *Nature Genetics* 38, 203–208.

Zhang, R., Zhang, C., Yu, C., Dong, J. and Hu, J. (2022) Integration of multi-omics technologies for crop improvement: Status and prospects. *Frontiers in Bioinformatics* 2, 1027457.

7 Genomic Prediction Models for Cereal Crops

Ishrat Parween, Prerna Kumari and Binod Kumar Mahto*
University Department of Botany, Ranchi University, Ranchi, Jharkhand, India

Abstract

Genomic prediction (GP) models have revolutionized plant breeding by enabling accurate and rapid selection of superior genotypes based on genome-wide marker information. In cereal crops such as rice, wheat, maize, and barley, where global food security relies heavily on sustained genetic improvement, GP offers a powerful approach to accelerate genetic gain. This chapter provides a comprehensive overview of genomic prediction models applied to cereal crops, focusing on their statistical underpinnings, computational frameworks, and practical implementation in breeding programs. The integration of high-throughput genotyping with phenotypic and environmental data has led to significant advancements in predicting complex quantitative traits, including yield, stress tolerance, disease resistance, and quality parameters. We discuss various statistical and machine learning models, including genomic best linear unbiased prediction (GBLUP), Bayesian approaches, and nonlinear models such as random forest and deep learning. We also address key factors that influence prediction accuracy, such as training population size, marker density, the genetic architecture of traits, and genotype × environment interactions. Furthermore, we highlight recent developments in multi-trait and multi-environment GP models, which are particularly relevant for breeding climate-resilient cereals. Challenges such as model overfitting, computational costs, and data integration are also examined.

Keywords: Cereal crops, GBLUP, Genomic prediction, Genomic selection, Machine learning, Trait prediction, Marker-assisted breeding, Plant breeding, Statistical models

7.1 Introduction

Cereal crops are an essential component of the global diet, contributing approximately 50% of the total dietary energy intake. Among these, wheat, rice, maize, and barley dominate the world's arable lands, contributing approximately 2817 million tonnes to global production each year. However, sustaining and improving the productivity of these crops is becoming increasingly difficult due to rising challenges such as climate variability, diminishing soil fertility, water shortage, and decreasing availability of arable land (Reynolds and Ortiz, 2010). In parallel, the rapidly growing global population presents an urgent challenge to meet future food demands. It is projected that the world population will exceed nine billion by 2050, which will impose substantial pressure on current food production systems (Furbank and Tester, 2011; Narasimhan *et al.*, 2019). Meeting the growing demand for food,

*Corresponding author: binodkrmahto@ranchiuniversity.ac.in, krvinod09@gmail.com

 Plant Genomic Prediction: AI-Enhanced Models for Plant Genomics and Crop Improvement (ed. Jen-Tsung Chen)
DOI: 10.1079/9781836991328.0007

while simultaneously addressing rural poverty and nutritional insecurity, requires innovative strategies that can enhance the efficiency and effectiveness of crop improvement.

Traditional breeding approaches, which rely heavily on hybridization and phenotypic selection over multiple generations, are no longer sufficient to address these challenges. The time-intensive nature of conventional breeding, typically requiring 5–12 years to release a new variety, makes it ill-suited for the urgent demands of modern agriculture. Furthermore, phenotypic selection is often constrained by environmental noise, genotype-by-environment (G × E) interactions, and the low heritability of many complex traits (Tuberosa, 2012). As a result, the integration of genomics-based approaches has become critical to accelerate genetic gain (Montesinos-López *et al.*, 2021a).

One of the most promising innovations in this regard is genomic selection (GS). GS represents an integrated approach that combines traditional breeding methods with advanced genomic tools, the development of double haploid (DH) lines, optimized field designs, and quantitative genetics. Unlike earlier molecular breeding strategies, such as marker-assisted backcrossing (MABC) and marker-assisted recurrent selection (MARS) which focused on identifying and introgressing a few major quantitative trait loci (QTLs), GS enables genome-wide prediction of breeding values without requiring prior knowledge of specific loci (Meuwissen *et al.*, 2001; Varshney *et al.*, 2017). The concept of GS was initially developed for livestock breeding, where it demonstrated notable success in predicting the genetic merit of individuals using dense genome-wide marker data (Meuwissen *et al.*, 2001). Its early application in dairy cattle breeding led to substantial improvements in selection accuracy (Su *et al.*, 2010). The advent of cost-effective high-throughput genotyping technologies (particularly single-nucleotide polymorphism (SNP) arrays) ,combined with advances in statistical modeling, facilitated the widespread adoption of GS in plant breeding (Robertsen *et al.*, 2019).

At the core of GS lies the concept of genomic prediction (GP), wherein the breeding value of an individual is estimated based on genome-wide marker data and predictive statistical models. GP enables selection for complex polygenic traits, such as drought tolerance and yield, even in the absence of detailed knowledge of the underlying genetic architecture. This allows breeders to skip some of the time-consuming field trials typically required in conventional programs, thereby increasing the rate of genetic gain per unit time (Heffner *et al.*, 2009; García-Ruiz *et al.*, 2016; Yabe *et al.*, 2018). In a standard GS framework, individuals for whom both phenotypic and genotypic data are available constitute the training population (TRP). This dataset is used to train a statistical model that predicts the performance of a breeding population (BP) or test population (TSP), which are genotyped but not phenotyped. The effectiveness of GS is highly dependent on the quality of marker data, the size and diversity of the TRP, and the genetic relatedness between TRP and BP (Shamshad and Sharma, 2018). The overarching goal of GS is to combine accurate phenotypic selection with robust genome-wide prediction, thereby enhancing breeding efficiency for complex traits.

In recent years, the growing availability of genomic data has led to extensive research on the statistical and machine learning methods best suited for GS. Traditional linear mixed models, such as genomic best linear unbiased predictor (GBLUP) and Bayesian models (including BayesA, BayesB, BayesC, and Bayesian Lasso), have been widely applied in the prediction of additive genetic effects. However, the increasing complexity of trait architectures and the recognition of non-linear interactions have led to the adoption of machine learning approaches such as random forests (RFs), support vector machines (SVMs), gradient boosting machines (GBMs), and deep learning algorithms (Montesinos-López *et al.*, 2018, 2019a, b, 2021a).

These advanced modeling techniques offer the flexibility to handle diverse data structures, variable importance, non-additive genetic effects, and high-dimensional feature spaces. Machine learning tools are particularly advantageous when dealing with large genomic datasets and when traits exhibit complex relationships with their underlying markers. Yet, no single model universally outperforms others across all scenarios, a principle explained by the 'No Free Lunch Theorem' (Wolpert, 1996). Thus, selecting the most suitable prediction model requires empirical evaluation based on the specific breeding context, crop species, and trait architecture.

Despite the substantial progress made, there remain multiple avenues for enhancing the efficiency of GS. These include improving the density and quality of marker coverage, refining the design of training and testing populations, identifying optimal stages within the breeding pipeline for GS application, and continually updating prediction models with new data. Integrating GS with complementary genomic tools such as genome-wide association studies (GWAS) and high-throughput phenotyping can further enrich the selection process and enhance decision making (Crossa *et al.*, 2017; Montesinos-López *et al.*, 2021a). Ultimately, the adoption of genomic selection has revolutionized crop improvement by enabling faster, more accurate, and cost-effective selection. In the face of climatic uncertainties, dwindling resources, and an ever-growing population, GS offers a promising pathway to develop climate-resilient, high-yielding cereal varieties. Its integration into mainstream breeding programs represents a pivotal shift toward data-driven agriculture, ensuring that future food systems are both sustainable and robust.

7.2 Methods for the Collection of Data on Cereal Crops

Data collection on cereal crops involves integrated methodologies including field-based agronomic assessments, high-throughput phenotyping, remote sensing technologies, and molecular techniques such as genotyping and omics-based analyses (Fig. 7.1).

7.2.1 Phenotypic data collection methods

Experimental design

Field trials are usually laid out in replicated block designs to control for field heterogeneity. For late-stage (advanced) trials, breeders often use fully replicated randomized complete block designs (RCBDs) or incomplete-block designs such as alpha-lattice. In contrast, early-stage screening may use unreplicated or partially replicated (p-rep) or augmented designs with checks. For example, one recent wheat trial used a single-replicated augmented block design (with repeated checks) and an alpha-lattice (3 replicates) for preliminary evaluation (Herr and Carter, 2023). These designs ensure balanced comparisons of genotypes while accommodating large numbers of entries or limited seed.

Trait categories

Phenotypic traits span agronomic, morphological, physiological, and quality classes. Agronomic traits include grain yield and yield components (tiller/ear number, biomass, harvest index). Morphological traits include plant height, leaf area, root architecture, flowering time, etc. Physiological traits include photosynthetic rate, chlorophyll content, water-use traits (stomatal conductance, canopy temperature), and stress indices. Quality traits cover grain protein, oil or sugar content, cooking or milling quality, disease resistance ratings, etc. (for example, normalized difference vegetation index (NDVI) is often used as a proxy for chlorophyll or biomass, while direct sensors measure chlorophyll content or canopy temperature). Modern phenotyping emphasizes high-precision measurements of these traits, often non-destructively.

High-throughput phenotyping (HTP)

Advances in imaging and sensing enable rapid trait measurement on many plots.

GROUND SENSORS. Portable spectrometers (e.g. the CROPSCAN MSR16R) sample canopy reflectance over multiple bands (e.g. 430–970 nm across 16 broad bands) (Herr and Carter, 2023) to compute vegetation indices. Chlorophyll meters (SPAD) or handheld multispectral devices can measure leaf greenness or specific bands.

AERIAL PLATFORMS. Drones (UAVs) carry RGB, multispectral, hyperspectral, or thermal cameras to capture canopy images at the flight scale. These images are processed into indices

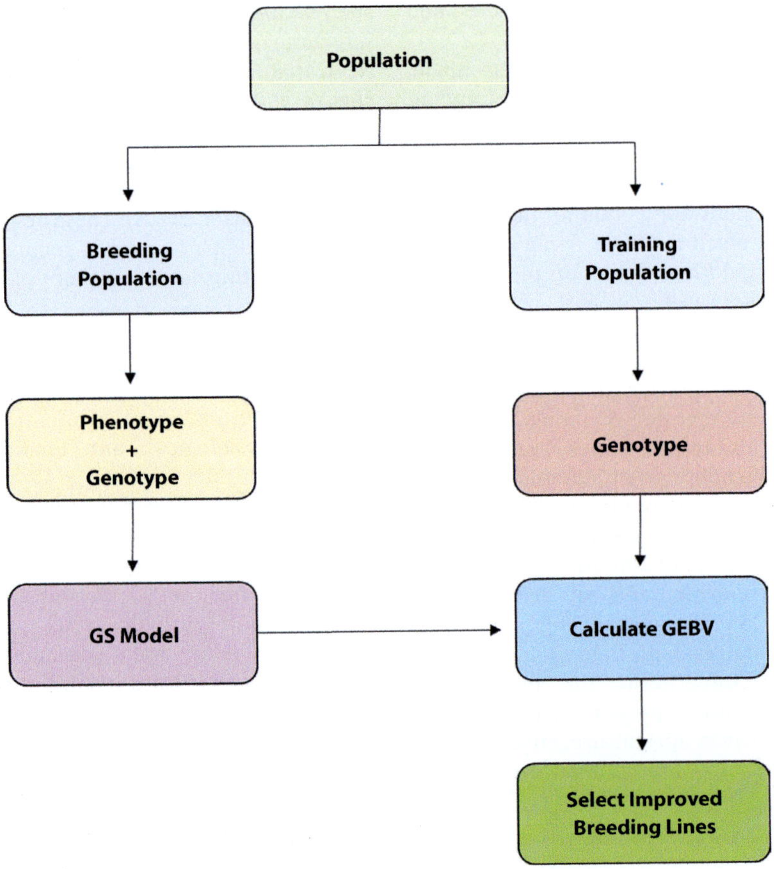

Fig. 7.1. Overview of the genomic selection process (Author's own image).

such as NDVI, NDRE (red-edge index), or water indices (e.g. NWI). NDVI in particular is widely used to estimate plant health or stress via the contrast of red vs near-infrared reflectance (Rouse *et al.*, 1973). For example, in wheat trials, NDVI and related indices have been shown to correlate with grain yield and chlorophyll content. Thermal infrared cameras (or thermal drones) measure canopy temperature as a proxy for water stress. Imaging LiDAR or structured-light scanners (e.g. PlantEye F600) can capture 3D canopy structure. Collectively, these HTP tools allow non-destructive, repeatable sampling of physiological and morphological traits at plot or canopy scales (Tayade *et al.*, 2022; Herr and Carter, 2023).

DATA CAPTURE TOOLS. Traditionally recorded on paper, phenotypic data are now collected via handheld devices. Mobile apps on tablets or smartphones (e.g. FieldBook, KDSmart, PhenoApp and newer tools like GridScore) provide guided, plot-by-plot entry of trait values. These apps can scan QR/barcodes for plot IDs, use GPS for geo-referencing, and accept multiple data types (numeric, categorical, photos, audio). For example, FieldBook is an Android app that steps the user through each plot, showing only one data field at a time to reduce errors (Raubach *et al.*, 2022). Such systems also allow quick syncing of data and reduce transcription errors. Open-source frameworks (e.g. BreedBase,

BrAPI-based apps) often integrate with databases to immediately link field data to genotype IDs.

7.2.2 Genotypic data collection methods

Tissue sampling and DNA extraction

Leaf tissue is typically sampled from each plot or genotype at a uniform growth stage. Young leaf samples may be frozen or desiccated in silica gel for transport. DNA is then extracted using CTAB-based protocols or commercial kits. Standard laboratory protocols (e.g. CTAB extraction) produce high-quality genomic DNA for genotyping (Kuo *et al.*, 2022). Quality control (e.g. gel electrophoresis or spectrophotometry) ensures that DNA is high molecular weight and free of contaminants before genotyping.

Genotyping platforms

Modern breeding uses high-density markers.

- **SNP arrays:** Many crops have custom SNP chip arrays (Illumina Infinium or Affymetrix Axiom) with thousands to millions of pre-selected SNPs. Arrays provide reproducible genotype calls at fixed loci at moderate cost per sample.
- **Genotyping-by-sequencing (GBS):** This reduced-representation method uses restriction enzymes to sample genomic fragments, then barcoded sequencing. It simultaneously discovers and scores SNPs without prior SNP panel design. For example, Elshire *et al.* (2011) showed that GBS can capture 25,000–200,000 markers in maize and barley by sequencing enzyme-digested fragments.
- **Whole-genome sequencing (WGS):** For some programs, low-pass or skim WGS is used to call genome-wide variants when costs allow.
- **Others:** Amplicon or targeted sequencing (e.g. exome capture), and KASP or simple sequence repeat (SSR) assays can be used for specific loci or smaller marker sets. The choice depends on cost, population size, and species genomic resources (Pavan *et al.*, 2020). In practice, breeders often outsource genotyping; they submit DNA samples and receive genotype files (e.g. variant call file (VCF) or matrix).

Genotypic data QC

Raw genotype calls are filtered to remove poor-quality data. Common filters include: removing SNPs or samples with high missingness (for example, requiring SNP call rate ≥95%) and filtering out rare variants (e.g. minor allele frequency (MAF) <1–5%). These thresholds may be tuned by dataset size. Markers with extreme heterozygosity or evidence of assay failure (e.g. poor clustering on arrays) are also dropped. If necessary, missing calls may be imputed. Tools like PLINK can apply these filters efficiently (Pavan *et al.*, 2020). After QC, the result is a cleaned genotype matrix (individuals × SNPs) ready for analysis.

Integration into genomic prediction models

To perform genomic prediction, the cleaned phenotypic and genotypic data are combined in statistical models. Typically, this follows a stage-wise analysis: phenotypic data from each trial are first analyzed (accounting for design and environment effects) to yield adjusted genotype means or BLUPs. These adjusted phenotypes (one per genotype per environment) become the response variable for genomic models.

7.3 Model Development (Genomic Prediction Model) for Cereal Crops

The development of genomic prediction models for cereal crops involves integrating high-density molecular marker data with phenotypic information using advanced statistical and machine learning algorithms to predict complex traits and enhance breeding efficiency accurately (Table 7.1).

7.3.1 Parametric model

A parametric model is a statistical approach that assumes a specific functional form for the relationship between variables, relying on a finite set

Table 7.1. Comparison table of model development (Author's own table).

Model	Type	Key feature/approach	Advantages/strengths	Limitations	References
RR-BLUP	Parametric	Linear mixed model (ridge regression); equal shrinkage	Simple, robust for additive traits; uses all markers equally	Assumes homogenous marker variances; may underperform if few loci have large effects	(Thavamanikumar *et al.*, 2015)
GBLUP	Parametric	Mixed model with genomic relationship matrix	Incorporates genomewide relatedness; widely used	Similar to RR-BLUP (additive); ignores nonadditive/epistatic effects	(Budhlakoti *et al.*, 2022)
Bayesian (A/B/C, BL, BRR, etc.)	Parametric	Bayesian MCMC with marker-specific priors	Models heterogeneous marker variances; selects large-effect SNPs	Computationally intensive; needs MCMC tuning; convergence issues are possible	(Thavamanikumar *et al.*, 2015; Budhlakoti *et al.*, 2022)
LASSO/Elastic Net	Parametric	Penalized regression (ℓ1, or ℓ1 + ℓ2)	Performs variable selection (LASSO) and handles correlated SNPs (ENet)	Requires penalty tuning; LASSO may ignore correlated markers, EN alleviates this	(Budhlakoti *et al.*, 2022)
Random forest	Non-parametric	Ensemble of decision trees (bagging)	Captures nonlinear/ epistatic effects; measures variable importance	Can overfit if not tuned; less interpretable; many hyperparameters (trees, depth)	(Montesinos López *et al.*, 2022)
Support vector machine (SVM)	Non-parametric	Kernel regression (e.g. RBF kernel)	Captures complex patterns via kernels; effective for high-dimensional SNP data	Kernel and hyperparameters (cost, gamma) must be selected carefully; scales poorly with huge data	(Mota *et al.*, 2024)
RKHS (kernel)	Non-parametric	Kernel-based regression (e.g. Gaussian kernel)	Flexible modeling of additive & epistatic variance; often higher accuracy when epistasis is present	Kernel choice critical; computationally heavy for large marker sets (matrix operations)	(Jiang and Reif, 2015)
Neural networks (ANN/DL)	Non-parametric	Multi-layer feedforward networks	Highly flexible, can model arbitrary nonlinear relationships	Requires large training data; risk of overfitting; 'black box' (low interpretability)	(Montesinos-López *et al.*, 2021b)

of parameters to predict phenotypic traits based on genomic data.

RR-BLUP

This is an R package based on BLUP. It is the most widely used mixed linear model for genomic prediction in animal and plant breeding. This package estimates the marker from the dataset, which is obtained by training the population, and ultimately estimates the GEBV for the selected candidates. The ridge regression (RR)-BLUP model assumes that all markers exhibit the same variance and applies uniform shrinkage to their effects equally to zero. Functionally, RR-BLUP is equal to BLUP, but it uses a realized relationship matrix estimated from the markers. Its simplicity and robustness are advantages, but the equal-variance assumption can be limiting when some loci have much larger effects (Endelman, 2011). In practice, this method is often implemented by using the RR-BLUP package. RR-BLUP is an open-source package available at https://CRAN.R-project.org/package=rrBLUP.

GBLUP

Genomic BLUP (GBLUP) is a mixed model approach equivalent to RR-BLUP under an infinitesimal model. GBLUP replaces the pedigree in BLUP and predicts breeding values directly by using a genomic relationship matrix $G=ZZ^T/p$. In fact, GBLUP predictions using the realized relationship matrix are mathematically equal to marker-RR-BLUP predictions (Zhang *et al.*, 2025). GBLUP treats all markers as contributing equally to genetic variance and typically uses REML to estimate variance components. GBLUP is widely used for its statistical efficiency; However, it shares RR-BLUP's limitation of primarily capturing additive variance and not modelling complex non-additive effects (Budhlakoti *et al.*, 2022).

Bayesian method

Bayesian methods (e.g. BayesA, BayesB, BayesCπ, BL, BRR, etc.) place prior distributions on marker variances and often use Markov chain Monte Carlo (MCMC) estimation. For example, BayesB assumes each marker has its own variance and that only a fraction (π) of markers have large effects, while others are zero. Bayesian models thus allow heterogeneous marker effects (some loci can have much larger influence) which can improve accuracy when trait genetics are oligogenic (Meuwissen *et al.*, 2001; Habier *et al.*, 2011; Thavamanikumar *et al.*, 2015). They can effectively perform variable selection (e.g. Bayesian LASSO, BayesB) at the cost of greater computational complexity (due to MCMC).

LASSO and elastic net

These are penalized regression methods. LASSO uses an ℓ1 penalty to induce sparsity, shrinking many marker coefficients to zero. It can select a subset of important markers, but tends to perform poorly when predictors are highly correlated (as often with linked SNPs) (Ogutu *et al.*, 2012). Elastic Net is the extension of LASSO, combining ℓ1 and ℓ2 penalties, providing a compromise that is more robust to correlated predictors (Friedman *et al.*, 2010). Both methods require tuning penalty strengths, but Elastic Net is especially useful in genomic data to handle linkage disequilibrium among SNPs.

7.3.2 Non-parametric models

Non-parametric models are data-driven approaches that do not assume a predefined functional form, enabling the capture of complex, non-linear relationships between genomic markers and phenotypic traits in cereal crop prediction.

Random forest (RF)

An ensemble machine learning method based on many decision trees (bagging), RF is highly flexible for genomic data because it automatically captures complex, nonlinear interactions among markers. It handles high-dimensional data effectively without requiring heavy pre-processing and can output variable importance measures. RF often achieves competitive accuracy in genomic prediction by implicitly modeling epistasis and marker interactions. Its drawbacks include the risk of overfitting (mitigated by tree averaging) and that it can be less interpretable than linear models (Montesinos López *et al.*, 2022).

Support vector machine (SVR)

Kernel-based regression (SVR) extends linear models by mapping inputs into high-dimensional feature spaces. In genomic prediction, SVR uses kernels (linear, polynomial, radial basis function, etc.) to capture nonlinear genotype–phenotype relationships. For example, the radial basis function (RBF) kernel can implicitly model complex epistatic effects. SVR is powerful for high-dimensional SNP data but requires careful choice and tuning of the kernel function. It can be computationally intensive on very large datasets, and its results depend on kernel hyperparameters (Mota *et al.*, 2024)

Reproducing kernel Hilbert space (RKHS)

A semi-parametric approach using kernel regression (often Gaussian kernels) to capture both additive and non-additive (epistatic) effects, RKHS models treat the relationship matrix implicitly via a kernel function, allowing flexible modeling of complex trait architectures. Studies have shown that RKHS regression can outperform linear models when epistasis or non-additive effects are substantial. The main consideration is choosing an appropriate kernel: for example, a Gaussian kernel is popular since it can implicitly encode multi-locus interactions. RKHS models can be computationally demanding for very large marker sets, due to the kernel matrix computations (Jiang and Reif, 2015).

Neural networks (ANN/deep learning)

These are highly flexible, non-parametric models composed of interconnected 'neurons' in layers. Multi-layer (deep) neural networks can, in principle, approximate any complex genotype–phenotype function by learning hierarchical features. They are especially useful if there are hidden nonlinear patterns or interactions in the data that are not pre-specified. However, neural networks typically require very large training datasets to avoid overfitting, and their black-box nature makes them less interpretable. Empirical studies suggest that deep learning can outperform linear models on traits with complex architectures, but performance can be mixed for small datasets (Montesinos-López *et al.*, 2021b).

7.4 Application on Major Cereal Crops

Genomic selection (GS) has revolutionized cereal crop improvement by enabling faster and more accurate prediction of complex traits compared with conventional breeding. Cereal crops such as wheat, rice, maize, and barley contribute nearly 50% of global dietary energy (WHO/FAO, 2003), and their productivity is increasingly threatened by climate change, evolving pathogens, and rising population pressures (Reynolds and Ortiz, 2010; Tester and Langridge, 2010). Developing high-yielding, resource-efficient, and stress-resilient varieties is critical but remains challenging, due to the polygenic nature of key traits, low heritability, and high environmental interactions (Mackay, 2001). GS addresses these limitations by using genome-wide marker data to predict breeding values, thereby expediting selection cycles and reducing dependence on extensive phenotyping (Bernardo and Yu, 2007; Lorenz *et al.*, 2011). GS has shown promising results in improving grain yield and its related traits, such as thousand-grain weight, panicle number, and filled grains, across various cereals, with prediction accuracies ranging from 0.28 to 0.78 (Yabe *et al.*, 2018; Rio *et al.*, 2019). Training population type and size, as well as statistical models used, significantly influence prediction accuracy, with recombinant inbred lines and doubled haploids outperforming natural populations (Fristche-Neto *et al.*, 2018; Liu *et al.*, 2018). Moreover, GS has enhanced hybrid prediction (VanRaden, 2008; Crossa *et al.*, 2017), and platforms have evolved from DArT to SNP-based systems for greater resolution and reliability (Crossa *et al.*, 2010; Poland *et al.*, 2012). For biotic stress resistance, GS has enabled the identification of resistant genotypes in wheat (e.g. rusts, powdery mildew, Fusarium head blight) with prediction accuracies up to 0.85 (Rutkoski *et al.*, 2012; Juliana *et al.*, 2017), and has been applied in rice for blast resistance and in maize for ear rot and Striga tolerance (Huang *et al.*, 2019). Abiotic stress tolerance, especially for drought and heat, has also benefited from GS with documented genetic gains in maize and rice using rapid-cycle schemes (Beyene *et al.*, 2015; Das *et al.*, 2020), and improved prediction accuracy via multi-trait and multi-environment models (Sun *et al.*, 2017; Sukumaran *et al.*, 2018). Additionally, GS is increasingly used for enhancing quality traits. In wheat, GS has improved protein content, dough properties, and milling quality, often with higher accuracy in multi-family populations (Heffner *et al.*, 2011; Michel *et al.*, 2019). Multi-trait GS and Bayesian models such as BRR have facilitated concurrent selection for yield and nutritional quality, achieving prediction accuracies up to 0.55 for micronutrient content (Joukhadar *et al.*, 2021). Similarly, rice and barley have shown accuracies ranging from 0.35 to 0.8 for grain dimensions and malting quality, respectively (Onogi *et al.*, 2015; Schmidt *et al.*, 2016). Collectively, GS offers an integrated and scalable approach to accelerate genetic gain, enhance stress resilience, and improve grain quality in cereals, making it an indispensable component of modern crop breeding programs. Table 7.2 compiles key genomic selection studies conducted in major cereal crops. It summarizes the targeted traits, prediction models used, genotyping strategies, and reported accuracies, providing a comparative perspective on GS implementation across diverse genetic and environmental contexts (Fig. 7.2).

7.5 Estimation of Genomic Prediction

Estimation of genomic prediction involves calculating the genomic estimated breeding values (GEBVs) using statistical models, machine learning that integrates genome-wide marker information with phenotypic data to predict the genetic potential of individuals (Table 7.3).

7.5.1 Statistical method

Genomic selection relies on accurately identifying and selecting individuals with desirable traits using genome-wide marker information. The process traditionally begins with a linear modeling approach, commonly known as the ordinary least squares (OLS) regression. The linear model is represented as:

$$Y = 1_n\mu + X\beta + \epsilon$$

where Y is an $n{\times}1$ vector of phenotypic observations, μ denotes the overall mean, β is a $p \times 1$ vector representing marker effects, and ε is the residual error term following a normal distribution, $\varepsilon \sim N(0, \sigma^2_e)$. The design matrix X of dimensions $n \times p$ contains genotype data, with rows corresponding to individuals and columns to molecular markers. A critical limitation of OLS in the context of GS arises from the fact that the number of predictors (markers, p) often vastly exceeds the number of observations (individuals, n), creating the classic 'large p, small n' scenario. This overparameterization hampers the stability and reliability of parameter estimation, especially in high-dimensional genomic datasets. To circumvent this, a common strategy involves using a subset of statistically significant markers. Meuwissen *et al.* (2001) proposed a modified least-squares method where each marker is analyzed independently using the model:

$$Y = X_j\beta_j + \epsilon$$

Here, X_j represents the jth marker column from the design matrix, and β_j is its corresponding genetic effect. The significance of each marker is assessed using a log-likelihood approach, and only those with substantial effects are retained for breeding value prediction.

However, this selection strategy might exclude informative markers, potentially compromising prediction accuracy. A more robust solution to the overparameterization issue is ridge regression (RR), a penalized regression technique that addresses both multi-collinearity and dimensionality by applying an ℓ^2 penalty to the model. Ridge regression shrinks the coefficients of correlated markers uniformly, thereby reducing variance in estimation and enhancing model stability (Meuwissen *et al.*, 2001). Another penalized method, the least absolute shrinkage and selection operator (LASSO), employs an ℓ^1 penalty, enabling it to produce sparse solutions by setting some coefficients exactly to zero (Tibshirani, 1996; Usai *et al.*, 2009). Despite its utility, LASSO performs sub-optimally in the presence of highly

Table 7.2. Genomic selection studies in cereals (Author's own table).

Crop	Trait(s)	Pop. Size	Marker Density/ Genotyping	Validation Strategy	Models Used	Trait Model Type	Accuracy	Country	Trait Category	Trait Heritability (Est.)	Reference
Oryza sativa (Rice)	Arsenic content	228	22,370 SNPs	CV & IV	GBLUP, Bayes A, RKHS	Single-trait	0.43–0.48	France	Abiotic stress	Moderate	Frouin *et al.*, 2019
Oryza sativa (Rice)	Days to heading	112	408,372 SNPs	CV & IV	GBLUP	Single and multi-trait	0.93–0.98	Japan	Phenology	High	Jarquin *et al.*, 2020
Oryza sativa (Rice)	Drought tolerance	280	215,000 SNPs	CV	GBLUP, RKHS	Single and multi-env.	0.22–0.80	France	Abiotic stress	Moderate–High	Bhandari *et al.*, 2019
Oryza sativa (Rice)	Grain weight	128	42,508 SNPs	CV	GBLUP, PLS	Single-trait	0.28–0.53	Japan	Yield component	Moderate	Yabe *et al.*, 2018
Oryza sativa (Rice)	Grain yield, plant height, flowering time	369	73,147 SNPs	CV	RR-BLUP	Single-trait	0.31, 0.34, 0.63	Philippines	Grain yield	Moderate	Spindel and Iwata, 2018
Oryza sativa (Rice)	Protein content, seed width	413	36,901 SNPs	CV	RR-BLUP	Single-trait	0.44, 0.84	China	Nutrition trait	Moderate–High	Guo *et al.*, 2014
Oryza sativa (Rice)	Grain yield-related traits	278	1,619 SNPs	Not specified	GBLUP, LASSO, SSVS	Single-trait	0.09–0.69	China	Grain yield	Not specified	Xu *et al.*, 2014
Oryza sativa (Rice)	Grain & morphological traits	575	3,299,150 SNPs	Not specified	GBLUP, MV-ADV	Single-trait	0.47–0.88	China	Morphological	Not specified	
Oryza sativa (Rice)	Grain yield (GY), Panicle weight	343	~7,650 DArT markers	540-fold CV	Bayesian LASSO	Single-trait	0.309, 0.327	France	Yield traits	Not reported	Grenier *et al.*, 2015
Oryza sativa (Rice)	Grain yield (GY)	363	73,147 GBS SNPs	5-fold CV	RR-BLUP	Single-trait	0.31	Philippines	Yield traits	Not specified	Spindel *et al.*, 2015
Oryza sativa (Rice)	Field grain weight	128	Illumina HiSeq 2000 SNPs	CV	GBLUP	Single-trait	0.28–0.30	Japan	Grain weight	Not specified	Yabe *et al.*, 2018
Oryza sativa (Rice)	Grain weight variance	128	Illumina HiSeq 4000/HiSeq X	CV	GBLUP	Single-trait	0.53	Japan	Variability	Not specified	Yabe *et al.*, 2018

Oryza sativa (Rice)	GY (~0.5), TGW (~0.28)	>500	GBS	CV	GBLUP, SVM, LASSO, PLS	Single-trait	~0.28–0.50	USA	Yield traits	Not specified	Xu *et al.*, 2018
Oryza sativa (Rice)	GY (0.54), Grain length (0.92)	Not specified	Illumina HiSeq 2000	CV	GBLUP	Single-trait	0.54, 0.92	China	Yield & grain traits	Not specified	Cui *et al.*, 2020
Oryza sativa (Rice)	Panicle weight, Nitrogen balance index	Not specified	GBS SNPs	CV	GBLUP, RKHS, Bayes B	Single-trait	0.21–0.30	Tunisia	Physiological traits	Not specified	Ben Hassen *et al.*, 2018
Oryza sativa (Rice)	GY (0.39), TGW (0.88)	Not specified	SNP array	CV	RR-BLUP, GBLUP	Single-trait	0.39, 0.88	China	Yield traits	Not specified	Wang *et al.*, 2018
Oryza sativa (Rice)	Blast resistance	Germplasm panel	GBS	CV	RR-BLUP, GBLUP	Single-trait	0.17–0.73	China	Biotic stress	Not specified	Huang *et al.*, 2019
Oryza sativa (Rice)	Drought tolerance	Germplasm panel	962K SNPs	CV / multi-env	GBLUP, RKHS	Single/multi-env	0.226–0.809	France	Abiotic stress	Not specified	Bhandari *et al.*, 2019
Triticum aestivum (Wheat)	Grain yield	1,325	9,290 SNPs	CV & IV	GBLUP	Single and multi-trait	0.18–0.31	Denmark	Grain yield	Low to moderate	Tsai *et al.*, 2020
Triticum aestivum (Wheat)	Rust resistance	1,744	15,000 SNPs	CV	GBLUP	Single-trait	0.16–0.50	Germany	Biotic stress	Moderate	Beukert *et al.*, 2020
Triticum aestivum (Wheat)	Grain yield & protein traits	282	7,426 SNPs	CV & IV	RKHS, RF, RRBLUP	Single-trait	0.17–0.70	USA	Yield & Quality	Moderate	Hu *et al.*, 2019
Hordeum vulgare (Barley)	DON resistance	–	Illumina GoldenGate	CV	RRBLUP	Single-trait	0.72	–	Biotic stress	High	Sallam *et al.*, 2015
Hordeum vulgare (Barley)	FHB resistance	–	Illumina GoldenGate	CV	RRBLUP	Single-trait	0.74	–	Biotic stress	High	Sallam *et al.*, 2015
Hordeum vulgare (Barley)	Thousand kernel weight (TKW)	–	GBS	CV	GBLUP, RKHS	Single-trait	0.67	–	Yield component	Moderate to high	Abed *et al.*, 2018

Continued

Table 7.2. Continued

Crop	Trait(s)	Pop. Size	Marker Density/ Genotyping	Validation Strategy	Models Used	Trait Model Type	Accuracy	Country	Trait Category	Trait Heritability (Est.)	Reference
Sorghum bicolor (Sorghum)	Yield, plant height, stay-green, flowering time	2,645	4,781 SNPs	CV	GBLUP	Multi-trait	0.36	–	Multi-trait	Moderate	Ortiz *et al.*, 2019
Sorghum bicolor (Sorghum)	Biofuel traits	369	61,976 SNPs	CV	Bayesian model	Single-trait	0.59	–	Bioenergy	Moderate to high	Habyarimana *et al.*, 2020
Secale cereale (Rye)	Grain yield, protein content	201	394 SNPs	CV	GBLUP	Single-trait	0.23	–	Yield & Quality	Moderate	Schulthess *et al.*, 2016
Secale cereale (Rye)	Plant height, heading stage, disease	465	7,728 SNPs	CV	rrBLUP	Multi-trait	0.72	–	Multi-trait	High	Gaikpa *et al.*, 2020

Fig. 7.2. Recent advances highlight the diverse applications of genomic selection in enhancing crop traits (Author's own image).

correlated predictors, which is often the case with single-nucleotide polymorphisms (SNPs) due to linkage disequilibrium (Ogutu *et al.*, 2012). To address this limitation, the elastic net (ENET) model combines the strengths of both ℓ^1 and ℓ^2 penalties, offering robustness against high multicollinearity (Zou and Hastie, 2005; Friedman *et al.*, 2010). While ridge regression assumes equal variance contributions from all markers, this does not reflect the true genetic architecture of complex traits. To accommodate variable marker contributions, Bayesian models have been introduced. These models incorporate prior distributions for marker effects and use posterior inference to estimate model parameters. Key Bayesian methods include BayesA, BayesB, BayesCπ, and BayesDπ, with further developments such as Bayesian ridge regression (BRR) and Bayesian LASSO (Meuwissen *et al.*, 2001; Habier *et al.*, 2011). In parallel, the Best Linear Unbiased Prediction (BLUP) framework, originally proposed by Henderson *et al.* (1959), has been adapted for genomic contexts. Variants such as genomic BLUP (GBLUP), ridge regression BLUP (RR-BLUP), single-step GBLUP (ssGBLUP), and kernel-based extensions like rrGBLUP are

Table 7.3. Statistical tools for genomic selection in cereal crops (Author's own table).

Tool Name (Platform)	Crops Benefited	Features	Statistical Methods	Reference
GMStool (R-based)	Soybean (flowering time prediction)	GWAS-based marker selection, ML/DL model, data preparation, marker selection, prediction	GWAS, ML, DL, Validation	Jeong *et al.*, 2020
SolGS (Linux-based)	Cassava (NEXTGEN)	Trait/trial/custom datasets, downloadable outputs	RR-BLUP, NLME	Endelman, 2011; Pinheiro *et al.*, 2017
RR-BLUP (R-based)	Maize, rice, wheat	Widely used, estimates GEBV, supports GBLUP	Ridge regression, BLUP, GBLUP	Endelman, 2011; Spindel *et al.*, 2015
BWGS (R-based)	Wheat (French breeding)	Imputation, dimension reduction, GEBV estimation, 15 models	Parametric and non-parametric models	Charmet *et al.*, 2020
BGLR (R-based)	Maize, wheat (genomic prediction accuracy)	Bayesian regression supports continuous & categorical traits	BayesA, BayesB, BRR	Gianola, 2013; Pérez and de los Campos, 2014
GSelection (R-based)	Rice, chickpea (multi-environment traits)	Integrates additive and non-additive models	BLUP, LASSO	Majumdar *et al.*, 2019
lme4GS (R-based)	Maize, sugar-beet (yield and drought stress)	Mixed models with user-defined covariance, combine pedigree + genomic	Linear Mixed Models	Vazquez *et al.*, 2010; Ziyatdinov *et al.*, 2018
STGS (R-based)	Rice, wheat (drought and nutrient traits)	Single trait GS, includes RR, BLUP, LASSO, SVM, ANN, RF	RR, BLUP, LASSO, SVM, ANN, RF	Budhlakoti *et al.*, 2019a
MTGS (R-based)	Wheat, sorghum	Multi-trait GS includes MRCE, MLASSO, and KMLASSO	MRCE, MLASSO, KMLASSO	Budhlakoti *et al.*, 2019b

widely applied. Unlike traditional BLUP, which relies on pedigree-based relationship matrices, GBLUP incorporates marker-derived genomic relationships to estimate genomic estimated breeding values (GEBVs) (Meuwissen *et al.*, 2001). These parametric approaches perform effectively for traits governed predominantly by additive genetic variance. However, their predictive performance diminishes when non-additive effects such as epistasis play a major role. To capture complex genetic architectures, Gianola *et al.* (2006) introduced non-parametric and semi-parametric techniques. This led to the application of statistical learning methods capable of modeling both additive and epistatic interactions (Xu, 2007; Cai *et al.*, 2011; Legarra and Reverter, 2018). Several non-parametric approaches have since been explored in GS, including the Nadaraya–Watson (NW) kernel estimator, reproducing kernel Hilbert spaces (RKHS), support vector machines (SVM) (Maenhout *et al.*, 2007; Long *et al.*, 2011), artificial neural networks (ANNs) (Gianola *et al.*, 2011), and random forests (RFs) (Holliday *et al.*, 2012). Among these, SVM, ANN, and RF are firmly rooted in the machine learning paradigm and provide flexible tools for genomic prediction under complex trait architectures. Historically, most of these methods have been employed under the assumption of single-trait genomic selection (STGS), wherein each trait is predicted independently. However, STGS can be sub-optimal when dealing with traits under pleiotropic control, where a single gene influences multiple traits. In such cases, single-trait models fail to utilize shared genetic information across traits, potentially leading to reduced prediction accuracy, especially for low-heritability traits. To overcome these constraints, multi-trait genomic selection (MTGS) has emerged as a more powerful approach. MTGS allows for simultaneous modeling of multiple correlated traits, thereby improving the accuracy of GEBV predictions. Methodologies within this framework include multivariate mixed models (Jia and Jannink, 2012; Klápště *et al.*, 2020), Bayesian multi-trait models (Jia and Jannink, 2012; Cheng *et al.*, 2018), multivariate regression with covariance estimation (MRCE) (Rothman *et al.*, 2010), and conditional Gaussian graphical models (cGGM) (Chiquet *et al.*, 2017). Jia and Jannink (2012) compared three multivariate linear models (GBLUP, BayesA, and BayesCπ) to their single-trait counterparts, illustrating the superiority of MTGS for pleiotropic traits. A comprehensive evaluation of both STGS and MTGS frameworks has also been provided by Budhlakoti *et al.* (2019a, b). Given the diversity of statistical methodologies described above, ranging from penalized regression and Bayesian inference to non-parametric and multi-trait models, several computational tools and software packages have been developed to facilitate the implementation of genomic selection in crop breeding. These tools integrate advanced algorithms with high-throughput genotypic and phenotypic data, allowing breeders to compute genomic estimated breeding values efficiently. Table 7.3 summarizes widely adopted open-source tools that support GS in crops, along with their statistical foundations, computational platforms, and validated crop-specific applications.

Statistical tools for genomic selection in cereal crops are shown in Table 7.3.

7.5.2 Machine learning

Over the past century, the application of statistical methods has played a transformative role in crop yield improvements. With the pioneering work of R.A. Fisher (Fisher, 1934, 1935) introducing techniques such as analysis of variance (ANOVA) and randomized control trials, plant breeding evolved into a quantitative science. These foundational methodologies led to the widespread adoption of models like ridge regression (RR), best linear unbiased prediction (BLUP), and genomic BLUP (GBLUP), especially within the genomic selection framework (Meuwissen *et al.*, 2001). However, the pace of crop yield increases remains inadequate in the face of mounting global food demands and the disruptive impacts of climate change.

Recent advancements in machine learning (ML) and deep learning (DL) offer powerful tools to address these challenges. ML approaches enable the modeling of complex, nonlinear relationships between genotype, phenotype, and the environment. Beyond statistical interference, ML excels in prediction, pattern recognition, and integration of high-dimensional datasets. As the genomic data has grown in volume and complexity, there has been a need for more adaptable and powerful methods to make sense of it. While classical models like RR-BLUP and GBLUP have laid the foundation for genomic selection, they often struggle to capture the intricate relationship between genotype, phenotype, and the environment, especially when these relationships are nonlinear or involve interactions among many variables. This is where machine learning has begun to play a central role. By allowing models to learn patterns directly from the data without relying on strict assumptions about genetic structure, machine learning brings a new level of flexibility and precision to genomic prediction. Among the most widely used machine learning techniques in genomic prediction are tree-based ensemble methods, which are especially valued for their ability to model nonlinear relationships and variable interactions. Random forest (RF) works by constructing a multitude of decision trees using random subsets of data and aggregating their outputs, making it robust to noise and overfitting.

In contrast, XG-Boost (Extreme Gradient Boosting) builds trees sequentially, optimizing for errors at each step to improve overall predictive accuracy. XG-Boost has demonstrated superior performance over both deep learning and traditional linear methods in several studies, particularly in soybean yield prediction (Montesinos-López *et al.*, 2019a, b). Another powerful class of algorithms, support vector machines (SVMs), transforms input features into higher-dimensional spaces using a kernel function, allowing the separation of complex nonlinear patterns. While effective for both classification and regression tasks, SVMs can be computationally intensive, especially when applied to large genomic datasets. In recent years, deep learning approaches, particularly convolutional neural networks (CNNs) and recurrent neural networks (RNNs), have gained attention for their ability to learn hierarchical and temporal features from raw genotype data. For example, Deep GS, a CNN-based framework, was among the earliest DL applications in wheat genomic selection and outperformed RR-BLUP for key yield traits (Ma *et al.*, 2018). RNNs, on the other hand, have shown promise in sequential data across multiple growing seasons. To further enhance prediction accuracy, ensemble and hybrid models that integrate outputs from multiple algorithms have emerged as a promising direction. Azodi *et al.* (2019) demonstrated that such integrated models combining DL with XG-Boost provided consistently higher prediction accuracies across a range of phenotypic traits, highlighting the strength of blending complementary methods.

The growing application of machine learning across various crop species has yielded notable successes, particularly in genomic prediction and trait improvement. In wheat, deep learning models such as Deep GS have demonstrated clear advantages over traditional statistical models like RR-BLUP by capturing high-dimensional hierarchical patterns in genomic data, resulting in improved prediction of traits such as grain yield and heading data. Similarly, in soybean, XG-Boost outperformed deep learning models across 13 out of 14 traits, including oil and protein content, while CNN-based models identified key SNPs contributing to phenotypic variation (Montesinos-López *et al.*, 2019a, b). In maize, combining deep learning with genotype-by-environment models improved prediction for nitrogen-use efficiency, with candidate genes validated through RNA-seq and XG-Boost (Li *et al.*, 2022).

Orphan crops have also benefited from these advances, with AgroNT enabling genome annotation in pigeonpea and quinoa, while in chia (*Salvia hispanica*) the use of large language models to transfer orthologous information from well-characterized species enabled the successful identification of nearly 94% of genes via BUSCO analysis. Despite these advances, several limitations continue to constrain the full potential of machine learning in plant breeding. The most pressing issue is the lack of high-quality, annotated genomic datasets for under-researched species, which restricts model training and validation. Deep learning models, while powerful, are prone to overfitting when applied to small datasets, and interpretability remains a challenge necessitating advanced techniques like saliency maps or SHAP values to understand feature contributions. Moreover, there remains a critical training gap, as many plant breeders lack the computational background to implement and interpret these models effectively. This highlights the need for interdisciplinary education programs, as previously emphasized by Cobb *et al.* (2019), that bridge bioinformatics, machine learning, and plant biology.

Looking forward, several emerging strategies hold potential to address these challenges and expand the impact of ML in genomic prediction. Federated learning offers a framework for collaborative model training across institutions without requiring sensitive genotype data to be centralized. Incorporating biological priors such as known QTLs and gene networks directly into ML models could enhance interpretability and biological relevance. With climate resilience becoming a key breeding target, machine learning is increasingly being used to identify genomic regions linked to traits like drought tolerance, heat stress response, and nutrient use efficiency. In parallel, fine-tuning foundational LLMs such as AgroNT and DNABERT for low-resource crops may unlock new opportunities for gene discovery and trait annotations where experimental data are scarce. Ultimately, while machine learning does not replace traditional expertise in plant breeding, it enriches the toolkit available to researchers, offering unprecedented precision in trait prediction and enabling the accelerated development of high-performing, resilient crop varieties suited to a changing climate.

7.6 Factors Affecting the Accuracy of Genomic Selection Models

The accuracy of genomic selection models is influenced by factors such as the size and diversity of the training population, marker density, trait heritability, linkage disequilibrium between markers and causal loci, and the statistical method employed for prediction.

7.6.1 Marker density

Marker density is one of the most fundamental parameters affecting the accuracy of genomic prediction. In general, higher marker densities improve prediction accuracy across a wide range of models, including BLUP, LASSO, and more recent machine learning or deep learning-based approaches. This is primarily because densely spaced markers are more likely to capture linkage disequilibrium (LD) with causal loci. However, in Bayesian models such as BayesA, BayesB, BayesCπ, and BayesDπ, excessive marker density can slow down model convergence due to the need for extensive Markov chain Monte Carlo (MCMC) iterations. Interestingly, even low-density marker sets ranging from a few hundred to a few thousand SNPs can yield high accuracy if strong LD exists among markers and trait-controlling loci. However, this is often trait-specific and varies with the underlying genetic architecture. Moreover, using extremely high-density marker platforms may not always be economically viable, especially in large-scale breeding programs. To make a balance between cost and performance, it is generally advisable to use a moderate number of markers, ideally around 2000 SNPs, without significantly compromising prediction accuracy. Genotyping costs can also be minimized by increasing the level of multiplexing in GBS platforms. For instance, in barley, using 96-plex GBS genotyping revealed over three times cheaper than using 9K or 50K SNP arrays, with no loss in accuracy. Additionally, the genomic positioning of SNPs also plays a role; intergenic SNPs have shown slightly better performance in capturing haplotypic diversity than genic SNPs, as they often span regulatory regions such as enhancers and promoters. Filtering markers based on quality metrics like minor allele frequency (MAF > 0.1) is also critical for ensuring stable and accurate predictions.

7.6.2 Population size

The size of the training population is another key determinant of genomic prediction accuracy. A small training set often results in reduced accuracy, because the estimation of marker effects becomes less reliable. The theoretical optimum training population size has been suggested to be $2 \times Ne \times L$, where Ne is the effective population size and L is the genome size in Morgans. This size is expected to deliver high prediction accuracy, approaching 0.9. If the training population is reduced to $1 \times Ne \times L$, the expected accuracy declines to around 0.7, particularly when the training and breeding populations are unrelated or separated by several generations. However, in most practical breeding programs, training and breeding populations share considerable genetic similarity. This relatedness often compensates for smaller training sizes, allowing breeders to achieve satisfactory accuracy levels with fewer individuals. Thus, while larger populations are generally preferred, a context-dependent approach, factoring in genetic relatedness, can optimize resource allocation and maintain prediction efficiency.

7.6.3 Trait heritability

Heritability of the target trait significantly influences the success of genomic selection. Traits with high heritability ($h^2 > 0.4$) typically show better genomic prediction accuracies, as the phenotypic variation is largely attributed to genetic components that the models can efficiently capture. In contrast, low-heritability traits ($h^2 < 0.4$) are more challenging to predict, due to greater environmental noise and reduced signal strength. In such cases, larger training populations are often required to achieve prediction accuracies comparable to those seen with moderate to high heritability traits. However, this presents cost-related constraints, particularly in developing-country contexts where resources are limited. Nevertheless, studies have demonstrated that genomic prediction models such as GBLUP, RR-BLUP, Bayesian approaches (Bayes A, B, Cπ, Dπ), and RKHS can still perform reasonably well for complex, low-heritability traits, especially when properly tuned. These models remain more robust compared with many nonlinear or parametric alternatives under such scenarios.

7.6.4 Trait architecture

The genetic architecture of a trait, defined by the number, effect size, and interactions of the underlying causal loci, also affects genomic prediction accuracy. Polygenic traits, which are controlled by a large number of small-effect loci, are generally more amenable to genomic selection, as GS models are specifically designed to handle such complex patterns. In contrast, oligogenic traits governed by a few large-effect loci may be better suited for traditional QTL mapping or marker-assisted selection. Additionally, the presence of non-additive genetic effects, gene-by-environment interactions, or pleiotropy may reduce the predictive power of standard GS models, necessitating the use of more advanced or customized frameworks.

7.6.5 Multi-trait models and missing observations

In breeding programs, missing phenotypic data and low trait heritability can often occur simultaneously, further complicating the prediction process. Multi-trait genomic selection (MTGS) models offer a viable solution by leveraging genetic correlations between traits. If a low-heritability trait is genetically correlated with one or more high-heritability traits, MTGS models can borrow statistical strength from the latter to improve the estimation of genomic estimated breeding values (GEBVs) for the former. This approach not only mitigates the effects of missing data but also enhances the precision of predictions across multiple traits, making it particularly valuable in multi-trait selection strategies in both plant and animal breeding.

7.7 Advantages of the Genomic Prediction Model

Genomic selection (GS) has revolutionized plant breeding by enabling the early and accurate prediction of complex traits through genome-wide marker information. Unlike conventional selection methods or marker-assisted selection (MAS), which typically focus on a limited number of markers associated with major-effect loci, GS captures the effects of both large and small-effect quantitative trait loci (QTLs), thereby improving the predictability of complex traits. Traits such as grain yield, drought tolerance, and disease resistance are often governed by polygenic architecture, making them difficult to dissect through traditional breeding or QTL mapping. GS addresses this challenge by simultaneously estimating the effects of thousands of markers across the genome, significantly enhancing selection accuracy. For instance, a study by Crossa *et al.* (2017) demonstrated that genomic prediction outperformed phenotypic selection for traits with low to moderate heritability in wheat, thus offering a more efficient tool for improving quantitative traits. One of the most pressing needs in modern agriculture is to develop climate-resilient crop varieties. Climate change introduces a spectrum of environmental uncertainties, including increased temperature fluctuations, drought, salinity, and emerging pest pressures. Genomic selection facilitates the incorporation of genotype-by-environment (G × E) interactions into predictive models, allowing breeders to develop cultivars that can maintain stable performance across diverse agro-climatic zones. A study by Jarquín *et al.* (2014) showed that incorporating G × E interactions into GS models enhanced prediction accuracy in maize under multiple stress environments. These models are particularly valuable for selecting genotypes suitable for variable conditions where phenotyping is often unreliable or resource-intensive. Thus, GS plays a pivotal role in climate-smart breeding programs aimed at addressing both biotic and abiotic stresses.

The economic viability of genomic selection has improved considerably with the advent of next-generation sequencing (NGS) technologies, particularly genotyping-by-sequencing (GBS). These advancements have reduced per-sample genotyping costs while simultaneously increasing marker density. As a result, GS is now feasible not only for globally important cereals like wheat, maize, and rice but also for underutilized crops and horticultural species. This cost-effectiveness, combined with higher selection intensity and reduced breeding cycle time, makes GS a valuable tool for both public and private breeding programs, especially in resource-limited settings.

Genomic selection also contributes to accelerated genetic gains per unit time, a key metric in evaluating the effectiveness of a breeding program. This acceleration results from improvements in all three components of the breeder's equation: selection intensity, selection accuracy, and cycle time Hayes *et al.* (2009). By enabling early-stage selection of candidates based on their genomic estimated breeding values (GEBVs), GS shortens the time required for selection cycles. Technow *et al.* (2021) reported that implementing rapid-cycle GS in tropical maize breeding increased grain yield under drought conditions and allowed for more frequent recombination, leading to an annual genetic gain of 0.135 t/ha/year. This efficiency is particularly valuable in the context of climate adaptation, where the timely delivery of improved cultivars is essential. Several crop-specific case studies emphasize the successful application of GS in developing climate-resilient varieties. In wheat, genomic selection has enabled the identification of lines with superior drought tolerance that maintain yield stability under water-limited conditions (Juliana *et al.*, 2019). In maize, multi-environment GS strategies have facilitated the improvement of grain yield under drought and waterlogging conditions that are increasingly frequent due to climate variability (Cairns *et al.*, 2021). Rice, another global staple, has benefited from GS-based prediction models for heat tolerance, where the inclusion of environmental covariates improved the selection of genotypes adapted to high-temperature conditions (Sandhu *et al.*, 2021). Barley has been improved for both drought and salinity tolerance using GS, supporting its cultivation in arid and marginal areas. Sorghum and pearl millet, key crops in semi-arid regions, have also been enhanced through GS for resilience traits such as high biomass yield, heat tolerance, and drought adaptation (Varshney *et al.*, 2021).

Thus, genomic selection presents a transformative advancement in plant breeding, offering multiple advantages, including efficient improvement of complex traits, enhanced climate resilience, cost-effectiveness, and accelerated genetic gains. Its application across a diverse range of crops underlines its versatility and effectiveness as a tool for future-ready agriculture. As breeding programs increasingly integrate high-throughput phenotyping and genomic data, GS is expected to play a central role in developing crop varieties that can sustainably meet the challenges of climate change and global food security.

7.8 Conclusion

Genomic prediction has transformed cereal crop breeding by enabling faster, more accurate selection for complex traits, including yield, stress tolerance, and quality. This chapter has contextualized a range of GP models – from traditional linear methods like RR-BLUP and GBLUP to advanced machine learning approaches such as random forests, SVMs, and deep learning – highlighting their strengths, limitations, and applications in crops like rice, wheat, maize, barley, and sorghum. Innovations in high-throughput phenotyping, affordable genotyping (e.g. GBS), and open-source tools have expanded GP's accessibility and effectiveness. However, challenges such as model interpretability, overfitting, and integration of complex datasets remain. Future efforts should focus on incorporating biological priors, federated learning, and climate-resilient frameworks to further strengthen GP's role in sustainable crop improvement.

Acknowledgements

The authors gratefully acknowledge the support of Jharkhand Council on Science, Technology and Innovation, Govt of Jharkhand, India, and Plant Physiology and Molecular Biology Laboratory, University Department of Botany, Ranchi University, Ranchi, Jharkhand, India, for providing the necessary facilities and resources to prepare this paper. BKM is supported by a Long-Term Research Project (Memo No. Estt.-02/2016 (Part-11) - 210 / 29.05.2025) granted by Jharkhand Council on Science, Technology and Innovation, Govt. of Jharkhand, India.

References

Abed, A., Pérez-Rodríguez, P., Crossa, J. and Belzile, F. (2018) When less can be better: How can we make genomic selection more cost-effective and accurate in barley. *Theoretical and Applied Genetics* 131, 1873–1890. DOI: 10.1007/S00122-018-3120-8.

Azodi, C.B., Bolger, E., McCarren, A., Roantree, M., de los Campos, G. *et al.* (2019) Benchmarking parametric and machine learning models for genomic prediction of complex traits. *G3: Genes, Genomes, Genetics* 9, 3691–3702.

Ben Hassen, M., Bartholomé, J., Valè, G., Cao, T.V. and Ahmadi, N. (2018) Genomic prediction accounting for genotype by environment interaction offers an effective framework for breeding simultaneously for adaptation to an abiotic stress and performance under normal cropping conditions in rice. *G3: Genes, Genomes, Genetics* 8(7), 2319–2332.

Bernardo, R. and Yu, J. (2007) Prospects for genome wide selection for quantitative traits in maize. *Crop Science* 47, 1082–1090. DOI: 10.2135/CROPSCI2006.11.0690.

Beukert, U., Thorwarth, P., Zhao, Y., Longin, C.F.H., Serfling, A. *et al.* (2020) Comparing the potential of marker-assisted selection and genomic prediction for improving rust resistance in hybrid wheat. *Frontiers in Plant Science* 11, 1650. DOI: 10.3389/fpls.2020.594113.

Beyene, Y., Semagn, K., Mugo, S., Tarekegne, A., Babu, R. *et al.* (2015) Genetic gains in grain yield through genomic selection in eight bi-parental maize populations under Drought Stress. *Crop Science* 55, 154–163. DOI: 10.2135/CROPSCI2014.07.0460.

Bhandari, A., Bartholomé, J., Cao-Hamadoun, T.V., Kumari, N., Frouin, J. *et al.* (2019) Selection of trait-specific markers and multi-environment models improve genomic predictive ability in rice. *PLoS ONE* 14, 1. DOI: 10.1371/JOURNAL.PONE.0208871.

Budhlakoti, N., Mishra, D.C., Rai, A. and Chaturvedi, K.K. (2019a) Package 'STGS' 1–11. Available at: https://cran.r-project.org/web/packages/STGS/STGS.pdf

Budhlakoti, N., Mishra, D.C. and Rai, A. (2019b) Package 'MTGS 1–6.

Budhlakoti, N., Kushwaha, A.K., Rai, A., Chaturvedi, K.K., Kumar, A. *et al.* (2022) Accelerating the efficiency of molecular genomic selection: A tool for breeding for development of climate-resilient crops. *Frontiers in Genetics* 13, 832153. DOI: 10.3389/fgene.2022.832153.

Cairns, J.E., Prasanna, B.M., Wang, N. and Crossa, J. (2021) Genomic prediction in maize breeding: Current status, challenges, and applications. *Frontiers in Genetics* 12, 657366. DOI: 10.3389/fgene.2021.657366.

Cai, X., Huang, A. and Xu, S. (2011) Fast empirical Bayesian LASSO for multiple quantitative trait locus mapping. *BMC Bioinformatics* 12, 211–213. DOI: 10.1186/1471-2105-12-211/FIGURES/5.

Charmet, G., Tran, L.-G., Auzanneau, J., Rincent, R. and Bouchet, S. (2020) BWGS: AR package for genomic selection and its application to a wheat breeding programme. *PLoS ONE* 15, e0222733. DOI: 10.1371/JOURNAL.PONE.0222733.

Cheng, H., Kizilkaya, K., Zeng, J., Garrick, D. and Fernando, R. (2018) Genomic prediction from multiple-trait bayesian regression methods using mixture priors. *Genetics* 209, 89–103. DOI: 10.1534/GENETICS.118.300650/-/DC1.

Chiquet, J., Mary-Huard, T., Robin, S. and Robin, S. (2017) Structured regularization for conditional Gaussian graphical models. *Statistics and Computing* 27, 789–804. DOI: 10.1007/s11222-016-9654-1.

Cobb, J.N., Juma, R.U., Biswas, P.S., Arbelaez, J.D., Rutkoski, J. *et al.* (2019) Enhancing the rate of genetic gain in public-sector plant breeding programs: Lessons from the breeder's equation. *Theoretical and Applied Genetics* 132(3), 627–645.

Crossa, J., de los Campos, G., Pérez, P., Gianola, D., Burgueño, J. *et al.* (2010) Prediction of genetic values of quantitative traits in plant breeding using pedigree and molecular markers. *Genetics* 186, 713–724. DOI: 10.1534/genetics.110.118521.

Crossa, J., Pérez-Rodríguez, P., Cuevas, J., Montesinos-López, O., Jarquín, D. *et al.* (2017) Genomic selection in plant breeding: Methods, models, and perspectives. *Trends in Plant Science* 22, 961–975. DOI: 10.1016/j.tplants.2017.08.011.

Cui, Z., Dong, H., Zhang, A., Ruan, Y., He, Y. *et al.* (2020) Assessment of the potential for genomic selection to improve husk traits in maize. *G3: Genes, Genomes, Genetics* 10, 3741–3749. DOI: 10.1534/G3.120.401600.

Das, R.R., Vinayan, M.T., Patel, M.B., Phagna, R.K., Singh, S.B. *et al.* (2020) Genetic gains with rapid-cycle genomic selection for combined drought and waterlogging tolerance in tropical maize (*Zea mays* L.). *Plant Genome* 13, 1–15. DOI: 10.1002/tpg2.20035.

Elshire, R.J., Glaubitz, J.C., Sun, Q., Poland, J.A., Kawamoto, K. *et al.* (2011) A robust, simple genotyping-by-sequencing (GBS) approach for high diversity species. *PLoS ONE* 6(5), e19379.

Endelman, J.B. (2011) Ridge regression and other kernels for genomic selection with R Package rrBLUP. *The Plant Genome* 4, 250–255. DOI: 10.3835/PLANTGENOME2011.08.0024.

Fisher, R.A. (1934) *Statistical Methods for Research Workers*, 5th edn. Oliver and Boyd, Edinburgh.

Fisher, R.A. (1935) *The Design of Experiments*. Oliver and Boyd, Edinburgh.

Friedman, J., Hastie, T. and Tibshirani, R. (2010) Regularization paths for generalized linear models via coordinate descent. *Journal of Statistical Software* 33, 1–22. DOI: 10.18637/jss.v033.i01.

Fristche-Neto, R., Akdemir, D. and Jannink, J.-L. (2018) Accuracy of genomic selection to predict maize single-crosses obtained through different mating designs. *Theoretical and Applied Genetics* 131, 1153–1162. DOI: 10.1007/s00122-018-3068-8.

Frouin, J., Labeyrie, A., Boisnard, A., Sacchi, G.A. and Ahmadi, N. (2019) Genomic prediction offers the most effective marker assisted breeding approach for ability to prevent arsenic accumulation in rice grains. *PLoS ONE* 14, e0217516. DOI: 10.1371/journal.pone.0217516.

Furbank, R.T. and Tester, M. (2011) Phenomics–technologies to relieve the phenotyping bottleneck. *Trends in Plant Science* 16(12), 635–644.

Gaikpa, D.S., Koch, S., Fromme, F.J., Siekmann, D., Würschum, T. *et al.* (2020) Genome-wide association mapping and genomic prediction of Fusarium head blight resistance, heading stage and plant height in winter rye (*Secale cereale*). *Plant Breeding* 139, 508–520. DOI: 10.1111/pbr.12810.

García-Ruiz, A., Cole, J.B., VanRaden, P.M., Wiggans, G.R., Ruiz-López, F.J. *et al.* (2016) Changes in genetic selection differentials and generation intervals in US Holstein dairy cattle as a result of genomic selection. *Proceedings of the National Academy of Sciences* 113(28), E3995–E4004.

Gianola, D. (2013) Priors in whole-genome regression: The Bayesian alphabet returns. *Genetics* 194(3), 573–596.

Grenier, C., Cao, T.-V., Ospina, Y., Quintero, C., Châtel, M.H. *et al.* (2015) Accuracy of genomic selection in a rice synthetic population developed for recurrent selection breeding. *PLoS ONE* 10, e0136594. DOI: 10.1371/JOURNAL.PONE.0136594.

Gianola, D., Fernando, R.L. and Stella, A. (2006) Genomic-assisted prediction of genetic value with semiparametric procedures. *Genetics* 173, 1761–1776. DOI: 10.1534/GENETICS.105.049510.

Gianola, D., Okut, H., Weigel, K.A. and Rosa, G.J. (2011) Predicting complex quantitative traits with Bayesian neural networks: A Case Study with Jersey cows and wheat. *BMC Genetics* 12, 87. DOI: 10.1186/1471-2156-12-87.

Guo, Z., Yang, W., Chang, X., Wang, Y. and Liu, X. (2014) Genome-wide association studies of flowering time, grain yield, and grain quality traits in rice. *BMC Genomics* 15, 543.

Habier, D., Fernando, R.L., Kizilkaya, K. and Garrick, D.J. (2011) Extension of the Bayesian alphabet for genomic selection. *BMC Bioinformatics* 12, 186–197. DOI: 10.1186/1471-2105-12-186/FIGURES/2.

Habyarimana, E., Lopez-Cruz, M. and Baloch, F.S. (2020) Genomic selection for optimum index with dry biomass yield, dry mass fraction of fresh material, and plant height in biomass sorghum. *Genes (Basel)* 11, 61. DOI: 10.3390/genes11010061.

Hayes, B.J., Bowman, P.J., Chamberlain, A.J. and Goddard, M.E. (2009) Invited: Genomic selection in dairy cattle: Progress and challenges. *Journal of Dairy Science* 92, 433–443. DOI: 10.3168/JDS.2008-1646.

Heffner, E.L., Sorrells, M.E. and Jannink, J.-L. (2009) Genomic selection for crop improvement. *Crop Science* 49, 1–12. DOI: 10.2135/cropsci2008.08.0512.

Heffner, E.L., Jannink, J.-L., Sorrells, M.E., Heffner, E.L., Sorrells, M.E. *et al.* (2011) Genomic selection accuracy using multifamily prediction models in a wheat breeding program. *The Plant Genome* 4, 65–75. DOI: 10.3835/plantgenome.

Henderson, C.R., Kempthorne, O., Searle, S.R. and von Krosigk, C.M. (1959) The estimation of environmental and genetic trends from records subject to culling. *Biometrics* 15, 192. DOI: 10.2307/2527669.

Herr, A.W. and Carter, A.H. (2023) Remote sensing continuity: A comparison of HTP platforms and potential challenges with field applications. *Frontiers in Plant Science* 14, 1233892.

Holliday, J.A., Wang, T. and Aitken, S. (2012) Predicting adaptive phenotypes from multilocus genotypes in Sitka spruce (*Picea sitchensis*) using random forest. *G3: Genes Genomes Genetics* 2, 1085–1093. DOI: 10.1534/g3.112.002733.

Hu, X., Carver, B.F., Powers, C., Yan, L., Zhu, L. *et al.* (2019) Effectiveness of genomic selection by response to selection for winter wheat variety improvement. *Plant Genome* 12, 180090. DOI: 10.3835/plantgenome2018.11.0090.

Huang, M., Balimponya, E.G., Mgonja, E.M., McHale, L.K., Luzi-Kihupi, A. *et al.* (2019) Use of genomic selection in breeding rice (Oryza *sativa L.*) for resistance to rice blast (*Magnaporthe oryzae*). *Molecular Breeding* 39, 1–16. DOI: 10.1007/S11032-019-1023-2.

Jarquin, D., Kajiya-Kanegae, H., Taishen, C., Yabe, S., Persa, R. *et al.* (2020) Coupling day length data and genomic prediction tools for predicting time-related traits under complex scenarios. *Scientific Reports* 10, 1. DOI: 10.1038/s41598-020-70267-9.

Jarquín, D., Crossa, J., Lacaze, X., Cheyron, P.D., Daucourt, J. *et al.* (2014) A reaction norm model for genomic selection using high dimensional genomic and environmental data. *Theoretical and Applied Genetics* 127, 595–607. DOI: 10.1007/s00122-013-2243-1.

Jeong, S., Kim, J.-Y. and Kim, N. (2020) GMStool: GWAS-based marker selection tool for genomic prediction from genomic data. *Scientific Reports* 10, 1–12. DOI: 10.1038/s41598-020-76759-y.

Jia, Y. and Jannink, J.-L. (2012) Multiple-trait genomic selection methods increase genetic value prediction accuracy. *Genetics* 192, 1513–1522. DOI: 10.1534/GENETICS.112.144246.

Jiang, Y. and Reif, J.C. (2015) Modeling epistasis in genomic selection. *Genetics* 201(2), 759–768.

Joukhadar, R., Thistlethwaite, R., Trethowan, R.M., Hayden, M.J., Stangoulis, J. *et al.* (2021) Genomic selection can accelerate the biofortification of spring wheat. *Theoretical and Applied Genetics* 134, 3339–3350. DOI: 10.1007/S00122-021-03900-4.

Juliana, P., Singh, R.P., Singh, P.K., Crossa, J., Rutkoski, J.E. *et al.* (2017) Comparison of models and whole-genome profiling approaches for genomic-enabled prediction of *Septoria tritici* blotch, *Stagonospora nodorum* blotch, and tan spot resistance in wheat. *Plant Genome* 10, 1–16. DOI: 10.3835/PLANTGENOME2016.08.0082.

Juliana, P., Singh, R.P., Poland, J.A., Mondal, S., Crossa, J. *et al.* (2019) Genomic selection for grain yield in CIMMYT spring wheat breeding programs. *Theoretical and Applied Genetics* 132(2), 1–17. DOI: 10.1007/s00122-018-3200-z.

Klápště, J., Dungey, H.S., Telfer, E.J., Suontama, M., Graham, N.J. *et al.* (2020) Marker selection in multivariate genomic prediction improves accuracy of low heritability traits. *Frontiers in Genetics* 11, 499094. DOI: 10.3389/FGENE.2020.499094/FULL.

Kuo, P., Henderson, I.R. and Lambing, C. (2022) CTAB DNA extraction and genotyping-by-sequencing to map meiotic crossovers in plants. In: Lambing, C. (ed.) *Plant Gametogenesis: Methods and Protocols*. Humana, New York, pp. 43–53.

Legarra, A. and Reverter, A. (2018) Semi-parametric estimates of population accuracy and bias of predictions of breeding values and future phenotypes using the LR method. *Genetics Selection Evolution* 50, 53–18. DOI: 10.1186/S12711-018-0426-6/FIGURES/3.

Li, C., Jia, Y., Zhou, R., Liu, L., Cao, M. *et al.* (2022) GWAS and RNA-seq analysis uncover candidate genes associated with alkaline stress tolerance in maize (*Zea mays* L.) seedlings. *Frontiers in Plant Science* 13, 963874. DOI: 10.3389/fpls.2022.963874.

Liu, X., Wang, H., Wang, H., Guo, Z., Xu, X. *et al.* (2018) Factors affecting genomic selection revealed by empirical evidence in maize. *The Crop Journal* 6(4), 341–352.

Long, N., Gianola, D., Rosa, G.J.M. and Weigel, K.A. (2011) Application of support vector regression to genome-assisted prediction of quantitative traits. *Theoretical and Applied Genetics* 123, 1065–1074. DOI: 10.1007/S00122-011-1648-Y.

Lorenz, A.J., Chao, S., Asoro, F.G., Heffner, E.L., Hayashi, T. *et al.* (2011) Genomic selection in plant breeding. *Advanced Agronomics* 110, 77–123. DOI: 10.1016/B978-012-385531-2.00002-5.

Majumdar, S.G., Rai, A. and Mishra, D.C. (2019) Package 'GSelection', 1–14. Available at: https://rdrr.io/cran/GSelection/man/GSelection-package.html

Ma, W., Qiu, Z., Song, J., Li, J., Cheng, Q. *et al.* (2018) A deep convolutional neural network approach for predicting phenotypes from genotypes. *Planta* 248, 1307–1318.

Mackay, T.F.C. (2001) The genetic architecture of quantitative traits. *Annual Review of Genetics* 35, 303–339.

Maenhout, S., De Baets, B., Haesaert, G. and Van Bockstaele, E. (2007) Support vector machine regression for the prediction of maize hybrid performance. *Theoretical and Applied Genetics* 115, 1003–1013. DOI: 10.1007/s00122-007-0627-9.

Meuwissen, T.H.E., Hayes, B.J. and Goddard, M.E. (2001) Prediction of total genetic value using genome-wide dense marker maps. *Genetics* 157, 1819–1829. DOI: 10.1093/GENETICS/157.4.1819.

Michel, S., Löschenberger, F., Ametz, C., Pachler, B., Sparry, E. *et al.* (2019) Combining grain yield, protein content and protein quality by multi-trait genomic selection in bread wheat. *Theoretical and Applied Genetics* 132, 2767–2780. DOI: 10.1007/S00122-019-03386-1.

Montesinos López, O.A., Montesinos López, A. and Crossa, J. (2022) Random forest for genomic prediction. In: *Multivariate Statistical Machine Learning Methods for Genomic Prediction*. Springer, pp. 633–681.

Montesinos-López, O.A., Montesinos-López, A., Crossa, J., Gianola, D., Hernández-Suárez, C.M. *et al.* (2018) Multi-trait, multi-environment deep learning modeling for genomic-enabled prediction of plant traits. *G3: Genes, Genomes, Genetics* 8(12), 3829–3840.

Montesinos-López, O.A., Martín-Vallejo, J., Crossa, J., Gianola, D., Hernández-Suárez, C.M. *et al.* (2019a) A benchmarking between deep learning, support vector machine and Bayesian threshold best linear unbiased prediction for predicting ordinal traits in plant breeding. *G3: Genes, Genomes, Genetics* 9(2), 601–618.

Montesinos-López, O.A., Martín-Vallejo, J., Crossa, J., Gianola, D., Hernández-Suárez, C. *et al.* (2019b) New deep learning genomic-based prediction model for multiple traits with binary, ordinal, and continuous phenotypes. *G3: Genes, Genomes, Genetics* 9, 1545–1556.

Montesinos-López, O.A., Montesinos-López, A., Mosqueda-González, B.A., Bentley, A.R., Lillemo, M. *et al.* (2021a) A new deep learning calibration method enhances genome-based prediction of continuous crop traits. *Frontiers in Genetics* 12, 798840.

Montesinos-López, O.A., Montesinos-López, A., Pérez-Rodríguez, P., Barrón-López, J.A., Martini, J.W. *et al.* (2021b) A review of deep learning applications for genomic selection. *BMC Genomics* 22(1), 19.

Mota, L.F., Arikawa, L.M., Santos, S.W., Fernandes Júnior, G.A., Alves, A.A. *et al.* (2024) Benchmarking machine learning and parametric methods for genomic prediction of feed efficiency-related traits in Nellore cattle. *Scientific Reports* 14(1), 6404.

Narasimhan, V.M., Patterson, N., Moorjani, P., Rohland, N., Bernardos, R. *et al.* (2019) The formation of human populations in South and Central Asia. *Science* 365(6457), eaat7487.

Ogutu, J.O., Schulz-Streeck, T. and Piepho, H.P. (2012) Genomic selection using regularized linear regression models: Ridge regression, Lasso, Elastic Net and their extensions. *BMC Proceedings* 6, S10. DOI: 10.1186/1753-6561-6-S2-S10.

Onogi, A., Ideta, O., Inoshita, Y., Ebana, K., Yoshioka, T. *et al.* (2015) Exploring the areas of applicability of whole-genome prediction methods for Asian rice (*Oryza sativa* L.). *Theoretical and Applied Genetics* 128, 41–53. DOI: 10.1007/S00122-014-2411Y.

Ortiz, R., Gowda, M., Zhao, Y., Velazco, J.G., Van Eeuwijk, F.A. *et al.* (2019) Genomic prediction of grain yield and drought adaptation capacity in Sorghum is enhanced by multi-trait analysis. *Frontiers in Plant Science* 10, 997. DOI: 10.3389/fpls.2019.00997.

Pavan, S., Delvento, C., Ricciardi, L., Lotti, C., Ciani, E. *et al.* (2020) Recommendations for choosing the genotyping method and best practices for quality control in crop genome-wide association studies. *Frontiers in Genetics* 11, 447.

Pérez, P. and de los Campos, G. (2014) Genome-wide regression and prediction with the BGLR Statistical Package. *Genetics* 198, 483–495. DOI: 10.1534/genetics.114.164442.

Pinheiro, J., Bates, D., DebRoy, S., Sarkar, D., Heisterkamp, S. *et al.* (2017) Package 'nlme'. Available at: https://cran.r-project.org/web/packages/nlme/nlme.pdf

Poland, J., Endelman, J., Dawson, J., Rutkoski, J., Wu, S. *et al.* (2012) Genomic selection in wheat breeding using genotyping-by-sequencing. *Plant Genome* 5, 1–11. DOI: 10.3835/PLANTGENOME2012.06.0006.

Raubach, S., Schreiber, M. and Shaw, P.D. (2022) GridScore: A tool for accurate, cross-platform phenotypic data collection and visualization. *BMC Bioinformatics* 23(1), 214.

Reynolds, M.P. and Ortiz, R. (2010) Adapting crops to climate change: A summary. In: Reynolds, M.P. (ed.) *Climate Change and Crop Production*. CAB International, Wallingford, UK, pp. 1–8. DOI: 10.1079/9781845936334.0001.

Rio, S., Mary-Huard, T., Moreau, L. and Charcosset, A. (2019) Genomic selection efficiency and a priori estimation of accuracy in a structured dent maize panel. *TAG. Theoretical and Applied Genetics* 132(1), 81–96.

Robertsen, C.D., Hjortshøj, R.L. and Janss, L.L. (2019) Genomic selection in cereal breeding. *Agronomy* 9(2), 95.

Rothman, A.J., Levina, E. and Zhu, J. (2010) Sparse multivariate regression with covariance estimation. *Journal of Computational and Graphical Statistics* 19, 947–962. DOI: 10.1198.JCGS.2010.09188.

Rouse Jr, J.W., Haas, R.H., Schell, J. and Deering, D. (1973) *Monitoring the Vernal Advancement and Retrogradation (Green Wave Effect) of Natural Vegetation*. Texas A&M University Remote Sensing Center, College Station, Texas.

Rutkoski, J., Benson, J., Jia, Y., Brown-Guedira, G., Jannink, J.-L. *et al.* (2012) Evaluation of genomic prediction methods for Fusarium head blight resistance in wheat. *The Plant Genome* 5, 51–61. DOI: 10.3835/PLANTGENOME2012.02.0001.

Sallam, A.H., Endelman, J.B., Jannink, J.L. and Smith, K.P. (2015) Assessing genomic selection prediction accuracy in a dynamic barley breeding population. *Plant Genome* 8, 20. DOI: 10.3835/PLANTGENOME2014.05.0020.

Sandhu, N., Subedi, S., Yadaw, R.B., Variar, M., Vijayalakshmi, M. *et al.* (2021) Harnessing genomic selection for improving heat tolerance in rice. *Frontiers in Plant Science* 12, 661559. DOI: 10.3389/fpls.2021.661559.

Schmidt, M., Kollers, S., Maasberg-Prelle, A., Großer, J., Schinkel, B. *et al.* (2016) Prediction of malting quality traits in barley based on genome wide marker data to assess the potential of genomic selection. *Theoretical and Applied Genetics* 129, 203–213. DOI: 10.1007/s00122-015-2639-1.

Schulthess, A.W., Wang, Y., Miedaner, T., Wilde, P., Reif, J.C. *et al.* (2016) Multiple- trait- and selection indices-genomic predictions for grain yield and protein content in rye for feeding purposes. *Theoretical and Applied Genetics* 129, 273–287. DOI: 10.1007/s00122-015-2626-6.

Shamshad, M. and Sharma, A. (2018) The usage of genomic selection strategy in plant breeding. In: Yelda Özden, C. (ed.) *Next Generation Plant Breeding*. IntechOpen, London.

Spindel, J. and Iwata, H. (2018) Genomic selection in rice breeding. In: Sasaki, T. and Ashikari, M. (eds) *Rice Genomics, Genetics and Breeding*. Springer, Singapore, pp. 473–496. DOI: 10.1007/978-981-10-7461-524.

Spindel, J., Begum, H., Akdemir, D., Virk, P., Collard, B. *et al.* (2015) Correction: Genomic selection and association mapping in rice (*Oryza sativa*): Effect of trait genetic architecture, training population composition, marker number and statistical model on accuracy of rice genomic selection in elite, tropical rice breeding lines. *PLOS Genetics* 11, e1005350. DOI: 10.1371/JOURNAL.PGEN.1005350.

Su, G., Guldbrandtsen, B., Gregersen, V. and Lund, M. (2010) Preliminary investigation on reliability of genomic estimated breeding values in the Danish Holstein population. *Journal of Dairy Science* 93(3), 1175–1183.

Sukumaran, S., Jarquin, D., Crossa, J. and Reynolds, M. (2018) Genomic-enabled prediction accuracies increased by modeling genotype × environment interaction in durum wheat. *Plant Genome* 11, 170112. DOI: 10.3835/PLANTGENOME2017.12.0112.

Sun, J., Rutkoski, J.E., Poland, J.A., Crossa, J., Jannink, J.L. *et al.* (2017) Multitrait, random regression, or simple repeatability model in high throughput phenotyping data improve genomic prediction for wheat grain yield. *Plant Genome* 10, 1–12. DOI: 10.3835/PLANTGENOME2016.11.0111.

Tayade, R., Yoon, J., Lay, L., Khan, A.L., Yoon, Y. *et al.* (2022) Utilization of spectral indices for high-throughput phenotyping. *Plants* 11(13), 1712.

Technow, F., Messina, C.D., Totir, L.R. and Cooper, M. (2021) Integrating crop growth models with whole genome prediction through approximate Bayesian computation. *PLoS ONE* 16(3), e0248433. DOI: 10.1371/journal.pone.0248433.

Tester, M. and Langridge, P. (2010) Breeding technologies to increase crop production in a changing world. *Science* 327, 818–822. DOI: 10.1126/SCIENCE.1183700.

Thavamanikumar, S., Dolferus, R. and Thumma, B.R. (2015) Comparison of genomic selection models to predict flowering time and spike grain number in two hexaploid wheat doubled haploid populations. *G3: Genes, Genomes, Genetics* 5(10), 1991–1998.

Tibshirani, R. (1996) Regression shrinkage and selection via the lasso. *Journal of the Royal Statistical Society, Series B (Methodological)* 58, 267–288. DOI: 10.1111/J.2517-6161.1996.TB02080.X.

Tsai, H.-Y., Cericola, F., Edriss, V., Andersen, J.R., Orabi, J. *et al.* (2020) Use of multiple traits genomic prediction, genotype by environment interactions and spatial effect to improve prediction accuracy in yield data. *PLoS ONE* 15, e0232665. DOI: 10.1371/journal.pone.0232665.

Tuberosa, R. (2012) Phenotyping for drought tolerance of crops in the genomics era. *Frontiers in Physiology* 3, 347.

Usai, M.G., Goddard, M.E. and Hayes, B.J. (2009) LASSO with cross-validation for genomic selection. *Genetic Research* 91, 427–436. DOI: 10.1017/S0016672309990334.

VanRaden, P.M. (2008) Efficient methods to compute genomic predictions. *Journal of Dairy Science* 91(11), 4414–4423.

Varshney, R.K., Roorkiwal, M. and Sorrells, M.E. (2017) Genomic selection for crop improvement: An introduction. In: Varshney, R.K., Roorkiwal, M. and Sorrells, M.E. (eds) *Genomic Selection for Crop Improvement: New Molecular Breeding Strategies for Crop Improvement*. Springer Nature, Cham, Switzerland, pp. 1–6.

Varshney, R.K., Bohra, A., Yu, J., Graner, A., Zhang, Q. *et al.* (2021) Designing future crops: Genomics-assisted breeding comes of age. *Trends in Plant Science* 26(6), 631–649. DOI: 10.1016/j.tplants.2021.02.010.

Vazquez, A.I., Bates, D.M., Rosa, G.J.M., Gianola, D. and Weigel, K.A. (2010) Technical note: An R package for fitting generalized linear mixed models in animal breeding. *Journal of Animal Science* 88, 497–504. DOI: 10.2527/JAS.2009-1952.

Wang, X., Xu, Y., Hu, Z. and Xu, C. (2018) Genomic selection methods for crop improvement: Current status and prospects. *Crop Journal* 6, 330–340. DOI: 10.1016/j.cj.2018.03.001.

WHO/FAO (2003) *Diet, Nutrition and the Prevention of Chronic Diseases: Recommendations for Preventing Excess Weight Gains and Obesity*. WHO, Geneva, Switzerland.

Wolpert, D.H. (1996) The lack of a priori distinctions between learning algorithms. *Neural Computation* 8(7), 1341–1390.

Xu, S. (2007) An empirical Bayes method for estimating epistatic effects of quantitative trait loci. *Biometrics* 63, 513–521. DOI: 10.1111/J.15410420.2006.00711.X.

Xu, S., Zhu, D. and Zhang, Q. (2014) Predicting hybrid performance in rice using genomic best linear unbiased prediction. *Proceedings of the National Academy of Sciences of the United States of America* 111(34), 12456–12461.

Xu, Y., Wang, X., Ding, X., Zheng, X., Yang, Z. *et al.* (2018) Genomic selection of agronomic traits in hybrid rice using an NCII population. *Rice (NY)* 11, 32–10. DOI: 10.1186/S12284-018-0223-4/FIGURES/5.

Yabe, S., Yoshida, H., Kajiya-Kanegae, H., Yamasaki, M., Iwata, H. *et al.* (2018) Description of grain weight distribution leading to genomic selection for grain-filling characteristics in rice. *PLoS ONE* 13, e0207627. DOI: 10.1371/journal.pone.0207627.

Zhang, D., Yang, F., Li, J., Liu, Z., Han, Y. *et al.* (2025) Progress and perspectives on genomic selection models for crop breeding. *Technology in Agronomy* 5(1).

Ziyatdinov, A., Ve1azquez-Santiago, M., Brunel, H., Martinez-Perez, A., Aschard, H. *et al.* (2018) LME4qtl: Linear mixed models with flexible covariance structure for genetic studies of related individuals. *BMC Bioinformatics* 19, 68. DOI: 10.1186/s12859-018-2057-x.

Zou, H. and Hastie, T. (2005) Regularization and variable selection via the Elastic Net. *Journal of the Royal Statistical Society B* 67, 301–320. DOI: 10.1111/j.1467-9868.2005.00503.x.

8 Genomic Prediction Models for Fruit and Vegetable Crops

Saniya[1] and Sarfraz Ahmad[2*]

[1]Department of Horticulture, ICAR-National Research Centre for Grapes, Pune, Maharashtra, India; [2]Department of Genetics and Plant Breeding, College of Agriculture Mandawa, Swami Keshwanand Rajasthan Agriculture University, Bikaner, Rajasthan, India

Abstract

Genomic prediction (GP) has emerged as a revolutionary approach to accelerate genetic gains in crop improvement, leveraging dense genome-wide marker data and advanced modeling techniques to predict the genetic merit of individuals. While cereal and legume crops have been at the forefront of GP applications, fruit and vegetable crops present unique opportunities and challenges due to their biological complexity, diverse reproductive systems, and high-value traits such as yield, quality, and stress resilience. This chapter provides a comprehensive overview of the theoretical foundations, genotyping resources, and statistical frameworks underpinning GP, with a focus on its application in fruit and vegetable breeding pipelines. Key topics include clonal propagation, polyploidy, long juvenile phases, and trait complexity, all of which necessitate tailored prediction strategies. We highlight commonly used statistical and machine learning models such as genomic BLUP, Bayesian methods, kernel-based approaches, and deep learning, detailing their strengths, limitations, and crop-specific applications. Case studies across tomato, potato, grapevine, apple, cucumber, and strawberry illustrate how GP can be integrated into practical breeding programs to improve efficiency and precision. The chapter further explores enabling resources, including high-throughput genotyping, phenotyping platforms, and multi-omics integration, which enhance the accuracy of prediction models. Finally, we discuss the challenges of deploying GP in fruit and vegetable crops, including data scarcity, genotype-by-environment interactions, and the need for computational infrastructure, while highlighting future perspectives for sustainable breeding. Overall, GP is poised to transform fruit and vegetable improvement, fostering climate resilience, nutritional quality, and consumer-preferred cultivars in the coming decades.

Keywords: Climate-resilient crops, Fruit breeding, Genomic prediction, High-throughput phenotyping, Machine learning, Multi-omics integration, Polyploidy, Vegetable breeding

8.1 Introduction

Fruit and vegetable crops represent a cornerstone of global food systems, providing essential micronutrients, dietary fiber, and bioactive compounds that directly contribute to human health and nutrition. Unlike cereals, which primarily supply calories, fruits and vegetables are indispensable sources of vitamins (e.g. vitamins C, A, K, folates), minerals, and antioxidants. However, despite their importance, the breeding of these crops faces several challenges: long juvenile

*Corresponding author: sarfraz_firoz@rediffmail.com

 Plant Genomic Prediction: AI-Enhanced Models for Plant Genomics and Crop Improvement (ed. Jen-Tsung Chen)
DOI: 10.1079/9781836991328.0008

periods, high heterozygosity, clonal propagation, polyploidy, and the need to improve highly complex quality-related traits such as taste, texture, aroma, and nutritional composition. These complexities slow the development of new cultivars and increase the cost of breeding programs compared with annual grain crops.

In the past decade, genomic prediction (GP), an advanced statistical and computational framework, has emerged as a promising solution to accelerate genetic gain in fruit and vegetable crops. Genomic prediction, first popularized in animal breeding (Chafai *et al.*, 2023), leverages genome-wide markers to predict the genetic potential of individuals for complex traits, without relying solely on the identification of major-effect loci. Unlike marker-assisted selection (MAS), which targets only a limited number of quantitative trait loci (QTLs), GP uses all available marker information simultaneously to estimate the genomic estimated breeding values (GEBVs) of individuals. This genome-wide approach makes it particularly effective for traits controlled by many small-effect loci, which are common in fruit and vegetable crops.

The utility of GP in horticultural crops has gained momentum in recent years due to technological advances in high-throughput genotyping (e.g. single-nucleotide polymorphism (SNP) arrays, genotyping-by-sequencing, whole-genome resequencing) and phenotyping platforms (e.g. imaging systems, drones, metabolomic profiling). These developments have enabled the collection of large-scale datasets required for accurate prediction modeling. Furthermore, the integration of multi-omics data (e.g. transcriptomics, metabolomics, proteomics, epigenomics) into GP frameworks is opening new avenues for predicting complex traits such as flavor, disease resistance, and post-harvest performance (Belay and Caleb, 2022).

From an applied breeding perspective, GP is particularly attractive for fruit and vegetable crops because it enables early selection, reducing the number of years breeders must wait for phenotypic evaluation. For instance, apple and citrus trees require several years to reach reproductive maturity, making traditional breeding cycles extremely long. By applying GP, breeders can identify elite seedlings at early stages, thereby accelerating cultivar development. Similarly, in clonally propagated crops such as potato and strawberry, GP can be used to predict and select superior genotypes before large-scale multiplication and field testing. For vegetable crops like tomato, cucumber, and pepper, where fruit quality traits are difficult to measure in early generations, GP provides a cost-effective approach for pre-screening breeding populations.

Recent empirical studies highlight the growing adoption of GP in horticulture. For example, GP models in apple have achieved high predictive ability for fruit firmness, soluble solids content, and scab resistance (Mendoza *et al.*, 2011). In tomato, genomic prediction has been applied to fruit weight, shape, and metabolite profiles, with machine learning models outperforming traditional approaches in some cases (Gerakari *et al.*, 2025). In sweet potato, a polyploid crop, dosage-aware models that account for allele dosage effects have demonstrated improved predictive accuracy (Batista *et al.*, 2022). These case studies illustrate how GP can be tailored to diverse crop-specific challenges.

Another critical driver of GP in fruit and vegetable crops is the increasing demand for climate-resilient cultivars. Climate change is altering pest and disease pressures, temperature regimes, and water availability, posing significant threats to horticultural production systems. Genomic prediction can help breeders design varieties with improved resilience by incorporating genotype × environment (G × E) interactions and enviromic data into predictive models (Song *et al.*, 2025). For instance, integrating environmental covariates with genomic data can improve the prediction of traits such as drought tolerance, fruit set under heat stress, and resistance to emerging pathogens.

Despite its promise, GP in fruit and vegetable crops faces several challenges and limitations. The relatively small size of breeding programs compared with cereals often constrains the size of training populations, reducing prediction accuracy. Additionally, many fruit and vegetable crops exhibit complex ploidy levels and high heterozygosity, complicating model development. The collection of high-quality phenotypic data remains a bottleneck, particularly for traits like flavor and nutritional quality that require destructive and costly analyses. Moreover, there is a need for breeder-friendly software and decision-support tools to facilitate the adoption of GP in practical programs.

In summary, genomic prediction represents a paradigm shift in the breeding of fruit and vegetable crops. It offers the potential to shorten breeding cycles, accelerate the delivery of improved cultivars, and address both productivity and quality-related challenges in horticulture. By integrating genomic data with multi-omics, phenomics, and environmental information, GP provides a holistic framework for predictive breeding. This chapter aims to: (i) review the theoretical foundations of GP; (ii) discuss crop-specific challenges and opportunities; (iii) summarize genotyping and phenotyping resources; (iv) evaluate statistical and machine learning models; (v) highlight case studies across fruit and vegetable crops; and (vi) explore future directions for GP-enabled horticultural breeding.

8.2 Theoretical Foundations of Genomic Prediction

8.2.1 Historical background and conceptual shift

The concept of genomic prediction (GP) originates from the seminal work of Meuwissen *et al.* (2001), who introduced the idea of estimating genetic merit using genome-wide dense marker data. Prior to this breakthrough, marker-assisted selection (MAS) was the dominant paradigm in molecular breeding. MAS focused on identifying and utilizing major-effect quantitative trait loci (QTLs) or single genes associated with traits of interest. While MAS was successful for traits controlled by a few loci (e.g. disease-resistance genes in tomato or grapevine), it often failed for polygenic traits, which are influenced by numerous loci of small effect. Most yield- and quality-related traits in fruit and vegetable crops fall into this polygenic category.

The transition from MAS to GP marked a paradigm shift: rather than attempting to identify and select only significant markers, GP uses all available marker information simultaneously, regardless of the individual effect sizes. The output is the genomic estimated breeding value (GEBV) of each individual, which represents the expected genetic contribution of that individual to the next generation. GEBVs can be estimated even without phenotypic evaluation of the selection candidates, provided that a training population linking markers to traits is available.

8.2.2 Statistical foundation of GP

The theoretical basis of GP lies in quantitative genetics, where the phenotype (y) of an individual can be decomposed as:

$$Y = \mu + g + e$$

where μ is the overall mean, g is the genetic effect, and e is the environmental effect. In GP, the genetic effect is modeled as a function of genome-wide marker data.

Predictive accuracy in GP depends on several factors:

- **Marker density:** Higher density improves linkage disequilibrium (LD) coverage but increases computational cost.
- **Trait heritability:** High-heritability traits are predicted with greater accuracy.
- **Training population size and structure:** Large, diverse, and representative training sets improve prediction.
- **Genetic architecture:** Prediction is more challenging for traits controlled by many small-effect loci or those with strong non-additive effects.

8.2.3 Core GP models

Several statistical models have been developed for GP, each with distinct assumptions and strengths:

Genomic best linear unbiased prediction (GBLUP)

- Uses a genomic relationship matrix (GRM) derived from marker data.
- Assumes all marker effects are normally distributed with equal variance.
- Effective for traits with highly polygenic architectures.

- Used in Persian walnuts due to its robustness and simplicity (Bernard *et al.*, 2025).

Bayesian models

- Models such as BayesA, BayesB, BayesCπ, and Bayesian LASSO assign prior distributions to marker effects.
- Allow for variable shrinkage, making them effective when a few large-effect QTLs coexist with many small-effect loci.
- Particularly relevant for traits such as disease resistance in tomato or pepper, where major genes play important roles.

Reproducing kernel Hilbert space (RKHS) regression

- A non-parametric method capable of modeling nonlinear relationships and epistatic interactions.
- Useful for crops like potato, where non-additive and dominance effects are significant in clonal propagation systems.

Machine learning and deep learning models

- Algorithms such as random forests, gradient boosting, support vector machines, and neural networks have been adapted for GP.
- These methods can capture complex, nonlinear interactions and integrate multi-omics data.
- Recent studies in tomato and strawberry suggest that deep learning can outperform traditional statistical models for metabolite prediction (Shawky *et al.*, 2025).

8.2.4 Training population and validation strategies

The training population (TP) forms the foundation of GP models. It consists of individuals with both genotype and phenotype data, used to estimate marker effects. Prediction is then applied to a validation population, which is genotyped but not phenotyped.

Key considerations include:

- **TP size:** Larger populations generally yield higher prediction accuracy.
- **Genetic relatedness:** TPs closely related to the breeding population enhance prediction power.
- **Cross-validation schemes:** Common strategies include k-fold cross-validation, leave-one-family-out, and temporal validation, depending on the breeding program design.

8.2.5 Prediction accuracy and model evaluation

Prediction accuracy in GP is often measured as the correlation between predicted GEBVs and observed phenotypes. Other metrics include mean squared error, regression coefficients, and bias. In fruit and vegetable crops, reported accuracies vary widely:

- Moderate to high (0.4–0.7) for traits such as fruit firmness in apple or soluble solids content in grapevine.
- Lower (0.2–0.4) for traits with strong environmental influence, such as flavor metabolites in tomato.

Accuracy depends on the alignment of the model with the underlying trait architecture and the representativeness of the training population.

8.2.6 Extensions of the GP framework

Beyond the classical models, several extensions have been proposed to address challenges specific to horticultural crops:

- **Genotype × environment (G×E) models:** Integrating environmental covariates to improve prediction under variable conditions (Heslot *et al.*, 2014).
- **Multi-trait genomic prediction:** Leveraging correlations between traits (e.g. yield and fruit quality) to improve accuracy.
- **Ploidy-aware models:** Adapting statistical frameworks for autopolyploids such as potato and strawberry, where allele dosage must be considered.
- **Multi-omics integration:** Incorporating transcriptomics, metabolomics, and

epigenomics for enhanced prediction of complex phenotypes (Ahmad *et al.*, 2025).

8.2.7 Relevance to fruit and vegetable crops

The theoretical foundations of GP provide a strong rationale for its application in fruit and vegetable breeding. These crops often display:

- Long breeding cycles (e.g. apple, citrus) → GP reduces time to selection.
- High heterozygosity (e.g. grapevine, strawberry) → GP can accommodate clonal propagation.
- Complex polygenic traits (e.g. tomato flavor, pepper pungency) → GP captures genome-wide small effects.
- Polyploidy (e.g. potato) → specialized models improve accuracy.

Thus, GP represents a versatile tool capable of addressing the inherent biological and breeding complexities of horticultural crops.

8.3 Breeding Features of Fruit and Vegetable Crops

The successful application of genomic prediction (GP) in fruit and vegetable crops requires a deep understanding of their unique biological and breeding features. Unlike major cereals and legumes, which are typically annuals with large breeding programs and standardized genomic resources, fruit and vegetable crops exhibit extraordinary biological diversity. This diversity encompasses variation in reproductive systems, ploidy levels, breeding strategies, and trait architectures. Collectively, these features present both opportunities and challenges for the adoption of GP in horticultural breeding.

8.3.1 Biological diversity in fruit and vegetable crops

Fruit and vegetable crops represent hundreds of species across diverse plant families, including Solanaceae (tomato, pepper, eggplant), Cucurbitaceae (cucumber, melon, squash), Rosaceae (apple, peach, strawberry), Vitaceae (grapevine), Rutaceae (citrus), Brassicaceae (broccoli, cabbage), and Apiaceae (carrot, celery). This taxonomic diversity translates into wide-ranging reproductive systems (selfing vs outcrossing vs clonal propagation), ploidy levels (diploid vs polyploid), and genetic architectures.

From a breeding perspective, the heterogeneity of crop biology complicates the establishment of universal GP pipelines. For example:

- **Tomato** is largely autogamous and diploid, with relatively short generation times, enabling rapid cycling.
- **Apple and citrus** are long-lived perennials with extended juvenile periods, which delay phenotypic evaluation.
- **Strawberry and potato** are polyploid and clonally propagated, requiring specialized dosage-aware models.
- **Cucumber and pepper** exhibit high genetic diversity and complex traits such as flavor, which are difficult to phenotype accurately.

Therefore, while GP theory is generalizable, implementation must be crop-specific, taking into account the unique reproductive biology and trait complexity.

8.3.2 Clonal propagation and heterozygosity

Clonal propagation is common in many fruit crops (apple, grapevine, banana, citrus) and several vegetables (potato, sweet potato, garlic). Unlike sexually propagated crops, where recombination reshuffles alleles each generation, clonal propagation fixes heterozygous genotypes across cycles. This strategy preserves favorable allele combinations but complicates traditional breeding.

For clonally propagated crops:

- Heterozygosity is extremely high, meaning that non-additive genetic effects (dominance and epistasis) play a major role in trait expression.

- Traditional selection relies heavily on phenotypic performance of clones across multiple environments, which is costly and time-consuming.
- GP can accelerate clonal breeding by predicting the value of seedlings before clonal multiplication, thereby reducing the number of candidates advanced to field testing.

For instance, in potato, which is an autotetraploid, GP models incorporating allele dosage and dominance effects have significantly improved prediction accuracies for tuber yield and quality traits (Stich and Van Inghelandt, 2018). Similarly, in grapevine and apple, where cultivar development may take 15–20 years, GP enables early-stage selection and helps reduce cycle length.

8.3.3 Polyploidy and dosage effects

Polyploidy is prevalent among fruit and vegetable crops. Autopolyploid species such as potato and strawberry contain multiple homologous chromosome sets, while allopolyploids such as *Brassica* species combine genomes from different progenitors. Polyploidy complicates breeding and prediction for several reasons:

- **Allele dosage:** In tetraploids, an allele can exist in 0–4 copies, creating complex genotype-to-phenotype relationships.
- **Segregation patterns:** Polyploid meiosis can involve multivalent pairing and double reduction, leading to deviations from diploid inheritance models.
- **Modeling challenges:** Standard GP models assume diploidy; adapting them to polyploid contexts requires dosage-aware marker coding and algorithms.

Recent advances in polyploid-specific GP models have shown promise. For instance, in autotetraploid potato, models that account for allele dosage outperform diploidized approaches. In octoploid strawberry, multi-allelic GP models have improved prediction for fruit size, flavor, and firmness (Navarro, 2023). These results underscore the importance of tailoring GP models to the cytogenetic context of each crop.

8.3.4 Long juvenile periods in perennials

Perennial fruit crops such as apple, peach, grapevine, and citrus require years before reproductive maturity is reached, delaying phenotypic evaluation. Traditional breeding cycles for apple, for example, can span 20–25 years, from initial crosses to commercial cultivar release (Laurens *et al.*, 2018).

GP addresses this challenge by enabling early prediction of breeding values in juvenile seedlings. Once genotyped, seedlings can be selected based on predicted performance rather than waiting for full phenotypic evaluation. This drastically shortens breeding cycles, increasing the rate of genetic gain per unit time. Moreover, GP can help breeders focus limited field-testing resources on the most promising candidates, improving cost-efficiency.

8.3.5 Complex trait architectures

Fruit and vegetable crops are often selected for traits beyond yield, including quality, shelf life, stress resilience, and consumer appeal. These traits typically exhibit polygenic architectures and strong environmental sensitivity. Examples include:

- **Tomato flavor and aroma:** Determined by the interaction of sugars, acids, and volatile metabolites, controlled by many loci of small effect.
- **Grapevine wine quality traits:** Influenced by polygenic metabolic pathways involving terpenes, flavonoids, and tannins.
- **Apple texture and firmness:** Controlled by multiple QTL and environmental conditions during ripening.
- **Brassica vegetable flavor compounds:** Derived from glucosinolates, which are affected by both genetics and environment.

For such traits, GP is more effective than MAS because it integrates information across the entire genome, capturing small-effect alleles. Furthermore, multi-trait GP models can leverage correlations between traits (e.g. soluble solids and flavor volatiles in tomato) to enhance prediction accuracy.

8.3.6 Selfing vs outcrossing systems

Breeding approaches differ significantly between self-pollinating vegetables (e.g. tomato, pepper, cucumber) and outcrossing perennials (e.g. apple, grapevine).

- **Selfing crops:** Inbred lines are used to produce F1 hybrids. GP can accelerate line development and optimize hybrid prediction by estimating combining abilities.
- **Outcrossing crops:** Populations are highly heterozygous. GP can predict breeding values for selection within populations, clonal evaluation, or controlled crosses.

These differences necessitate customized GP pipelines, with hybrid prediction models being increasingly important for vegetable crops where hybrid seeds dominate the market.

8.3.7 Environmental sensitivity and G×E interactions

Fruit and vegetable crops often exhibit strong genotype × environment (G × E) interactions, particularly for traits such as yield, quality, and stress resilience. For example:

- Tomato flavor varies widely across environments due to climate, soil, and cultivation practices.
- Apple firmness and acidity are influenced by growing season temperatures and storage conditions.
- Grapevine metabolite profiles fluctuate strongly with terroir (soil, climate, management).

GP models incorporating G × E interactions (so-called enviromic models) improve predictive accuracy in multi-environment trials (Crossa *et al.*, 2022). This approach is particularly critical for global breeding programs that aim to develop widely adapted cultivars.

8.3.8 Implications for GP adoption

The breeding features of fruit and vegetable crops have direct implications for GP:

- **Clonal crops:** Require models capturing dominance and epistasis.
- **Polyploids:** Need dosage-aware GP frameworks.
- **Perennials:** Benefit from early selection strategies enabled by GP.
- **Complex traits:** Demand genome-wide prediction over MAS.
- **G × E-prone crops:** Require integration of environmental covariates.

Despite these challenges, the diversity of breeding systems also presents unique opportunities. For example, clonal propagation allows direct capture of superior genotypes identified by GP, avoiding the dilution of favorable combinations through recombination. Similarly, self-pollinating vegetables with short cycles provide rapid feedback for model improvement.

8.4 Genotyping Resources for Genomic Prediction in Fruit and Vegetable Crops

The success of genomic prediction in fruit and vegetable crops is critically dependent on the availability and quality of genomic resources. These resources provide the foundational information for developing molecular markers, constructing high-density genetic maps, and generating the large-scale genotype datasets needed for training statistical and machine learning models. Unlike major cereals where standardized genotyping platforms are well established, fruit and vegetable crops present unique challenges due to their diverse biology, clonal propagation, polyploidy, and the economic importance of traits like flavor, texture, nutritional value, and shelf life. This section reviews the major genotyping resources available for fruit and vegetable crops, with emphasis on marker systems, sequencing technologies, reference genome assemblies, pangenome initiatives, and crop-specific databases that underpin GP applications.

8.4.1 Marker systems for genomic prediction

Early genomic prediction studies in horticultural crops relied heavily on simple sequence repeats

(SSRs) due to their co-dominant inheritance and high informativeness. However, their low genome coverage limited predictive ability in polygenic traits. The advent of single-nucleotide polymorphisms (SNPs) revolutionized genotyping by enabling high-throughput, genome-wide marker discovery and cost-effective genotyping. SNP arrays, such as the Axiom Apple 480K SNP array (Bianco *et al.*, 2016), the Vitis18k SNP chip for grapevine (Vervalle *et al.*, 2022), and the Potato SolCAP array (Yousaf *et al.*, 2023), have become cornerstone resources for GP in fruit and vegetable crops. These arrays provide dense, reproducible, and species-specific marker sets suitable for genomic selection across breeding programs.

However, SNP arrays have limitations in terms of ascertainment bias and inability to capture structural variation, which can be particularly relevant in species with high heterozygosity, clonal propagation, or polyploidy (Balagué-Dobón *et al.*, 2022). Consequently, there has been a shift toward genotyping-by-sequencing (GBS) and restriction-site associated DNA sequencing (RAD-seq), which enable *de novo* SNP discovery and genotyping at relatively low cost. These approaches have been widely used in crops like cucumber, tomato, and strawberry, providing large datasets for genomic selection models.

8.4.2 Next-generation sequencing (NGS) and whole-genome resources

Advances in NGS platforms such as Illumina and long-read sequencing technologies (PacBio HiFi, Oxford Nanopore) have greatly facilitated reference genome assembly and annotation in horticultural crops. High-quality reference genomes now exist for many fruit and vegetable crops, including tomato (*Solanum lycopersicum*) (Hosmani *et al.*, 2019), cucumber (*Cucumis sativus*) (Osipowski *et al.*, 2020), grapevine (*Vitis vinifera*) (Shi *et al.*, 2023), banana (*Musa acuminata*) (Droc *et al.*, 2013), and apple (*Malus domestica*) (Peace *et al.*, 2019). These assemblies have provided unprecedented opportunities to anchor SNPs, identify gene models, and integrate multi-omics information into GP models.

Importantly, ongoing efforts in pan-genome construction are addressing the limitations of single reference genomes. For example, the tomato pan-genome project revealed significant presence/absence variations (PAVs) and structural variations linked to traits such as fruit weight, sugar content, and disease resistance (Gao *et al.*, 2019). Similar initiatives in brassicas, potato, and pepper are uncovering novel alleles not present in reference cultivars, thereby improving genomic coverage for GP.

8.4.3 Polyploid and clonally propagated crops

Many economically important fruit and vegetable crops such as potato, strawberry, sweet potato, and banana are polyploids, which complicates genotyping, due to allele dosage effects, complex inheritance patterns, and difficulties in distinguishing homologous from homeologous loci. Specialized genotyping platforms and algorithms have been developed for polyploid crops. For instance, GBS approaches with dosage-sensitive SNP calling (Olukolu and Yencho, 2024) and polyploid-specific SNP arrays (e.g. the Axiom IStraw90 for cultivated strawberry) have improved marker resolution for genomic prediction in polyploid crops.

Clonally propagated crops such as potato, banana, and cassava present additional challenges because recombination is often suppressed, and elite cultivars are maintained through vegetative propagation. This reduces effective population sizes and complicates the creation of representative training populations for GP. Nonetheless, dense SNP marker sets combined with haplotype-based methods have proven effective in capturing genetic variance for predictive modeling in clonal crops.

8.4.4 High-density genetic maps and linkage disequilibrium (LD) patterns

High-density genetic maps are essential for evaluating the distribution of linkage disequilibrium (LD) across genomes, which in turn influences the resolution of genomic prediction. In many fruit crops, LD decays slowly due to clonal propagation or narrow genetic bottlenecks, as observed in apple and grapevine, where LD extends over hundreds of kilobases (Nybom and Lacis, 2021). In contrast, outcrossing vegetable species like cucumber display faster LD decay, necessitating denser SNP coverage for effective prediction.

The development of integrated genetic maps combining multiple biparental populations or diversity panels has improved marker density and facilitated cross-population prediction. These maps also help breeders design optimized training populations by accounting for LD structure, thereby enhancing the transferability of GP models across germplasm pools.

8.4.5 Crop-specific genotyping platforms

Several crop-specific genotyping platforms have been developed to accelerate GP implementation:

- **Tomato:** SolCAP SNP array (Sauvage *et al.*, 2020), complemented by resequencing and pan-genome projects.
- **Potato:** SolCAP 20K array, updated with dosage-aware SNP calling pipelines for autotetraploids.
- **Apple:** 20K and 480K SNP arrays, enabling large-scale genomic selection in international breeding consortia.
- **Strawberry**: IStraw90 and Axiom IStraw35 arrays for octoploid genotyping.
- **Grapevine**: 18k SNP array widely used for mapping and selection.
- **Pepper and cucumber**: GBS-based resources integrated with draft genome assemblies.

These resources have enabled predictive modeling of traits such as fruit size, disease resistance, flowering time, and yield across fruit and vegetable crops.

8.4.6 Emerging technologies: Pan-genomics, structural variation, and multi-omics integration

While SNPs remain the backbone of genotyping for GP, emerging research highlights the importance of structural variants (SVs), copy number variations (CNVs), and presence/absence variants (PAVs) in shaping complex traits. Long-read sequencing and Hi-C scaffolding have improved the detection of these variants, which are often overlooked in SNP-based studies. For example, in tomato, large-effect structural variants in the *fw2.2* locus contribute significantly to fruit size variation (Clevenger, 2012).

Moreover, integration of epigenomic and transcriptomic data with genotyping provides deeper insights into regulatory variants affecting fruit ripening, stress responses, and metabolite accumulation. This multi-omics integration is increasingly being explored in GP models to enhance predictive accuracy.

8.4.7 Databases and bioinformatics tools

The rapid accumulation of genotyping data necessitates robust databases and bioinformatics tools for storage, management, and analysis. Prominent databases include:

- **Genome Database for Rosaceae (GDR)** for apple, peach, and related species.
- **Sol Genomics Network (SGN)** for tomato, potato, and pepper.
- **Cucurbit Genomics Database (CuGenDB)** for cucumber, melon, and watermelon.
- **Grape Genome Database (CRIBI, URGI)** for grapevine.

These platforms provide access to SNP datasets, genome browsers, genetic maps, and functional annotations, making them indispensable for breeders implementing GP pipelines.

8.5 Statistical and Machine Learning Models for Genomic Prediction in Fruit and Vegetable Crops

The effectiveness of genomic prediction (GP) depends not only on the quality of genotypic and phenotypic data but also on the statistical models and computational frameworks used to translate these data into predictive breeding values. Over the past two decades, a wide range of models have been applied to GP, ranging from linear mixed models rooted in quantitative genetics to Bayesian regression, kernel-based methods, and modern machine learning and deep learning approaches. In fruit and vegetable crops, the diversity of reproductive systems, genetic architectures, and trait complexities necessitates careful selection and tailoring of models. This section reviews the principal statistical and machine learning models used in GP, with emphasis on their applications, strengths, and limitations in fruit and vegetable crop breeding.

8.5.1 Best linear unbiased prediction (BLUP) and genomic BLUP (GBLUP)

The BLUP framework, originally developed for animal breeding, estimates breeding values using phenotypic and pedigree data. With the advent of molecular markers, the approach was extended to genomic BLUP (GBLUP), which replaces pedigree relationships with realized genomic relationships derived from genome-wide SNP data (Misztal *et al.*, 2020). GBLUP assumes that all markers contribute equally to the genetic variance with normally distributed small effects.

- **Applications:** In fruit crops, GBLUP has been successfully applied in apple for predicting disease resistance and fruit quality traits (Kumar *et al.*, 2012), in grapevine for yield-related traits (Bharati *et al.*, 2023), and in tomato for shelf life and soluble solid content (Yamamoto *et al.*, 2017).
- **Strengths:** Simple to implement, computationally efficient, and robust with moderate-sized datasets.
- **Limitations:** Reduced accuracy when traits are controlled by few large-effect loci or when marker density is insufficient to capture linkage disequilibrium (LD).

GBLUP remains widely used in horticultural GP because it provides a reliable baseline against which more complex models can be compared.

8.5.2 Ridge regression BLUP (RR-BLUP)

RR-BLUP is closely related to GBLUP but applies a ridge penalty on marker effects rather than modeling genetic relationships directly. This penalization reduces overfitting in high-dimensional datasets where the number of markers far exceeds the number of individuals.

- **Applications:** In **potato**, RR-BLUP has been used to predict tuber yield and disease resistance under multi-environment trials (Gerakari *et al.*, 2025). In pepper, it has shown moderate success in predicting fruit size and shape traits (Jaiswal *et al.*, 2022).
- **Strengths:** Effective for highly polygenic traits, scalable to large marker sets.
- **Limitations:** Like GBLUP, assumes constant marker effect variance and may underperform for traits with major QTLs.

8.5.3 Bayesian regression models

Bayesian methods extend linear models by allowing flexible prior distributions on marker effects, making them suitable for traits with diverse genetic architectures. The most common models include:

- **BayesA:** Assumes all markers have effects with variances drawn from scaled-t distributions.
- **BayesB:** Allows some markers to have zero effect, making it well-suited for traits influenced by a few large QTLs.
- **BayesCπ:** Similar to BayesB but with a shared variance for non-zero marker effects.
- **Bayesian LASSO**: Shrinks marker effects based on double-exponential priors.
- **Applications:**
  In **strawberry**, Bayesian models outperformed GBLUP for predicting fruit firmness and soluble solid content due to the presence of major QTLs (Gezan *et al.*, 2017).
- **Strengths**: Flexibility to capture oligogenic and polygenic architectures.
- **Limitations**: Computationally demanding, requiring careful tuning of hyperparameters.

Bayesian models are increasingly favored in fruit and vegetable GP due to the common presence of large-effect loci linked to disease resistance and quality traits.

8.5.4 Kernel methods: reproducing kernel Hilbert space (RKHS)

Kernel-based approaches such as RKHS regression allow the modeling of nonlinear relationships between markers and phenotypes by projecting data into higher-dimensional feature spaces. This makes them suitable for traits influenced by dominance, epistasis, or genotype-by-environment (G × E) interactions.

- **Applications**: RKHS models have been applied in tomato for predicting fruit yield under stress conditions (Hernández-Bautista *et al.*, 2020) and in solanaceous crops for modeling complex flavor-related traits with significant epistatic interactions (Gerakari *et al.*, 2025).
- **Strengths**: Captures nonlinear genetic effects, increases accuracy in traits with non-additive variance.
- **Limitations**: Higher computational cost, sensitive to kernel choice, risk of overfitting with small datasets.

Kernel methods are promising for crops like pepper, cucumber, and grapevine, where flavor, texture, and quality traits often involve epistatic gene interactions.

8.5.5 Machine learning models

Machine learning (ML) approaches such as random forests (RF), support vector machines (SVMs), and gradient boosting machines (GBMs) have been tested in GP pipelines. Unlike linear models, ML methods can capture complex marker–trait relationships without explicit genetic assumptions.

- **Random forests (RF)**: Effective for variable selection and modeling non-additive effects but can suffer from reduced predictive power when marker numbers are extremely high relative to sample size.
- **Support vector machines (SVMs)**: Capture nonlinear relationships using kernel tricks but require parameter optimization.
- **Gradient boosting machines (GBMs)**: Combine weak learners (decision trees) into strong predictors, improving prediction accuracy.
- **Applications**: RF and SVMs have been tested in cucumber for disease resistance prediction with moderate success (Mia *et al.*, 2021). In tomato, GBLUP showed improved performance for predicting fruit traits (Yeon *et al.*, 2024).
- **Strengths**: Flexibility, ability to capture interactions, and suitability for multi-trait prediction.
- **Limitations**: Computationally intensive, risk of overfitting, lack of interpretability.

8.5.6 Deep learning approaches

Deep learning (DL) has recently been applied in GP to model highly complex genotype–phenotype relationships using architectures such as convolutional neural networks (CNNs), recurrent neural networks (RNNs), and deep fully connected networks (DNNs).

- **CNNs** are particularly effective for structured genomic data, capturing local LD blocks and haplotype effects.
- **RNNs** are suited for sequential data, such as time-series phenotyping linked to fruit development stages.
- **Autoencoders** can reduce dimensionality of genotype data while preserving informative features.
- **Applications**: In sweet pepper, CNNs have been shown to outperform GBLUP for predicting yield-related traits under varying environments (Maleki *et al.*, 2025).
- **Strengths**: Capture high-order interactions, integrate heterogeneous data (genomics, transcriptomics, phenomics).
- **Limitations**: Require large training datasets, high computational resources, and careful tuning to avoid overfitting.

Despite these challenges, DL is considered one of the most promising directions for GP in fruit and vegetable crops, particularly when paired with high-throughput phenotyping and multi-omics integration.

8.5.7 Model comparison and selection

The relative performance of GP models depends on trait architecture, population structure, and available data:

- **GBLUP and RR-BLUP**: Baseline models for polygenic traits with moderate heritability.
- **Bayesian methods**: More effective for traits with major QTLs (disease resistance, color, specific metabolites).
- **Kernel methods (RKHS)**: Useful when epistasis or dominance contributes substantially to genetic variance.
- **Machine learning**: Flexible for multi-trait and nonlinear prediction but requires large, diverse datasets.
- **Deep learning**: Offers the greatest potential for complex traits and integrative prediction but is data-hungry.

Model ensembles that combine linear and nonlinear methods are emerging as powerful alternatives. For example, hybrid frameworks combining GBLUP and DL models have been proposed to maximize prediction accuracy while maintaining interpretability (Li *et al.*, 2025).

To integrate genotypic and phenotypic information efficiently, genomic prediction models are embedded into structured breeding pipelines (Table 8.1). A conceptual workflow of genomic prediction models for fruit and vegetable crops is illustrated in Fig. 8.1, highlighting the steps from raw data acquisition to model training and final breeding decisions. This framework demonstrates how linear, Bayesian, kernel, and machine learning/deep learning models can be flexibly applied depending on trait architecture, dataset size, and breeding objectives.

8.6 Applications of Genomic Prediction in Fruit and Vegetable Crops

Genomic prediction (GP) has moved from theoretical exploration to practical application across a wide spectrum of fruit and vegetable crops. These species present unique challenges, due to their biological diversity, polyploidy, clonal propagation, long juvenile phases, and complex trait architectures, yet they are also high-value crops where even small improvements in yield, disease resistance, or quality can have significant economic impacts. This section highlights crop-specific applications of GP, illustrating advances in tomato, potato, grapevine, apple, cucumber, and strawberry, which represent both vegetatively and seed-propagated crops. Table 8.2 summarizes the applications of genomic prediction (GP) in fruit and vegetable crops, with traits, models used, and reported prediction accuracies.

8.6.1 Tomato

Tomato (*Solanum lycopersicum*) is a model crop for GP research owing to its well-annotated genome and availability of high-density genotyping arrays. GP has been applied to predict fruit quality traits, including firmness, soluble solid content, flavor-related volatiles, and shelf life. For instance, Adunola *et al.* (2024) reported that Bayesian and RKHS models outperformed GBLUP for predicting fruit shelf life, reflecting the oligogenic nature of some postharvest traits in blueberry. GP has also been used in disease resistance breeding, with moderate success in predicting resistance to *Fusarium oxysporum* under greenhouse and field conditions (Kaushal *et al.*, 2020).

8.6.2 Potato

Potato (*Solanum tuberosum*) is an autotetraploid crop where GP is highly relevant due to the complexity of allele dosage effects and the difficulties in conventional selection. Ortiz *et al.* (2022) demonstrated that GP models accounting for polyploidy, such as Bayesian models and GBLUP with dosage effects, improved prediction of tuber yield, starch content, and late blight resistance. GP is increasingly being implemented in commercial potato breeding programs, where prediction-based selection accelerates cycle time and reduces reliance on multi-year clonal trials.

Table 8.1. Comparison of genomic prediction models in fruit and vegetable crops with applications (Author's own table)

Model	Key features	Strengths	Weaknesses	Crop examples
GBLUP (genomic best linear unbiased prediction)	Uses genomic relationship matrix; assumes equal, small marker effects	Simple, robust, computationally efficient; good baseline model	Underestimates effects of large QTLs; limited for oligogenic traits	Apple – fruit firmness and disease resistance (Kumar *et al*., 2012); Tomato – quality traits (Yamamoto *et al*., 2017); Grapevine – yield-related traits (Bharati *et al*., 2023)
RR-BLUP (ridge regression BLUP)	Penalized regression; shrinks all marker effects equally	Handles large marker datasets; good for polygenic traits	Assumes constant marker variance; less accurate for major-effect loci	Potato – tuber yield and late blight resistance (Gerakari *et al*., 2025); Pepper – fruit size and shape (Jaiswal *et al*., 2022)
Bayesian models (BayesA, BayesB, BayesCπ, Bayesian LASSO)	Flexible priors on marker effects; allow some loci to have large or zero effects	Effective for traits with major QTLs; captures sparse genetic architecture	Computationally intensive; requires tuning	Strawberry – firmness, soluble solids (Gezan *et al*., 2017)
RKHS (kernel-based methods)	Captures nonlinear relationships via kernel functions	Models epistasis, dominance, and G × E; higher accuracy in complex traits	High computational cost; risk of overfitting	Tomato – yield under stress (Hernández-Bautista *et al*., 2020); Solanaceous crops – flavor-related traits (Gerakari *et al*., 2025)
Random forest (RF)	Ensemble decision trees; variable importance ranking	Captures non-additive effects; interpretable variable selection	Reduced performance with very high marker density; risk of overfitting	Cucumber – disease resistance (Mia *et al*., 2021); Tomato – vegetative growth traits (Yeon *et al*., 2024)
Deep learning (CNNs)	Neural architectures capturing high-order interactions	Integrates multi-omics; handles complex LD and trait networks	Data-hungry; computationally expensive; interpretability issues	Sweet pepper – yield traits under different environment (Maleki *et al*., 2025)

8.6.3 Grapevine

Grapevine (*Vitis vinifera*) breeding is hindered by long juvenile phases and strong genotype-by-environment (G × E) interactions. GP has emerged as a powerful tool to accelerate improvement for disease resistance (powdery and downy mildew), fruit yield, and wine quality traits. Brault (2021) showed that GBLUP and kernel methods achieved moderate prediction accuracies for complex traits such as berry weight and anthocyanin concentration. More recently, Zhang *et al*. (2025) applied deep learning models integrating SNP and transcriptomic data, which improved predictive ability for aroma-related secondary metabolites. Importantly, GP in grapevine has demonstrated potential to shorten breeding cycles by early selection of seedlings before vineyard establishment.

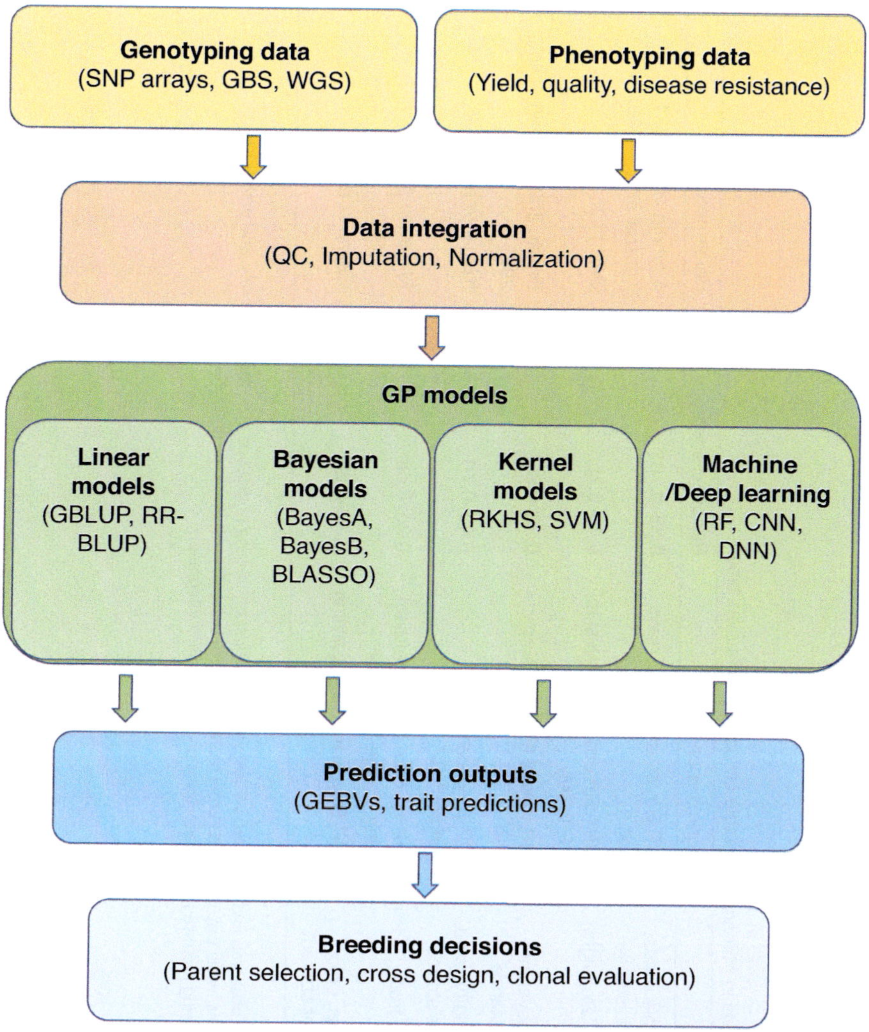

Fig. 8.1. Conceptual workflow of genomic prediction models in fruit and vegetable breeding pipelines. The workflow begins with phenotypic and genotypic data collection (e.g. SNPs, multi-omics, phenomics), followed by preprocessing (quality control, normalization, and imputation). Various modeling approaches are then applied, ranging from linear models (GBLUP, RR-BLUP), Bayesian approaches, and kernel methods (e.g. RKHS, SVM) to advanced machine learning (RF, GBM) and deep learning architectures (CNNs, RNNs, DNNs). The outputs of these models, expressed as genomic estimated breeding values (GEBVs), are used to inform selection, crossing, and breeding strategies in fruit and vegetable crops (Author's own image).

8.6.4 Apple

Apple (*Malus domestica*) is characterized by long juvenile phases and high heterozygosity, making it a prime target for GP. Early work by Kumar *et al.* (2012) demonstrated that GP could predict fruit firmness, acidity, and sugar content with accuracies sufficient for pre-field selection. More recent applications have integrated multi-environment trials into GP frameworks, improving accuracy for traits with strong G × E such as yield and storage performance. Additionally, GP

Table 8.2. Applications of genomic prediction models in fruit and vegetable crops (Author's own table)

Crop	Target traits	GP models applied	Reported accuracy (r)	References
Tomato (*Solanum lycopersicum*)	Fruit firmness, soluble solids, shelf life, disease resistance	GBLUP, Bayesian, RKHS, deep learning	0.35–0.70	Adunola *et al.*, 2024; Kaushal *et al.*, 2020
Potato (*Solanum tuberosum*)	Tuber yield, starch, late blight resistance	GBLUP (with dosage), Bayesian, ML (RF, GBM)	0.40–0.75	Ortiz *et al.*, 2022
Grapevine (*Vitis vinifera*)	Berry weight, anthocyanins, aroma, mildew resistance	GBLUP, Kernel (RKHS), deep learning	0.30–0.65	Brault, 2021; Zhang *et al.*, 2025
Apple (*Malus domestica*)	Firmness, sugar, acidity, yield, storage, disease resistance	GBLUP, Bayesian, multi-environment GP	0.40–0.70	Kumar *et al.*, 2012;
Cucumber (*Cucumis sativus*)	Disease resistance, fruit shape, yield	GBLUP, random forest, gradient boosting, image-based GP	0.35–0.68	Mia *et al.*, 2021
Strawberry (*Fragaria × ananassa*)	Sweetness, firmness, color, shelf life, antioxidant content	GBLUP, Bayesian, deep learning	0.30–0.60	Gezan *et al.*, 2017

Notes:
- Prediction accuracy values (Pearson correlation between observed and predicted values) vary depending on training population size, marker density, and environment.
- Linear models (GBLUP, RR-BLUP) provide stable baselines, while Bayesian and kernel models perform better for oligogenic or nonlinear traits.
- Machine learning and deep learning approaches are showing promise for complex traits and multi-omics integration.

has been applied for disease resistance, particularly against apple scab (*Venturia inaequalis*) and fire blight (*Erwinia amylovora*), with Bayesian models showing advantages when resistance loci of major effect are present. GP has now been embedded in several European and North American apple breeding consortia.

8.6.5 Cucumber

Cucumber (*Cucumis sativus*) is a seed-propagated diploid vegetable crop where GP has been tested primarily for disease resistance and yield-related traits. Mia *et al.* (2021) reported that machine learning models such as random forests and gradient boosting achieved higher accuracies than GBLUP for resistance to *Pseudoperonospora cubensis* (downy mildew). GP has also been applied for fruit shape and size, which are controlled by multiple small-effect loci.

8.6.6 Strawberry

Strawberry (*Fragaria* × *ananassa*), an octoploid crop, poses significant challenges for genomic prediction due to polyploid complexity and clonal propagation. Nevertheless, GP has shown promise for fruit quality traits such as sweetness, firmness, and color. Gezan *et al.* (2017) reported moderate predictive ability for fruit firmness using GBLUP. The deployment of GP in strawberry breeding programs is expected to reduce the number of years required for clonal evaluation and accelerate release of superior cultivars.

8.6.7 Summary of applications

The collective evidence across fruit and vegetable crops demonstrates that GP has transitioned from a research tool to a practical breeding strategy. While linear models (e.g. GBLUP, RR-BLUP) remain robust baselines, Bayesian, kernel, and machine learning approaches are increasingly applied depending on the trait genetic architecture. Multi-omics integration, high-throughput phenotyping, and deep learning are emerging as transformative approaches to further enhance GP accuracy. Importantly, the deployment of GP in perennial fruit crops (e.g. grapevine, apple) has the potential to drastically reduce breeding cycle times, whereas in annual vegetables (e.g. tomato, cucumber), GP facilitates rapid pyramiding of quality and disease resistance traits (Table 8.2).

8.7 Challenges and Future Perspectives

8.7.1 Challenges in applying GP to fruit and vegetable crops

Despite the significant advances of genomic prediction (GP) in major cereals, translating these models into fruit and vegetable crops presents unique challenges. Unlike annual field crops, most fruit and vegetable species exhibit complex biological and breeding characteristics, such as clonality, polyploidy, perenniality, and strong genotype × environment (G × E) interactions, which complicate model performance and generalization.

Limited and unbalanced training populations

Training populations for GP are often small in perennial fruits (e.g. apple, grapevine, citrus) due to long juvenile periods, limited funding, and breeding cycle constraints. Small datasets reduce the statistical power of models, especially for complex traits like yield stability and disease resistance.

Genetic complexity and polyploidy

Many fruit and vegetable crops, including strawberry (octoploid), potato (tetraploid), and sweet potato (hexaploid), possess polyploid genomes with multiple allelic interactions and dosage effects (Batista *et al.*, 2022). This genomic complexity limits the effectiveness of standard diploid GP models, necessitating specialized algorithms that account for allele dosage and polysomic inheritance.

Phenotyping bottlenecks

Phenotyping remains a major bottleneck, as GP models rely on high-quality phenotypic records. For traits such as fruit flavor, nutritional content, or shelf life, phenotyping is costly and often destructive, leading to incomplete datasets. High-throughput phenotyping (HTP) using imaging, spectroscopy, and sensor-based technologies is still underutilized in fruit and vegetable crops compared with cereals (Angidi *et al.*, 2025).

Environment × genotype (G×E) interactions

Fruit and vegetable crops are grown across diverse agro-ecological zones, making G × E interactions highly relevant. Prediction models trained in one environment often fail to transfer well to another. Multi-environment GP models exist, but they require dense environmental covariates and sophisticated statistical frameworks (Malosetti *et al.*, 2013).

Computational and data integration challenges

The integration of large-scale genotyping, phenotyping, and multi-omics datasets poses computational challenges. Handling heterogeneous data types, including genomic, transcriptomic, metabolomic, and imaging data, requires robust pipelines and machine learning models capable of integrating multiple modalities (Ballard *et al.*, 2024).

8.7.2 Future perspectives

Despite these challenges, several opportunities can enhance the adoption of GP in fruit and vegetable crops.

Multi-omics integration

Integrating genomics with transcriptomics, metabolomics, and epigenomics has the potential to improve the prediction of complex traits such as flavor, nutritional value, and disease resistance (Zhang *et al.*, 2022). For example, combining genomic markers with metabolomic data has improved predictions for fruit quality traits in tomato and grapevine (Savoi *et al.*, 2022).

Deep learning and AI approaches

Deep learning models, including convolutional neural networks (CNNs) and graph neural networks (GNNs), offer enhanced capacity to capture nonlinearities, epistasis, and multi-locus interactions. These models, when coupled with high-throughput phenomics, may outperform traditional linear and Bayesian GP approaches in complex trait prediction.

High-throughput phenotyping (HTP) and remote sensing

Coupling GP with HTP platforms (such as drones, hyperspectral imaging, and automated greenhouse systems) can generate dense, multi-temporal datasets that capture dynamic trait expression. This integration enhances the capacity to model G × E interactions in fruit and vegetable breeding (Xu *et al.*, 2022).

Advances in polyploid GP models

New statistical frameworks are emerging for polyploid GP, including algorithms that incorporate allele dosage, double reduction, and dominance effects (Meirmans *et al.*, 2018). These advances are particularly important for potato, strawberry, and sweet potato, where diploid models underestimate genetic variance.

Genomic selection in pre-breeding and wild germplasm

Fruit and vegetable crops possess diverse wild relatives that harbor alleles for disease resistance, abiotic stress tolerance, and nutritional traits. GP can accelerate introgression breeding, helping to predict the value of wild alleles in hybrid backgrounds (Pratap *et al.*, 2021). This is particularly relevant for crops like tomato, cucumber, and pepper, where pre-breeding is ongoing.

Toward climate-resilient breeding

Climate change is reshaping the production environments for fruit and vegetable crops. GP combined with envirotyping and predictive modeling could accelerate the development of climate-resilient cultivars with improved water use efficiency, heat tolerance, and pest resistance (Cvejić *et al.*, 2022).

8.8 Conclusion

Genomic prediction (GP) has evolved into a cornerstone of modern plant breeding, offering a paradigm shift in evaluating genetic potential. For fruit and vegetable crops, GP reduces breeding cycles, enhances selection accuracy, and facilitates prediction of complex traits such as yield, quality, and stress tolerance. These crops pose unique challenges, including polyploid genomes, clonal propagation, long generation times, and multi-dimensional quality attributes. Nonetheless, advances in linear mixed models, Bayesian frameworks, kernel methods, and deep learning have shown promising results, with case studies in tomato, potato, grapevine, apple, cucumber, and strawberry achieving moderate to high predictive accuracies. Integration of polyploid-focused models, machine learning, multi-omics, and high-throughput phenotyping further strengthens GP applications. Looking forward, GP will increasingly shape breeding pipelines by combining with genomic selection, envirotyping, and pre-breeding, thereby enabling climate-smart, nutrition-sensitive, and market-oriented cultivar development essential for future global food security.

References

Adunola, P., Ferrão, L.F.V., Benevenuto, J., Azevedo, C.F. and Munoz, P.R. (2024) Genomic selection optimization in blueberry: Data-driven methods for marker and training population design. *The Plant Genome* 17(3), e20488.

Ahmad, S., Shah, M.H., Saniya, Raghib, M.G., Belwal, V. *et al.* (2025) High-throughput metabolomics for plant phenotyping. In: *High-Throughput Plant Metabolomics*. CAB International, Wallingford, UK, pp. 236–251.

Angidi, S., Madankar, K., Tehseen, M.M. and Bhatla, A. (2025) Advanced high-throughput phenotyping techniques for managing abiotic stress in agricultural crops – a comprehensive review. *Crops* 5(2), 8.

Balagué-Dobón, L., Cáceres, A. and González, J.R. (2022) Fully exploiting SNP arrays: A systematic review on the tools to extract underlying genomic structure. *Briefings in Bioinformatics* 23(2), bbac043.

Ballard, J.L., Wang, Z., Li, W., Shen, L. and Long, Q. (2024) Deep learning-based approaches for multi-omics data integration and analysis. *BioData Mining* 17(1), 38.

Batista, L.G., Mello, V.H., Souza, A.P. and Margarido, G.R. (2022) Genomic prediction with allele dosage information in highly polyploid species. *Theoretical and Applied Genetics* 135(2), 723–739.

Belay, Z.A. and Caleb, O.J. (2022) Role of integrated omics in unravelling fruit stress and defence responses during postharvest: A review. *Food Chemistry: Molecular Sciences* 5, 100118.

Bernard, A., Bénéjam, J., Roth, M., Lheureux, F. and Dirlewanger, E. (2025) Genomic prediction in persian walnut: Optimization levers according to genetic architecture of complex traits. *The Plant Genome* 18(2), e70047.

Bharati, R., Sen, M.K., Severová, L., Svoboda, R. and Fernández-Cusimamani, E. (2023) Polyploidization and genomic selection integration for grapevine breeding: A perspective. *Frontiers in Plant Science* 14, 1248978.

Bianco, L., Cestaro, A., Linsmith, G., Muranty, H., Denancé, C. *et al.* (2016) Development and validation of the Axiom® Apple480K SNP genotyping array. *The Plant Journal* 86(1), 62–74.

Brault, C. (2021) Grapevine breeding optimization with genomic and phenomic predictions. Doctoral dissertation, Montpellier SupAgro.

Chafai, N., Hayah, I., Houaga, I. and Badaoui, B. (2023) A review of machine learning models applied to genomic prediction in animal breeding. *Frontiers in Genetics* 14, 1150596.

Clevenger, J. (2012) Metabolic and genomic analysis of elongated fruit shape in tomato (*Solanum lycopersicum*). MSc Thesis, Ohio State University.

Crossa, J., Montesinos-López, O.A., Pérez-Rodríguez, P., Costa-Neto, G., Fritsche-Neto, R. *et al.* (2022) Genome and environment based prediction models and methods of complex traits incorporating genotype × environment interaction. In: Ahmadi, N. and Bartholome, J. (eds) *Genomic Prediction of Complex Traits: Methods and Protocols*. Humana, New York, pp. 245–283.

Cvejić, S., Jocić, S., Mitrović, B., Bekavac, G., Mirosavljević, M. *et al.* (2022) Innovative approaches in the breeding of climate-resilient crops. In: Benkeblia, N. (ed.) *Climate Change and Agriculture: Perspectives, Sustainability and Resilience*. John Wiley & Sons, pp. 111–156.

Droc, G., Lariviere, D., Guignon, V., Yahiaoui, N., This, D. *et al.* (2013) The banana genome hub. *Database* 2013, bat035.

Gao, L., Gonda, I., Sun, H., Ma, Q., Bao, K. *et al.* (2019) The tomato pan-genome uncovers new genes and a rare allele regulating fruit flavor. *Nature Genetics* 51(6), 1044–1051.

Gerakari, M., Katsileros, A., Kleftogianni, K., Tani, E., Bebeli, P.J. *et al.* (2025) Breeding of solanaceous crops using AI: Machine learning and deep learning approaches – a critical review. *Agronomy* 15(3), 757.

Gezan, S.A., Osorio, L.F., Verma, S. and Whitaker, V.M. (2017) An experimental validation of genomic selection in octoploid strawberry. *Horticulture Research* 4.

Hernández-Bautista, A., Lobato-Ortiz, R., García-Zavala, J.J., Cruz-Izquierdo, S., Chávez-Servia, J.L. *et al.* (2020) Prediction accuracy of genomic selection models for earliness in tomato. *Chilean Journal of Agricultural Research* 80(4), 505–514.

Heslot, N., Akdemir, D., Sorrells, M.E. and Jannink, J.L. (2014) Integrating environmental covariates and crop modeling into the genomic selection framework to predict genotype by environment interactions. *Theoretical and Applied Genetics* 127(2), 463–480.

Hosmani, P.S., Flores-Gonzalez, M., van de Geest, H., Maumus, F., Bakker, L.V. *et al.* (2019) An improved de novo assembly and annotation of the tomato reference genome using single-molecule sequencing, Hi-C proximity ligation and optical maps. *Biorxiv* 767764.

Jaiswal, V., Gahlaut, V., Chhapekar, S.S. and Ramchiary, N. (2022) Genomic selection in solanaceae: Status, opportunities and future prospects. In: Elias, A.A. and Goel, S. (eds) *Genomic Selection in Plants*. CRC Press, pp. 111–131.

Kaushal, A., Sadashiva, A.T., Ravishankar, K.V., Singh, T.H., Prasanna, H.C. *et al.* (2020) A rapid disease resistance breeding in tomato (*Solanum lycopersicum* L.). In: Gosal, S.S. and Wanai, S.H. (eds) *Accelerated Plant Breeding, Volume 2: Vegetable Crops*. Springer International, Cham, pp. 17–55.

Kumar, S., Chagné, D., Bink, M.C., Volz, R.K., Whitworth, C. *et al.* (2012) Genomic selection for fruit quality traits in apple (Malus × domestica Borkh.). *PLoS ONE* 7(5), e36674.

Laurens, F., Aranzana, M.J., Arus, P., Bassi, D., Bink, M. *et al.* (2018) An integrated approach for increasing breeding efficiency in apple and peach in Europe. *Horticulture Research* 5, 11.

Li, R., Zhang, D., Han, Y., Liu, Z., Zhang, Q. *et al.* (2025) Hybrid deep learning approaches for improved genomic prediction in crop breeding. *Agriculture* 15(11), 1171.

Maleki, H.H., Darvishzadeh, R. and Azad, N. (2025) Sweet pepper yield modeling via deep learning and selection of superior genotypes using GBLUP and MGIDI. *Scientific Reports* 15(1), 1–12.

Malosetti, M., Ribaut, J.M. and van Eeuwijk, F.A. (2013) The statistical analysis of multi-environment data: Modeling genotype-by-environment interaction and its genetic basis. *Frontiers in Physiology* 4, 44.

Meirmans, P.G., Liu, S. and van Tienderen, P.H. (2018) The analysis of polyploid genetic data. *Journal of Heredity* 109(3), 283–296.

Mendoza, F., Lu, R., Ariana, D., Cen, H. and Bailey, B. (2011) Integrated spectral and image analysis of hyperspectral scattering data for prediction of apple fruit firmness and soluble solids content. *Postharvest Biology and Technology* 62(2), 149–160.

Meuwissen, T.H., Hayes, B.J. and Goddard, M.E. (2001) Prediction of total genetic value using genome-wide dense marker maps. *Genetics* 157(4), 1819–1829.

Mia, M.J., Maria, S.K., Taki, S.S. and Biswas, A.A. (2021) Cucumber disease recognition using machine learning and transfer learning. *Bulletin of Electrical Engineering and Informatics* 10(6), 3432–3443.

Misztal, I., Lourenco, D. and Legarra, A. (2020) Current status of genomic evaluation. *Journal of Animal Science* 98(4), skaa101.

Navarro, A.T. (2023) Harvesting data from polyploid plants: Developing tools for genetic analysis in strawberry. Doctoral dissertation, Wageningen University and Research.

Nybom, H. and Lacis, G. (2021) Recent large-scale genotyping and phenotyping of plant genetic resources of vegetatively propagated crops. *Plants* 10(2), 415.

Olukolu, B.A. and Yencho, C. (2024) Evolution of molecular marker use in cultivated sweetpotato. In: Yench, G.C., Olukolu, B.A. and Isobe, S. (eds) *The Sweetpotato Genome*. Springer, Cham, pp. 27–44.

Ortiz, R., Crossa, J., Reslow, F., Perez-Rodriguez, P. and Cuevas, J. (2022) Genome-based genotype × environment prediction enhances potato (*Solanum tuberosum* L.) improvement using pseudo-diploid and polysomic tetraploid modeling. *Frontiers in Plant Science* 13, 785196.
Osipowski, P., Pawełkowicz, M., Wojcieszek, M., Skarzyńska, A., Przybecki, Z. *et al.* (2020) A high-quality cucumber genome assembly enhances computational comparative genomics. *Molecular Genetics and Genomics* 295(1), 177–193.
Peace, C.P., Bianco, L., Troggio, M., Van de Weg, E., Howard, N.P. *et al.* (2019) Apple whole genome sequences: Recent advances and new prospects. *Horticulture Research* 6, 59.
Pratap, A., Das, A., Kumar, S. and Gupta, S. (2021) Current perspectives on introgression breeding in food legumes. *Frontiers in Plant Science* 11, 589189.
Sauvage, C., Arnoux, S. and Causse, M. (2020) Population genomics of tomato. In: *Population Genomics: Crop Plants*. Springer International, Cham, pp. 533–571.
Savoi, S., Santiago, A., Orduna, L. and Matus, J.T. (2022) Transcriptomic and metabolomic integration as a resource in grapevine to study fruit metabolite quality traits. *Frontiers in Plant Science* 13, 937927.
Shawky, E., Zhu, W., Tian, J., Abu El-Khair, R.A. and Selim, D.A. (2025) Metabolomics-driven prediction of vegetable food metabolite patterns: Advances in machine learning approaches. *Food Reviews International* 41(4), 1051–1080.
Shi, X., Cao, S., Wang, X., Huang, S., Wang, Y. *et al.* (2023) The complete reference genome for grapevine (*Vitis vinifera* L.) genetics and breeding. *Horticulture Research* 10(5), uhad061.
Song, L., Ye, J. and Lu, Z. (2025) The integration of genomic and environmental data: Intelligent breeding for enhancing crop resilience and yield improvement. *Advances in Resources Research* 5(1), 167–185.
Stich, B. and Van Inghelandt, D. (2018) Prospects and potential uses of genomic prediction of key performance traits in tetraploid potato. *Frontiers in Plant Science* 9, 159.
Vervalle, J.A., Costantini, L., Lorenzi, S., Pindo, M., Mora, R. *et al.* (2022) A high-density integrated map for grapevine based on three mapping populations genotyped by the Vitis 18K SNP chip. *Theoretical and Applied Genetics* 135(12), 4371–4390.
Xu, Y., Zhang, X., Li, H., Zheng, H., Zhang, J. *et al.* (2022) Smart breeding driven by big data, artificial intelligence, and integrated genomic-enviromic prediction. *Molecular Plant* 15(11), 1664–1695.
Yamamoto, E., Matsunaga, H., Onogi, A., Ohyama, A., Miyatake, K. *et al.* (2017) Efficiency of genomic selection for breeding population design and phenotype prediction in tomato. *Heredity* 118(2), 202–209.
Yeon, J., Nguyen, T.T.P. and Kim, M. (2024) Prediction accuracy of genomic estimated breeding values for fruit traits in cultivated tomato (*Solanum lycopersicum* L.). *BMC Plant Biology* 24, 222.
Yousaf, M.F., Zia, M.A.B. and Naeem, M. (2023) Application of SolCAP genotyping in potato (*Solanum tuberosum* L.) association mapping. In: *Plant Genotyping: Methods and Protocols*. Springer, New York, pp. 415–435.
Zhang, H.M., Lyu, X.J., Sun, Z.Y., Sun, Q., Wang, Y.C. *et al.* (2025) GWAS identifies a molecular marker cluster associated with monoterpenoids in grapes. *Horticulture Research* uhaf144.
Zhang, R., Zhang, C., Yu, C., Dong, J. and Hu, J. (2022) Integration of multi-omics technologies for crop improvement: Status and prospects. *Frontiers in Bioinformatics* 2, 1027457.

9 Genomic Prediction Models for Trees

Anas Hamdani[1]*, Abdellatif Boutagayout[2]*, Mohamed Kouighat[3], Brahim Ouahzizi[4] and Atman Adiba[5]

[1]National Agricultural Research Institute, Meknes, Morocco; [2]Environment and Valorization of Microbial and Plant Resources Unit, Faculty of Sciences, Moulay Ismail University, Meknes, Morocco; [3]Research Unit of Plant Breeding and Plant Genetic Resources Conservation, National Institute of Agricultural Research, Regional Agricultural Research Center of Meknes, Meknes, Morocco; [4]Biochemistry of Natural Resources Team, Faculty of Sciences and Techniques, Moulay Ismail University, Errachidia, Morocco; [5]Regional Agricultural Research Center of Tadla, National Institute of Agricultural Research, Avenue Ennasr, Rabat, Morocco

Abstract

Genomic prediction of trees, an essential branch of plant breeding, focuses on predicting phenotypic traits such as growth, wood quality, and disease resistance. By leveraging genomic data and integrating genetics, molecular biology, and data science, this method transforms genetic improvement through early and accurate selection, bypassing lengthy phenotypic evaluations while enhancing tree species' performance. Tree breeding poses significant challenges because of the long lifespans of trees. However, genomic advancements have demonstrated the potential of using genetic information to optimize desirable traits. Various statistical models, often based on linear regression with predefined assumptions, have been developed for genomic predictions. Studies have indicated that while differences in predictive accuracy among genomic prediction models are minor, they consistently outperform traditional pedigree-based methods. This chapter explores the core principles of genomic prediction in trees, emphasizing modern statistical and algorithmic models while advocating for their integration as a cornerstone strategy in tree breeding programs.

Keywords: Genomic prediction, Genotyping, Phenotyping, Statistical models, Tree improvement

9.1 Introduction

Historically, plant breeding was based mainly on the phenotypic evaluation of traits of interest, observed directly in the field. This process generally requires several growth cycles in order to reliably identify the superior genotypes to be retained as progenitors for subsequent generations (Crossa *et al.*, 2024). These phenotypic traits, though, are heavily affected by the conditions in the environment, which means that it is hard to make a precise estimate of the real genetic merit of the individuals. The emergence of broadband technologies for genotyping has enabled new approaches to selection to emerge, providing higher precision and more robustness against the effects of the environment. These approaches are designed to minimize costs,

*Corresponding authors: hamdanianasse@gmail.com and a.boutagayout@edu.umi.ac.ma

 Plant Genomic Prediction: AI-Enhanced Models for Plant Genomics and Crop Improvement (ed. Jen-Tsung Chen)
DOI: 10.1079/9781836991328.0009

accelerate the selection procedure and produce a large volume that can be used for decision making. Of these, genomic prediction seems to be an effective and innovative strategy (Ejima *et al.*, 2023). The genetic value of a genotype is assessed on the basis of its genomic profile, instead of awaiting the full expression of its traits on the farm (Barreto *et al.*, 2024). This is based on the development of statistical models on reference populations, where genotypic data (i.e. genetic information derived from molecular markers discovered on DNA) and phenotypic data (i.e. perceptual traits like height, diameter, or wood density) are available simultaneously. When validated, such models could be applied to predict the genetic potential of genotyped but non-phenotyped individuals (Gholami *et al.*, 2021). This genomic prediction is increasingly being used by forest geneticists and tree breeders as a widely applied method for plant improvement (Ejima *et al.*, 2023). Using genomic prediction has since been largely adopted in plant development. This method, which does not require the tree to be fully developed or phenotyped *in situ*, makes it possible to assess a tree's genetic value simply on the information provided by its genome. What works well in one environment can be less effective in others, and certain genotypes that were originally ruled out can give excellent performance in conditions that have not yet been evaluated. The phenotypic representation that genotypes express may fluctuate significantly according to their environmental conditions (Barreto *et al.*, 2024). Furthermore, the ability to carry out fair and comprehensive tests in various contexts may be impeded by the high cost of testing and the high level of genotypes to be investigated in preliminary studies (Jarquin *et al.*, 2020). In order to surmount the limits of lengthy selection cycles, genomic prediction (GP) is proving promising as a way of identifying the genetic merit of individuals that have not been subject to phenotypic evaluation (Dias *et al.*, 2018). Considering the extended life cycles of trees, which can often require years or even several decades in order to characterize interesting qualities such as growth, wood quality, or disease resistance, this is a particularly important method for forest improvement (Sáenz-Romero *et al.*, 2016) Establishing sustainable, productive forests that are adaptive to changing environments may be aided by incorporating genomic prediction within forest breeding programs, which can significantly enhance the precision and effectiveness of breeding decision making (Lin *et al.*, 2018). This technique enables the most promising genotypes to be spotted at an extremely early stage, thereby considerably shortening breeding cycles in genetic improvement programs. It is particularly effective for complex variables such as yield, tolerance to abiotic stress, and disease resistance, where heritability is low or moderate. In brief, genomic prediction is now an essential way of enhancing plant breeding by spending less, gaining time, and improving the precision of genetic selections (Lenz *et al.*, 2020).

The objective of this chapter is to provide a thorough overview of genomic prediction models for tree crops, highlighting their underlying science and methodological developments to date. To enhance genetic selection for diverse and complex traits such as growth, wood quality, and tolerance to biotic and abiotic factors, this chapter examines conventional and new methodologies, including the integration of multi-omic data and the use of machine learning algorithms. In addition to examining the main issues related to the complexity of tree genetics and the technological limitations that exist, this chapter also gives practical applications in the field of forestry. Finally, it offers promising opportunities for genomic prediction to help make forestry stronger, more productive, and sustainable in light of climate change.

9.2 Scientific Foundations of Genomic Prediction in Trees

9.2.1 Essential concepts of quantitative genetics: Heritability, genetic and phenotypic variability

The phenotypic traits of plants were conventionally evaluated by hand, which represented a laborious, time-consuming, and frequently devastating procedure that was also subject to researcher bias. This is because, during their long lives, trees are subject to alternating periods of dormancy and development, and too many stresses, such as autumn and late spring frosts (Lebedev *et al.*, 2020). Inadequate comprehension of the phenotype formation mechanisms may lead to unforeseeable behavior on the part of certain genotypes under specific environmental conditions (Moraes *et al.*, 2018). For this reason, the phenotyping of trees needs to be performed in various environments, because the phenotype is affected by both genetic and environmental factors. Genomic prediction is no different, and phenotyping trees is as important as their genotyping, as much for trees to predict their genetic potential as for reference populations. Whereas genotyping was once the bottleneck, the spread of modern technologies has lowered genotyping costs and phenotyping is now the limiting factor (Li *et al.*, 2017). This has allowed extensive genetic analysis to be used, particularly at the genome and genetic marker level, to achieve this genomic prediction. This has been made possible by the advent of statistical methods that allow accurate prediction of marker effects (Robertsen *et al.*, 2019). The majority of tree traits – yield, resistance to biotic and abiotic stress, and wood density – have economic importance, multiple genes, and a complex genetic architecture that involves several genes with different levels of phenotypic effect. A major benefit of genomic prediction (GP) compared with conventional breeding is its capacity to effectively evaluate genotypes for such genetically complex multigenic traits, by taking into consideration the large number of genes with low effect (Jung *et al.*, 2020). Generally speaking, genetic complexity and heritability (h2) are directly linked to the number of genes and their interaction to control a trait. Traits controlled by a few genes with a large effect generally show greater heritability than traits controlled by many genes with varying levels of genotypic effect. Thus, heritability influences the GP and impacts its accuracy (Bernardo, 2016). All three concepts of heritability, genotypic variability, and phenotypic variability are vital to understanding how complex traits are passed on from one generation to the next, and to improve them within the framework of selection programs or adaptation studies.

9.2.2 Key concepts of genomic prediction: Presentation of the main models (BLUP, GBLUP, Bayesian, etc.)

Genomic prediction aims to estimate phenotypes from unexpressed genotypic data, after learning the relationship between genotype and phenotype (Fig. 9.1). This relationship is often complex, due to the simultaneous contribution of numerous genetic markers to phenotypic variation, and requires the use of appropriate models (Crossa *et al.*, 2024). The most commonly used statistical approaches, which form the historical basis, include the best linear unbiased

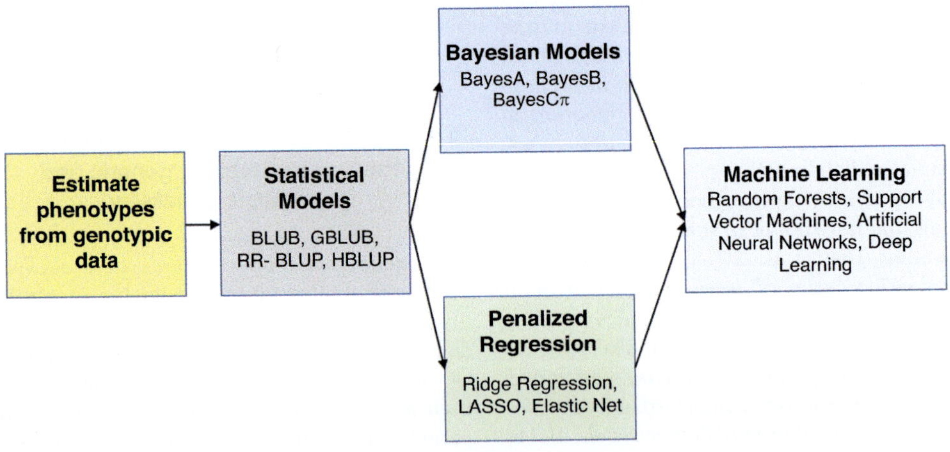

Fig. 9.1. Approach to genomic prediction based on genetic data (Author's own image).

Table 9.1. Comparison of the main models used in genomic prediction (Author's own table).

Model	Type	Main hypothesis	Principal advantage	Major limit
BLUP	Linear	Pedigree parentage	Simple, robust	No genomic info
GBLUP	Linear	SNPs effects	Uses genomic markers	Assumes equivalent effects
Bayesian	Linear	SNPs with variable effects	Flexibility, complex architecture	Complex, expensive
LASSO	Linear	Penalization of coefficients	Variable selection	Can ignore epistasis
Machine learning	Nonlinear	No *a priori* distribution	Capturing complex interactions	Interpretation, overlearning

prediction (BLUP) model and its genomic versions GBLUP, RR-BLUP, and HBLUP (Table 9.1) which assume that all marker influences are normally distributed, that these markers have equal variation with a small but non-zero effect, and that relatedness between individuals can be described by a pedigree matrix. This model can use genealogical information to estimate breeding values, but it does not consider genomic information from molecular markers. Thus, it can be exploited even when pedigrees are simple or absent, making it a preferred choice for forest tree breeding programs without a long history of commercial cultivation (Lebedev *et al.*, 2020). Similarly, Bayesian models such as BayesA, BayesB, and BayesCπ allow for a more nuanced consideration of variable marker effects that assume the Student's t distribution, unlike the normal distribution in BLUP, which outperforms them for disease resistance traits controlled by a limited number of loci (Lin *et al.*, 2014). These Bayesian models are better suited to traits where marker effects are modeled to follow a prior distribution. In addition to these classical models, penalized regression approaches such as ridge regression (equivalent to GBLUP), LASSO, or Elastic Net are also used. They allow for the processing of a large number of markers by reducing the dimensionality of the statistical problem while controlling overfitting. LASSO is particularly useful for variable selection, as it can constrain certain coefficients to zero, thus reducing the number of markers considered significant (Legarra *et al.*, 2011). Furthermore, machine learning methods, such as random forests (RFs), support vector machines (SVMs), artificial neural networks (ANNs), and deep learning (DL) models, offer the capacity to model nonlinear and complex interactions between markers (Montesinos-López *et al.*, 2021). The choice of model depends on the nature of the data available, the complexity of the traits studied, and the objectives of genomic selection (Cuevas *et al.*, 2024).

9.2.3 Particularities of genomic prediction in trees compared with annual crops

The domestication and selection of forest trees, through modern methods, have pushed researchers to solve the problems related to the adaptation of trees to climate change and to meet the high demands for wood for construction, fuel, and paper production (Isik *et al.*, 2015). This is why trees are considered an important crop, due to their use in the agri-food and industrial sectors (Jamnadass *et al.*, 2016). Therefore, the application of genomic prediction, which is a modern approach to selection and genetic improvement, seems to be more effective for trees to overcome the problems of yield and adaptability, particularly because they are characterized by a long life cycle and low-selection intensity (Seyum *et al.*, 2022). This has been endorsed by many studies, namely Bouvet *et al.* (2016) on eucalyptus, Minamikawa *et al.* (2018) on citrus, and McElroy *et al.* (2018) on cocoa, who all mentioned that genomic prediction showed high accuracy compared with conventional methods in predicting the genetic potential of non-phenotypic trees. Despite this, very few GP studies have been reported on fruit tree species compared with annual crops (Sun et al., 2024). Therefore, advances should recommend research on genomic prediction for trees to improve both the prediction and selection of more fruit crops.

9.3 Advanced Methods and Techniques

9.3.1 Genomic data generation technologies

Advanced sequencing platforms

The rapid development of sequencing technologies has revolutionized forest genomics, enabling the analysis of the genomes of tree species that are often complex, polyploid, heterozygous, and previously neglected (Lebedev *et al.*, 2020). High-throughput sequencing platforms, known as next-generation sequencing (NGS), have considerably reduced costs while increasing the speed and depth of genetic data analysis (Kalendar *et al.*, 2022). These advances have paved the way for the widespread use of genomics in breeding programs and the development of genomic prediction models (Zhang *et al.*, 2023).

Among the main approaches used, whole-genome sequencing (WGS) is the most exhaustive, providing complete coverage of the nuclear genome (Hong *et al.*, 2023). It enables the identification of all forms of genetic variation, single-nucleotide polymorphisms (SNPs), insertions/deletions (indels), and copy number variations (CNVs) (Wang *et al.*, 2025a). Although highly informative, this approach remains costly and requires significant computational resources, particularly for species with large genomes (Bourgeois and Warren, 2021). In contrast, genotyping-by-sequencing (GBS) stands out for its efficiency and low cost (Torkamaneh *et al.*, 2018). This method relies on enzymatic DNA digestion followed by targeted sequencing, which generates a large number of SNPs randomly distributed throughout the genome. It is well-suited to genetic diversity studies and associative analyses, even if it does not cover the entire genome (Sharma *et al.*, 2024). A similar approach, restriction site-associated DNA sequencing (RAD-seq), uses restriction enzymes to target specific regions of the genome (Kumar *et al.*, 2021). It is particularly useful for non-model species lacking a reference genome, notably for QTL mapping, the study of population genetic structure, and the establishment of markers for selection (Ramasamy *et al.*, 2025). RNA-seq, although derived from transcriptomics, plays an increasingly important role in genetic analysis. It enables the detection of SNPs in expressed genes (exonic SNPs), making it a powerful complementary tool for predictive models, particularly when examining physiological, adaptive, or stress-response traits (Gogolev *et al.*, 2021). Finally, targeted sequencing (such as exome capture) enables analysis to focus on regions of interest, such as exons or candidate genes. This strategy significantly reduces costs while maximizing the biological relevance of the data obtained, particularly for functional studies or targeted validations (Burridge *et al.*, 2022).

The choice of sequencing platform will depend on a number of factors: the available budget, complexity of the species genome, existence or non-existence of a reference genome, research objectives, and nature of the traits to be predicted. Thus, each technology represents a compromise between the depth of coverage, cost, and applicability, and must be carefully selected in the context of tree genomic prediction strategies (Table 9.2).

Genetic marker

One of the key stages in the use of data generated by sequencing technologies is the extraction of genetic markers, which are polymorphic elements of the genome that enable genetic variation to be linked to phenotypic variation within the tree populations (Mahajan and Gupta, 2012; Nam *et al.*, 2020). Molecular markers are polymorphic DNA fragments used to detect genetic variations between individuals. They play a central role in marker-assisted selection as well as in modern genomic prediction approaches by establishing a link between genotype and phenotype. There are two main categories: genomic markers based on DNA sequences, such as restriction fragment length polymorphism (RFLP), random amplified polymorphic DNA (RAPD), amplified fragment length polymorphism (AFLP), microsatellites (SSRs), indels insertions and deletions (indels), and single-nucleotide polymorphisms (SNPs);

Table 9.2. Benefits and limitations of the major sequencing platforms used in forest genomics (Author's own table).

Sequencing platform	Benefits	Limits
Whole-genome Sequencing (WGS)	Full coverage of the genome, allowing a comprehensive identification of genetic variations.	High cost, especially for species with large genomes.
	Identifies SNPs, indels and CNVs.	Requires an important data processing and storage infrastructure.
	Ideal for comprehensive studies of genetic variation.	High complexity of bioinformatics analysis.
Genotyping-by-Sequencing (GBS)	Low cost and fast, generating a large number of SNPs.	Covers only a limited portion of the genome.
	Ideal for genetic diversity and associative studies.	Does not allow a complete analysis of genetic variations.
	Easy to apply even for non-model species.	SNPs are often distributed randomly, which can make interpretation more difficult.
RAD-seq	Targeting specific regions of the genome with a small amount of data required.	Covers only a fraction of the genome and may leave unsampled regions.
	Particularly useful for species without a reference genome.	Less accurate in highly polypoid or structured regions.
	Effective tool for QTL mapping and genetic diversity analysis.	Requires a good knowledge of the restriction sites for good targeting.
RNA-seq	Allows analysis of gene expression and identification of expressed SNPs (exonic SNPs).	Requires high-quality RNA samples, which can be difficult to obtain in trees.
	Useful for the study of genes involved in physiological or adaptive traits.	Does not detect unexpressed variations or regulatory variations (non-coding).
	Validates candidate genes for genomic selection and prediction.	The complexity of transcriptome analysis can make interpretation difficult.
Targeted sequencing (exome capture)	Allows for focus on specific regions of interest (genes, exons, etc.), thus reducing costs.	Limited to targeted regions, which may leave out important genetic areas for some traits.
	Generates rich data for candidate genes or functional regions.	Does not cover the entire genome, which may limit the scope of global studies.
	Reduces the volume of data to be processed and simplifies bioinformatics analysis.	May not be representative of overall genetic variation if poorly targeted.

Table 9.3. Advantages and limitations of the different genetic markers used in forest genomics (SNPs, indels, SSRs, etc.) (Author's own table).

Genetic marker	Benefits	Limits
SNPs (single-nucleotide polymorphisms)	Very abundant and widely distributed in the genome.	Require high genomic coverage for accurate detection.
	Stable and co-dominant, allowing reliable analysis of genetic variation.	Susceptible to sequencing errors, which can lead to false positives.
	Suitable for large-scale analysis and quantitative genetics studies.	The cost of sequencing can be high for very large genomes.
Indels (Insertions/deletions)	May have functional effects, especially in coding regions (genes).	More sensitive to alignment errors than SNPs.
	Practices for the study of structural variations and their impact on phenotype.	Their detection is often more complex than that of SNPs.
	Can be used to detect rare events in the genome.	Calling variants is more difficult for small indels, which can result in incomplete coverage.
SSRs (simple sequence repeats)	Highly polymorphic and powerful for the study of genetic diversity and population structure.	Difficult typing on a large scale and more complex than for SNPs.
	Easy to use in population-level diversity studies.	Can be biased by amplification errors, making it difficult to use them for large-scale analyses.
	Excellent for kinship studies and genetic mapping.	Their frequency in the genome is lower than that of SNPs, limiting their ability to cover a wide range of genetic variation.
CNVs (copy number variations)	Useful for detecting variations in the number of copies of genes or specific regions of the genome.	Their detection requires advanced techniques (e.g. high-throughput DNA sequencing).
	May have important functional effects, especially in gene regulation.	Detection is often less accurate, especially for small variations.
	May be involved in complex traits or diseases.	Sensitive to sample quality and data alignment.

and transcriptomic markers derived from gene expression (by RNA-seq) (Bidyananda *et al.*, 2024). Molecular markers such as SNPs, indels, and microsatellites (SSRs) are at the heart of genomic prediction models used for breeding and the study of genetic diversity (Table 9.3). However, markers such as RFLP, AFLP, and SSR, although historically important, are rarely used for genomic prediction because of their low resolution, high cost, and the need for more accurate prediction (Kumar *et al.*, 2024). SNPs are the most abundant form of genetic variation and are widely used in genomic selection studies (Delfini *et al.*, 2021). These variations consist of the substitution of a single nucleotide base at a given position in the genome. SNPs represent the reference standard because of their high density, abundance, stability, homogeneous distribution,

Table 9.4. Software tools for variant detection and their application in tree genomics (Author's own table).

Software	Main function of the software	Tree species	Reference
GATK	Detection and annotation of variants (SNPs, indels) from high-speed sequencing data	*Eucalyptus robusta*	Silva Junior *et al*. (2015)
		Pear (*Pyrus* spp.)	Zhang *et al*. (2021)
		Litchi (*Litchi chinensis* Sonn.)	Deng *et al*. (2022)
SAMtools + BCFtools	Alignment file handling (SAM/BAM) and variant call (VCF/BCF)	Henry chinkapin (*Castanea henryi*)	Scarcelli *et al*. (2016)
		Populus (*Populus simonii* var. *przewalskii*)	Mousavi *et al*. (2016)
MUM & Co	Detection of structural variants from comparisons between genome assemblies	The chocolate tree (*Theobroma cacao*)	Hämälä *et al*. (2021)
SyRi	Comparison of genomes and detection of structural variants (insertions, deletions, duplications, inversions, etc.)	Japanese chestnut (*Castanea crenata*)	Wang *et al*. (2022)
Sniffles_v2	Detection of structural variants from long readings (PacBio, Nanopore)	Pear (*Pyrus* spp.)	Liu *et al*. (2022)
Manta	High sensitivity detection of structural variants (insertions, deletions, duplications, inversions, translocations)	Papaya (*Carica papaya* L.) (often considered a small tree)	Liao *et al*. (2021)
		Peach (*Prunus persica* L.)	Li *et al*. (2021)
Delly	Detection of structural variants via matching algorithms	Apple (*Malus pumila* Mill.)	Shen *et al*. (2022)
		Butternut (*Juglans cinerea* L.)	Chen *et al*. (2022)
CNVnator	Detection of variations in the number of copies (CNVs) from coverage data	Peach (*Prunus persica* L.)	Cao *et al*. (2024)
GRIDSS	Detection of structural variants by combining local alignment and paired reading	Peach (*Prunus persica* L.)	Guo *et al*. (2020)

and compatibility with high-throughput technologies, whereas transcriptomic markers are increasingly integrated into multi-omic approaches to enhance predictive accuracy (Kumar *et al.*, 2024). Their identification requires the use of bioinformatics pipelines, where sequence alignment is performed on a reference genome (where available), using software such as GATK, SAMtools + BCFtools, SyRi, Manta, or Dellyare to detect variants (Table 9.4) (Li *et al.*, 2021; Deng *et al.*, 2022; Wang *et al.*, 2022).

Indels, which refer to insertions or deletions in small DNA sequences, represent another important type of genetic variation. These variations can be functionally significant, particularly when they occur in coding regions of the genome (Zhou *et al.*, 2024). The detection of indels follows a process similar to that of SNPs, but this task can be more prone to alignment errors, requiring special attention during analysis (Song *et al.*, 2024). In addition to SNPs and indels, other types of markers, such as simple sequence repeats (SSRs), also known as microsatellites, have been used in studies of genetic diversity and population structure (Choudhury *et al.*, 2023). These markers, which consist of repeats of short DNA motifs, are useful for analyzing genetic variability; however, their use in predictive models is less frequent because of the difficulties associated with their automated typing, which can prove more complex (Alves *et al.*, 2024). The extraction of these genetic markers was followed by rigorous filtering to ensure the quality of the data used for the prediction models. This filtering involves several criteria, such as the complete genotyping rate for each marker, minimum allelic frequency (MAF) to eliminate uninformative rare variants, and compliance with the Hardy-Weinberg equilibrium according to the study objective. Excessive heterozygosity may also indicate genotyping errors and is therefore also taken into account (Kanaka *et al.*, 2023). After these quality control steps, validated markers are used to build genotype matrices, which form the essential basis of genomic prediction models.

9.3.2 Genomic prediction methods

Classical statistical models (BLUP, GBLUP, Bayesian models 'BayesA, B, Cπ')

Classical statistical models play a fundamental role in genomic prediction, offering robust and efficient tools for estimating the genetic value of individuals from their genotypes. These models are particularly popular in the early stages of plant and tree genomic selection, as they enable population structure and relationships between individuals to be considered (Montesinos López *et al.*, 2022a; Sandhu *et al.*, 2022). One of the most widely used models in this field is the best linear unbiased prediction (BLUP) model, which was adapted into the genomic BLUP (GBLUP) model for genomic selection. GBLUP incorporates a genomic similarity matrix calculated from SNPs and estimates the genetic values of individuals using the following equation (Duarte *et al.*, 2024):

$$y = X\beta + Zu + \epsilon$$

where y is the vector of observed phenotypes, X and Z are the incidence matrices for fixed and random effects, respectively, β is the vector of fixed effects (e.g. the site effect), u is the vector of random additive genetic values, and ϵ represents the residual error. The random effect u was modeled using the genomic relatedness matrix G calculated from the SNPs. GBLUP assumes that all loci contribute equally to genetic variation, which can be a limitation when major quantitative trait loci (QTLs) are present. Bayesian models have been introduced to overcome some of the limitations of GBLUP, notably the assumption of constant variance for all markers. These models allow for more flexible modeling of loci effects using specific *a priori* distributions. Among the most widely used Bayesian models are BayesA, where each SNP has its own variance following an inverse-chi-squared *a priori* distribution, and BayesB, which assumes that some SNPs have zero effect (sparse model), whereas others follow a distribution similar to Bayes A (McGowan *et al.*, 2021). Other variants include BayesCπ and Bayesian LASSO, which estimate the proportion of uninformative SNPs and apply effect shrinkage, respectively. These Bayesian models are particularly well-suited to complex traits, controlled by a limited number of major-effect loci, and are effective in situations involving multi-collinearity between markers (Gianola and Fernando, 2020). In the forestry field, these models have proven effective in predicting traits of interest, such as growth, wood quality, and resistance to biotic and abiotic stresses. Their ability to manage complex family structures and process phenotypes measured over long periods is a valuable tool for accelerating selection cycles in tree-breeding programs (Lebedev *et al.*, 2020).

Machine learning and artificial intelligence (random forest, SVM, neural networks, deep learning)

With the increase in massive and complex genomic data, machine learning and artificial intelligence (AI) approaches are gradually being integrated into genomic prediction models (Esposito *et al.*, 2019). These methods offer a significant advantage in capturing nonlinear relationships and complex interactions between genetic markers, thereby improving prediction accuracy, particularly for traits with medium to low heritability (Xu *et al.*, 2022). Unlike classical statistical models, which often rely on strict parametric assumptions, machine learning algorithms can learn structures directly from data without the need for *a priori* assumptions (Mahood *et al.*, 2020). One of the most widely used algorithms is the random forest (RF), which is an ensemble model based on decision trees. Random forest constructs several trees from random subsamples of the data and then aggregates their predictions. This method has several advantages, including its ability to model nonlinear relationships and complex interactions between SNPs without the need to specify them explicitly (Eraslan *et al.*, 2019). It is also robust to overlearning and tolerant to noisy data, making it a valuable tool for genomic prediction. However, random forest (RF) is less effective on very large datasets and has limited interpretability compared with classical linear models (Montesinos López *et al.*, 2022b). Artificial neural networks (ANNs), inspired by the work of the human brain, represent another powerful approach to machine learning. These networks, particularly deep neural networks, can capture complex patterns in genomic data, including high-level interactions and epistatic effects (Xu *et al.*, 2022). Neural networks are particularly well-suited for predicting complex quantitative traits, such as tree growth, resistance, and yield, and enable the integration of multi-omics data, such as genomic, transcriptomic, and environmental data (Cembrowska-Lech *et al.*, 2023). However, their efficiency requires large volumes of data, which can make them difficult to use in studies with smaller datasets (Zang *et al.*, 2024). In addition to these popular methods, other artificial intelligence techniques show great potential for genomic prediction. These include gradient boosting machines (such as XGBoost and LightGBM), which are effective for classification and regression tasks, and support vector machines (SVMs), which are used for nonlinear classification and regression (Yan *et al.*, 2021). K-nearest neighbors (k-NN) are also used in simpler situations or as a comparative method. Autoencoders are used for dimensionality reduction or imputation of missing data (Huang *et al.*, 2023). Table 9.5 highlights the main strengths and weaknesses of each approach, making it easier to decide which model to use, depending on the specifics of the data and the objectives of genomic prediction. Current trends in genomic research are moving towards the hybridization of classical and modern approaches (Chakrabarty and Külheim, 2025). For example, the use of RF feature selection, followed by Bayesian modeling or the integration of neural network layers into statistical frameworks, makes it possible to combine the advantages of each method (Ao *et al.*, 2019). These hybrid approaches offer particular potential for modeling the phenotypic plasticity of trees, integrating spatiotemporal environmental variables, and predicting complex traits that are difficult to measure on a large scale, thus offering promising prospects for the genetic improvement of trees in complex environments (Lebedev *et al.*, 2020).

9.3.3 Towards integrated genomics

Multi-omics integration

Multi-omics integration represents a significant advancement in modern predictive approaches. It involves combining different biological data sources at several levels of regulation, including genomics, transcriptomics, proteomics, metabolomics, and phenotypic and environmental factors (Fig. 9.2) (Wang *et al.*, 2023). This integration aims to improve our understanding and prediction of complex traits governed by multiple biological mechanisms. In the context of trees, whose genomes are often large, heterozygous, and poorly annotated, multi-omics integration offers an effective solution for overcoming the limitations of genetic data based solely on SNPs. This allows the exploitation of the richness of molecular responses related to traits such as growth, stress resistance, and environmental adaptation (Abdullah-Zawawi *et al.*, 2022). The transcriptomic data obtained by RNA-seq allows the analysis of gene expression in specific tissues or at different stages of development (Xanthopoulou *et al.*, 2021). The

Table 9.5. Advantages and limitations of classical statistical models, machine learning, and artificial intelligence (AI) approach to genomic prediction (Author's own table).

Model	Advantages	Limits
Classical statistical models	Robustness and reliability of results. Interpretability of models (clear, fixed, and random effects). Less data required.	Strict parametric assumptions. Limited in managing nonlinear relationships and complex interactions.
BLUP (best linear unbiased prediction)	Widely used and proven in genomic selection. Good management of the genetic structure of populations.	Does not accurately model the major QTLs. Assumption of equal effects for all loci.
GBLUP (genomic BLUP)	Integrates genomic data via a kinship matrix. Ability to manage complex populations.	Less suitable for characters controlled by a small number of major QTLs. Limited accuracy for complex strokes.
Bayesian models	Flexibility in modelling the effects of loci. Ability to deal with heterogeneous effects between markers.	Requires an appropriate selection of a priori distributions. Computation-intensive and more complex to interpret.
Machine learning	Ability to detect non-linear relationships and complex interactions. Can handle large amounts of data.	Requires large data sets. High computational complexity and low interpretability.
Random forest	Robust to overlearning. Modeling of complex interactions between SNPs without pre-selection. Good performance on noisy data.	Less efficient on very large datasets. Less interpretable than conventional models.
Artificial neural networks	Excellent ability to model complex relationships. Possible integration of multi-omics data. Adapted to complex quantitative characters.	Requires large volumes of data. Very high computational complexity. Difficulty in biological interpretation.
Gradient boosting machines	High performance on nonlinear datasets. Good for classification and regression.	Long drive on big data. Less effective on very noisy data.
Support vector machines (SVMs)	Effective for nonlinear classification. Good for unbalanced data management.	Sensitive to the choice of parameters. Less effective on large datasets.

Continued

Table 9.5. Continued

Model	Advantages	Limits
k-Nearest neighbors (k-NN)	Simple to understand and apply. Used as a comparative method in simple situations.	Not effective on very large databases. Sensitive to noise in data.
Autoencoders	Good dimensionality reduction capability. Useful for the imputation of missing data.	Requires large amounts of data to be effective. Can be difficult to interpret.

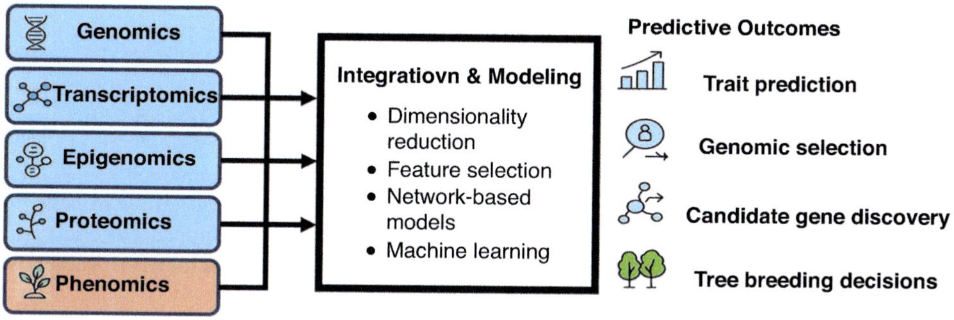

Fig. 9.2. Integrating multi-omics data into modern predictive approaches for trees. The figure illustrates the vertical integration of different omics layers – genomics, transcriptomics, epigenomics, proteomics, metabolomics, and phenotypics – into a unified analytical framework. These heterogeneous data are combined using statistical methods and artificial intelligence algorithms (machine learning, gene networks, Bayesian models, etc.) to predict complex traits of interest (growth, wood quality, stress resistance) and optimize genomic selection in tree improvement programs (Author's own image).

integration of these data with genotypes opens up several possibilities. It allows, for example, the identification of key regulatory genes, weight of SNPs located in or near differentially expressed genes, and improves the accuracy of phenotypic predictions by adding functional dimensions (Riehl *et al.*, 2023). Approaches such as the Transcriptome-Wide Association Study (TWAS) are used to associate gene expression with traits of interest (Ding *et al.*, 2022). These methods are sometimes combined with whole-genome association studies (GWAS) to provide a more complete and integrated view of the genetic and functional factors affecting phenotypes (Abdullah-Zawawi *et al.*, 2022). Metabolomic data provide an overview of biological activity by revealing the final metabolites that result from both gene expression and interaction with the environment. In trees, these data are particularly useful for characterizing wood quality by analyzing components, such as lignin, tannins, and essential oils. They also provide a better understanding of pathogen defense mechanisms and explore metabolic signatures associated with complex phenotypes (Rodríguez *et al.*, 2018). Proteomics, although less commonly used than transcriptomics or metabolomics, represents a direct functional link between genotypes and phenotypes. The integration of proteomics with other types of omics data can refine predictions and contribute to the identification of specific functional targets (Abril *et al.*, 2011). Several integration strategies for omics data are currently being explored. Horizontal integration is a combination of data of the same type from different sources or species, which allows for a wider range of analyses (Richardson *et al.*, 2016). In contrast, vertical integration merges different types of omics data for a single sample, such as the combination of genomic, transcriptomic, and phenotypic data (Cembrowska-Lech *et al.*, 2023). Network-based approaches, which construct gene–metabolite or gene–phenotype interaction networks, reveal functional links between different data. Finally, deep learning models, such as multi-input neural networks, offer a unified approach for integrating multi-omics and environmental matrices into increasingly complex and powerful predictive frameworks (Crossa *et al.*, 2024).

Genome editing and functional validation

The emergence of genome editing technologies offers new perspectives for the functional validation of genomic predictions and targeted improvement of traits of interest in trees. In addition to the traditional approaches of marker-assisted selection or genomic prediction, these tools allow for direct intervention of the genome to test, confirm, or modify the function of candidate genes (Pak and Li, 2022). This opens up important opportunities for refining and accelerating the genetic improvement of trees. CRISPR/Cas technology is a precise and efficient genome editing tool derived from the bacterial immune system, enabling the modification or regulation of specific genes (Fernandez i Marti and Dodd, 2018). It is available in several types, of which the most widely used in tree genomics are Cas9 (type II), widely used for gene deletion or modification, and Cas12a (type V), which is appreciated for its flexibility in DNA targeting and insertion (Zhao *et al.*, 2022; Martín-Valmaseda *et al.*, 2023). CRISPR/Cas9 allows for the precise, rapid, and targeted modification of genetic sequences through an endonuclease (Cas9) guided by a specific RNA (gRNA) that directs the cut of the DNA at the target site (Thapliyal *et al.*, 2023). This method allows genetic insertions, deletions, or substitutions. In the tree context, genomic editing has been successfully applied to several model and commercial species, such as poplar and eucalyptus, to modify genes related to wood growth or quality, improve tolerance to drought or pathogens, and study the function of genes identified by omics approaches or genome-wide association studies (GWAS) (Klápště *et al.*, 2024; Cao *et al.*, 2024). CRISPR variants, such as CRISPR/Cas12, CRISPR base editors (which allow double-strand editing without a break), or prime editing, represent additional advances, offering finer genetic modifications without the insertion of exogenous material (Sen *et al.*, 2025). Genome editing is also a key tool for the functional validation of genes identified using genomic prediction approaches, multi-omics analyses, or association studies. It allows, for example, confirmation of the direct involvement of a gene in a given phenotypic trait through approaches such as 'knock-out' (deletion of the gene) or 'knock-in' (insertion of a specific gene). Furthermore, genome editing offers the possibility of studying the pleiotropic effects of genes and their interactions with the environment, thus facilitating a better understanding of the molecular mechanisms that cause observed variations (Kumar *et al.*, 2021). These functional validation capabilities are essential for moving from statistical signals to biological proofs and for developing improved genotypes in a targeted and accelerated manner (Kamali and Singh, 2023).

9.4 Practical Applications

9.4.1 Improvement of priority phenotypic traits

Phenotype describes the physical, physiological, and biochemical characteristics of organisms that are determined or influenced by genes and the environment. Accurate extraction of phenotypic data is a prerequisite for the selection of forest plantations. Phenotypic traits can be extracted directly by observation or indirectly predicted by inversion models, helping to provide information on structural changes and their response to changing environments in tree growth and development. Combined with genetic trait assessments, forest phenotyping models will help to accelerate the plant breeding process. In forest trees, simulations based on deterministic models and experimental data have shown that genomic selection would have a considerable impact in terms of improving the efficiency of genetic gain per unit of time when the duration of the selection cycle is reduced and the intensity of selection is increased. Reports are available on tree species, including studies focusing on the relationship between genetic variation and environmental adaptation in *Eucalyptus grandis*, *Picea abies*, and *Pinus pinea*, as well as phenotypic plasticity or adaptation to local environments (He *et al.*, 2019). In the context of tree breeding, Niu *et al.* (2023) demonstrated the potential of gradient-boosting machine models for the accurate estimation of canopy nitrogen and non-structural carbohydrate contents in Eucalyptus. Floral traits are essential for the ornamental value of horticultural plants, but the long juvenile period of woody plants complicates their selection. A study explored the genetic architecture and genome prediction for flower pattern and color in *Camellia japonica* (Fan *et al.*, 2024). The results confirmed that the combination of GWAS-associated markers and machine learning technology can predict flower traits. In the study by Beaulieu *et al.*, 2014, high accuracy was obtained in genomic selection models to predict wood property traits in fir. In Pinus pinaster Aiton, Isik *et al.* (2015) reported that even with a relatively low marker density, the ability to predict growth traits remained high. The predictive ability of GP models can be increased by using different methodological strategies. (Cappa *et al.*, 2017) used the single-step GBLUP (ssGBLUP) method for growth and wood quality traits in Eucalyptus, which consists of simultaneously considering genotyped and non-genotyped trees. Genomic prediction models are an important determining factor in the selection of polygenic qualitative traits. Several studies have used prediction models to evaluate qualitative traits in fruit trees such as apple, peach, or citrus. In apple, genomic selection has only been partially addressed for fruit texture via predictions of classical fruit firmness measures and sensory evaluations. Using best linear unbiased predictors (BLUPs), Roth *et al.* (2020) successfully predicted fruit texture in apples, facilitating genomic selection for this trait. A study by Minamikawa *et al.* (2018) evaluated the use of genomic prediction models for fruit quality traits such as harvest time, fruit weight, flesh firmness, sugar content, acid content, fruit skin color, pre-harvest fruit drop, core rot, and watery heart in Japanese pear (*Pyrus pyrifolia* Nakai). Other studies have evaluated the effectiveness of predictive models in coffee (*Coffea canephora*) to predict key traits such as bean production, leaf rust resistance, and green bean yield. All this shows the potential of using predictive models in genomic selection to optimize tree selection, especially by reducing the duration of selection cycles and increasing genetic gain per unit time.

9.4.2 Case studies: Successes and advances in arboriculture

Tree breeding relies on long and costly selection cycles due to its long life cycle and the complexity of the traits of interest. The integration of genomic prediction models into forest breeding programs offers a promising alternative by allowing an early and efficient selection of the best-performing genotypes. Different models, such as GBLUP (genomic best linear unbiased prediction), Bayesian approaches (BayesA, BayesB, BayesCπ, Bayesian LASSO, Bayesian ridge regression), and reproducing kernel Hilbert spaces (RKHS) methods, have been applied to various forest species to predict traits related to growth, wood quality, and resistance to biotic and abiotic stresses (Table 9.6).

9.4.3 Towards sustainable and resilient forestry

The use of genetic prediction models represents a major advance in the development of sustainable and resilient forestry. These models make it possible to anticipate tree performance based on its genetic characteristics, thus optimizing the selection of individuals best suited to environmental challenges, including climate change, pests and diseases, and soil degradation. The integration of advanced statistical methods, such as Bayesian models and neural networks, refines the prediction of essential traits such as growth, resistance to biotic and abiotic stresses, and wood quality. Applying a forest dynamics model (LUBERON2) to combine the study of forest growth and demography with the analysis of genetic diversity and heredity processes can help assess the impact of silvicultural interventions on genetic variability and the resilience of forests in the face of future climate changes. In addition, genetic prediction

Table 9.6. Applications of genomic prediction models for improving traits of interest in different forest tree species (Author's own table).

Species	Traits Predicted	Model	Reference
Cryptomeria japonica D	Tree height, wood rigidity, and number of male flowers	BayesA, BayesB, BayesCπ, Bayesian lasso, Bayesian ridge regression (BRR).	Ejima *et al.* (2023)
Eucalyptus globulus Labill	Volume, stem straightness, diameter at breast height, branch-cluster frequency	Bayesian ridge regression, Bayesian LASSO, BayesB, BayesC, reproducing kernel Hilbert spaces, GBLUP, BRNN	Maldonado *et al.* (2020)
Eucalyptus grandis	Breast height, stem straightness	GBLUP multitrait	Cappa *et al.* (2017)
Eucalyptus pellita F. Muell	Diameter at breast height, tree height, volume	GBLUP, single-step GBLUP, BayesB, BayesC, Bayesian LASSO, Bayesian ridge regression	Müller *et al.* (2017)
Eucalyptus robusta Sm.	Volume, lignans, holo-cellulose	Reproducing kernel Hilbert spaces, GBLUP, EN	Rambolarimanana *et al.* (2018)
Eucalyptus nitens (H.Deane & Maiden) Maiden	Diameter at breast height, stem straightness, tangential air-dry shrinkage, wood-related traits	GBLUP, BayesC, BA, BayesB, Bayesian LASSO, Bayesian ridge regression	Klápště *et al.* (2017)
Fraxinus excelsior L.	Resistance to *Hymenoscyphus fraxineus*	GBLUP, BRR, BayesB. GBLUP with RR and GAUSS	Meger *et al.* (2024)
Hevea brasiliensis Muell. Arg	Rubber production, circumference at breast height	RKHS, RRBLUP with the variable selection procedure, Bayesian LASSO	Cros *et al.* (2019)
Picea glauca (Moench) Voss	Diameter at breast height, volume, acoustic velocity, pungenol concentration in needles	GBLUP	Lenz *et al.* (2020)
Pinus contorta Douglas ex Loudon	Microfibril angle, tree height,	GBLUP, BayesC	Ukrainetz (2019)
Pinus mariana (Mill.)	Britton, Sterns & Poggenb diameter at breast height, microfibril angle	GBLUP	Lenz *et al.* (2017)
Pinus radiata D. Don	Stem straightness, diameter at breast height, modulus of elasticity	GBLUP	Klápště *et al.* (2020)

Continued

Table 9.6. Continued

Species	Traits Predicted	Model	Reference
Pinus sylvestris Thunb.	Diameter at breast height, microfibril angle, modulus of elasticity,	GBLUP, Bayesian ridge regression, Bayesian LASSO	Calleja-Rodriguez *et al.* (2020)
Populus nigra L.	Tree height, circumference at breast height, bud flush, bud set, resistance to rust	GBLUP, Bayesian LASSO	Rincent *et al.* (2018)
Pseudotsuga menziesii (Mirb.)	Juvenile height	RR-BLUP with the variable selection procedure, BayesB	He *et al.* (2019)

models offer valuable insights into anticipating the influence of environmental factors on the growth and reproduction of forest species. For example, the different development phases of trees, such as flowering and seed production, are sensitive to temperature variations and can be modeled to predict their response to thermal stress. Unlike conventional approaches, which are often focused on monogenic traits, genetic prediction models offer a more comprehensive approach by identifying polygenic traits that influence tolerance to environmental stresses and diseases. This approach provides a sustainable solution for improving forest species. Integrating environmental data into machine learning algorithms also enables the assessment of the impact of extreme weather events on forest productivity, facilitating the adaptation of silvicultural strategies to future conditions. Predictive genomic selection plays a key role in identifying genetic markers associated with resistance to pathogens and abiotic stresses, paving the way for genetic improvement programs to enhance the resilience of forest species. Finally, by simulating the performance of new tree varieties under future climate conditions, these models actively contribute to forest resilience in the face of environmental changes. They also leverage emerging technologies, such as IoT sensors, drones, and weather stations, to provide real-time recommendations for forest management and adaptation to changing conditions. The integration of these digital tools and predictive analytics into forest management optimizes practices, minimizes environmental impact, and ensures the long-term sustainability of forest ecosystems.

9.5 Challenges and Limitations

9.5.1 Genetic complexity of trees: High variability, polyploidy, and genome size

Germplasm conservation can conserve genetic diversity, which is the degree of genetic heterogeneity among individuals of a variety, population, or species (Brown, 1983). Plant genetic diversity is widely seen as a crucial topic, as increasing urban populations and shrinking arable land are the main causes of food insecurity in the developing world. Agricultural scientists discovered that plant genetic resources, such as gene banks, DNA libraries, and so on, can be used to capture and retain genetic variety. These resources are kept in biorepositories, which maintain genetic material for extended periods (Govindaraj *et al.*, 2015). Another important characteristic of plants that usually does not change within species is genome size (Carta *et al.*, 2020). The DNA content of an unreplicated holoploid genome with n chromosomes is measured by the 1C-value, which is frequently used to symbolize it. In polyploid species, the size of the haploid genome is effectively shown by the 1Cx-value, which is the DNA content of an unreplicated haploid genome with x chromosomes (Greilhuber *et al.*, 2005). The 1C-value and 1Cx-value are the same in species that do not have polyploid reproduction (Greilhuber *et al.*, 2005). Most plant genomes have experienced polyploidization at some point, and it is a crucial aspect of plant evolution (Van de Peer *et al.*, 2017; Simonin and Roddy, 2018). Although autopolyploidization, or entire genome duplications within a species, is possible, allopolyploids are believed to be more prevalent. After hybridization between divergent lineages, allopolyploids are created by duplicating their genomes (Otto and Whitton, 2000; Mallet, 2007). The induction of polyploidy in crop species often creates opportunities and changes in breeding systems. Triploid breeding programs can be particularly beneficial; however, there have been only a few instances of implementation and evaluation in key forestry species, for example *Populus* (Kang and Wei, 2022), *Acacia* (Nghiem *et al.*, 2018), *Betula* (Mu *et al.*, 2012a), and *Eucalyptus urophylla* S.T.Blake (Yang *et al.*, 2018). Numerous plant breeding studies have focused on genomic prediction, and numerous breeding companies have been incorporating it into their breeding operations (Hickey *et al.*, 2017). However, the development and deployment of genomic prediction in polyploid crops have lagged behind, with the majority of these efforts concentrating on diploid crops (Wilson *et al.*, 2021). The intricate genetic makeup of polyploids is the cause of this delay. With more than two homologous copies of each chromosome, polyploids differ from diploids in that they have a higher degree of heterozygosity, a greater number of potential allelic dosages, a greater variety of genetic classes, and the potential for multivalent pairing (Ramsey and Schemske, 2002; Dufresne *et al.*, 2014; Lloyd and Bomblies, 2016). These features introduce substantial analytical and computational challenges, which have hindered the development and practical implementation of genomic prediction methods in polyploid species (Endelman *et al.*, 2018; De Bem Oliveira *et al.*, 2019). A lot of work has been done in the past few decades to understand and apply GP in breeding programs. Nevertheless, it is questionable if the majority of current techniques can be directly applied to polyploid crops, because they are based on diploid crops. Selecting the appropriate genomic prediction model is not always simple, because a variety of factors influence the prediction's accuracy. There is more genetic variance to capture when a model has a high heritability, which will often improve its predictive power. Nevertheless, Meher *et al.* (2022) discovered that not all models experience this gain equally, with variable selection models typically benefiting more from an increase in heritability than shrinkage models would.

9.5.2 Costs and technical challenges associated with genomic technologies

Genetic breeding, also known as breeding by genetic marker-assisted selection, has been restricted to a small number of traits, mostly those associated with oligogenic control and disease resistance (Gessler and Pertot, 2012; Zuriaga *et al.*, 2013). The availability of genome sequences and next-generation sequencing (NGS) methods, however, might offer fresh approaches to solving the majority of the issues with tree breeding. High-throughput DNA sequencing technology has made genome sequences and phenotypic variation data available, allowing for the dissection of complicated features in many significant tree species. These methods make it easier to analyze population diversity and provide shortcuts for finding genes associated with particular traits. Genome-wide association studies (GWASs) and association mapping (AM) make it easier to identify molecular markers for aided breeding and selection by facilitating the link between genetic markers and phenotypic traits in complex and unstructured populations. Thousands of markers are produced using genotyping-by-sequencing (GBS) techniques (Elshire *et al.*, 2011; Poland *et al.*, 2012), which enable the identification of genomic areas implicated in desired traits. For forest tree species with big genomes, like oak, eucalyptus, and loblolly pine, NGS techniques have been essential in delivering new genome sequence releases and better assembly of older ones. These technologies are accessible to possibilities for detailed expression analyses, mutation mapping, polymorphism discovery, studies on non-coding regions, and other applications that could influence how quickly questions can be answered and outcomes achieved in plant breeding, as they can economically generate millions of DNA sequence reads in a single run (Mardis, 2008).

9.5.3 Lack of diversified and representative genomic data

The lengthy generation times and large genomes of forest trees, the scarcity of well-characterized mutations for reverse genetic techniques, and the lack of funding have all impeded progress in research in this field. However, because of the efforts of a few committed researchers, forest tree genetics has advanced over the past 20 years (Neale and Kremer, 2011). In contrast to other plant genomics research communities that focus on species like *Arabidopsis thaliana*, maize, rice, wheat, and tomato, where hundreds or even thousands of researchers work on a single species, the forest tree genomics research community only consists of a thousand or so people worldwide, and research is conducted on dozens of species. At least 100,000 species of forest trees exist, but most genomics research has focused on a small number of highly domesticated, temperate species. The category of tree species with the most variety, i.e. non-domesticated tropical species, has not been included in genomics research. Each of the about a dozen species with advanced genomics research programs has significant economic value (Neale and Kremer, 2011).

9.5.4 Integrating environmental factors into genomic approaches

For the successful use of genomic selection (GS) based on breeding values, genomic-enabled prediction models are crucial. In contrast to animal breeding, plant breeding incorporates a large amount of data from multi-environment and multi-year field trials. Therefore, genotype × environment (G × E) interaction should be incorporated into genomic-enabled prediction models, since it typically improves prediction performance when line responses vary from environment to environment (Crossa *et al.*, 2022). Therefore, the employment of contemporary genomics-assisted prediction techniques, particularly for use in a variety of contexts, is hampered by enviromics. Novel approaches have integrated field trial data with DNA sequences using different sources of enviromics, such as linear and nonlinear reaction-norm models (Jarquín *et al.*, 2014; Morais-Júnior *et al.*, 2018; Millet *et al.*, 2019; Monteverde *et al.*, 2019; Costa-Neto *et al.*, 2020a), crop growth model (CGM) outputs (Rincent *et al.*, 2019), CGM integrated with GS (Cooper *et al.*, 2016; Robert *et al.*, 2022) and historical weather records to predict cultivars in years to come (De los Campos *et al.*, 2020). For example, De los Campos *et al.* (2020) suggested a method that accounts for model parameter uncertainty while evaluating genomic × environment (G × E) patterns discovered through field trials. This method not only forecasts a cultivar's expected performance in a given environment but also analyzes the expected distribution of that cultivar's performance under various potential weather conditions. This is a novel approach to multi-environment trial analysis that can expedite the evaluation of grain yield stability and adaptation.

9.6 Future Prospects and Opportunities

9.6.1 Improving prediction accuracy through deep learning

In recent years, the application of deep learning (DL) techniques has gained considerable momentum in the field of genomic prediction (GP), particularly in complex biological systems such as perennial tree species (Aono *et al.*, 2022; Mora-Poblete *et al.*, 2024). Unlike traditional statistical models that primarily rely on linear assumptions, deep learning methods offer the ability to capture nonlinear, hierarchical, and high-order interactions within genomic datasets (Huang *et al.*, 2021). This is especially valuable in forestry genetics, where phenotypic traits are often governed by intricate gene networks and environmental interactions (Feng *et al.*, 2025). The increasing availability of high-throughput genomic and phenotypic data, combined with advances in computational infrastructure, has enabled the implementation of sophisticated DL architectures such as convolutional neural networks (CNNs), recurrent neural networks (RNNs), and transformers tailored to improve prediction accuracy across diverse traits and breeding populations (Fritsche-Neto *et al.*, 2024; Li *et al.*, 2025). These models not only enhance predictive performance but also provide opportunities for improved biological interpretation, thus offering a promising direction for next-generation tree breeding programs (Montesinos-López *et al.*, 2021). Among these architectures, long short-term memory (LSTM) networks, a subclass of RNNs, have demonstrated particular efficacy in capturing temporal dependencies in sequential biological data (Sethi *et al.*, 2023). Hybrid frameworks combining LSTM with classical machine learning models such as random forest (RF) and support vector regression (SVR) have yielded superior prediction accuracies for traits such as yield and stem diameter in perennial crops, outperforming conventional regression-based approaches (Rashid *et al.*, 2021; Jelovica *et al.*, 2024). Additionally, deep belief networks (DBNs) and deep denoising autoencoders (DDAs), both unsupervised deep learning models, have been employed for noise reduction and automated feature extraction, particularly from image-based phenotypic datasets (Fu *et al.*, 2015; Balasubramaniam *et al.*, 2024). These methods have achieved impressive classification accuracies (ranging from 96% to 98%) in the detection of foliar diseases and pest infestations (Ngugi *et al.*, 2024). Furthermore, hybrid architectures that integrate deep feed-forward neural networks with gradient-boosted decision trees have emerged as powerful tools for improving trait prediction, especially in scenarios involving heterogeneous data types or limited sample sizes, characteristics often observed in tree genomic datasets (Montesinos-López *et al.*, 2024). In this sense, several researches revealed that deep learning techniques have emerged as transformative tools in plant genomic prediction, particularly in enhancing the accuracy of phenotype forecasting across various plant species, including trees such as the olive (Mendes *et al.*, 2024), apple (Cembrowska-Lech *et al.*, 2023), plum (Rodríguez *et al.*, 2018), and pistachio (Heidary-Sharifabad *et al.*, 2021). These efforts underscore the utility of DL frameworks in decoding complex genotype–phenotype relationships and advancing precision breeding in woody perennials (Niazian and Niedbała, 2020).

9.6.2 Building global genomic databases specifically for trees

A pivotal breakthrough in plant genomics in recent years has been the emergence of meticulously curated genomic databases specifically designed for forest and fruit tree species (Ramasamy *et al.*, 2025). These platforms serve as comprehensive repositories, integrating a wide array of high-throughput sequencing resources, including reference genomes (Falk *et al.*, 2018), transcriptomic datasets (Li *et al.*, 2012), genetic linkage maps (Palmer and Cardon, 2005), and extensive variant catalogs (Falk *et al.*, 2018). Importantly, many of these databases are enriched with corresponding phenotypic measurements and environmental metadata, thereby enabling multidimensional analyses that link genotype to phenotype across diverse ecological contexts (Opgenoorth *et al.*, 2021). Several genomic databases for trees were available for the users. In this sense, the TreeGenes database was one of the most important platforms for the trees. It was created in 1990 and currently represents more than 1790 tree species and over 1500 registered users (Falk *et al.*, 2018). To support this goal, TreeGenes has been integrated into the federation of Tripal databases, enhancing its capacity to serve the specialized needs of the forest tree research community (Malakar *et al.*, 2024). Through this integration, TreeGenes has implemented advanced modules designed to facilitate seamless cross-site data querying, efficient data transfer, and direct connectivity to a suite of analytical tools (Falk *et al.*, 2018). These developments enable researchers to access and analyze forest tree genomic and phenotypic data more effectively, fostering collaboration and accelerating discoveries in forest biology and conservation (Isabel *et al.*, 2020). TreeGenes currently offers an extensive repository of genetic and genomic resources, tailored to a variety of data types (Wegrzyn *et al.*, 2008). Its collection includes more than 2142 BioProjects, 19 fully sequenced reference genomes, around 188,982 cDNA sequences, over 15 million dbSNP records, and 466 GEO studies, all sourced from GenBank (Falk *et al.*, 2018). This wealth of data supports in-depth research and analysis within the forest tree research community. In addition to its extensive genetic resources, the TreeGenes platform also integrates curated phenotypic data, primarily obtained through direct submissions from contributing researchers (Wegrzyn *et al.*, 2020). On the other hand, the Tree Barcode of Life (TreeBOL) was considered as the second most important database, globally coordinated effort spearheaded by institutions such as the New York Botanical Garden, in collaboration with the International Barcode of Life (iBOL) consortium (Kress *et al.*, 2005; Hollingsworth *et al.*, 2009). TreeBOL aims to generate standardized DNA barcodes typically using chloroplast gene regions for thousands of tree species worldwide (Chac and Thinh, 2023). This global endeavor is designed to facilitate accurate species identification, resolve taxonomic ambiguities, and support biodiversity assessments and conservation planning across forest ecosystems (Hollingsworth *et al.*, 2011). A critical component of this initiative is the integration of DNA sequence data into the Barcode of Life Data Systems (BOLD), which enables researchers to match unidentified specimens to reference barcodes and access comprehensive metadata, including voucher specimen information, geographic distribution, and ecological context (Ratnasingham and Hebert, 2007). By focusing on floristically rich and taxonomically underrepresented regions, particularly tropical forests, the TreeBOL addresses longstanding challenges in tree taxonomy, while contributing significantly to global conservation genomics, ecosystem monitoring, and the documentation of arboreal biodiversity (Kress *et al.*, 2005; Fazekas *et al.*, 2008).

9.6.3 Increasing contribution of genomic prediction to sustainable forestry and the fight against climate change

While tree planting is often promoted as a nature-based solution to climate change, it is not a panacea. To maximize the long-term impact of reforestation and conservation efforts, it is essential to move beyond simply increasing forest cover and toward enhancing the adaptive capacity and genetic resilience of forest ecosystems (Kumar *et al.*, 2024; Wang *et al.*, 2025b). A deeper understanding of the genetics and genomics of forest trees can significantly improve the success of these initiatives (Holliday *et al.*, 2017). Genomic tools enable the identification and selection of tree genotypes better suited to future climatic conditions, improving survival, growth, and carbon sequestration potential (Aitken and Bemmels, 2016; Holliday *et al.*, 2017). Moreover, they provide insight into local adaptation processes and help anticipate how populations might respond to shifting environmental pressures (Fitzpatrick and Keller, 2015; Capblancq *et al.*, 2020). Integrating genomic knowledge into reforestation strategies is thus key to building climate-resilient forests and ensuring the long-term sustainability of forest-based climate mitigation efforts (Ahmad *et al.*, 2023). Genomic prediction (GP), which uses genome-wide markers to predict the genetic value of individuals, has emerged as a transformative tool in forestry breeding programs (Ramasamy *et al.*, 2025). Its integration into sustainable forest management contributes to enhancing productivity, accelerating breeding cycles, and increasing the resilience of forest trees to climate-related stressors such as drought, heat, and pests (Kumar *et al.*, 2024; Ramasamy *et al.*, 2025). These advances support the global effort to mitigate and adapt to climate change (Bay *et al.*, 2017; Ramasamy *et al.*, 2025). Recent studies have emphasized the role of GP in improving key adaptive and economically important traits in forest trees. In this sense, Isik (2014) highlighted the successful application of genomic selection in *Pinus taeda* and *Eucalyptus* species, leading to significant gains in growth rate and wood quality while reducing breeding cycle time. Similarly, Budhlakoti *et al.* (2022) demonstrated that GP can capture complex trait architectures influenced by genotype-by-environment interactions, crucial for selecting climate-resilient genotypes. In southern Brazil forestry, Tambarussi *et al.* (2025) applied GP in *Picea abies* (Norway spruce), showing improved predictions of height and diameter growth under future climate scenarios. This supports the use of GP for selecting genotypes adapted to warming temperatures and longer growing seasons. Moreover, Sharma *et al.* (2024) emphasized that GP can facilitate the deployment of improved genotypes at large scales, increasing forest carbon sequestration potential, an essential climate mitigation strategy. The integration of GP with phenotyping technologies and environmental data, sometimes referred to as genomic–environmental prediction frameworks, further enhances its utility (Fritsche-Neto *et al.*, 2024). Trudić *et al.* (2024) proposed an integrated approach combining GP and remote sensing for predicting drought tolerance in *Populus* species, underscoring the potential for selecting climate-resilient trees with reduced field trial costs. Additionally, Voss-Fels *et al.* (2019) argued that

GP also aids in maintaining genetic diversity in breeding populations, enabling selection without narrowing the genetic base. This is vital for sustaining the long-term adaptability of forests in the face of unpredictable climatic changes. These findings collectively indicate that genomic prediction is increasingly vital not only for enhancing forest productivity but also for fostering ecological resilience and mitigating climate change impacts (Feng *et al.*, 2025). Its integration into forest management and policy frameworks will be essential to meet global sustainability targets (Wang *et al.*, 2025a).

9.7 Conclusion

This chapter examines the underlying methods of tree genomic prediction (GP) and demonstrates its potential for transforming plant breeding. This approach is central to various stages of breeding, identifying the most likely individual candidates for further crosses, and reducing or even eliminating the requirement for extensive phenotyping over many generations of successive breedings. Using vast genomic and phenotypic datasets, genomic prediction can predict genetic values for specific traits, thus avoiding lengthy and resource-intensive trials. This rationalized process accelerates the selection process, facilitating the discovery of superior germplasm possessing desirable traits such as higher yields, resistance to pathogens and insect infestation, and adaptability to changing environmental conditions, which are often exacerbated by global warming. In contrast to conventional selection methods, which are largely based on phenotypic observations affected by both genetic and environmental conditions, genomic prediction looks directly at the genetic makeup of plants, providing a more precise and reliable evaluation of their potential performance. This simplifies the identification of favorable alleles but also enables gene-based interactions and environmental impacts to be incorporated into specific target traits, resulting in the creation of more robust and resilient germplasms. This improved genetic pool makes the selection and distribution of the targeted cultivars more efficient. As breeding value GP advances, its incorporation with machine learning and artificial intelligence is a promising new frontier in crop genetic enhancement research. The synergy between vast genomic datasets and advanced computational models is helping to reveal previously neglected, subtle genetic patterns and interactions. In broad terms, studies on GP in forest trees have shown that it can reduce the duration of the selection cycle and enhance genetic gain. Lower genotyping and phenotyping expenses should also favor the wider use of this method in forest selection.

Genomic prediction is proving to be a promising tool for tree selection, as long as the predictions are strictly validated and correctly explained, bearing in mind the specific characteristics of the targeted traits in the population under study and the genetic context.

Acknowledgments

The authors sincerely thank the researchers cited in this chapter for their valuable contributions to research in the field of genomic prediction.

References

Abdullah-Zawawi, M.R., Govender, N., Harun, S., Muhammad, N.A.N., Zainal, Z. *et al.* (2022) Multi-omics approaches and resources for systems-level gene function prediction in the plant kingdom. *Plants* 11(19), 2614. DOI: 10.3390/plants11192614.

Abril, N., Gion, J.M., Kerner, R., Müller-Starck, G., Cerrillo, R.M.N. *et al.* (2011) Proteomics research on forest trees, the most recalcitrant and orphan plant species. *Phytochemistry* 72(10), 1219–1242. DOI: 10.1016/j.phytochem.2011.01.005.

Ahmad, M.M., Imtiaz, M.A., Rasheed, F., Jameel, M.A., Asif, M. *et al.* (2023) The crucial role of forests in combating climate change. *Trends in Biotechnology and Plant Sciences* 1(1), 1–6.

Aitken, S.N. and Bemmels, J.B. (2016) Time to get moving: Assisted gene flow of forest trees. *Evolutionary Applications* 9(1), 271–290.

Alves, S.I.A., Dantas, C.W.D., Macedo, D.B. and Ramos, R.T.J. (2024) What are microsatellites and how to choose the best tool: A user-friendly review of SSR and 74 SSR mining tools. *Frontiers in Genetics* 15, 1474611. DOI: 10.3389/fgene.2024.1474611.

Ao, Y., Li, H., Zhu, L., Ali, S. and Yang, Z. (2019) The linear random forest algorithm and its advantages in machine learning assisted logging regression modeling. *Journal of Petroleum Science and Engineering* 174, 776–789. DOI: 10.1016/j.petrol.2018.11.067.

Aono, A.H., Francisco, F.R., Souza, L.M., Gonçalves, P. de S., Scaloppi Junior, E.J. *et al.* (2022) A divide-and-conquer approach for genomic prediction in rubber tree using machine learning. *Scientific Reports* 12(1), 18023.

Balasubramaniam, S., Prasanth, A., Kumar, K.S. and Kavitha, V. (2024) Medical image analysis based on deep learning approach for early diagnosis of diseases. In: Murugeswari, K., Sundaravadivazhagan, B., Poonkuntran, S. and Puyalnithi, T. (eds) *Deep Learning for Smart Healthcare*. Auerbach Publication, Boca Raton, Florida, pp. 54–75.

Barreto, C.A.V., das Graças Dias, K.O., de Sousa, I.C., Azevedo, C.F., Nascimento, A.C.C. *et al.* (2024) Genomic prediction in multi-environment trials in maize using statistical and machine learning methods. *Scientific Reports* 14(1), 1062.

Bay, R.A., Rose, N.H., Logan, C.A. and Palumbi, S.R. (2017) Genomic models predict successful coral adaptation if future ocean warming rates are reduced. *Science Advances* 3(11), e1701413.

Beaulieu, J., Doerksen, T.K., MacKay, J., Rainville, A. and Bousquet, J. (2014) Genomic selection accuracies within and between environments and small breeding groups in white spruce. *BMC Genomics* 15, 1048. DOI: 10.1186/1471-2164-15-1048.

Bernardo, R. (2016) Bandwagons I, too, have known. *Theoretical and Applied Genetics* 129, 2323–2332. DOI: 10.1007/s00122-016-2772-5.

Bidyananda, N., Jamir, I., Nowakowska, K., Varte, V., Vendrame, W.A. *et al.* (2024) Plant genetic diversity studies: Insights from DNA marker analyses. *International Journal of Plant Biology* 15(3), 607–640. DOI: 10.3390/ijpb15030046.

Bourgeois, Y.X. and Warren, B.H. (2021) An overview of current population genomics methods for the analysis of whole-genome resequencing data in eukaryotes. *Molecular Ecology* 30(23), 6036–6071. DOI: 10.1111/mec.15989.

Bouvet, J.-M., Makouanzi, G., Cros, D. and Vigneron, P. (2016) Modeling additive and non-additive effects in a hybrid population using genome-wide genotyping: Prediction accuracy implications. *Heredity* 116, 146–157. DOI: 10.1038/hdy.2015.78.

Brown, W. (1983) Genetic diversity and genetic vulnerability—an appraisal. *Economic Botany* 37, 4–12. DOI: 10.1007/BF02859301.

Budhlakoti, N., Kushwaha, A.K., Rai, A., Chaturvedi, K.K., Kumar, A. *et al.* (2022) Genomic selection: A tool for accelerating the efficiency of molecular breeding for development of climate-resilient crops. *Frontiers in Genetics* 13, 832153.

Burridge, A.J., Winfield, M.O., Wilkinson, P.A., Przewieslik-Allen, A.M., Edwards, K.J. *et al.* (2022) The use and limitations of exome capture to detect novel variation in the hexaploid wheat genome. *Frontiers in Plant Science* 13, 841855. DOI: 10.3389/fpls.2022.841855.

Calleja-Rodriguez, A., Pan, J., Funda, T., Chen, Z., Baison, J. *et al.* (2020) Evaluation of the efficiency of genomic versus pedigree predictions for growth and wood quality traits in Scots pine. *BMC Genomics* 21(1), 796. DOI: 10.1186/s12864-020-07188-4.

Cao, H.X., Michels, D., Vu, G.T.H. and Gailing, O. (2024) Applications of CRISPR technologies in forestry and molecular wood biotechnology. *International Journal of Molecular Sciences* 25(21), 11792. DOI: 10.3390/ijms252111792.

Capblancq, T., Fitzpatrick, M.C., Bay, R.A., Exposito-Alonso, M. and Keller, S.R. (2020) Genomic prediction of (mal) adaptation across current and future climatic landscapes. *Annual Review of Ecology, Evolution, and Systematics* 51(1), 245–269.

Cappa, E.P., El-Kassaby, Y.A., Muñoz, F., Garcia, M.N., Villalba, P.V., Klápště, J. and Marcucci Poltri, S.N. (2017) Improving accuracy of breeding values by incorporating genomic information in spatial-competition mixed models. *Molecular Breeding* 37(10), 125.

Carta, A., Bedini, G. and Peruzzi, L. (2020) A deep dive into the ancestral chromosome number and genome size of flowering plants. *New Phytologist* 228(3), 1097–1106. DOI: 10.1111/nph.16668.

Cembrowska-Lech, D., Krzemińska, A., Miller, T., Nowakowska, A., Adamski, C. *et al.* (2023) An integrated multi-omics and artificial intelligence framework for advance plant phenotyping in horticulture. *Biology* 12(10), 1298.

Chac, L.D. and Thinh, B.B. (2023) Species identification through DNA barcoding and its applications: A review. *Biology Bulletin* 50(6), 1143–1156.

Chakrabarty, S. and Külheim, C. (2025) Trends in tree improvement methods: From classical breeding to genomic technologies. *Tree Genetics & Genomes* 21(3), 12.

Chen, Y., Miao, Y., Bai, W., Lin, K. and Pang, E. (2022) Characteristics and potential functional effects of long insertions in Asian butternuts. *BMC Genomics* 23(1), 732. DOI: 10.1186/s12864-022-08961-3.

Choudhury, D.R., Kumar, R., Maurya, A., Semwal, D.P., Rathi, R.S. *et al.* (2023) SSR and SNP marker-based investigation of indian rice landraces in relation to their genetic diversity, population structure, and geographical isolation. *Agriculture* 13(4), 823. DOI: 10.3390/agriculture13040823.

Cooper, M., Technow, F., Messina, C., Gho, C. and Radu Totir, L. (2016) Use of crop growth models with whole-genome prediction: Application to a maize multienvironment trial. *Crop Science* 56, 2141–2156. DOI: 10.2135/cropsci2015.08.0512.

Costa-Neto, G., Fritsche-Neto, R. and Crossa, J. (2020a) Nonlinear kernels, dominance, and envirotyping data increase the accuracy of genome-based prediction in multi-environment trials. *Heredity (Edinburgh)* 126, 92–106. DOI: 10.1038/s41437-020-00353-1.

Cros, D., Mbo-Nkoulou, L., Bell, J.M., Oum, J., Masson, A. *et al.* (2019) Within-family genomic selection in rubber tree (*Hevea brasiliensis*) increases genetic gain for rubber production. *Industrial Crops and Products* 138, 111464.

Crossa, J., Montesinos-Lopez, O.A., Pérez-Rodríguez, P., Costa-Neto, G., Fritsche-Neto, R. *et al.* (2022) Genome and environment based prediction models and methods of complex traits incorporating genotype × environment interaction. In: Ahmadi, N. and Bartholomé, J. (eds) *Genomic Prediction of Complex Traits: Methods and Protocols*. Humana Press, New York, pp. 245–283. DOI: 10.1007/978-1-0716-2205-6_9.

Crossa, J., Martini, J.W., Vitale, P., Pérez-Rodríguez, P., Costa-Neto, G. *et al.* (2024) Expanding genomic prediction in plant breeding: Harnessing big data, machine learning, and advanced software. *Trends in Plant Science* 30(7), 756–774.

Cuevas, J., González-Diéguez, D., Dreisigacker, S., Martini, J.W.R., Crespo-Herrera, L. *et al.* (2024) Modeling within and between subgenomes epistasis of synthetic hexaploid wheat for genomeenabled prediction of diseases. *Genes (Basel)* 15, 262.

De Bem Oliveira, I., Resende, M.F.R., Ferrão, L.F.V., Amadeu, R.R., Endelman, J.B. *et al.* (2019) Genomic prediction of autotetraploids; influence of relationship matrices, allele dosage, and continuous genotyping calls in phenotype prediction. *G3 Genes| Genomes| Genetics* 9(4), 1189–1198. DOI: 10.1534/G3.119.400059.

De los Campos, G., Pérez-Rodríguez, P., Bogard, M., Gouache, D. and Crossa, J. (2020) A data-driven simulation platform to predict cultivars' performances under uncertain weather conditions. *Nature Communications* 11, 4876. DOI: 10.1038/s41467-020-18480-y.

Delfini, J., Moda-Cirino, V., dos Santos Neto, J., Ruas, P.M., Sant'Ana, G.C. *et al.* (2021) Population structure, genetic diversity and genomic selection signatures among a Brazilian common bean germplasm. *Scientific Reports* 11(1), 2964. DOI: 10.1038/s41598-021-82437-4.

Deng, Y., Liu, S., Zhang, Y., Tan, J., Li, X. *et al.* (2022) A telomere-to-telomere gap-free reference genome of watermelon and its mutation library provide important resources for gene discovery and breeding. *Molecular Plant* 15(8), 1268–1284. DOI: 10.1016/j.molp.2022.06.010.

Dias, K.O.D.G., Gezan, S.A., Guimarães, C.T., Nazarian, A., da Costa e Silve, A. *et al.* (2018) Improving accuracies of genomic predictions for drought tolerance in maize by joint modeling of additive and dominance effects in multi-environment trials. *Heredity* 121, 24–37.

Ding, X., Diao, S., Luan, Q., Wu, H.X., Zhang, Y. *et al.* (2022) A transcriptome-based association study of growth, wood quality, and oleoresin traits in a slash pine breeding population. *PLoS Genetics* 18(2), e1010017. DOI: 10.1371/journal.pgen.1010017.

Duarte, D., Jurcic, E.J., Dutour, J., Villalba, P.V., Centurión, C. *et al.* (2024) Genomic selection in forest trees comes to life: Unraveling its potential in an advanced four-generation *Eucalyptus grandis* population. *Frontiers in Plant Science* 15, 1462285. DOI: 10.3389/fpls.2024.1462285.

Dufresne, F., Stift, M., Vergilino, R. and Mable, B.K. (2014) Recent progress and challenges in population genetics of polyploid organisms: An overview of current state-of-the-art molecular and statistical tools. *Molecular Ecology* 23(1), 40–69. DOI: 10.1111/MEC.12581.

Ejima, A., Uchiyama, K., Mori, H. and Tsumura, Y. (2023) Genomic prediction of tree height, wood stiffness, and male flower quantity traits across two generations in selected individuals of *Cryptomeria japonica* D. Don (Japanese cedar). *Forests* 14(10), 2097.

Elshire, R.J., Glaubitz, J.C., Sun, Q., Poland, J.A., Kawamoto, K. *et al.* (2011) A robust, simple genotyping-by-sequencing (GBS) approach for high diversity species. *PLoS ONE* 6, e19379. DOI: 10.1371/journal.pone.0019379.

Endelman, J.B., Carley, C.A.S., Bethke, P.C., Coombs, J.J., Clough, M.E. *et al.* (2018) Genetic variance partitioning and genome-wide prediction with allele dosage information in autotetraploid potato. *Genetics* 209(1), 77–87. DOI: 10.1534/GENETICS.118.300685/-/DC1.

Eraslan, G., Avsec, Ž., Gagneur, J. and Theis, F.J. (2019) Deep learning: New computational modelling techniques for genomics. *Nature Reviews Genetics* 20(7), 389–403. DOI: 10.1038/s41576-019-0122-6.

Esposito, S., Carputo, D., Cardi, T. and Tripodi, P. (2019) Applications and trends of machine learning in genomics and phenomics for next-generation breeding. *Plants* 9(1), 34. DOI: 10.3390/plants9010034.

Falk, T., Herndon, N., Grau, E., Buehler, S., Richter, P. *et al.* (2018) Growing and cultivating the forest genomics database, TreeGenes. *Database* 2018, bay084.

Fan, M., Wei, X., Song, Z., Zhang, Y., Li, X. *et al.* (2024) GWAS-driven gene mining and genomic prediction of ornamental traits in flowering trees: A case study of *Camellia japonica*. *Horticultural Plant Journal*. DOI: 10.1016/j.hpj.2024.05.017.

Fazekas, A.J., Burgess, K.S., Kesanakurti, P.R., Graham, S.W., Newmaster, S.G. *et al.* (2008) Multiple multilocus DNA barcodes from the plastid genome discriminate plant species equally well. *PLoS ONE* 3(7), e2802.

Feng, C., Chen, B., Hofer, J., Shi, Y., Jiang, M. *et al.* (2025) Genomic and genetic insights into Mendel's pea genes. *Nature* 1–10.

Fernandez i Marti, A. and Dodd, R.S. (2018) Using CRISPR as a gene editing tool for validating adaptive gene function in tree landscape genomics. *Frontiers in Ecology and Evolution* 6, 76. DOI: 10.3389/fevo.2018.00076.

Fitzpatrick, M.C. and Keller, S.R. (2015) Ecological genomics meets community-level modelling of biodiversity: Mapping the genomic landscape of current and future environmental adaptation. *Ecology Letters* 18(1), 1–16.

Fritsche-Neto, R., Al-Nowibet, K., Montesinos-López, A. and Crossa, J. (2024) A review of multimodal deep learning methods for genomic-enabled prediction in plant breeding. *Genetics* 228, 4.

Fu, Y., Zhang, Y., Qiao, H., Li, D., Zhou, H. *et al.* (2015) Analysis of feature extracting ability for cutting state monitoring using deep belief networks. *Procedia CIRP* 31, 29–34.

Gessler, C. and Pertot, I. (2012) Vf scab resistance of Malus. *Trees* 26, 95–108. DOI: 10.1007/s00468-011-0618-y.

Gholami, M., Wimmer, V., Sansaloni, C., Petroli, C., Hearne, S.J. *et al.* (2021) A comparison of the adoption of genomic selection across different breeding institutions. *Frontiers in Plant Science* 12, 728567.

Gianola, D. and Fernando, R.L. (2020) A multiple-trait Bayesian Lasso for genome-enabled analysis and prediction of complex traits. *Genetics* 214(2), 305–331. DOI: 10.1101/852749.

Gogolev, Y.V., Ahmar, S., Akpinar, B.A., Budak, H., Kiryushkin, A.S. *et al.* (2021) OMICs, epigenetics, and genome editing techniques for food and nutritional security. *Plants* 10(7), 1423. DOI: 10.3390/plants10071423.

Govindaraj, M., Vetriventhan, M. and Srinivasan, M. (2015) Importance of genetic diversity assessment in crop plants and its recent advances: An overview of its analytical perspectives. *Genetics Research International (Hindawi Publishing Corp.)* 15, 431487. DOI: 10.1155/2015/431487.

Greilhuber, J., Dolezel, J., Lysa, K.M. and Bennett, M. (2005) The origin, evolution and proposed stabilization of the terms 'genome size' and 'c-value' to describe nuclear DNA contents. *Annals of Botany* 95(1), 255–260. DOI: 10.1093/aob/mci019.

Guo, J., Cao, K., Deng, C., Li, Y., Zhu, G. *et al.* (2020) An integrated peach genome structural variation map uncovers genes associated with fruit traits. *Genome Biology and Evolution* 21(1), 258. DOI: 10.1186/s13059-020-02169-y.

Hämälä, T., Wafula, E.K., Guiltinan, M.J., Ralph, P.E., dePamphilis, C.W. *et al.* (2021) Genomic structural variants constrain and facilitate adaptation in natural populations of *Theobroma cacao*, the chocolate tree. *Proceedings of the National Academy of Sciences USA* 118, 35. DOI: 10.1073/pnas.2102914118.

He, L., Xiao, J., Rashid, K.Y., Jia, G., Li, P. *et al.* (2019) Evaluation of genomic prediction for pasmo resistance in flax. *International Journal of Molecular Sciences* 20(2), 359.

Heidary-Sharifabad, A., Zarchi, M.S., Emadi, S. and Zarei, G. (2021) An efficient deep learning model for cultivar identification of a pistachio tree. *British Food Journal* 123(11), 3592–3609.

Hickey, J.M., Chiurugwi, T., Mackay, I. and Powell, W. (2017) Genomic prediction unifies animal and plant breeding programs to form platforms for biological discovery. *Nature Genetics* 49(9), 1297–1303. DOI: 10.1038/ng.3920.

Holliday, J.A., Aitken, S.N., Cooke, J.E.K., Fady, B., González Martínez, S.C. *et al.* (2017) *Advances in Ecological Genomics in Forest Trees and Applications to Genetic Resources Conservation and Breeding*. Wiley Online Library.

Hollingsworth, C.P.W.G., Hollingsworth, P.M., Forrest, L.L., Spouge, J.L., Hajibabaei, M. *et al.* (2009) A DNA barcode for land plants. *Proceedings of the National Academy of Sciences* 106(31), 12794–12797.

Hollingsworth, P.M., Graham, S.W. and Little, D.P. (2011) Choosing and using a plant DNA barcode. *PLoS ONE* 6(5), e19254.

Hong, K., Radian, Y., Manda, T., Xu, H. and Luo, Y. (2023) The development of plant genome sequencing technology and its conservation and application in endangered gymnosperms. *Plants* 12(23), 4006. DOI: 10.3390/plants12234006.

Huang, K., Xiao, C., Glass, L.M., Critchlow, C.W., Gibson, G. *et al.* (2021) Machine learning applications for therapeutic tasks with genomics data. *Patterns* 2, 10.

Huang, L., Song, M., Shen, H., Hong, H., Gong, P. *et al.* (2023) Deep learning methods for omics data imputation. *Biology* 12(10), 1313. DOI: 10.3390/biology12101313.

Isabel, N., Holliday, J.A. and Aitken, S.N. (2020) Forest genomics: Advancing climate adaptation, forest health, productivity, and conservation. *Evolutionary Applications* 13(1), 3–10.

Isik, F. (2014) Genomic selection in forest tree breeding: The concept and an outlook to the future. *New Forests* 45, 379–401.

Isik, F., Kumarx, S., Martínez-García, P.J., Iwatajj, H. and Yamamoto, T. (2015) Acceleration of forest and fruit tree domestication by genomic selection. *Advances in Botanical Research* 74, 93–124.

Jamnadass, R., McMullin, S., Iiyama, M., Dawson, I.K., Powell, B. *et al.* (2016) Understanding the roles of forests and tree-based systems in food provision. In: Mansourian, S., Vira, B. and Wildburger, C. (eds) *Forests and Food: Addressing Hunger and Nutrition Across Sustainable Landscapes*. OBP Collection, Open Book Publishers, Cambridge, pp. 29–72.

Jarquin, D., Howard, R., Crossa, J., Beyene, Y., Gowda, M. *et al.* (2020) Genomic prediction enhanced sparse testing for multi-environment trials. *G3: Genes, Genomes, Genetics* 10, 2725–2739.

Jarquín, D., Crossa, J., Lacaze, X., Du Cheyron, P., Daucour, J. *et al.* (2014) A reaction norm model for genomic selection using high-dimensional genomic and environmental data. *Theoretical and Applied Genetics* 127, 595–607. DOI: 10.1007/s00122-013-2243-1.

Jelovica, B., Erkinaro, J., Orell, P., Kløve, B., Haghighi, A.T. *et al.* (2024) Atlantic salmon habitat-abundance modeling using machine learning methods. *Ecological Indicators* 160, 111832.

Jung, M., Roth, M., Aranzana, M.J., Auwerkerken, A., Bink, M. *et al.* (2020) The apple REFPOP-a reference population for genomics-assisted breeding in apple. *Horticultural Research* 7(1), 189. DOI: 10.1038/s41438-020-00408-8.

Kalendar, R., Hunter, C. and Orbovic, V. (2022) Innovative applications of sequencing technologies in plant science. *Frontiers in Plant Science* 13, 1058347. DOI: 10.3389/fpls.2022.1058347.

Kamali, S. and Singh, A. (2023) Genomic and transcriptomic approaches to developing abiotic stress-resilient crops. *Agronomy* 13(12), 2903. DOI: 10.3390/agronomy13122903.

Kanaka, K.K., Sukhija, N., Goli, R.C., Singh, S., Ganguly, I. *et al.* (2023) On the concepts and measures of diversity in the genomics era. *Current Plant Biology* 33, 100278. DOI: 10.1016/j.cpb.2023.100278.

Kang, X. and Wei, H. (2022) Breeding polyploid populus: Progress and perspective. *Forestry Research* 2(4). DOI: 10.48130/fr-2022-0004.

Klápště, J., Suontama, M., Telfer, E., Graham, N., Low, C. *et al.* (2017) Exploration of genetic architecture through sib-ship reconstruction in advanced breeding population of *Eucalyptus nitens*. *PLoS ONE* 12(9), e0185137.

Klápště, J., Dungey, H.S., Telfer, E.J., Suontama, M., Graham, N.J. *et al.* (2020) Marker selection in multivariate genomic prediction improves accuracy of low heritability traits. *Frontiers in Genetics* 11, 499094.

Klápště, J., Silva-Duque, A H. and Cortés, A.J. (2024) Applied genomics in Poplar: Innovative breeding. In: The Poplar Genome. Springer International Publishing, Cham, pp. 191–224. DOI: 10.1007/978-3-031-50787-8_10.

Kress, W.J., Wurdack, K.J., Zimmer, E.A., Weigt, L.A. and Janzen, D.H. (2005) Use of DNA barcodes to identify flowering plants. *Proceedings of the National Academy of Sciences* 102(23), 8369–8374.

Kumar, D., Pandey, V. and Dixit, S. (2024) Agronomic strategies for enhancing forest resilience to climate change. In: *Forests and Climate Change: Biological Perspectives on Impact, Adaptation, and Mitigation Strategies*. Springer, pp. 385–420.

Kumar, P., Choudhary, M., Jat, B.S., Kumar, B., Singh, V. *et al*. (2021) Skim sequencing: An advanced NGS technology for crop improvement. *Journal of Genetics* 100, 1–10. DOI: 10.1007/s12041-021-01285-3.

Lebedev, V.G., Lebedeva, T.N., Chernodubov, A.I. and Shestibratov, K.A. (2020) Genomic selection for forest tree improvement: Methods, achievements and perspectives. *Forests* 11(11), 1190.

Legarra, A., Robert-Granie, C., Croiseau, P., Guillaume, F. and Fritz, S. (2011) Improved LASSO for genomic selection. *Genetics Research* 93, 77–87.

Lenz, P.R.N., Beaulieu, J., Mansfield, S.D., Clément, S., Desponts, M. *et al*. (2017) Factors affecting the accuracy of genomic selection for growth and wood quality traits in an advanced breeding population of black spruce (*Picea mariana*). *BMC Genomics* 18, 335.

Lenz, P.R.N., Nadeau, S., Azaiez, A., Gerardi, S., Deslauriers, M. *et al*. (2020) Genomic prediction for hastening and improving efficiency of forward selection in conifer polycross mating designs: An example from white spruce. *Heredity* 124, 562–578.

Li, J., Dai, X., Liu, T. and Zhao, P.X. (2012) LegumeIP: An integrative database for comparative genomics and transcriptomics of model legumes. *Nucleic Acids Research* 40(D1), D1221–D1229.

Li, R., Zhang, D., Han, Y., Liu, Z., Zhang, Q. *et al*. (2025) Hybrid deep learning approaches for improved genomic prediction in crop breeding. *Agriculture* 15(11), 1171.

Li, Y., Suontama, M., Burdon, R.D. and Dungey, H.S. (2017) Genotype by environment interactions in forest tree breeding: Review of methodology and perspectives on research and application. *Tree Genetics and Genomes* 13, 60.

Li, Y., Cao, K., Li, N., Zhu, G., Fang, W. *et al*. (2021) Genomic analyses provide insights into peach local adaptation and responses to climate change. *Genome Research* 31(4), 592–606. DOI: 10.1101/gr.261032.120.

Liao, Z., Zhang, X., Zhang, S., Lin, Z., Zhang, X. *et al*. (2021) Structural variations in papaya genomes. *BMC Genomics* 22(1), 335. DOI: 10.1186/s12864-021-07665-4.

Lin, Y.T., Whitman, W.B., Coleman, D.C. and Chiu, C.Y. (2018) Effects of reforestation on the structure and diversity of bacterial communities in subtropical low mountain forest soils. *Frontiers in Microbiology* 9, 1–10. DOI: 10.3389/fmicb.2018.01968.

Lin, Z., Hayes, B.J. and Daetwyler, H.D. (2014) Genomic selection in crops, trees and forages: A review. *Crop & Pasture Science* 65, 1177–1191.

Liu, Y., Zhang, M., Wang, R., Li, B., Jiang, Y. *et al*. (2022) Comparison of structural variants detected by PacBio-CLR and ONT sequencing in pear. *BMC Genomics* 23(1), 830. DOI: 10.1186/s12864-022-09074-7.

Lloyd, A. and Bomblies, K. (2016) Meiosis in autopolyploid and allopolyploid Arabidopsis. *Current Opinion in Plant Biology* 30, 116–122. DOI: 10.1016/J.PBI.2016.02.004.

Mahajan, R. and Gupta, P. (2012) Molecular markers: Their use in tree improvement. *Journal of Forest Science* 58(3), 137–144.

Mahood, E.H., Kruse, L.H. and Moghe, G.D. (2020) Machine learning: A powerful tool for gene function prediction in plants. *Applications in Plant Sciences* 8(7), e11376. DOI: 10.1002/aps3.11376.

Malakar, A., Chandra, G. and Barthwal, S. (2024) Tree genome databases: A new era in the development of cyber-infrastructures for forest trees. In: Genomics Data Analysis for Crop Improvement. Springer Nature Singapore, Singapore, pp. 285–308.

Maldonado, C., Mora-Poblete, F., Contreras-Soto, R.I., Ahmar, S., Chen, J.T. *et al*. (2020) Genome-wide prediction of complex traits in two outcrossing plant species through Deep learning and Bayesian Regularized Neural Network. *Frontiers in Plant Science* 11, 593897.

Mallet, J. (2007) Hybrid speciation. *Nature* 446, 279–283. DOI: 10.1038/nature05706.

Mardis, E.R. (2008) The impact of next-generation sequencing technology on genetics. *Trends in Genetics* 24, 133–141. DOI: 10.1016/j.tig.2007.12.007.

Martín-Valmaseda, M., Devin, S.R., Ortuño-Hernández, G., Pérez-Caselles, C., Mahdavi, S.M.E. *et al.* (2023) CRISPR/Cas as a genome-editing technique in fruit tree breeding. *International Journal of Molecular Sciences* 24(23), 16656. DOI: 10.3390/ijms242316656.

McElroy, M.S., Navarro, A.J.R., Mustiga, G., Stack, C., Gezan, S. *et al.* (2018) Prediction of cacao (*Theobroma cacao*) resistance to *Moniliophthora* spp. diseases via genome-wide association analysis and genomic selection. *Frontiers in Plant Science* 9, 343. DOI: 10.3389/fpls.2018.00343.

McGowan, M., Wang, J., Dong, H., Liu, X., Jia, Y. *et al.* (2021) Ideas in genomic selection with the potential to transform plant molecular breeding: A review. *Plant Breeding Reviews* 45, 273–319. DOI: 10.1002/9781119828235.ch7.

Meger, J., Ulaszewski, B., Pałucka, M., Kozioł, C. and Burczyk, J. (2024) Genomic prediction of resistance to *Hymenoscyphus fraxineus* in common ash (*Fraxinus excelsior* L.) populations. *Evolutionary Applications* 17(5), e13694.

Meher, P.K., Rustgi, S. and Kumar, A. (2022) Performance of Bayesian and BLUP alphabets for genomic prediction: Analysis, comparison and results. *Heredity* 128(6), 519–530. DOI: 10.1038/S41437-022-00539-9.

Mendes, J., Lima, J., Costa, L., Rodrigues, N. and Pereira, A.I. (2024) Deep learning networks for olive cultivar identification: A comprehensive analysis of convolutional neural networks. *Smart Agricultural Technology* 8, 100470.

Millet, E.J., Kruijer, W., Coupel-Ledru, A., Alvarez Prado, S., Cabrera Bosquet, L. *et al.* (2019) Genomic prediction of maize yield across European environmental conditions. *Nature Genetics* 51, 952–956. DOI: 10.1038/s41588-019-0414-y.

Minamikawa, M.F., Takada, N., Terakami, S., Saito, T., Onogi, A. *et al.* (2018) Genome-wide association study and genomic prediction using parental and breeding populations of Japanese pear (*Pyrus pyrifolia* Nakai). *Science Reports* 8, 11994. DOI: 10.1038/s41598-018-30154-w.

Montesinos López, O.A., Montesinos López, A. and Crossa, J. (2022a) *Multivariate Statistical Machine Learning Methods for Genomic Prediction*. Springer Nature, Cham. DOI: 10.1007/978-3-030-89010-0.

Montesinos López, O.A., Montesinos López, A. and Crossa, J. (2022b) Random forest for genomic prediction. In: *Multivariate Statistical Machine Learning Methods for Genomic Prediction*. Springer International Publishing, Cham, pp. 633–681. DOI: 10.1007/978-3-030-89010-0_15.

Montesinos-López, O.A., Chavira-Flores, M., Kismiantini,Crespo-Herrera, L., Saint Piere, C. *et al.* (2024) A review of multimodal deep learning methods for genomic-enabled prediction in plant breeding. *Genetics* 228(4), iyae161.

Montesinos-López, O.A., Montesinos-López, A., Pérez-Rodríguez, P., Barrón-López, J.A., Martini, J.W.R. *et al.* (2021) A review of deep learning applications for genomic selection. *BMC Genomics* 22, 1–23.

Monteverde, E., Gutierrez, L., Blanco, P., Pérez de Vida, F., Rosas, J.E. *et al.* (2019) Integrating molecular markers and environmental covariates to interpret genotype by environment interaction in rice (*Oryza sativa* L) grown in subtropical areas. *G3 Genes Genomes Genetics* 9, 1519–1531. DOI: 10.1534/g3.119.400064.

Moraes, B.F.X., dos Santos, R.F., de Lima, B.M., Aguiar, A.M., Missiaggia, A.A. *et al.* (2018) Genomic selection prediction models comparing sequence capture and SNP array genotyping methods. *Molecular Breeding* 38, 115.

Morais-Júnior, O.P., Duarte, J.B., Breseghello, F., Coelho, A.S.G. and Magalhães, A.M., Jr (2018) Single-step reaction norm models for genomic prediction in multienvironment recurrent selection trials. *Crop Science* 58, 592–607. DOI: 10.2135/cropsci2017.06.0366.

Mora-Poblete, F., Mieres-Castro, D., do Amaral Júnior, A.T., Balach, M. and Maldonado, C. (2024) Integrating deep learning for phenomic and genomic predictive modeling of Eucalyptus trees. *Industrial Crops and Products* 220, 119151.

Mousavi, M., Tong, C., Liu, F., Tao, S., Wu, J. *et al.* (2016) De novo SNP discovery and genetic linkage mapping in poplar using restriction site associated DNA and whole-genome sequencing technologies. *BMC Genomics* 17(1), 656. DOI: 10.1186/s12864-016-3003-9.

Mu, H., Jiang, J., Li, H. and Liu, G. (2012a) Seed vigor, photosynthesis and early growth of saplings of different triploid Betula families. *Dendrobiology* 68, 11–20.

Müller, B.S.F., Neves, L.G., Filho, J.E.D.A., Resende, M.F.R. and Muñoz, P.R. (2017) Genomic prediction in contrast to a genome-wide association study in explaining heritable variation of complex growth traits in breeding populations of *Eucalyptus*. *BMC Genomics* 18, 524.

Nam, V.T., Hang, P.L.B., Linh, N.N., Ly, L.H., Hue, H.T.T. *et al.* (2020) Molecular markers for analysis of plant genetic diversity. *Vietnam Journal of. Biotechnology* 18, 589–608.

Neale, D.B. and Kremer, A. (2011) Forest tree genomics: Growing resources and applications. *Nature Reviews Genetics* 12(2), 111–122. DOI: 10.1038/nrg2931.

Nghiem, C.Q., Griffin, R.A., Harbard, J.L., Harwood, C.E., Le, S. *et al.* (2018) Reduced fertility in triploids of *Acacia auriculiformis* and its hybrid with *A. mangium*. *Euphytica* 214, 77. DOI: 10.1007/s10681-018-2157-8.

Ngugi, H.N., Ezugwu, A.E., Akinyelu, A.A. and Abualigah, L. (2024) Revolutionizing crop disease detection with computational deep learning: A comprehensive review. *Environmental Monitoring and Assessment* 196(3), 302.

Niazian, M. and Niedbała, G. (2020) Machine learning for plant breeding and biotechnology. *Agriculture* 10(10), 436.

Niu, X., Song, Z., Xu, C., Wu, H., Luan, Q. *et al.* (2023) Prediction of needle physiological traits using UAV imagery for breeding selection of slash pine. *Plant Phenomics* 5, 0028. DOI: 10.34133/plantphenomics.0028.

Opgenoorth, L., Dauphin, B., Benavides, R., Heer, K., Alizoti, P. *et al.* (2021) The GenTree Platform: Growth traits and tree-level environmental data in 12 European forest tree species. *GigaScience* 10(3), giab010.

Otto, S.P. and Whitton, J. (2000) Polyploid incidence and evolution. *Annual Review of Genetics* 34(1), 401–437. DOI: 10.1146/annurev.genet.34.1.401.

Pak, S. and Li, C. (2022) Progress and challenges in applying CRISPR/Cas techniques to the genome editing of trees. *Forestry Research* 2, 6. DOI: 10.48130/FR-2022-0006.

Palmer, L.J. and Cardon, L.R. (2005) Shaking the tree: Mapping complex disease genes with linkage disequilibrium. *The Lancet* 366(9492), 1223–1234.

Poland, J.A., Brown, P.J., Sorrells, M.E. and Jannink, J.L. (2012) Development of high-density genetic maps for barley and wheat using a novel two-enzyme genotyping-by-sequencing approach. *PLoS ONE* 7, e32253. DOI: 10.1371/journal.pone.0032253.

Ramasamy, Y., Arunachalam, S., Dasgupta, M. and Nambiar-Veetil, M. (2025) Genetic markers, genomics and genetic modification in forest trees: Current status and prospects. In: Mandal, A.K. and Nicodemus, A. (eds) *Textbook of Forest Science*. Springer, Singapore, pp. 717–737. DOI: 10.1007/978-981-97-8289-5_33.

Rambolarimanana, T., Ramamonjisoa, L., Verhaegen, D., Leong Pock Tsy, J.M., Jacquin, L. *et al.* (2018) Performance of multi-trait genomic selection for *Eucalyptus robusta* breeding program. *Tree Genetics & Genomes* 14(5), 71.

Ramsey, J. and Schemske, D.W. (2002) Neopolyploidy in flowering plants. *Annual Review of Ecology and Systematics* 33, 589–639. DOI: 10.1146/annurev.ecolsys.33.010802.150437.

Rashid, M., Bari, B.S., Yusup, Y., Kamaruddin, M.A. and Khan, N. (2021) A comprehensive review of crop yield prediction using machine learning approaches with special emphasis on palm oil yield prediction. *IEEE Access* 9, 63406–63439.

Ratnasingham, S. and Hebert, P.D.N. (2007) BOLD: The barcode of life data system. *Molecular Ecology Notes* 7(3), 355–364. Available at: http://www.barcodinglife.org

Richardson, S., Tseng, G.C. and Sun, W. (2016) Statistical methods in integrative genomics. *Annual Review of Statistics and Its Application* 3(1), 181–209. DOI: 10.1146/annurev-statistics-041715-033506.

Riehl, J.F., Cole, C.T., Morrow, C.J., Barker, H.L., Bernhardsson, C. *et al.* (2023) Genomic and transcriptomic analyses reveal polygenic architecture for ecologically important traits in aspen (*Populus tremuloides* Michx.). *Ecology and Evolution* 13(10), e10541. DOI: 10.1002/ece3.10541.

Rincent, R., Charpentier, J.P., Faivre-Rampant, P., Paux, E., Le Gouis, J. *et al.* (2018) Phenomic selection is a low-cost and high-throughput method based on indirect predictions: Proof of concept on wheat and poplar. *G3: Genes, Genomes, Genetics* 8(12), 3961–3972. DOI: 10.1534/g3.118.200760.

Rincent, R., Malosetti, M., Ababaei, B., Touzy, G., Mini, A. *et al.* (2019) Using crop growth model stress covariates and AMMI decomposition to better predict genotype-by-environment interactions. *Theoretical and Applied Genetics* 132, 3399–3411. DOI: 10.1007/s00122-019-03432-y.

Robert, P., Brault, C., Rincent, R. and Segura, V. (eds) (2022) Phenomic selection: A new and efficient alternative to genomic selection. In: *Methods in Molecular Biology*, Vol. 2467. Springer Nature, pp. 397–420.

Robertsen, C.D., Hjortshøj, R.L. and Janss, L.L. (2019) Genomic selection in cereal breeding. *Agronomy* 9, 95. DOI: 10.3390/agronomy9020095.

Rodríguez, F.J., García, A., Pardo, P.J., Chávez, F. and Luque-Baena, R.M. (2018) Study and classification of plum varieties using image analysis and deep learning techniques. *Progress in Artificial Intelligence* 7, 119–127.

Roth, M., Muranty, H., Di Guardo, M., Guerra, W., Patocchi, A. *et al.* (2020) Genomic prediction of fruit texture and training population optimization towards the application of genomic selection in apple. *Horticulture Research* 7, 148. DOI: 10.1038/s41438-020-00370-5.

Sáenz-Romero, C., Lindig-Cisneros, R.A., Joyce, D.G., Beaulieu, J., Bradley, J.S. *et al.* (2016) Assisted migration of forest populations for adapting trees to climate change. *Rev Chapingo, Ser Ciencias For y Del Ambient* 22, 303–323. DOI: 10.5154/r.rchscfa.2014.10.052.

Sandhu, K.S., Shiv, A., Kaur, G., Meena, M.R., Raja, A.K. *et al.* (2022) Integrated approach in genomic selection to accelerate genetic gain in sugarcane. *Plants* 11(16), 2139. DOI: 10.3390/plants11162139.

Scarcelli, N., Mariac, C., Couvreur, T.L.P., Faye, A., Richard, D. *et al.* (2016) Intra-individual polymorphism in chloroplasts from NGS data: Where does it come from and how to handle it? *Molecular Ecology Resources* 16(2), 434–445. DOI: 10.1111/1755-0998.12462.

Sen, M.K., Mondal, S.K., Bharati, R., Severova, L. and Šrédl, K. (2025) Multiplex genome editing for climate-resilient woody plants. *Frontiers in Forests and Global Change* 8, 1542459. DOI: 10.3389/ffgc.2025.1542459.

Sethi, B.K., Singh, D., Rout, S.K. and Panda, S.K. (2023) Long short-term memory–deep belief network-based gene expression data analysis for prostate cancer detection and classification. *IEEE Access* 12, 1508–1524.

Seyum, E.G., Bille, N.H., Abtew, W.G., Munyengwa, N., Bell, J.M. *et al.* (2022) Genomic selection in tropical perennial crops and plantation trees: A review. *Molecular Breeding* 42(10), 58.

Sharma, U., Sankhyan, H.P., Kumari, A., Thakur, S., Thakur, L. *et al.* (2024) Genomic selection: A revolutionary approach for forest tree improvement in the wake of climate change. *Euphytica* 220(1), 9.

Shen, F., Bianco, L., Wu, B., Tian, Z., Wang, Y. *et al.* (2022) A bulked segregant analysis tool for out-crossing species (BSATOS) and QTL-based genomics-assisted prediction of complex traits in apple. *Journal of Advanced Research* 42, 149–162. DOI: 10.1016/j.jare.2022.03.013.

Silva Junior, O.B., Faria, D.A. and Grattapaglia, D. (2015) A flexible multi species genome wide 60K SNP chip developed from pooled resequencing of 240 Eucalyptus tree genomes across 12 species. *New Phytologist* 206(4), 1527–1540. DOI: 10.1111/nph.13322.

Simonin, K.A. and Roddy, A.B. (2018) Genome downsizing, physiological novelty, and the global dominance of flowering plants. *PLoS Biology* 16(1), e2003706. DOI: 10.1371/journal.pbio.2003706.

Song, B., Buckler, E.S. and Stitzer, M.C. (2024) New whole-genome alignment tools are needed for tapping into plant diversity. *Trends in Plant Science* 29(3), 355–369. DOI: 10.1016/j.tplants.2023.08.013.

Sun, M., Zhang, M., Kumar, S., Qin, M., Liu, Y. *et al.* (2024) Genomic selection of eight fruit traits in pear. *Horticultural Plant Journal* 10(2), 318–326.

Tambarussi, E.V., Shalizi, M.N., Grattapaglia, D., Hodge, G., Isik, F. *et al.* (2025) Genome-wide SNP-based relationships improve genetic parameter estimates and genomic prediction of growth traits in a large operational breeding trials of *Pinus taeda* L. *Forestry* 98(5), cpaf004.

Thapliyal, G., Bhandari, M.S., Vemanna, R.S., Pandey, S., Meena, R.K. *et al.* (2023) Engineering traits through CRISPR/Cas genome editing in woody species to improve forest diversity and yield. *Critical Reviews in Biotechnology* 43(6), 884–903. DOI: 10.1080/07388551.2022.2092714.

Torkamaneh, D., Boyle, B. and Belzile, F. (2018) Efficient genome-wide genotyping strategies and data integration in crop plants. *Theoretical and Applied Genetics* 131, 499–511. DOI: 10.1007/s00122-018-3056-z.

Trudić, B., Stojnić, S., Avramidou, E.V., Malliarou, E., Ivezić, A. *et al.* (2024) Recent global (bio) technological advancements in the service of forest genetic resources protection under climate change. *Časopis Topola* 213, 17–31.

Ukrainetz, T.A. (2019) Sketch and Speak: An expository intervention using note-taking and oral practice for children with language-related learning disabilities. *Language, Speech, and Hearing Services in Schools* 50(1), 53–70.

Van de Peer, Y., Mizrachi, E. and Marchal, K. (2017) The evolutionary significance of polyploidy. *Nature Reviews Genetics* 18(7), 411–244. DOI: 10.1038/nrg.2017.26.

Voss-Fels, K.P., Cooper, M. and Hayes, B.J. (2019) Accelerating crop genetic gains with genomic selection. *Theoretical and Applied Genetics* 132, 669–686.

Wang, J., Hong, P., Qiao, Q., Zhu, D., Zhang, L. *et al.* (2022) Chromosome-level genome assembly provides new insights into Japanese chestnut (*Castanea crenata*) genomes. *Frontiers in Plant Science* 13, 1049253. DOI: 10.3389/fpls.2022.1049253.

Wang, M., Li, R. and Zhao, Q. (2023) Multi-omics techniques in genetic studies and breeding of forest plants. *Forests* 14(6), 1196. DOI: 10.3390/f14061196.

Wang, L., Wei, F., Tagesson, T., Fang, Z. and Svenning, J.C. (2025a) Transforming forest management through rewilding: Enhancing biodiversity, resilience, and biosphere sustainability under global change. *One Earth* 8(3).

Wang, T., Zuo, Y., Manda, T., Hwarari, D. and Yang, L. (2025b) Harnessing artificial intelligence, machine learning and deep learning for sustainable forestry management and conservation: Transformative potential and future perspectives. *Plants* 14(7), 998.

Wegrzyn, J.L., Lee, J.M., Tearse, B.R. and Neale, D.B. (2008) TreeGenes: A forest tree genome database. *International Journal of Plant Genomics* 2008(1), 412875.

Wegrzyn, J.L., Falk, T., Grau, E., Buehler, S., Ramnath, R. *et al.* (2020) Cyberinfrastructure and resources to enable an integrative approach to studying forest trees. *Evolutionary Applications* 13(1), 228–241.

Wilson, S., Zheng, C., Maliepaard, C., Mulder, H.A., Visser, R.G.F. *et al.* (2021) Understanding the effectiveness of genomic prediction in tetraploid potato. *Frontiers in Plant Science* 12, 672417. DOI: 10.3389/fpls.2021.672417.

Xanthopoulou, A., Montero-Pau, J., Picó, B., Boumpas, P., Tsaliki, E. *et al.* (2021) A comprehensive RNA-seq-based gene expression atlas of the summer squash (*Cucurbita pepo*) provides insights into fruit morphology and ripening mechanisms. *BMC Genomics* 22(1), 341. DOI: 10.1186/s12864-021-07683-2.

Xu, Y., Zhang, X., Li, H., Zheng, H., Zhang, J. *et al.* (2022) Smart breeding driven by big data, artificial intelligence, and integrated genomic–enviromic prediction. *Molecular Plant* 15(11), 1664–1695. DOI: 10.1016/j.molp.2022.09.001.

Yan, J., Xu, Y., Cheng, Q., Jiang, S., Wang, Q. *et al.* (2021) LightGBM: Accelerated genomically designed crop breeding through ensemble learning. *Genome Biology* 22, 1–24. DOI: 10.1186/s13059-021-02492-y.

Yang, J., Wang, J., Liu, Z., Xiong, T., Lan, J. *et al.* (2018) Megaspore chromosome doubling in *Eucalyptus urophylla* S.T. Blake induced by colchicine treatment to produce triploids. *Forests* 9(11), 728. DOI: 10.3390/f9110728.

Zang, Z., Xu, Y., Duan, C., Wu, J., Li, S.Z. *et al.* (2024) A review of artificial intelligence based biological-tree construction: Priorities, methods, applications and trends. *arXiv Preprint* 2410.04815. DOI: 10.48550/arXiv.2410.04815.

Zhang, C., Jiang, S., Tian, Y., Dong, X., Xiao, J. *et al.* (2023) Smart breeding driven by advances in sequencing technology. *Modern Agriculture* 1(1), 43–56. DOI: 10.1002/moda.8.

Zhang, M.Y., Xue, C., Hu, H., Li, J., Xue, Y. *et al.* (2021) Genome-wide association studies provide insights into the genetic determination of fruit traits of pear. *Nature Communications* 12(1), 1144. DOI: 10.1038/s41467-021-21378-y.

Zhao, Y., Tian, Y., Sun, Y. and Li, Y. (2022) The development of Forest genetic breeding and the application of genome selection and CRISPR/Cas9 in Forest breeding. *Forests* 13(12), 2116. DOI: 10.3390/f13122116.

Zhou, X., Li, J., Chen, L., Guo, M., Liang, R. *et al.* (2024) The genomic pattern of insertion/deletion variations during rice improvement. *BMC Genomics* 25(1), 1263. DOI: 10.1186/s12864-024-11178-1.

Zuriaga, E., Soriano, J.M., Zhebentyayeva, T., Romero, C., Dardick, C. *et al.* (2013) Genomic analysis reveals MATH gene(s) as candidate(s) for Plum pox virus (PPV) resistance in apricot (*Prunus armeniaca* L.). *Molecular Plant Pathology* 14, 663–677. DOI: 10.1111/mpp.12037.

10 High-Throughput and Precision Phenotyping for Plant Genomic Prediction

A.K.M. Aminul Islam*, T. Anam and F.A. Shraboni
Department of Genetics and Plant Breeding, Gazipur Agricultural University, Bangladesh

Abstract
High-throughput phenotyping (HTP) and precision phenotyping are revolutionizing plant breeding by enabling rapid and large-scale trait assessment at the early growth stage for genomic prediction. HTP provides distinct benefits that allow for rapid, non-destructive, high-throughput detection and conquer the drawbacks of traditional and manual phenotyping. This method helps the efficient selection of genotypes through connecting phenotypes to genotypes. High-throughput and precision phenotyping involves a quick and precise technique of measuring numerous characteristics of plants on a large scale, which is crucial for genomic prediction. Researchers can accelerate breeding programs and improve the prediction accuracy of plant performance by integrating high-throughput phenotyping data with high-density genomic data. This integration allows breeders to select the best genotypes with targeted traits more precisely and efficiently for effective crop improvement. This chapter reviews different aspects of high-throughput and precision phenotyping, genomic prediction, and their potential in plant breeding. It also explores how HTP data can be incorporated into genomic selection models to improve the precision of prediction and accelerate genetic gain, and the chapter highlights their advancement in the areas of crop improvement.

Keywords: Crop improvement, Genomic data, Genomic selection, Phenomics, Plant speed breeding

10.1 Introduction

The most effective means to boost the productivity of key food and feed crops globally is by enhancing plant genetics and breeding. However, the field of crop improvement must adapt to face the growing challenges due to climate change. Phenotyping is a crucial component of plant science that focuses on measuring and analyzing observable plant traits and their response to various factors related to genes and environments (Fiorani and Schurr, 2013). This has led to the process becoming integral in modern agricultural research, generating fundamental data to promote programs of breeding, crop production, and resilience against environmental stresses (Harfouche *et al.*, 2019). The emergence of global challenges (for example, climate change, food security, and loss of biodiversity) has fueled the demand for advanced phenotyping technologies (Fu and Jiang, 2022).

The phenotyping process is the key to connecting the genetic potential of plant genotypes with their phenotypic outcome in plant breeding

*Corresponding author: aminulgpb@bsmrau.edu.bd

 Plant Genomic Prediction: AI-Enhanced Models for Plant Genomics and Crop Improvement (ed. Jen-Tsung Chen)
DOI: 10.1079/9781836991328.0010

(Cobb *et al.*, 2013). By measuring traits like yield and drought tolerance, as well as diseases and patterns of growth, researchers can identify the plants with the best characteristics for breeding. Phenotyping used to be a slow, laborious process, but advances in high-throughput technologies have revolutionized it by aiding the fast and precise collection of data on a large scale (Li *et al.*, 2021). In section 10.2, the various types of phenotyping and how phenotyping technologies have changed over time are described, and also highlight how high-throughput phenotyping (HTP) fits into modern plant breeding. The subsequent sections cover traditional phenotyping, molecular phenotyping, and HTP, describing the major tools and techniques involved in these processes, key platforms used for conducting high-throughput studies, and challenges associated with these advances.

10.2 Phenotyping: Origin, Concept, and Importance

Initial studies of genetics and plant breeding resulted in the development of phenotyping. In the early 20th century, plant scientists would physically assess and record characteristics like height, leaf size, or flower color to select superior genotypes for breeding (Cobb *et al.*, 2013). With the explosive advancements in genetics, there emerged a need to bridge genotypes with phenotypes, and phenotyping began including molecular tools to allow more precise measurements of traits (Schlichting and Smith, 2002). Phenotyping is the observable expression of an organism's genome. Phenotype encompasses more than just visible traits; it includes a diverse range of physiological, biochemical, and morphological attributes (Varshney *et al.*, 2021). Genomic technologies and high-throughput phenotyping, combined with environmental data analysis, can enhance and expand this notion of phenotype into physical characteristics (e.g. structure) or molecular traits in plants. In plant breeding and crop improvement programs, phenotyping is the most important part. This allows scientists to compare field germplasm and determine the inherent foundations of key complex traits like yield potential, and tolerance to abiotic and biotic stresses (Chawade *et al.*, 2019). Therefore, as crop improvement tilts towards precision breeding, where the focus is on one or few targeted traits, high-throughput phenotyping complements such approaches by providing data that can lessen the time of the breeding cycle while also improving selection accuracy. Within a climate change context, phenotyping also enables the exploration of core areas such as adaptation and sustainability by developing new plant varieties with improved resistance to disease and pests, and with tolerance to extreme heat, drought, salinity, and other abiotic stressors (Chapman *et al.*, 2012).

10.3 Phenotyping Types and Technologies

Phenotyping can be generally grouped into three major categories: traditional phenotyping; molecular phenotyping; and high-throughput phenotyping. Each of these categories has its advantages and disadvantages, and the combination of different forms of phenotyping delivers a broader view of plant traits.

10.3.1 Traditional phenotyping

Traditional phenotyping depends on visual observation and the physical measurements of observable traits (Fig. 10.1) (He *et al.*, 2024). This process, which has been used for centuries, includes the measurement of height, leaf area, flower color, and seed size, among other traits (Lamichhane and Thapa, 2022). These methods can be applied to screen a handful of traits in smaller populations, yet they are labor-intensive and very much susceptible to human error. In addition, traditional phenotyping does not identify complex or hidden characters like the architecture of the root system or plant response to stress, which play a vital role in the modern field of crop breeding. Traditional phenotyping is still necessary for evaluating some qualitative traits but has limitations, including the scale and accuracy, or precision, particularly when applied to high-density or large-scale breeding applications (Sun *et al.*, 2024).

Fig. 10.1. Traditional phenotyping practices in rice fields (Author's own image).

10.3.2 Molecular phenotyping

Molecular phenotyping is a modern approach that goes beyond external traits to analyze biochemical and genetic characteristics of plants. This type of phenotyping uses molecular markers like DNA, RNA, proteins, and metabolites to study the plant's genetic blueprint (Varshney *et al.*, 2021). Techniques like genomics, proteomics, transcriptomics, and metabolomics help to determine the genes and their pathways associated with traits of interest (Fig. 10.2). Molecular phenotyping plays an important role in analyzing traits that cannot be observed visually, such as tolerance to abiotic stress, resistance to diseases and pests, and nutrient absorption (Hammer *et al.*, 2016). This also enables marker-assisted selection (MAS), in which specific genomic markers are used to identify desirable traits more efficiently than traditional methods. Yet, molecular phenotyping is costly, needing laboratories and technical professionals in the field, limiting its availability for breeding programs to some extent (Ribaut *et al.*, 2010).

10.3.3 High-throughput phenotyping (HTP)

The quick and large-scale data on plant traits can be recorded by using advanced technologies like high-throughput phenotyping (HTP) (Li *et al.*, 2021). This allows for capturing vast datasets with high speed and accuracy, which facilitates more efficient plant breeding. HTP integrates multiple techniques like high-throughput imaging, sensors, remote sensing, robotics, and artificial intelligence (AI) for capturing and analyzing phenotypic data (Watt *et al.*, 2020).

High-resolution phenotyping

High-resolution phenotyping is the application of sophisticated cameras (3D imaging, infra-red, and hyperspectral) to take measurements of

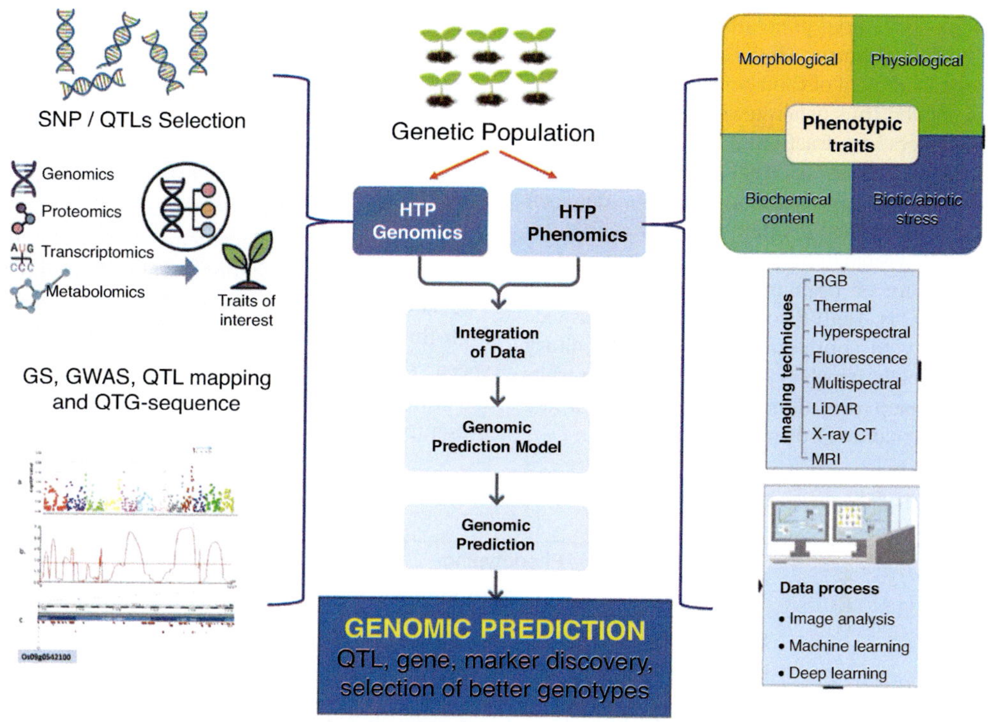

Fig. 10.2. High-throughput genotyping (HTG) and phenotyping (HTP) for gene or marker discovery for genomic prediction of better genotypes (Author's own image).

numerous traits at once (Smith *et al.*, 2021). These can differentiate visible traits (e.g. leaf area) from non-visible early changes in growth (e.g. internal water or root architecture) at high spatial and temporal resolution.

Sensor-based phenotyping

Various types of sensors are used for determining physiological and environmental variables in a process called sensor-based phenotyping. These sensors measure different parameters, including soil moisture, plant temperature, chlorophyll content, and transpiration rate (Xie and Yang, 2020). These types of measurements are crucial to understanding the plant's response to the stresses due to environmental change and can be monitored in real-time, allowing faster detection of stresses like drought or disease (Abebe *et al.*, 2023). In spite of the developments in HTP, many drawbacks still exist, including high equipment costs, storage of huge datasets, and the difficulty in analyzing data (Mir *et al.*, 2019). Developing cost-effective and scalable HTP solutions, improving data integration methods, and overcoming technical limitations of sensor resolution are crucial steps in making HTP more accessible and effective both for research and real-world utilization. However, standardization of protocols and data management practices is critical to ensuring that phenotyping results are consistent and comparable across platforms and environments (Krajewski *et al.*, 2015).

10.4 Traditional Phenotyping versus High-Throughput Phenotyping

Traditional phenotyping, while still valuable for certain traits, has limitations that are increasingly addressed by high-throughput phenotyping (HTP). The primary advantage of HTP over traditional phenotyping is its speed and scalability (Li *et al.*, 2021). Traditional approaches are labor-intensive and time-consuming, particularly when dealing with large population sizes or multiple traits. HTP platforms, on the other hand, facilitate the measurement of thousands of plants simultaneously and thus generate much larger datasets for breeders in a fraction of the time (Chawade *et al.*, 2019). In addition, HTP technologies provide more detailed and accurate data on plant traits needed to characterize ones that are not easily visible to humans, such as root system architecture or responses to abiotic stress (He *et al.*, 2024). High-throughput systems provide the ability to continuously monitor hereditary plant growth, which is vital for monitoring traits that develop over time (e.g. tolerance under drought conditions, disease resistance).

Although HTP has clear advantages over the traditional phenotyping approach, the latter continues to be adopted by breeders for initial screening of large populations and qualitative traits that do not warrant high-tech solutions (Kock, 2021). However, the increasing affordability of HTP technologies will lead to their incremental replacement of traditional methods for most applications in plant breeding. The key differences between traditional and high-throughput phenotyping, highlighting speed, scale, precision, cost-effectiveness, and advantages of HTP over traditional methods, are presented in Table 10.1.

Table 10.1. Differences between key aspects of traditional vs high-throughput phenotyping (Author's own table)

Aspect	Phenotyping		References
	Traditional	High-throughput	
Speed	Slow and manual processes.	Rapid and automated advanced systems.	Fiorani and Schurr (2013); Li *et al.* (2021); Abebe *et al.* (2023)
Scale	Applicable for limited samples only.	Large-scale phenotyping with high capacity.	Furbank and Tester (2011); Araus and Cairns (2014)
Precision	Low precision level and human error create an impact.	High precision level with sensor and imaging technologies.	Yang *et al.* (2020); Sun *et al.* (2024)
Resolution	The individual's visualization focused on macro traits like height.	High-resolution imaging enables micro- and macro-trait analysis.	Fiorani and Schurr (2013)
Trait coverage	Limited to visible, easily measurable traits.	Captures complex traits (e.g. photosynthesis efficiency)	Araus and Cairns (2014); Mir *et al.* (2019)
Environmental data	Minimum integration with environmental factors.	Monitors stress responses to environmental conditions.	Furbank and Tester (2011)
Technology	Simple tools like rulers and calipers.	Sophisticated tools like drones, hyperspectral cameras.	Furbank and Tester (2011)
Data Management	Limited data storage, manual data logging.	Generates big data, requires AI-based analytics.	Yang *et al.* (2020)
Cost	Low initial cost but labor-intensive.	High initial investment but cost-effective for longer term.	Fiorani and Schurr (2013); Abebe *et al.* (2023)
Application	Suitable for small-scale or targeted studies.	Essential for breeding, genomic studies, and stress analytics.	Araus and Cairns (2014)

10.5 Genomic Prediction: Concepts

Genomic prediction (GP) refers to forecasting of trait performance based on genomic data. Nowadays, high-throughput genotyping technology permits breeders to produce a vast number of genomic datasets that can be used in the genomic prediction of traits through genome-wide association studies (GWAS) (Meuwissen *et al.*, 2001). Genomic prediction, also known as genomic selection (GS), uses genomic data to predict an individual plant's trait performance, leveraging machine learning and deep learning methods. Machine learning (ML) provides nonlinear predictive models for predicting the genetic merits of individuals at an early stage with maximum prediction accuracy (Crossa *et al.*, 2025). The machine learning approach can accelerate genetic gain to extraordinary heights.

10.5.1 Prediction of genotypes

Prediction of genotypes involves understanding how genes are passed from parents to offspring. Punnett Square is a simple tool to visualize and predict the probabilities of different genotypes resulting from a cross in Mendelian inheritance. In case of complex traits inheritance and large populations, methods of statistics and machine learning models are used for analyzing large genetic datasets generated through high-throughput genotyping for prediction of phenotypes based on genotypes and environment (Danilevicz *et al.*, 2022). Prediction of genotypes uses genetic markers (for example, SNPs) throughout the genome to predict genetic values for selecting the best genotypes at an early stage in a breeding program. High-throughput genotyping and machine learning, including deep learning, are used increasingly to predict phenotypes of plants from the genotypes, where interaction between genes and environment plays a significant role. Therefore, understanding of environmental influence in the expression of genes is crucial for accurate prediction of genotypes to phenotypes of plants (Guo and Li, 2023).

10.5.2 Prediction of phenotype

Prediction of phenotype is a method of forecasting a plant's phenotype (observable traits) based on its genetic makeup (genotype) and environmental interaction. Phenotype prediction is a complex challenge in any crop improvement program, but high-throughput phenotyping, machine learning, and additional computational approaches are increasingly used to make accurate predictions (Kaya, 2025). Phenotypic prediction links a plant's genotype (DNA sequence) and its environment to its phenotype. This prediction is important in crop breeding to select plants having targeted characters for improved yield and resilience. Phenotypic prediction methods include machine learning to analyze large datasets, GWAS analysis to select desirable heritable variations, deep learning to predict phenotypes, metabolic modeling to understand the mechanism behind genotype-to-phenotype association, QTL-mapping to identify genomic regions influencing traits, and the Bayesian method to find complex relationships of genotype–phenotype–environment. These methods and models of phenotypic prediction face challenges of technological complexity, problems in data integration, model interpretation, and generalized ability, as they are not generalizing well to all datasets or genetic populations (Washburn *et al.*, 2021; Annan *et al.*, 2023).

10.5.3 Estimated breeding value (EBV)

EBV is an estimate of an individual plant's breeding value, which is the genetic potential it can transmit to its progeny. EBVs are numerical predictions of an individual plant's genetic merit for a specific trait, used in plant breeding to select superior plants (Bernardo, 2010). These values help breeders to predict how offspring will perform, facilitating more accurate selection. EBV, either pedigree-based or genomics-based, helps breeders to predict trait performance and select superior individuals for breeding. Pedigree-based estimated breeding value (PEBV) is calculated using the statistical method 'best linear unbiased prediction' (BLUP) and genomic estimated breeding value (GEBV) by single-step genomic BLUP (ssGBLUP). GEBV incorporates DNA information in the process of estimation and gives more

Table 10.2. Key differences and comparison between pedigree-based estimated breeding value (PEBV) and genomic estimated breeding value (GEBV) (Author's own table)

Features	PEBV	GEBV
Information used	Pedigree, own performance	Pedigree, own performance, and DNA data
Accuracy	Low	High
Cost	Low	High
Genetic gain	Low	High
Genetic progress	Slower	Faster

accurate predictions (Table 10.2). Both PEBV and GEBV are valuable tools for genetic improvement. The choice of these two methods depends on the trait being evaluated, resource availability, and the desired level of accuracy.

10.5.4 Genomic selection

Genomic selection (GS) is a breeding method that utilizes DNA information to predict the breeding value of individuals, allowing for more precise and faster identification of better traits. It differs from traditional methods by using markers across the entire genome, rather than a few known markers to predict genetic values (Bhat *et al.*, 2016). Genomic selection approach significantly speeds genetic gain of breeding programs by reducing breeding cycles and the cost associated with evaluation or phenotyping of a large number of breeding materials. GS can significantly shorten the breeding cycle by helping early generation selection based on DNA information, and decrease the time needed for the development of a new variety. It can capture a more comprehensive picture of the genetic merit of an individual by considering markers across the entire genome, leading to more accurate predictions. By aiding quicker and precise selection, GS can lead to improved genetic gain over generations. The approach of genomic selection could be effective when applied to traits that are difficult to phenotype using conventional methods, and additionally, these traits are often found to be influenced by multiple genes, e.g. quantitative traits (Budhlakoti *et al.*, 2022). Currently, GS is increasingly used to improve traits like yield, stress tolerance, and disease resistance in plant breeding programs.

10.6 High-Throughput Phenotyping: Tools and Technologies

HTP uses a diverse set of tools and technologies to precisely measure plant traits on a large scale (Fig. 10.3). Such tools can be roughly categorized into ground-based and aerial-based platforms, each offering unique advantages for different types of phenotyping applications.

10.6.1 High-throughput platforms

Ground-based platforms

Ground-based platforms, such as automated phenotyping systems, robotic systems, and sensor arrays, are primarily used in controlled environments like greenhouses or laboratory settings (Rui *et al.*, 2024). These systems are designed to monitor plant growth, health, and physiological processes in the long term. For example, field phenotyping robots can survey large plant populations and quantify plant traits by means of attached cameras or other sensors to capture real-time data on plant traits (Xu and Li, 2022).

Aerial-based platforms

Drones and unmanned aerial vehicles (UAVs) are increasingly being harnessed for field-based phenotyping as an aerial platform (Jin *et al.*, 2020). Multispectral or hyperspectral camera-drones are capable of flying over large areas and acquiring high-resolution plant imagery, leading to potential cost-efficient data collection at scale without resorting to manual labor (Feng *et al.*, 2021). One implementation, for instance, is the use of aerial-based systems that can monitor large fields or diverse environments where ground-based platforms may be impractical.

Fig. 10.3. A high-throughput phenotyping (HTP) platform featuring two distinct areas: a modern ground-based and an aerial-based setup (Author's own image).

10.6.2 Sensors and imaging

HTP relies heavily on sensor technologies and imaging systems. These systems comprise multi-spectral and hyper-spectral cameras, and thermal infrared sensors, as well as environmental sensors that quantify different plant characteristics and environmental factors (Zhang and Zhang, 2018). These sensors provide critical data on the physiological traits of the plant, like chlorophyll content, water status, and temperature, which are crucial to understanding the pattern of plant growth, development, and health (Damm *et al.*, 2018). Imaging technologies have a wide range of advanced applications, and particularly allow for high-resolution phenotyping to capture fine detail on plant structures, such as leaf morphology and canopy architecture (Fig. 10.4). These technologies also allow 3D modeling of plant growth that yields insight into structures and developmental strategies that cannot be captured with traditional phenotyping methods (Li *et al.*, 2014).

10.6.3 Data acquisition and analysis

In HTP, large datasets are collected by using sensors, imaging systems, and environmental data sources during the stage of data acquisition (Fig. 10.5). Data analysis is the next challenge to derive some insights from these data (Adão *et al.*, 2017). Newer computational approaches, including artificial intelligence (AI), and machine learning (ML), can be applied to explore phenotypic data in ways that cannot be revealed by conventional studies (Bongomin *et al.*, 2024). Using this information, these tools can be utilized for the prediction of plant performance, identifying required traits, and optimizing plant breeding strategies.

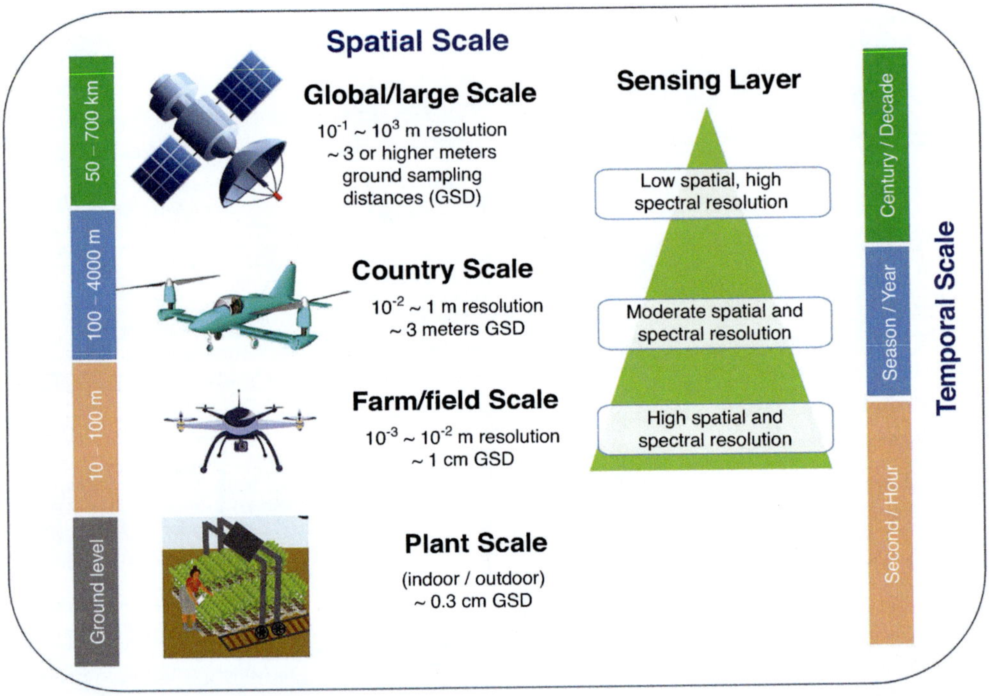

Fig. 10.4. Using drones for versatile crop phenotyping: different scales and sensing levels (Modified from Guo *et al.*, 2021)

10.7 High-Throughput Phenotyping for Genomic Prediction

High-throughput phenotyping (HTP) is revolutionizing plant breeding by enabling quick and precise assessment of a vast number of plant characters that are used in genomic prediction models to select superior plant genotypes. This approach combines advanced technologies like imaging, sensors, and automation to collect extensive phenotypic data, accelerating the identification of plants with desired characters such as yield, stress tolerance, and disease resistance (Nguyen *et al.*, 2025).

10.7.1 Increase data volume and accuracy

HTP platforms generate a vast amount of detailed phenotypic datasets, such as morphological, physiological, and biochemical characters that were previously impossible with traditional methods (Xiao *et al.*, 2022). High-throughput phenotyping techniques allow quick and accurate estimation of plant characteristics on a large scale, leading to a rise in data volume and enhanced data accuracy in plant breeding. This is accomplished by integrating advanced technologies like imaging, automation, machine learning, etc. (Sheikh *et al.*, 2024). High-throughput phenotyping platforms (robotics and sensors) can collect vast amounts of data automatically and continuously, exceeding data collection through conventional methods. This technique also allows simultaneous measurement of numerous plant characteristics (e.g. growth, stress response) on a large number of samples, generating vast datasets (Casto *et al.*, 2021). HTP platforms also capture data over time (e.g. hourly images of plant growth), which can further expand the data volume and enable dynamic analysis (Chang *et al.*, 2020). HTP platforms often utilize non-destructive methods

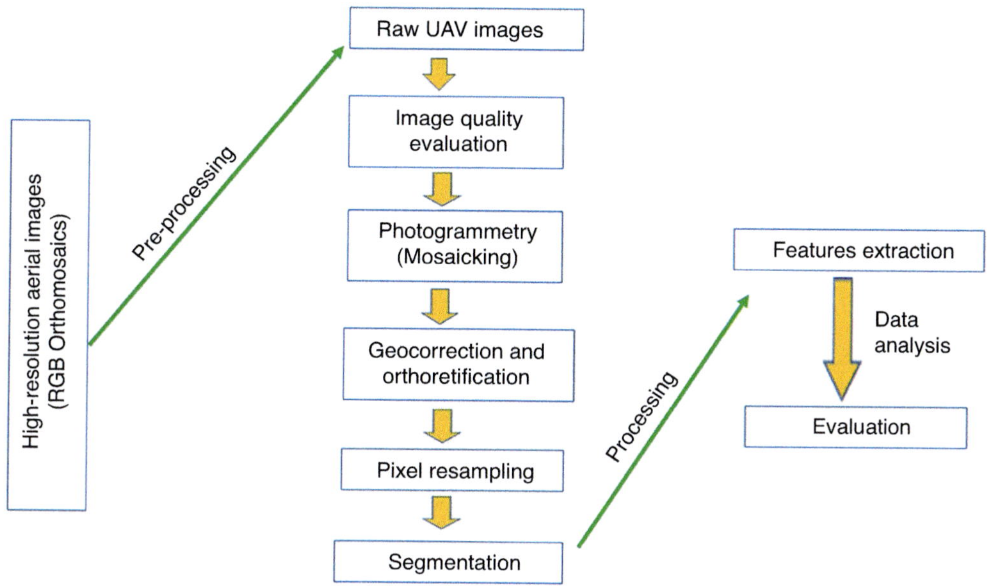

Fig. 10.5. General approach to unmanned aerial vehicle (UAV) image processing and analytics (Modified from Bongomin *et al*., 2024).

like imaging that reduce the risk of human error and sample damage associated with destructive measurements. Automation also ensures consistent application of measurement protocols, minimizing variability and improving data reliability. Integration of machine learning can analyze the large datasets generated by HTP platforms, identify subtle patterns, and extract valuable information with high accuracy. Technologies like hyperspectral and thermal imaging can detect plant health and stress response, and improve the accuracy of measurements. This platform often includes preprocessing and correction steps to minimize noise and improve the quality of data (Gill *et al*., 2022).

10.7.2 Improve trait prediction

HTP can significantly improve trait prediction in plant breeding by providing detailed, non-destructive measurements of multiple traits across various growth stages. HTP-generated data combined with genomic information and machine learning can be used in genomic prediction models that accurately forecast the genetic potential of plants for various traits, even before the plant is fully grown or in specific environments (Persa *et al*., 2021). The performance of untested lines in different environments can be predicted through this model, enabling precise predictions of genetic values and accelerating the process of superior plant selection (Kaushal *et al*., 2024). HTP data allows for the identification of traits at an early stage, which is highly associated with desired traits in later stages, enabling breeders to select superior plants at the early stage of the breeding cycle, reducing the time and costs required for evaluation. It also enables simultaneous measurement of multiple traits, allowing for the development of multi-trait genomic prediction models (Baba *et al*., 2020).

10.7.3 Enhance breeding efficiency

With integration of HTP data into genomic selection, breeders can select better plants with higher efficiency, and accelerate the breeding of improved crop varieties with desired traits (Angidi *et al*., 2025). HTP significantly enhances breeding efficiency through allowing rapid, vast-scale evaluation of plant characters, leading to faster selection of superior plants. This platform allows breeders to collect large numbers of phenotypic data, including growth, disease resistance, and stress tolerance, with high spatial and temporal resolution. Integration of collected datasets with genomic information and artificial intelligence accelerates the selection of suitable characters in early generations, their correlation with genetic markers boosting the breeding progress. HTP also helps to measure traits that are impossible or difficult to assess manually, like stress tolerance and disease responses. Integration of HTP in the breeding program can save costs and time through reducing labor costs, faster breeding progress by shortening breeding cycles, and optimizing resource utilization (He *et al.*, 2024).

10.7.4 Overcoming bottlenecks in breeding

Traditionally, plant breeding is a slow method, due to the time-consuming behavior of evaluating of large number of plants for various traits. This 'phenotypic bottleneck' limits the rate at which new and improved crop varieties can be developed. HTP addresses the bottleneck in traditional plant breeding by providing rapid and accurate phenotyping, which is crucial for genomic selection (Song *et al.*, 2021; Ninomiya, 2022). HTP overcomes this bottleneck through automation, non-destructive measurement, imaging, and data analysis. HTP utilizes automated systems such as robotics, drones, etc., to collect data on plant traits, reducing the need for labor and increasing the speed of data acquisition. It also allows the collection of data without damaging the plants, enabling repeated measurement from the same plant samples, assessing a large number of plants and traits simultaneously to increase the efficiency of the breeding process. Machine learning and image analysis techniques are often used to produce meaningful evidence from data generated through HTP platforms (Gill *et al.*, 2022).

10.7.5 Applications beyond yield

HTP is not limited to improving yield; it can be useful for a wide range of traits like stress tolerance, disease resistance, and other functional traits of crops, ultimately contributing to more resilient and sustainable agriculture (Islam *et al.*, 2024). HTP can be used to select plants having inherent disease resistance, allow breeders to select plants with resistant genes, and incorporate these genes into new varieties, which reduces the use of chemical pesticides (Nguyen *et al.*, 2025). HTP also assesses the ability of plants to respond against abiotic stresses like drought, extreme temperature, and salinity, which is difficult by conventional methods. Nutritional composition of crops can be analyzed by HTP, allowing breeders to select plants with higher levels of essential vitamins, minerals, and other beneficial elements (Casto *et al.*, 2021). HTP assists in selecting plants with traits that reduce weed competition, such as rapid early growth or allelopathic effects, which leads to effective weed management and reduces the use of herbicides. HTP enables continuous monitoring of the growth and development of plants, and delivers important observations on plant health and stress responses, leading to timely and targeted interventions, which is nearly impossible by conventional methods (He *et al.*, 2024).

10.8 Data-Driven Genomic Prediction

Genomic prediction (GP) is an innovative and novel breeding approach, also known as genomic selection (GS), that leverages statistical and machine learning algorithms integrated with genotypic and phenotypic datasets to predict breeding values in crops (Fig. 10.6). Marker-assisted selection (MAS) is a widespread method to select intended genes of interest accurately based on genetic markers. However, the application of MAS breeding poses challenges, as minor-effect loci control several complex agronomic traits, while the mapping of quantitative trait loci (QTLs) can identify loci with only major effects (Desta and Ortiz, 2014). Genomic selection can overcome this constraint by identifying QTLs with even minimal effects that MAS cannot trace. Genomic selection builds prediction models using phenotypic and genomic data, enabling the prediction of

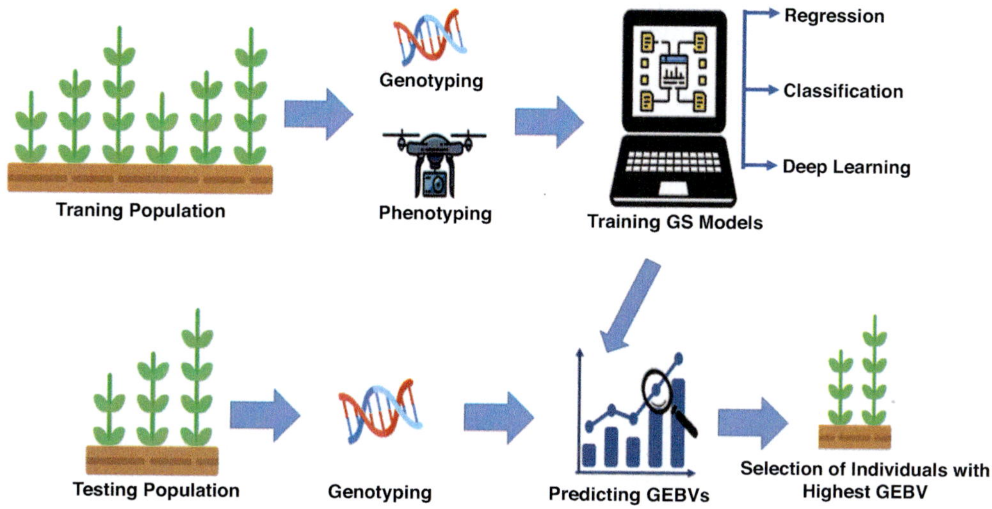

Fig. 10.6. Simple overview of genomic selection (genomic selection, GS, and genomic estimated breeding value, GEBV).

previously unrecognized phenotypic traits in genes and facilitating the selection of genotypes for breeding purposes. Thus, genomic selection reduces the time required to complete breeding cycles for developing new varieties by half.

Different platforms designed for high-throughput phenotyping could facilitate enhanced genomic selection in plants. Employing a UAV equipped with a remote-sensing unit featuring either a near-infrared or RGB camera, the height of sorghum plants has been efficiently assessed, and multiple genomic prediction models have been executed (Watanabe *et al.*, 2017). Rutkoski *et al.* (2016) concluded that the utilization of secondary trait information, for example, red and green NDVI, and canopy temperature obtained through HTP, could aid in improving the prediction of grain yield and GS in wheat. Additionally, the International Maize and Wheat Improvement Centre (CIMMYT) observed that the genomic selection techniques could enable crop growers to identify elite, high-yielding varieties with greater efficiency and accelerate genetic improvement. They reached this conclusion after investigating the integration of genotyping-by-sequencing (GBS) data with HTP-generated phenotypic data in advanced lines of wheat (Crain *et al.*, 2018).

Machine learning approaches have been integrated with high-throughput phenotyping to develop genomic selection models. The DeepGS technique introduced by Ma *et al.* (2018) employs convolutional neural networks (CNNs) to analyze and forecast grain-related traits in wheat. Five traits in soybeans were identified using CNNs and showed improved performance (Liu *et al.*, 2019). A genomic prediction model based on multi-traits has been developed for leveraging field-HTP and machine learning to build a predictive model for the plant height and yield of maize (Adak *et al.*, 2024). Montesinos-López *et al.* (2019) utilized deep learning in the prediction of multi-trait phenotypes in classification scenarios in wheat. The combination of GS and HTP data helps to develop a more robust model that accounts for both genomic and environmental factors affecting yield. Integrating GS and HTP data represents a transformative method for yield prediction in crop breeding. Utilizing the strengths of both technologies, plant breeders can take accurate and data-driven decisions, ultimately improving the sustainability, resilience, and productivity of crops to face the environmental challenges.

10.9 Challenges of High-Throughput and Precision Phenotyping

High-throughput and precision phenotyping face several challenges, including equipment cost, integration complexity of HTP data from various sources, and the need for robust data management systems. Moreover, genotype–environment interactions and environmental variability complicate the selection of traits and analysis. The equipment of advanced phenotyping platforms, including sensors, robotics, and imaging systems, can be expensive to acquire and maintain. The robust volume of the dataset generated by HTP phenotyping requires robust data management systems, including standardization, storage, and accessibility, adhering to principles like FAIR (findable, accessible, interoperable, and reusable). Uncontrolled environmental conditions, like fluctuating light, temperature, and humidity, may create variation in phenotypic data, especially in field-based studies. Genotype-by-environment interaction changes gene expression, making it challenging to identify stable traits across different locations and seasons. The genotypic and phenotypic data integration is a critical step for understanding the genetic basis of plant characters, but this integration is a complex method and requires specialized expertise. Analysis of a large dataset generated through high-throughput phenotyping requires advanced computational resources and expertise in the field of machine learning and analysis of captured images. Adoption of HTP in breeding programs is too slow, due to high cost, complexity, and the requirement for special expertise. Security and proper access to sensitive phenotypic data are crucial when dealing with large datasets and remote data access. HTP and genomic prediction also face similar challenges, including high cost, data management complexities, and necessity of robust and field-deployed technologies. Integration of HTP with genomic prediction faces hurdles related to data standardization, environmental variability, and the development of suitable statistical models.

10.10 Conclusion and Prospects

HTP and genomic prediction in plant breeding involve the integration of data generated by high-throughput phenotyping with genomic information to increase the efficiency and precision of selection of plants having desirable traits. This integration ultimately hastens the cycle of plant breeding and yield improvement. Combining HTP data with genomic data leads to more accurate predictions of plant performance, enabling breeders to identify superior plants more effectively and accelerate the breeding cycle. Integrating machine learning techniques with HTP and GP can further explore the improved prediction accuracy and efficiency of the complex relationships between large data, genomic information, and agronomic traits. Genomic prediction by considering multiple traits simultaneously also shows promise in improving prediction accuracy for complex traits. Through the integration of HTP and GP, breeders can develop more precise and targeted breeding strategies for the improvement of suitable characters that are desired. It can also assist in improving climate-resilient crop varieties' resistance to diseases and tolerance to abiotic stresses. More investigations are required to comprehend the HTP–GP-derived indices and prediction models for unraveling the complex relationships between traits and environments for effective selection.

References

Abebe, A.M., Kim, Y., Kim, J., Kim, S.L. and Baek, J. (2023) Image-based high-throughput phenotyping in horticultural crops. *Plants* 12, 10.

Adak, A., DeSalvio, A.J., Arik, M.A. and Murray, S.C. (2024) Field-based high-throughput phenotyping enhances phenomic and genomic predictions for grain yield and plant height across years in maize. *G3: Genes, Genomes, Genetics* 14(7), jkae092. DOI: 10.1093/g3journal/jkae092.

Adão, T., Hruška, J., Pádua, L., Bessa, J., Peres, E. *et al.* (2017) Hyperspectral imaging: A review on UAV-based sensors, data processing and applications for agriculture and forestry. *Remote Sensing* 9(11), 1110.

Angidi, S., Madankar, K., Tehseen, M.M. and Bhatla, A. (2025) Advanced high-throughput phenotyping techniques for managing abiotic stress in agricultural crops—a comprehensive review. *Crops* 5(2), 8. DOI: 10.3390/crops5020008.

Annan, R., Qingge, L. and Yang, P. (2023) Machine learning models for phenotype prediction from genotype. In: *23rd International Conference on Bioinformatics and Bioengineering (BIBE)*, IEEE 2023, Dayton, Ohio, USA, pp. 81–86. DOI: 10.1109/BIBE60311.2023.00021.

Araus, J.L. and Cairns, J.E. (2014) Field high-throughput phenotyping: The new crop breeding frontier. *Trends in Plant Science* 19(1), 52–61.

Baba, T., Momen, M., Campbell, M.T., Walia, H. and Morota, G. (2020) Multi-trait random regression models increase genomic prediction accuracy for a temporal physiological trait derived from high-throughput phenotyping. *PLOS ONE* 15(2), e0228118. DOI: 10.1371/journal.pone.0228118.

Bernardo, R. (2010) *Breeding for Quantitative Traits in Plants*, 2nd edn. Stemma Press, Woodbury, Minnesota, USA, p. 390 .

Bhat, J.A., Ali, S., Salgotra, R.K., Mir, Z.A., Dutta, S. *et al.* (2016) Genomic selection in the era of next generation sequencing for complex traits in plant breeding. *Frontiers in Genetics* 7, 221. DOI: 10.3389/fgene.2016.00221.

Bongomin, O., Lamo, J., Guina, J.M., Okello, C., Ocen, G.G. *et al.* (2024) UAV image acquisition and processing for high-throughput phenotyping in agricultural research and breeding programs. *The Plant Phenome Journal* 7(1), e20096.

Budhlakoti, N., Kushwaha, A.K., Rai, A., Chaturvedi, K.K., Kumar, A. *et al.* (2022) Genomic selection: A tool for accelerating the efficiency of molecular breeding for development of climate-resilient crops. *Frontiers in Genetics* 13, 832153. DOI: 10.3389/fgene.2022.832153.

Casto, A.L., Schuhl, H., Tovar, J.C., Wang, Q., Bart, R.S. *et al.* (2021) Picturing the future of food. *Plant Phenome Journal* 4, e20014. DOI: 10.1002/ppj2.20014.

Chang, S., Lee, U., Hong, M.J., Jo, Y.D. and Kim, J.-B. (2020) High-throughput phenotyping (HTP) data reveal dosage effect at growth stages in *Arabidopsis thaliana* irradiated by gamma rays. *Plants* 9(5), 557. DOI: 10.3390/plants9050557.

Chapman, S.C., Chakraborty, S., Dreccer, M.F. and Howden, S.M. (2012) Plant adaptation to climate change—opportunities and priorities in breeding. *Crop and Pasture Science* 63(3), 251–268.

Chawade, A., van Ham, J., Blomquist, H., Bagge, O., Alexandersson, E. *et al.* (2019) High-throughput field-phenotyping tools for plant breeding and precision agriculture. *Agronomy* 9(5), 258.

Cobb, J.N., DeClerck, G., Greenberg, A., Clark, R. and McCouch, S. (2013) Next-generation phenotyping: Requirements and strategies for enhancing our understanding of genotype–phenotype relationships and its relevance to crop improvement. *Theoretical and Applied Genetics* 126, 867–887.

Crain, J., Mondal, S., Rutkoski, J., Singh, R.P. and Poland, J. (2018) Combining high-throughput phenotyping and genomic information to increase prediction and selection accuracy in wheat breeding. *The Plant Genome* 11(1), 170043. DOI: 10.3835/plantgenome2017.05.0043.

Crossa, J., Martini, J.W.R., Vitale, P., Pérez-Rodríguez, P., Costa-Neto, G. *et al.* (2025) Expanding genomic prediction in plant breeding: Harnessing big data, machine learning, and advanced software. *Trends in Plant Science* 30(7), 756–774. DOI: 10.1016/j.tplants.2024.12.009.

Damm, A., Paul-Limoges, E., Haghighi, E., Simmer, C., Morsdorf, F. *et al.* (2018) Remote sensing of plant-water relations: An overview and future perspectives. *Journal of Plant Physiology* 227, 3–19.

Danilevicz, M.F., Gill, M., Anderson, R., Batley, J., Bennamoun, M. *et al.* (2022) Plant genotype to phenotype prediction using machine learning. *Frontiers in Genetics* 13, 822173. DOI: 10.3389/fgene.2022.822173.

Desta, Z.A. and Ortiz, R. (2014) Genomic selection: Genome-wide prediction in plant improvement. *Trends in Plant Science* 19(9), 592–601. DOI: 10.1016/j.tplants.2014.05.006.

Feng, L., Chen, S., Zhang, C., Zhang, Y. and He, Y. (2021) A comprehensive review on recent applications of unmanned aerial vehicle remote sensing with various sensors for high-throughput plant phenotyping. *Computers and Electronics in Agriculture* 182, 106033. DOI: 10.1016/j.compag.2021.106033.

Fiorani, F. and Schurr, U. (2013) Future scenarios for plant phenotyping. *Annual Review of Plant Biology* 64(1), 267–291.

Furbank, R.T. and Tester, M. (2011) Phenomics–technologies to relieve the phenotyping bottleneck. *Trends in Plant Science* 16(12), 635–644.

Fu, X. and Jiang, D. (2022) High-throughput phenotyping: The latest research tool for sustainable crop production under global climate change scenarios. In: Liu, F., Li, X., Hogy, P., Jian, D., Brestic, M. et al. (eds) *Sustainable Crop Productivity and Quality Under Climate Change*. Academic Press, pp. 313–381.

Gill, T., Gill, S.K., Saini, D.K., Chopra, Y., de Koff, J.P. *et al.* (2022) A comprehensive review of high throughput phenotyping and machine learning for plant stress phenotyping. *Phenomics* 2(3), 156. DOI: 10.1007/s43657-022-00048-z.

Guo, T. and Li, X. (2023) Machine learning for predicting phenotype from genotype and environment. *Current Opinion in Biotechnology* 79, 102853. DOI: 10.1016/j.copbio.2022.102853.

Guo, W., Carroll, M.E., Singh, A., Swetnam, T.L., Merchant, N. *et al.* (2021) UAS-based plant phenotyping for research and breeding applications. *Plant Phenomics* 9840192.

Hammer, G., Messina, C., van Oosterom, E., Chapman, S., Singh, V. *et al.* (2016) Molecular breeding for complex adaptive traits: How integrating crop ecophysiology and modelling can enhance efficiency. In: Yin, X. and Struik, P. (eds) *Crop Systems Biology: Narrowing the Gaps between Crop Modelling and Genetics*. Springer, Cham, Switzerland, pp. 147–162.

Harfouche, A.L., Jacobson, D.A., Kainer, D., Romero, J.C., Harfouche, A.H. *et al.* (2019) Accelerating climate resilient plant breeding by applying next-generation artificial intelligence. *Trends in Biotechnology* 37(11), 1217–1235.

He, S., Li, X., Chen, M., Xu, X., Tang, F. *et al.* (2024) Crop HTP technologies: Applications and prospects. *Agriculture* 14(5), 723. DOI: 10.3390/agriculture14050723.

Islam, M.M., Rahman, M.M., Sarker, S.S., Islam, M.N., Bhuiyan, F.H. *et al.* (2024) Beyond yield: Unveiling farmer perceptions and needs regarding weed management in Bangladesh. *Frontiers in Bioengineering and Biotechnology* 12, 1410128. DOI: 10.3389/fbioe.2024.1410128.

Jin, X., Li, Z., Feng, H., Ren, Z. and Li, S. (2020) Estimation of maize yield by assimilating biomass and canopy cover derived from hyperspectral data into the AquaCrop model. *Agricultural Water Management* 227, 105846. DOI: 10.1016/j.agwat.2019.105846.

Kaushal, S., Gill, H.S., Billah, M.M., Khan, S.N., Halder, J. *et al.* (2024) Enhancing the potential of phenomic and genomic prediction in winter wheat breeding using high-throughput phenotyping and deep learning. *Frontiers in Plant Science* 15, 1410249. DOI: 10.3389/fpls.2024.1410249.

Kaya, C. (2025) Optimizing crop production with plant phenomics through high-throughput phenotyping and ai in controlled environments. *Food Energy Security* 14, e70050. DOI: 10.1002/fes3.70050.

Kock, M.A. (2021) Open intellectual property models for plant innovations in the context of new breeding technologies. *Agronomy* 11(6), 1218.

Krajewski, P., Chen, D., Ćwiek, H., van Dijk, A.D., Fiorani, F. *et al.* (2015) Towards recommendations for metadata and data handling in plant phenotyping. *Journal of Experimental Botany* 66(18), 5417–5427.

Lamichhane, S. and Thapa, S. (2022) Advances from conventional to modern plant breeding methodologies. *Plant Breeding and Biotechnology* 10(1), 1–14.

Li, D., Quan, C., Song, Z., Li, X., Yu, G. *et al.* (2021) High-throughput plant phenotyping platform (HT3P) as a novel tool for estimating agronomic traits from the lab to the field. *Frontiers in Bioengineering and Biotechnology* 8, 623705.

Li, L., Zhang, Q. and Huang, D. (2014) A review of imaging techniques for plant phenotyping. *Sensors* 14(11), 20078–20111.

Liu, Y., Wang, D., He, F., Wang, J., Joshi, T. *et al.* (2019) Phenotype prediction and genome-wide association study using deep convolutional neural network of soybean. *Frontiers in Genetics* 10, 1091. DOI: 10.3389/fgene.2019.01091.

Ma, W., Qiu, Z., Song, J., Li, J., Cheng, Q. *et al.* (2018) A deep convolutional neural network approach for predicting phenotypes from genotypes. *Planta* 248, 1307–1318. DOI: 10.1007/s00425-018-2976-9.

Meuwissen, T.H.E., Hayes, B.J. and Goddard, M.E. (2001) Prediction of total genetic value using genome-wide dense marker maps. *Genetics* 157(4), 1819–1829. DOI: 10.1093/genetics/157.4.1819.

Mir, R.R., Reynolds, M., Pinto, F., Khan, M.A. and Bhat, M.A. (2019) High-throughput phenotyping for crop improvement in the genomics era. *Plant Science* 282, 60–72.

Montesinos-López, O.A., Montesinos-López, A., Tuberosa, R., Maccaferri, M., Sciara, G. *et al.* (2019) Multi-trait, multi-environment genomic prediction of durum wheat with genomic best linear unbiased predictor and deep learning methods. *Frontiers in Plant Science* 10, 1311. DOI: 10.3389/fpls.2019.01311.

Nguyen, H.T., Khan, M.A.R., Nguyen, T.T., Pham, N.T., Nguyen, T.T.B. *et al.* (2025) Advancing crop resilience through high-throughput phenotyping for crop improvement in the face of climate change. *Plants* 14(6), 907. DOI: 10.3390/plants14060907.

Ninomiya, S. (2022) High-throughput field crop phenotyping: Current status and challenges. *Breeding Science* 72(1), 3–18. DOI: 10.1270/jsbbs.21069.

Persa, R., de Oliveira Ribeiro, P.C. and Jarquin, D. (2021) The use of high-throughput phenotyping in genomic selection context. *Crop Breeding and Applied Biotechnology* 21(S), e385921S6. DOI: 10.1590/1984-70332021v21Sa19.

Ribaut, J.M., De Vicente, M.C. and Delannay, X. (2010) Molecular breeding in developing countries: Challenges and perspectives. *Current Opinion in Plant Biology* 13(2), 213–218.

Rui, Z., Zhang, Z. and Chen, T. (2024) A review of field plant phenotyping platform, sensing and navigation technology. In: Zhang, Z., Jian, Y., Wen, C., Men, S. and Zhang, Y. (eds) *Transforming Technologies in Smart Agriculture: New Applications During the Past Decade*, Vol. 8. Smart Agriculture, Springer, Singapore, pp. 1–46. DOI: 10.1007/978-981-97-4141-0_1.

Rutkoski, J., Poland, J., Mondal, S., Autrique, E., Pérez, L.G. *et al.* (2016) Canopy temperature and vegetation indices from high-throughput phenotyping improve accuracy of pedigree and genomic selection for grain yield in wheat. *G3: Genes, Genomes, Genetics* 6(9), 2799–2808. DOI: 10.1534/g3.116.032888.

Schlichting, C.D. and Smith, H. (2002) Phenotypic plasticity: Linking molecular mechanisms with evolutionary outcomes. *Evolutionary Ecology* 16, 189–211.

Sheikh, M., Iqra, F., Ambreen, H., Pravin, K.A., Ikra, M. *et al.* (2024) Integrating artificial intelligence and high-throughput phenotyping for crop improvement. *Journal of Integrative Agriculture* 23(6), 1787–1802. DOI: 10.1016/j.jia.2023.10.019.

Smith, D.T., Potgieter, A.B. and Chapman, S.C. (2021) Scaling up high-throughput phenotyping for abiotic stress selection in the field. *Theoretical and Applied Genetics* 134(6), 1845–1866.

Song, P., Wang, J., Guo, X., Yang, W. and Zhao, C. (2021) High-throughput phenotyping: Breaking through the bottleneck in future crop breeding. *The Crop Journal* 9(3), 633–645. DOI: 10.1016/j.cj.2021.03.015.

Sun, L., Lai, M., Ghouri, F., Nawaz, M.A., Ali, F. *et al.* (2024) Modern plant breeding techniques in crop improvement and genetic diversity: From molecular markers and gene editing to artificial intelligence—a critical review. *Plants* 13(19), 2676.

Varshney, R.K., Bohra, A., Roorkiwal, M., Barmukh, R., Cowling, W.A. *et al.* (2021) Fast-forward breeding for a food-secure world. *Trends in Genetics* 37(12), 1124–1136.

Washburn, J.D., Cimen, E., Ramstein, G., Reeves, T., O'Briant, P. *et al.* (2021) Predicting phenotypes from genetic, environment, management, and historical data using CNNs. *Theoretical and Applied Genetics* 134(12), 3997–4011. DOI: 10.1007/s00122-021-03943-7.

Watanabe, K., Guo, W., Arai, K., Takanashi, H., Kajiya-Kanegae, H. *et al.* (2017) High-throughput phenotyping of sorghum plant height using an unmanned aerial vehicle and its application to genomic prediction modeling. *Frontiers in Plant Science* 8. DOI: 10.3389/fpls.2017.00421.

Watt, M., Fiorani, F., Usadel, B., Rascher, U., Muller, O. *et al.* (2020) Phenotyping: New windows into the plant for breeders. *Annual Review of Plant Biology* 71(1), 689–712.

Xiao, Q., Bai, X., Zhang, C. and He, Y. (2022) Advanced high-throughput plant phenotyping techniques for genome-wide association studies: A review. *Journal of Advanced Research* 35, 215–230. DOI: 10.1016/j.jare.2021.05.002.

Xie, C. and Yang, C. (2020) A review on plant high-throughput phenotyping traits using UAV-based sensors. *Computers and Electronics in Agriculture* 178, 105731.

Xu, R. and Li, C. (2022) A review of high-throughput field phenotyping systems: Focusing on ground robots. *Plant Phenomics* 16, 9760269. DOI: 10.34133/2022/9760269.

Yang, J., Du, L., Gong, W., Shi, S. and Sun, J. (2020) Estimating leaf nitrogen concentration based on the combination with fluorescence spectrum and first-derivative. *Royal Society Open Science* 7(2), 191941. DOI: 10.1098/rsos.191941.

Zhang, Y. and Zhang, N. (2018) Imaging technologies for plant high-throughput phenotyping: A review. *Frontiers of Agricultural Science and Engineering* 5(4), 406–419.

11 Validation Methods for Plant Genomic Prediction

Mohd Ali[1]*, Roohi Sharma[2], Parkirti[2], Anchita Singh[2], Alisha Sharma[2], Harshdeep Rana[3], Arun Dev Singh[4], Jaswinder Singh[1] and Puja Ohri[2]

[1]Post Graduate Department of Zoology, Khalsa College Charitable Society, Amritsar, Punjab, India; [2]Department of Zoology, Guru Nanak Dev University, Amritsar, Punjab, India; [3]BRIC-National Agri-Food and Biomanufacturing Institute (BRIC-NABI), Knowledge City, SAS Nagar, Mohali, Punjab, India; [4]Department of Botanical and Environmental Sciences, Guru Nanak Dev University, Amritsar, Punjab, India

Abstract

The genetic prediction of plant and animal phenotypes and breeding values has quickly grown into a distinct area of study. Due to differences in data simulation, incomplete descriptions of real or simulated data, and the failure to present all pertinent results, the results of genetic prediction studies are frequently hard to compare. Additionally, several novel techniques have only been compared in a small number of genetic designs, which could result in inaccurate conclusions. With the growing evidence that genomic prediction (GP) enhances genetic gains in plant breeding, it is appropriate to examine the critical elements that increase its effectiveness. In this chapter, we focus on the genomic prediction, characteristics of genomic prediction, and validation methods for plant genomic prediction. We outline how genomic prediction works in plants. Additionally, we examine the important findings and current limitations for genomic prediction in plants. Finally, we discuss the diagnostic features of deep learning (DL) models used in genomic prediction (GP).

Keywords: Accuracy, Deep learning, Genetic gains, Genomic prediction

11.1 Introduction

The rate of population growth worldwide is probably going to stay the same or even increase in the coming decades (Taagepera and Nemčok, 2024). While crop output has been reduced by a number of biotic and abiotic stresses made worse by anthropogenic climate change, the demand for food is predicted to rise by the same amount in order to feed the population (Oshunsanya *et al.*, 2019). In order to create new cultivars with increased yield, better quality, and tolerance or resistance to a variety of biotic and abiotic challenges, plant breeding is essential (Hussain, 2015). Genetic improvement of crops has traditionally depended on traditional cross-breeding techniques, in which genotypes are selected and bred only on the basis of phenotypic and pedigree performance (Begna, 2021). Improved crop cultivars have been developed in large part through careful parent evaluation for various traits, targeted crossing, generation advancement using the summer and winter shuttle breeding schemes to shorten the breeding cycle, key location evaluation of elite

*Corresponding author: xyaansari774@gmail.com

 Plant Genomic Prediction: AI-Enhanced Models for Plant Genomics and Crop Improvement (ed. Jen-Tsung Chen)
DOI: 10.1079/9781836991328.0011

germplasms, and efficient database management. But the rapid development of DNA-sequencing technology has given breeders access to extensive genetic data on crops, which is crucial for selection (Nadeem *et al.*, 2018; Ghosh *et al.*, 2024). Plant breeders now have access to a much greater variety of DNA markers thanks to the advent of many genotyping techniques based on DNA markers. Breeding has advanced through several phases during the past 40 years. It began with linkage-based mapping of quantitative trait loci (QTLs) in which individuals segregating with a specific characteristic were found to be associated with a QTL and utilized for marker-assisted selection (MAS) using a small number of DNA markers (Soller and Plotkin-Hazan, 1977; Amiteye, 2021). Due to its time-consuming nature, narrow allelic variation, poor resolution, and requirement for a set of segregating individuals derived from biparental crosses, the approach has little practical impact in plant breeding efforts. With the development of easily accessible genome-wide single-nucleotide polymorphisms (SNPs), genomic selection (GS), when applied to practical selection, has become a potent technique in plant breeding. GS was initially developed by Meuwissen *et al.* (2001) 20 years ago, in addition to early contributors (Lande and Thompson, 1990; Haley and Visscher, 1998; Whittaker, 2000). By proposing that genetic value prediction from marker profiles could significantly boost genetic gain in plant and animal breeding, especially when paired with reproductive techniques to reduce the generation interval, the authors of this ground-breaking study opened a new path in plant breeding. The two primary theme areas being intensively investigated in plant GP research are training population optimization and creating statistical machine-learning models. Their potential is the reason for this to increase the accuracy of predictions, as the current performance is far from ideal. This chapter starts with a condensed description of GP before delving into the most recent, extensively used cross-validation (CV) techniques in plant breeding. Additionally, the impact of the identified parameters on the accuracy of GP in various attributes is illustrated by the analysis of empirical study results using wheat, maize, and potato as examples of self-pollinating, cross-pollinating, and clonally propagated crops, respectively. Last but not least, a prime example of the use of GS is provided by current empirical research from both governmental and private breeding programs. In conclusion, helpful recommendations are sent to help plant breeding programs successfully apply GS.

11.2 Genomic Prediction

Genomic prediction is one of the most recent data-driven techniques that has gained widespread acceptance and is a useful instrument for accelerating genetic gain in plant breeding. GP uses sophisticated statistical machine learning models to choose individuals from a breeding population according to breeding values calculated from markers found throughout the genome. Data from a training population, including both genotypic and phenotypic information, is used in this selection process. These models use advanced statistical machine learning models to select individuals within a breeding population based on breeding values estimated from genome-wide markers. These models use genotypic data to predict breeding or phenotypic qualities for a specific population, following thorough training. However, before selecting prediction models, it is important to evaluate their performance using CV. This step in GP is crucial for evaluating prediction models and comparing them with various scenarios, including incorporating multiple traits, known major genes and marker-trait associations (QTLs), genotype 3 environments (G3E) interaction, and other omics data like transcriptomics, metabolomics, and proteomics.

GP defines an offspring's genomic estimated breeding value (GEBV) by taking into account the breeding values of the parental average and the Mendelian sampling deviation. This enables the method to be utilized for: (i) quick selection cycle with short breeding interval at early generations by predicting the additive effects (i.e. GS at the F2 level of a biparental cross); and (ii) selection of lines at late stages of selection by predicting the individual genotypic values, where the final commercial value of the line is determined by both additive and non-additive effects. GP is impacted by a variety of circumstances that can drastically lower its accuracy (Fig. 11.1). As a result, if left unchecked, they may make it more difficult to use GP effectively in plant breeding initiatives. During training-population optimization, the

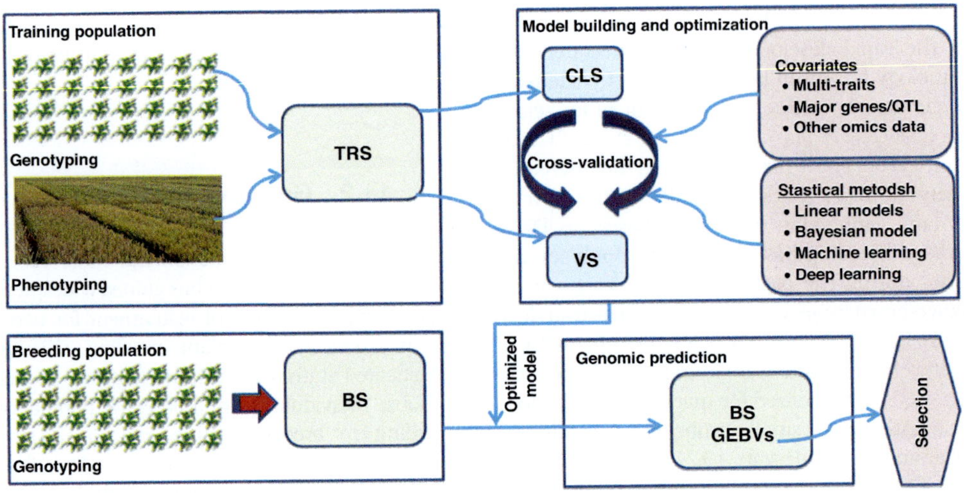

Fig. 11.1. Schematic representation of genomic prediction (Author's own image)

three main characteristics to focus on are the population size, genetic diversity, and genetic relatedness with the breeding population. Other significant characteristics that make GP in plant breeding even more difficult include the degree of linkage disequilibrium between QTLs and markers (in both the training and breeding (testing) population), genetic complexity and heritability of target traits, quality/precision phenotyping, statistical machine-learning models, G3E interaction, and other non-additive factors.

11.3 Genomic Prediction in Plants

Over the past two decades, the concept of genome-based prediction (GP) of genetic values has brought a quiet revolution in plant breeding (Meuwissen *et al.*, 2001; Jarquin *et al.*, 2020). Genome prediction works by using dense molecular markers spread across the entire genome to create a genomic relationship matrix (Dekkers *et al.*, 2021). This matrix becomes a powerful tool, allowing breeders to predict the performance of lines whose genotypes are known but phenotypes are yet to be measured (Isidro y Sánchez and Akdemir, 2021; Kumar *et al.*, 2023 . In other words, GP enables breeders to see the genetic potential of a plant before it is ever grown in the field. This predictive ability has been harnessed in multiple ways. For example, GP can boost selection accuracy by treating predictions as additional phenotypic data, such as those gathered over multiple years or locations (Jarquín *et al.*, 2014; Wu *et al.*, 2025). It can also shorten the breeding cycle by allowing certain testing stages to be skipped or by testing only a subset of genotypes, thus freeing resources to evaluate more candidates and increase selection intensity (Crossa *et al.*, 2017).

The genomic prediction process begins with the establishment of a training population (or reference set), comprising plants for which both genotype data (molecular marker profiles) and phenotype data (trait measurements) are available. This dataset is used to develop a statistical model that captures the relationship between specific genomic marker patterns and the observed trait performance (Jeon *et al.*, 2023). Once the model is trained, it can be applied to select candidate plants for which only genotype data are available to estimate their genomic estimated breeding values (GEBVs) (Selga *et al.*, 2022). Individuals with the highest predicted GEBVs are then prioritized for crossing or advanced field evaluations, thereby accelerating the breeding cycle and improving selection efficiency (Alemu *et al.*, 2024). However, several factors influence the accuracy of genomic selection, such as the genetic diversity and size of the breeding population, the heritability of the trait in question, genotype-by-environment (G×E) interactions, the density of genetic

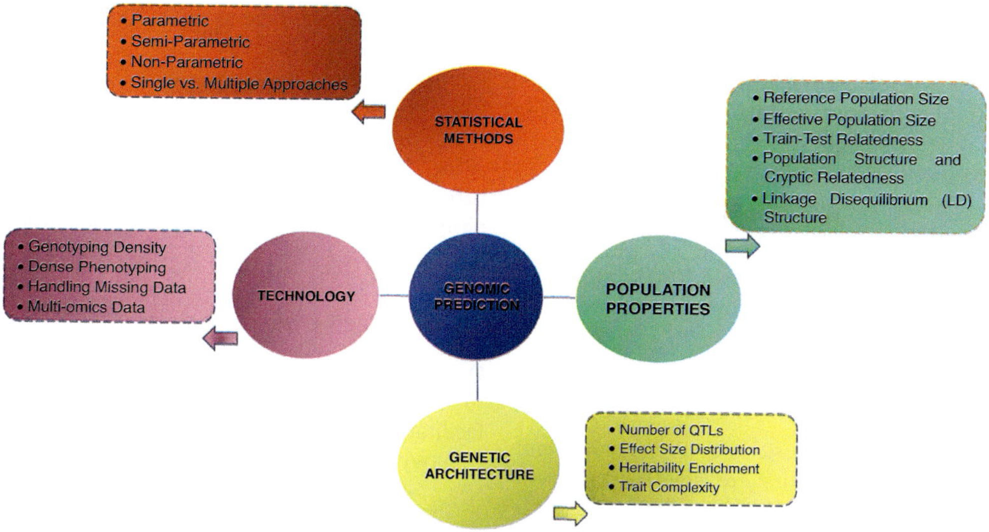

Fig. 11.2. Characteristics of genomic prediction.

markers, and the genetic relatedness between the training population and the breeding population or selection candidates (Abou Khater *et al.*, 2025)

11.4 Characteristics of Genomic Prediction

Genomic prediction (GP) is a potent statistical tool that forecasts genetic merit or breeding values of an individual using genome-wide molecular marker data. GP uses the complete genetic profile, which markedly makes it especially useful for complicated quantitative features, in contrast to marker-assisted selection, which mainly depends on a small number of important markers. The main factors that affect how well genomic prediction works are genetic architecture, accessible technologies, population properties, and statistical methods (Farooq *et al.*, 2023) (Fig. 11.2).

11.4.1 Genetic architecture

The number, size, and distribution of quantitative trait loci (QTLs) are referred to as the genetic architecture of traits. These QTLs and their interactions play a crucial role in influencing the accuracy of predictions.

Number of QTLs

Polygenic traits governed by numerous loci with minor to moderate effects are ideal for GP, as dense marker sets can capture the cumulative effect. In contrast, traits controlled by a small number of large-effect loci may not gain much from GP and could be better addressed through marker-assisted approaches.

Effect size distribution

A combination of big and small impact loci can be found in certain traits. In these situations, statistical techniques that permit variable shrinking of marker effects, such as Bayesian methods, frequently yield superior results (de Los Campos *et al.*, 2013).

Heritability enrichment

High heritability improves GP accuracy by allowing a greater percentage of phenotypic variance to be accounted for by genetic variance (Ren *et al.*, 2022).

Trait complexity

While qualities impacted by strong genotype × environment (G × E) interactions are more difficult to predict, purely genetic traits are much easier to predict. Improvements are frequently obtained by models that use G × E terms or multi-environment data (Crossa *et al.*, 2017).

11.4.2 Technology

The efficacy of genomic prediction is supported by technological developments in phenomics and genomics.

Genotyping density

Although high-density SNP arrays and sequencing techniques are able to capture a greater amount of genetic diversity, their high dimensionality can lead to problems, including multi-collinearity and computing difficulties. However, unless there are high linkage disequilibriums, low-density markers run the risk of missing significant QTLs (Kriaridou *et al.*, 2020).

Handling missing data

Imputation techniques have become essential for completing missing genotypes, enhancing model performance without appreciably raising costs.

Dense phenotyping

Rich trait data is produced by high-throughput phenotyping technology (such as imaging and remote sensing). However, autocorrelations are frequently present in these datasets and need to be carefully modelled.

Multi-omics data

To improve prediction, particularly for complex or low-heritability features, genomic data is increasingly being integrated with transcriptomics, metabolomics, or proteomics (Ye *et al.*, 2020).

11.4.3 Population properties

Predictive ability is strongly influenced by the structure and design of the target (prediction) population and the reference (training) population (Guo *et al.*, 2014).

Reference population size

A large size of the reference population captures more genetic variation and lowers sampling error, resulting in increased accuracy (Habier *et al.*, 2007).

Effective population size (Ne)

Overfitting risks rise in small populations due to increased linkage disequilibrium, which facilitates the capture of QTL effects by markers (Daetwyler *et al.*, 2010).

Train–test relatedness

The genetic relatedness between training and test individuals has a significant impact on prediction accuracy. When populations are distantly connected, accuracy drastically declines. (Pérez-Enciso and Zingaretti, 2019).

Population structure and cryptic relatedness

Predictions may be skewed by concealed relatedness or the existence of subpopulations. Reducing false linkages is facilitated by models that take these patterns into account.

Linkage disequilibrium (LD) structure

The ability of SNP markers to track QTL alleles is determined by LD. While weak LD decreases predictive capacity, strong LD increases it between markers and causal variations.

11.4.4 Statistical methods

The choice of statistical methodology crucially determines how marker information will be translated into successful predictions (Montesinos López *et al.*, 2020).

Parametric models

These models make explicit assumptions about the distribution of marker effects and usually treat the trait as a linear function of genotype (Howard *et al.*, 2014).

RR-BLUP (RIDGE REGRESSION BEST LINEAR UNBIASED PREDICTION). Assumes all markers have small, normally distributed effects with equal variance. Works well for highly polygenic traits.

GBLUP (GENOMIC BLUP). Uses a genomic relationship matrix (GRM) to model similarity among individuals. Computationally efficient and widely used in animal and plant breeding.

BAYESIAN METHODS ('BAYESIAN ALPHABET' MODELS). Advantage: Allows unequal shrinkage of marker effects. Good for traits with some large-effect loci alongside many small effects.

SEMI-PARAMETRIC MODELS. These methods combine parametric assumptions with more flexible components to capture non-linearities and epistasis.

REPRODUCING KERNEL HILBERT SPACE REGRESSION (RKHS). Good for traits where epistatic or nonlinear interactions between loci matter. Uses similarity functions to capture both additive and non-additive variance.

GAUSSIAN PROCESSES. Similar to RKHS but framed in probabilistic machine learning. Advantage: More flexible than BLUP; downside is higher computational cost.

Non-parametric models

These make minimal assumptions about data structure and can learn complex relationships directly. Machine learning approaches (e.g. random forests, support vector machines, neural networks) can model complex interactions without strong assumptions, but may require large datasets to avoid overfitting

Single vs multiple model approaches

- **Single models:** Apply one statistical method across all data. Examples: GBLUP, BayesB, RKHS. Advantage: simpler, faster, interpretable.
- **Multiple models:** Combine different models to leverage complementary strengths.
- **Ensembled models:** Average predictions from several methods.
- **Stacked models:** Use one model's outputs as inputs for another. Evidence suggests ensembles can improve prediction accuracy, especially across multiple traits or heterogeneous populations.

11.5 Types of Genomic Prediction Methods

Genomic prediction methods generally involve the statistical or machine learning algorithms employed to estimate breeding values from genome-wide marker data (Wang *et al.*, 2023). These models differ in their assumptions about the distribution and effect sizes of markers, as well as their capacity to capture linear or non-linear relationships between genotype and phenotype. Statistical methods are well-established in breeding, require fewer data, and are easier to explain, but they often assume simple genetic architectures, whereas machine learning methods can uncover hidden patterns and nonlinear interactions but usually demand more data and computational power, and they may lack transparency (Montesinos López *et al.*, 2022). An overall sketch of prediction models is described in Fig. 11.3.

11.5.1 Statistical models

Genomic best linear unbiased prediction (GBLUP)

Genomic best linear unbiased prediction assumes that all markers contribute equally and have small additive effects on the trait of interest. The method relies on a genomic relationship matrix to capture the genetic similarity among individuals and predict breeding values. Here, plant traits can be controlled by one gene (monogenic) or many genes (polygenic) (Staerk *et al.*, 2024). For example, a single resistance gene might control a disease, while yield per hectare is influenced by many genes, each with a

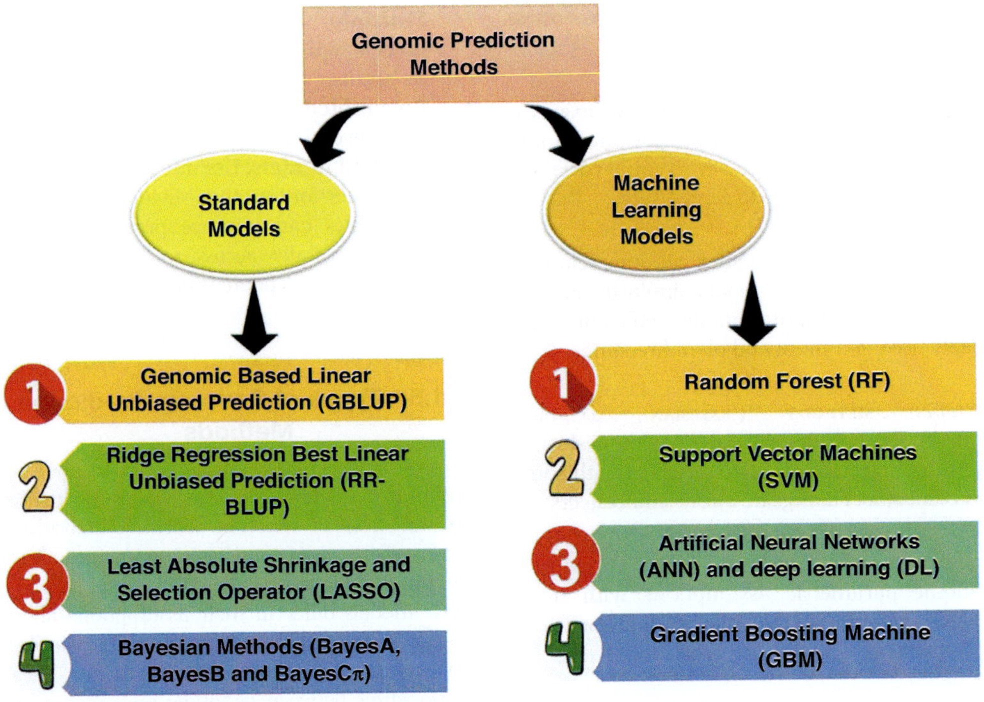

Fig. 11.3. An overview of genomic prediction models (Author's own image).

small effect. When only a few genes are involved, breeders can identify them through mapping and use marker-assisted selection (MAS). But GP is most useful for complex traits controlled by many genes (Crossa *et al.*, 2025). In GBLUP, the standard linear expression is described as:

$$y = \mu + Mu + \epsilon \tag{11.1}$$

where y = observed trait values, μ = overall average, M = marker data, u = effect of each marker, and ϵ = random error.

GBLUP uses all marker data at once to predict performance, especially for polygenic traits.

RR-BLUP (ridge regression best linear unbiased prediction)

Ridge regression best linear unbiased prediction uses the same basic linear model as GBLUP (Equation 11.1), but instead of working with a relationship matrix, it focuses directly on estimating the effects of individual markers. Since the number of markers is usually much larger than the number of genotypes, regularization is used to avoid overfitting. In RR-BLUP, this regularization penalizes the squared size of marker effects (u), which shrinks them toward zero without making them exactly zero. This assumes that all markers contribute small effects, fitting well for traits controlled by many genes (polygenic traits). Ridge regression BLUP and genomic BLUP are mathematically very similar; the main difference lies in how the penalty is applied and whether working with marker effects or genetic relationships (Crossa *et al.*, 2025). In RR-BLUP, marker effects u are estimated by minimizing the residual sum of squares plus an L_2 penalty:

$$\hat{u}RR = argmin\, u \left\{ (y - 1\mu - Zu)^T (y - 1\mu - Zu) + \lambda_{RR} u^T u \right\} \tag{11.2}$$

where y is the vector of phenotypes, 1 is a vector of ones, μ is the intercept, Z is the marker genotype matrix, u is the vector of marker effects, and λ_{RR} is the ridge penalty parameter controlling shrinkage (Riedelsheimer *et al.*, 2012).

LASSO (least absolute shrinkage and selection operator)

Least absolute shrinkage and selection operator is another version of the same linear model described in Equation 1, but it applies a different restriction on the marker effects (u). While RR-BLUP penalizes the sum of squared effect sizes, LASSO penalizes the sum of the absolute effect sizes. This difference means that LASSO tends to set many marker effects exactly to zero, effectively selecting only a subset of markers that contribute to the trait (Brault *et al.*, 2021). This property makes LASSO useful when the number of relevant loci is small and the trait's heritability is high (Crossa *et al.*, 2025). In the LASSO approach, the L_2 penalty is replaced by an L_1 penalty on the absolute size of the marker effects:

$$\hat{u}L = argmin\, u \left\{ \left(y - 1\mu - Zu\right)^T \left(y - 1\mu - Zu\right) + \lambda_L \sum_{j=1}^{p} |u_j| \right\} \quad (11.3)$$

where all terms are as defined above, and λ_L controls the degree of sparsity; larger values set more marker effects exactly to zero (Riedelsheimer *et al.*, 2012).

Bayesian methods (BayesA, BayesB, BayesCπ)

Several economically important livestock traits, such as litter size, calving difficulty, and disease resistance, are threshold traits (categorical or discrete). Traditional linear models often fail for these traits due to non-linearity, non-normality, and variance–mean dependence, which can be addressed using threshold models. Bayesian regression methods such as BayesA, BayesB, and their variants are well-suited for these cases because they can naturally incorporate prior distributions, handle non-normal data, and jointly estimate marker effects and threshold parameters (Wang *et al.*, 2013). These approaches are more flexible for traits influenced by both major genes and numerous minor-effect loci.

Bayesian regression models for genomic prediction can be grouped into one-group and two-group models, depending on how genetic markers are treated when estimating effects (Meuwissen *et al.*, 2001; Gianola *et al.*, 2009; Shi *et al.*, 2021). One-group models (e.g. BayesA; Meuwissen *et al.*, 2001) include all markers but shrink their effects toward zero through variable shrinkage. Two-group models (e.g. BayesB, BayesC; Meuwissen *et al.*, 2001; Habier *et al.*, 2011) assume that only a subset of markers have non-zero effects, while the rest have exactly zero effect.

BAYESA MODEL. In BayesA, all markers are included, but each has its own variance parameter, allowing differential shrinkage.

$$y = X\beta + \sum_{k=1}^{p} z_k a_k + e \quad (11.4)$$

where y = vector of phenotypes, $X\beta$ = fixed effects (e.g. overall mean), z_k = genotype covariate for SNP $k, a_k \sim N(0,\sigma_k^2)$ = SNP effect, each with its own variance, σ_k^2 drawn from an inverse chi-square prior, and e = residual error

BAYESB MODEL. This model performs both selection (deciding which SNPs matter) and shrinkage (controlling their effect sizes).

$$y = X\beta + \sum_{k=1}^{p} \delta_k z_k a_k + e \quad (11.5)$$

where δ_k is an indicator variable: $\delta_k=1$ with probability $1-\pi$ (marker has non-zero effect) and $\delta_k= 0$ with probability π (marker effect is zero). For markers with $\delta_k=1, a_k \sim N(0,\sigma_k^2)$ with its own inverse chi-square prior.

Multi-group models are extensions of this approach. Some models, such as BayesCp (Habier *et al.*, 2011), combine selection and shrinkage, first selecting markers and then shrinking some of their effects toward zero. BayesCp can also estimate the proportion of markers with non-zero effects, although it can face issues like poor convergence and mixing in some cases (Wolc *et al.*, 2011). To overcome the drawbacks of BayesA and BayesB, particularly the sensitivity to prior hyperparameters, two Bayesian methods, BayesCπ and BayesDπ, were developed for genomic prediction to treat the prior probability (π) that an SNP has zero effect as unknown (Habier *et al.*, 2011). The general statistical model is written as: e = residual error

Before analyzing the data, the effect of a SNP is zero with probability π and with probability (1–π) it follows a normal distribution with mean zero and a locus-specific variance. In BayesCπ, all SNPs share a common variance whose scale has an inverse chi-square prior, while in BayesDπ that variance scale is treated as unknown with its own prior, offering greater flexibility but at a higher computational cost.

11.5.2 Machine learning models

With the increasing availability of genomic and phenotypic data in plant breeding, machine learning (ML) has become indispensable for analyzing large, complex datasets efficiently (van Dijk *et al.*, 2021). High-throughput genomics and phenomics generate vast, multi-layered data, such as genomics, transcriptomics, and metabolomics, that ML methods can integrate to better understand complex traits controlled by multiple genetic and molecular pathways. Commonly used ML models in genomic prediction include random forests (RF), support vector machines (SVMs), artificial neural networks (ANNs) and deep learning (DL), and gradient boosting machines (GBMs) (Kalusivalingam *et al.*, 2021). These approaches excel at learning complex, non-linear relationships, often outperforming traditional statistical methods in capturing genotype-by-environment interactions and hierarchical genetic effects (Niazian and Niedbała, 2020).

- **Random forests (RFs)** are ensemble learning methods that build numerous decision trees and aggregate their outputs to improve robustness and accuracy, effectively handling noisy, high-dimensional genomic data. It effectively handles high-dimensional and imbalanced genomic data, making it ideal for classifying complex traits in plant breeding. The strength of RF in improving prediction accuracy by focusing on key markers and using balanced sampling has been demonstrated (Patra *et al.*, 2021).
- **Support vector machine (SVM)** is a supervised learning method used in genomic prediction to model complex, often nonlinear relationships between genetic markers and traits. By employing kernel functions, SVM identifies optimal hyperplanes that separate data points with the greatest margin, enabling accurate prediction even in high-dimensional datasets where markers outnumber samples. With proper kernel choice and parameter tuning, SVM achieves competitive accuracy compared with traditional prediction models (Zhao *et al.*, 2020).
- **Artificial neural networks (ANN) and deep learning (DL)** are machine learning techniques inspired by the human brain's structure. ANNs consist of layers of interconnected nodes that learn complex, nonlinear relationships between genetic markers and traits. Deep Learning, a subset of ANN, employs multiple hidden layers to capture hierarchical and intricate patterns, including epistatic interactions among markers. These methods improve prediction accuracy for complex traits by automatically extracting relevant features from high-dimensional genomic data. However, they require large datasets and significant computational power for effective training. DL and ANN provide flexible frameworks that complement traditional genomic prediction models by effectively modeling nonlinear genetic architectures (van Dijk *et al.*, 2021).
- **Gradient boosting machines (GBMs)** are powerful ensemble learning methods that build predictive models by sequentially adding decision trees. Each new tree focuses on correcting the errors made by the previous ones, which improves overall prediction accuracy. GBM effectively handles complex, nonlinear relationships in high-dimensional genomic and phenotypic data. Due to its iterative nature, GBM can capture subtle interactions among genetic markers, making it well-suited for genomic prediction in plant breeding. However, careful tuning of hyperparameters is essential to avoid overfitting and achieve optimal results (van Dijk *et al.*, 2021).

Together, these ML models provide flexible and powerful tools for genomic prediction, complementing traditional models such as genomic best linear unbiased prediction (GBLUP). Studies have demonstrated that non-parametric methods like RF and SVM can outperform parametric GS models, especially for complex traits and multi-environment datasets, by effectively capturing non-linear patterns and interactions (van Dijk *et al.*, 2021).

11.6 Validation Methods

11.6.1 Cross-validation (CV)

Cross-validation (CV) is a widely adopted statistical and machine learning technique used to evaluate the predictive performance and generalization ability of models. Its central aim is to estimate how well a model trained on a given dataset will perform on unseen data, thereby reducing overfitting, minimizing bias, and ensuring robustness (Roorkiwal *et al.*, 2018). In genomic prediction (GP) studies, CV plays a pivotal role in assessing models before their deployment for selecting candidate individuals in breeding programs. The fundamental concept of CV involves dividing the training population (training set (TRS)) into calibration and validation subsets (Tougui *et al.*, 2021). The calibration set is used to train the model, while the validation set is used to test its predictive accuracy, simulating its real-world application.

Types of cross-validation methods

K-FOLD CROSS-VALIDATION. One of the most common methods, K-fold CV divides the dataset into k equal-sized folds. For example, in a five-fold CV, the TRS is randomly split into five groups; in each round, four folds are used for calibration and one fold for validation. The process is repeated k times, so each fold serves as the validation set once, and the results are averaged to obtain the final accuracy. K-fold CV balances accuracy and computational efficiency, making it suitable for medium to large datasets (Nti *et al.*, 2021).

LEAVE-ONE-OUT CROSS-VALIDATION (LOOCV). Here, only one data point (or genotype in GP studies) is used for validation in each iteration, while the rest form the calibration set. This method is exhaustive and provides a near-unbiased accuracy estimate but is computationally expensive, especially with large datasets (Cheng *et al.*, 2017).

REPEATED K-FOLD CROSS-VALIDATION. Extends K-fold CV by repeating the process multiple times with different random splits, leading to more stable performance estimates and reduced variance.

STRATIFIED K-FOLD CV. Ensures that each fold preserves the proportion of target classes or trait categories, making it especially valuable in imbalanced datasets.

CROSS-VALIDATION IN MULTI-ENVIRONMENT GENOMIC PREDICTION. For multi-environment GP, CV designs also address genotype × environment (G × E) interactions like the following:

- **CV1:** Predicts the genomic estimated breeding values (GEBVs) of untested genotypes in environments that have already been evaluated,
- **CV2:** (Sparse Testing): Predicts the performance of genotypes tested in some environments based on data from other tested environments,
- **CV0:** Predicts the performance of tested genotypes in environments that have not been tested before,
- **CV00:** The most challenging design, predicting untested genotypes in completely untested environments (Crossa *et al.*, 2017).

ADDITIONAL ADVANCED CV VARIANTS.

- **Nested cross-validation:** Used when both model evaluation and hyperparameter tuning are required. The dataset is split into an outer loop for performance estimation and an inner loop for parameter optimization, preventing information leakage and overly optimistic accuracy estimates (Abdulaal *et al.*, 2018).
- **Blocked cross-validation:** Groups data based on spatial, temporal, or genetic relatedness before splitting, reducing bias when samples are correlated. This is particularly important in field trials where spatial effects

or kinship can inflate accuracy estimates (Valavi *et al.*, 2018).

- **Monte Carlo cross-validation (random subsampling):** Randomly splits the dataset into training and validation sets multiple times, averaging the results. It offers flexibility but may leave some data points untested (Nakatsu, 2023).

By choosing the appropriate CV strategy, researchers can more reliably estimate model performance, optimize model parameters, and ensure that genomic prediction models are applicable to real-world breeding scenarios (Seraj *et al.*, 2023). The choice of CV method depends on dataset size, computational resources, experimental objectives, and the complexity of G×E interactions. In modern breeding pipelines, CV has become an indispensable step to ensure that models are not only accurate but also stable, generalizable, and capable of guiding selection decisions effectively (Allgaier and Pryss, 2024).

11.7 Key Discoveries and Present Limitations for Plant Genomic Prediction

11.7.1 Important findings

Genomic prediction functions in various situations

The most significant finding from all of the research is that molecular marker-based performance prediction is accurate. In fact, even for features like grain yield, GEBV accuracy is comparatively good. Given the literature on rice, and particularly research utilizing breeding material, it is evident that the efficacy of genomic prediction, which worries many rice producers, is not warranted (Anilkumar *et al.*, 2023).

The accuracy of predictions can be improved

Breeders can manipulate several elements to improve forecast accuracy or lower implementation costs. In fact, higher accuracies can be achieved by enhancing the evaluation and composition of the training set, focusing on informative molecular markers (polymorphic with a medium to high minor allele frequency and distributed throughout the genome), or incorporating extra data (historical, environmental covariates, crop method, etc.) (Cooper *et al.*, 2014; Adhikari *et al.*, 2017). One of the key aspects influencing accuracy in Genomic Prediction for Rice Breeding 585 is the size and makeup of the training set, which determines the strength of the genetic association with the selection candidates. Consequently, algorithms for choosing the training set have been devised (Spindel *et al.*, 2015). In relation to molecular markers, various studies indicate that it is possible to decrease marker density to a certain degree without compromising prediction accuracy (Al-Samarai and Al-Kazaz, 2015)

In the context of hybrids, genomic prediction is efficient

The knowledge gained about model selection, training set identification, and marker concentrations is equally applicable to hybrid as well as inbred breeding plans. It is true that hybrid programs pose special difficulties because they allow for the application of forecasts that are not relevant to other breeding plans (Matsumi and Solove, 2025). The forecast of how males and females could be mixed to produce better hybrid combinations is noteworthy. There is some evidence that the convergence of additive genetics from the male and female lines drives hybrid performance in rice. It does not appear to assist in including no additive parameters in the prediction (Labroo *et al.*, 2021).

11.7.2 Current limitations

Despite the quantity and variety of research, the literature on some crops still leaves out several important details. They may hinder the ability to fully utilize genomic selection, depending on the situation (Lin *et al*., 2014).

Accuracy by itself is insufficient to evaluate genetic prediction's efficacy

Nearly all of the studies used prediction accuracy as a basis for evaluating genomic selection. Despite being a crucial component in evaluating the effectiveness of prediction models, accuracy does not reveal which persons are best chosen by the various techniques (Heslot *et al*., 2012).

The accuracy of within-family predictions is not adequately considered

The accuracy of within-family predictions using parental data or numerous biparental families as the training set has not been thoroughly examined in any crop study (Zhu *et al*., 2021).

References

Abdulaal, M.J., Casson, A.J. and Gaydecki, P. (2018) Performance of nested vs. non-nested SVM cross-validation methods in visual BCI: Validation study. In: *26th European Signal Processing Conference (EUSIPCO)*, IEEE, pp. 1680–1684.

Abou Khater, L., Joukhadar, R., Maalouf, F., Alsamman, A.M., Babiker, Z. *et al*. (2025) Genomic prediction for heat and herbicide tolerance in faba bean. *The Plant Genome* 18(3), e70076.

Adhikari, S., Saha, S., Biswas, A., Rana, T.S., Bandyopadhyay, T.K. and Ghosh, P. (2017) Application of molecular markers in plant genome analysis: A review. *The Nucleus* 60(3), 283–297.

Alemu, A., Åstrand, J., Montesinos-Lopez, O.A., Isidro y Sánchez, J.I., Fernandez-Gonzalez, J. *et al*. (2024) Genomic selection in plant breeding: Key factors shaping two decades of progress. *Molecular Plant* 17(4), 552–578.

Allgaier, J. and Pryss, R. (2024) Cross-validation visualized: A narrative guide to advanced methods. *Machine Learning and Knowledge Extraction* 6(2), 1378–1388.

Al-Samarai, F.R. and Al-Kazaz, A.A. (2015) Molecular markers: An introduction and applications. *European Journal of Molecular Biotechnology* 9(3), 118–130.

Amiteye, S. (2021) Basic concepts and methodologies of DNA marker systems in plant molecular breeding. *Heliyon* 7, 10.

Anilkumar, C., Muhammed Azharudheen, T.P., Sah, R.P., Sunitha, N.C., Devanna, B.N. *et al*. (2023) Gene based markers improve precision of genome-wide association studies and accuracy of genomic predictions in rice breeding. *Heredity* 130(5), 335–345.

Begna, T. (2021) Conventional breeding methods widely used to improve self-pollinated crops. *International Journal of Research* 7(1), 1–16.

Brault, C., Doligez, A., Cunff, L., Coupel-Ledru, A., Simonneau, T., Chiquet, J. *et al*. (2021) Harnessing multivariate, penalized regression methods for genomic prediction and QTL detection of drought-related traits in grapevine *G3: Genes, Genomes, Genetics* 11(9), jkab248.

Cheng, H., Garrick, D.J. and Fernando, R.L. (2017) Efficient strategies for leave-one-out cross validation for genomic best linear unbiased prediction. *Journal of Animal Science and Biotechnology* 8(1), 38.

Cooper, M., Messina, C.D., Podlich, D., Totir, L.R., Baumgarten, A. *et al*. (2014) Predicting the future of plant breeding: Complementing empirical evaluation with genetic prediction. *Crop and Pasture Science* (4), 311–336.

Crossa, J., Pérez-Rodríguez, P., Cuevas, J., Montesinos-López, O., Jarquín, D. *et al*. (2017) Genomic selection in plant breeding: Methods, models, and perspectives. *Trends in Plant Science* 22(11), 961–975.

Crossa, J., Martini, J.W., Vitale, P., Pérez-Rodríguez, P., Costa-Neto, G. *et al*. (2025) Expanding genomic prediction in plant breeding: Harnessing big data, machine learning, and advanced software. *Trends in Plant Science* 30(7), 756–774.

Daetwyler, H.D., Pong-Wong, R., Villanueva, B. and Woolliams, J.A. (2010) The impact of genetic architecture on genome-wide evaluation methods. *Genetics* 185(3), 1021–1031.

de Los Campos, G., Hickey, J.M., Pong-Wong, R., Daetwyler, H.D. and Calus, M.P. (2013) Whole-genome regression and prediction methods applied to plant and animal breeding. *Genetics* 193(2), 327–345.

Dekkers, J.C., Su, H. and Cheng, J. (2021) Predicting the accuracy of genomic predictions. *Genetics Selection Evolution* 53(1), 55.

Farooq, M., van Dijk, A.D., Nijveen, H., Mansoor, S. and de Ridder, D. (2023) Genomic prediction in plants: Opportunities for ensemble machine learning based approaches. *F1000Research* 11, 802.

Ghosh, S., Roy, A. and Dutta, S. (2024) Rapid generation advance methods to fast-track crop breeding: A review. *Agricultural Reviews* 45(4).

Gianola, D., de los Campos, G., Hill, W.G., Manfredi, E. and Fernando, R. (2009) Additive genetic variability and the Bayesian alphabet. *Genetics* 183, 347–363. DOI: 10.1534/genetics.109.103952.

Guo, Z., Tucker, D.M., Basten, C.J., Gandhi, H., Ersoz, E. *et al.* (2014) The impact of population structure on genomic prediction in stratified populations. *Theoretical and Applied Genetics* 127(3), 749–762.

Habier, D., Fernando, R.L. and Dekkers, J. (2007) The impact of genetic relationship information on genome-assisted breeding values. *Genetics* 177(4), 2389–2397.

Habier, D., Fernando, R.L., Kizilkaya, K. and Garrick, D.J. (2011) Extension of the Bayesian alphabet for genomic selection. *BMC Bioinformatics* 12(1), 186.

Haley, C.S. and Visscher, P.M. (1998) Strategies to utilize marker-quantitative trait loci associations. *Journal of Dairy Science* 81, 85–97.

Heslot, N., Yang, H.P., Sorrells, M.E. and Jannink, J.L. (2012) Genomic selection in plant breeding: A comparison of models. *Crop Science* 52(1), 146–160.

Howard, R., Carriquiry, A.L. and Beavis, W.D. (2014) Parametric and nonparametric statistical methods for genomic selection of traits with additive and epistatic genetic architectures. *G3: Genes, Genomes, Genetics* 4(6), 1027–1046.

Hussain, B. (2015) Modernization in plant breeding approaches for improving biotic stress resistance in crop plants. *Turkish Journal of Agriculture and Forestry* 39(4), 515–530.

Isidro y Sánchez, J. and Akdemir, D. (2021) Training set optimization for sparse phenotyping in genomic selection: A conceptual overview. *Frontiers in Plant Science* 12, 715910.

Jarquin, D., Howard, R., Crossa, J., Beyene, Y., Gowda, M. *et al.* (2020) Genomic prediction enhanced sparse testing for multi-environment trials. *G3: Genes, Genomes, Genetics* 10(8), 2725–2739.

Jarquín, D., Crossa, J., Lacaze, X., Cheyron, P.D., Daucourt, J. *et al.* (2014) A reaction norm model for genomic selection using high-dimensional genomic and environmental data. *Theoretical and Applied Genetics* 127, 595–607. DOI: 10.1007/s00122-013-2243-1.

Jeon, D., Kang, Y., Lee, S., Choi, S., Sung, Y. *et al.* (2023) Digitalizing breeding in plants: A new trend of next-generation breeding based on genomic prediction. *Frontiers in Plant Science* 14, 1092584.

Kalusivalingam, A.K., Sharma, A., Patel, N. and Singh, V. (2021) Leveraging deep learning and random forest algorithms for AI-driven genomics in personalized medicine. *International Journal of AI And* (3).

Kriaridou, C., Tsairidou, S., Houston, R.D. and Robledo, D. (2020) Genomic prediction using low density marker panels in aquaculture: Performance across species, traits, and genotyping platforms. *Frontiers in Genetics* 11, 124.

Kumar, M., Kumar, S., Sandhu, K.S., Kumar, N., Saripalli, G. *et al.* (2023) GWAS and genomic prediction for pre-harvest sprouting tolerance involving sprouting score and two other related traits in spring wheat. *Molecular Breeding* 43(3), 14.

Labroo, M.R., Studer, A.J. and Rutkoski, J.E. (2021) Heterosis and hybrid crop breeding: A multidisciplinary review. *Frontiers in Genetics* 12, 643761.

Lande, R. and Thompson, R. (1990) Efficiency of marker-assisted selection in the improvement of quantitative traits. *Genetics* 124(3), 743–756.

Lin, Z., Hayes, B.J. and Daetwyler, H.D. (2014) Genomic selection in crops, trees and forages: A review. *Crop and Pasture Science* 65(11), 1177–1191.

Matsumi, H. and Solove, D.J. (2025) The prediction society: AI and the problems of forecasting the future. *University of Illinois Law Review* 1, 62.

Meuwissen, T.H.E., Hayes, B.J. and Goddard, M.E. (2001) Prediction of total genetic value using genome-wide dense marker maps. *Genetics* 157(4), 1819–1829.

Montesinos López, O.A., Franco-Pérez, E., Luna-Vásquez, F.J., Salinas-Ruiz, J., Sandoval-Carrillo, S. *et al.* (2020) Benchmarking between item based collaborative filtering algorithm and genomic best linear unbiased prediction (GBLUP) model in terms of prediction accuracy for wheat and maize. *Biotecnia* 22(2), 136–146.

Montesinos López, O.A., Montesinos López, A. and Crossa, J. (2022) *Multivariate Statistical Machine Learning Methods for Genomic Prediction*. Springer Nature, p. 691.

Nadeem, M.A., Nawaz, M.A., Shahid, M.Q., Doğan, Y., Comertpay, G. *et al.* (2018) DNA molecular markers in plant breeding: Current status and recent advancements in genomic selection and genome editing. *Biotechnology & Biotechnological Equipment* 32(2), 261–285.

Nakatsu, R.T. (2023) Validation of machine learning ridge regression models using Monte Carlo, bootstrap, and variations in cross-validation. *Journal of Intelligent Systems* 32(1), 20220224.

Niazian, M. and Niedbała, G. (2020) Machine learning for plant breeding and biotechnology. *Agriculture* 10(10), 436.

Nti, I.K., Nyarko-Boateng, O. and Aning, J. (2021) Performance of machine learning algorithms with different K values in K-fold cross-validation. *International Journal of Information Technology and Computer Science* 13(6), 61–71.

Oshunsanya, S.O., Nwosu, N.J. and Li, Y. (2019) Abiotic stress in agricultural crops under climatic conditions. In: *Sustainable Agriculture, Forest and Environmental Management*. Springer Singapore, pp. 71–100.

Patra, S.S., Jena, O.P., Kumar, G., Pramanik, S., Misra, C. *et al.* (2021) Random forest algorithm in imbalance genomics classification. In: Satpathy, R., Choudhury, T., Satpathy, S., Mohanty, S.N. and Zhang, X. (eds) *Data Analytics in Bioinformatics: A Machine Learning Perspective*. Wiley, pp. 173–190.

Pérez-Enciso, M. and Zingaretti, L.M. (2019) A guide on deep learning for complex trait genomic prediction. *Genes* 10(7), 553.

Ren, D., Cai, X., Lin, Q., Ye, H., Teng, J. *et al.* (2022) Impact of linkage disequilibrium heterogeneity along the genome on genomic prediction and heritability estimation. *Genetics Selection Evolution* 54(1), 47.

Riedelsheimer, C., Technow, F. and Melchinger, A.E. (2012) Comparison of whole-genome prediction models for traits with contrasting genetic architecture in a diversity panel of maize inbred lines. *BMC Genomics* 13(1), 452.

Roorkiwal, M., Jarquin, D., Singh, M.K., Gaur, P.M., Bharadwaj, C. *et al.* (2018) Genomic-enabled prediction models using multi-environment trials to estimate the effect of genotype × environment interaction on prediction accuracy in chickpea. *Scientific Reports* 8(1), 11701.

Selga, C., Reslow, F., Pérez-Rodríguez, P. and Ortiz, R. (2022) The power of genomic estimated breeding values for selection when using a finite population size in genetic improvement of tetraploid potato. *G3: Genes, Genomes, Genetics* 12(1), jkab362.

Seraj, A., Mohammadi-Khanaposhtani, M., Daneshfar, R., Naseri, M., Esmaeili, M. *et al.* (2023) *Cross-validation*. In: *Handbook of Hydroinformatics*. Elsevier, pp. 89–105.

Shi, S., Li, X., Fang, L., Liu, A., Su, G. *et al.* (2021) Genomic prediction using Bayesian regression models with global–local prior. *Frontiers in Genetics* 12, 628205.

Soller, M. and Plotkin-Hazan, J. (1977) The use marker alleles for the introgression of linked quantitative alleles. *Theoretical and Applied Genetics* 51(3), 133–137.

Spindel, J., Begum, H., Akdemir, D., Virk, P., Collard, B. *et al.* (2015) Genomic selection and association mapping in rice (*Oryza sativa*): Effect of trait genetic architecture, training population composition, marker number and statistical model on accuracy of rice genomic selection in elite, tropical rice breeding lines. *PLoS Genetics* 11(2), e1004982.

Staerk, C., Klinkhammer, H., Wistuba, T., Maj, C. and Mayr, A. (2024) Generalizability of polygenic prediction models: How is the R2 defined on test data? *BMC Medical Genomics* 17(1), 132.

Taagepera, R. and Nemčok, M. (2024) World population growth over millennia: Ancient and present phases with a temporary halt in-between. *The Anthropocene Review* 11(1), 163–183.

Tougui, I., Jilbab, A. and El Mhamdi, J. (2021) Impact of the choice of cross-validation techniques on the results of machine learning-based diagnostic applications. *Healthcare Informatics Research* 27(3), 189–199.

Valavi, R., Elith, J., Lahoz-Monfort, J.J. and Guillera-Arroita, G. (2018) BlockCV: An R package for generating spatially or environmentally separated folds for k-fold cross-validation of species distribution models. *Biorxiv* 357798.

van Dijk, A.D.J., Kootstra, G., Kruijer, W. and de Ridder, D. (2021) Machine learning in plant science and plant breeding. *Iscience* 24(1).

Wang, C.L., Ding, X.D., Wang, J.Y., Liu, J.F., Fu, W.X. *et al.* (2013) Bayesian methods for estimating GEBVs of threshold traits. *Heredity* 110(3), 213–219.

Wang, K., Abid, M.A., Rasheed, A., Crossa, J., Hearne, S. *et al.* (2023) DNNGP, a deep neural network-based method for genomic prediction using multi-omics data in plants. *Molecular Plant* 16(1), 279–293.

Whittaker, R.J. (2000) Scale, succession and complexity in island biogeography: Are we asking the right questions? *Global Ecology and Biogeography* 9(1), 75–85.

Wolc, A., Stricker, C., Arango, J., Settar, P., Fulton, J.E. *et al.* (2011) Breeding value prediction for production traits in layer chickens using pedigree or genomic relationships in a reduced animal model. *Genetics Selection Evolution* 43, 5. DOI: 10.1186/1297-9686-43-5.

Wu, H., Jin, S., Xiang, C., Tang, J., Xian, J. *et al.* (2025) GPS: Harnessing data fusion strategies to improve the accuracy of machine learning-based genomic and phenotypic selection. *Plant Communications* 6(8), 101416.

Ye, S., Li, J. and Zhang, Z. (2020) Multi-omics-data-assisted genomic feature markers preselection improves the accuracy of genomic prediction. *Journal of Animal Science and Biotechnology* 11(1), 109.

Zhao, W., Lai, X., Liu, D., Zhang, Z., Ma, P. *et al.* (2020) Applications of support vector machine in genomic prediction in pig and maize populations. *Frontiers in Genetics* 11, 598318.

Zhu, X., Leiser, W.L., Hahn, V. and Würschum, T. (2021) Training set design in genomic prediction with multiple biparental families. *The Plant Genome* 14(3), e20124.

12 Genomic Prediction for Developing Climate Change-Ready Crops

Sarfraz Ahmad[1]* and Saniya[2]

[1]*Department of Genetics and Plant Breeding, College of Agriculture Mandawa, Swami Keshwanand Rajasthan Agriculture University, Bikaner, Rajasthan, India;*
[2]*Department of Horticulture, ICAR-National Research Centre for Grapes, Pune, Maharashtra, India*

Abstract

Climate change poses unprecedented challenges to global agriculture, demanding the rapid development of crop varieties resilient to multiple abiotic and biotic stresses. Genomic prediction (GP) has emerged as a transformative tool in plant breeding, leveraging genome-wide marker information, phenotypic data, and environmental variables to predict the performance of untested genotypes. Unlike conventional selection methods, GP accelerates breeding cycles by enabling early selection and reducing dependence on extensive field trials. This chapter explores the theoretical foundations of GP, its applications in breeding for climate resilience, and its integration into modern breeding pipelines. We discuss how GP models capture complex polygenic traits such as drought tolerance, heat resilience, and submergence survival, which are central to crop adaptation under variable climates. Case studies in maize, wheat, and rice demonstrate the potential of GP in addressing genotype × environment interactions and targeting multiple stress-tolerance traits simultaneously. Advances in high-throughput phenotyping, speed breeding, and envirotyping further enhance GP accuracy and utility. Moreover, integrating GP with digital platforms and decision-support systems is reshaping breeding into a more data-driven, predictive science. Despite its promise, challenges remain, including data scarcity, infrastructure gaps, interpretability issues, and inequitable access in resource-limited regions. Addressing these requires multi-omics integration, AI-driven modeling, participatory breeding, and international collaborations. In conclusion, GP represents a paradigm shift toward climate-smart breeding ecosystems. Accelerating genetic gains and enabling precise targeting of resilience traits, it provides a cornerstone for developing climate change-ready crops and ensuring future food and nutritional security.

Keywords: Abiotic stress, Climate change-ready crops, Climate resilience, Data-driven breeding, Genomic prediction, High-throughput phenotyping, Multi-environment trials, Stress tolerance

12.1 Introduction

Agriculture faces an unprecedented challenge in the 21st century as climate change threatens global food security and the sustainability of crop production systems. Rising temperatures, shifting rainfall patterns, increasing frequency of extreme weather events, soil degradation, and the emergence of novel pests and pathogens collectively impose enormous stress on agricultural productivity (Singh *et al.*, 2023). For staple crops such as rice, wheat, maize, and soybean,

*Corresponding author: sarfraz_firoz@rediffmail.com

 Plant Genomic Prediction: AI-Enhanced Models for Plant Genomics and Crop Improvement (ed. Jen-Tsung Chen)
DOI: 10.1079/9781836991328.0012

projections suggest yield declines of 10–25% by 2050 under high-emission climate scenarios if no adaptation measures are implemented (Muller and Robertson, 2014). At the same time, global population growth is expected to drive food demand upward by 50% or more, creating a dual challenge of producing more food on less arable land under increasingly uncertain environmental conditions (Daszkiewicz, 2022).

Conventional plant breeding has historically been successful in enhancing crop yields and improving tolerance to some environmental stresses. However, its efficiency is constrained by long breeding cycles, reliance on field trials across multiple environments, and limited ability to capture complex polygenic traits such as drought tolerance, heat resilience, and nutrient use efficiency. These limitations are particularly critical under climate change, where crop performance is shaped by multifactorial stresses and dynamic genotype × environment × management (G × E × M) interactions (Cooper *et al.*, 2022). Consequently, traditional selection strategies are often too slow to meet the urgent need for climate-ready cultivars.

Genomic prediction (GP), also referred to as genomic selection, has emerged as a transformative approach for accelerating crop improvement. First proposed by Meuwissen *et al.* (2001) in the context of animal breeding, GP leverages dense genome-wide marker information and statistical models to predict the genetic merit of untested individuals. Unlike marker-assisted selection, which focuses on a limited number of major-effect loci, GP captures the contributions of thousands of small-effect loci simultaneously, thereby improving predictive accuracy for complex quantitative traits (Veerendrakumar *et al.*, 2024). This makes GP particularly powerful for breeding climate change-ready crops, where adaptive traits are typically polygenic and influenced by complex G × E dynamics.

The integration of GP into crop breeding programs offers several advantages relevant to climate resilience. First, it enables early selection in breeding cycles, reducing reliance on multi-season phenotyping and thereby shortening the time to variety release (Heffner *et al.*, 2009). Second, GP allows for the prediction of performance across diverse environments, especially when combined with multi-environment trial data and environmental covariates (Jarquín *et al.*, 2014). Third, it facilitates the optimization of crossing designs by identifying parental combinations with the highest potential to generate climate-adapted progeny (Feng *et al.*, 2024). Finally, GP is compatible with complementary technologies such as high-throughput phenotyping, speed breeding, and genome editing, creating integrated pipelines for rapid development of climate-smart cultivars (Naqvi *et al.*, 2024).

Applications of GP in climate change adaptation are expanding across major crops. In maize, genomic prediction models have been employed to improve drought tolerance and grain yield stability under water-limited conditions in sub-Saharan Africa (Beyene *et al.*, 2016). In rice, GP has been integrated with trait introgression strategies to accelerate the development of submergence-tolerant and drought-resilient varieties (Lokeshkumar *et al.*, 2025). Wheat breeding programs in South Asia and Australia are increasingly using GP to predict performance under terminal heat and salinity stress conditions (Sukumaran *et al.*, 2021). These successes highlight the role of GP as a cornerstone of modern breeding pipelines aimed at mitigating climate risks.

Despite its promise, GP faces challenges in widespread deployment for climate-ready crop development. The accuracy of prediction models depends on the size and diversity of the training population, the density and quality of marker data, and the representation of relevant environmental variation (Akdemir and Isidro-Sánchez, 2019). Moreover, the integration of G × E into GP frameworks remains complex, as it requires combining genomic, phenotypic, and environmental data at scale (Cooper *et al.*, 2022). Developing countries, which are disproportionately affected by climate change, often lack the resources for large-scale genotyping and phenotyping, raising concerns about equitable access to GP technologies (Sukumaran *et al.*, 2021).

Nevertheless, advances in sequencing technologies, machine learning, high-throughput phenotyping, and global data-sharing initiatives are steadily lowering these barriers. Emerging approaches such as reaction-norm models, deep learning architectures, and multi-omics integration promise to enhance the predictive power of GP in the face of environmental complexity (Kunwar *et al.*, 2025). Coupled with participatory breeding and socio-economic considerations, GP can be harnessed not only to increase productivity but also to improve the resilience and sustainability of agricultural systems.

In this chapter, we explore the role of genomic prediction in developing climate change-ready crops. We begin with a review of its historical and theoretical foundations, followed by an overview of models and methods. We then examine its applications for specific climate-relevant traits such as drought, heat, salinity, and flood tolerance. Case studies highlight its implementation across major crops, while later sections address challenges, limitations, and emerging innovations such as integration with artificial intelligence and gene editing. The chapter concludes by discussing future directions for embedding GP into global breeding pipelines to secure food production in a warming world.

12.2 Historical Context and Evolution of Genomic Prediction

The concept of predicting the genetic potential of individuals has long been central to plant and animal breeding. Traditionally, breeders relied on phenotypic selection, where individuals were chosen based on observed traits, and pedigree-based selection, where the performance of relatives was used to estimate breeding values. The latter was formalized through best linear unbiased prediction (BLUP), introduced in animal breeding in the 1970s, which provided a statistical framework for estimating breeding values by leveraging pedigree relationships (Henderson, 1975). While pedigree BLUP increased selection accuracy compared with pure phenotypic selection, it was limited by the fact that pedigree relationships are only an approximation of the true underlying genomic relationships.

12.2.1 Marker-assisted selection (MAS) and its limitations

The molecular genetics revolution in the late 20th century introduced DNA-based markers such as restriction fragment length polymorphisms (RFLPs), simple sequence repeats (SSRs), and later single-nucleotide polymorphisms (SNPs). These markers made it possible to directly track alleles associated with desirable traits. Marker-assisted selection (MAS) became widely adopted in the 1990s and 2000s, particularly for traits controlled by major genes or quantitative trait loci (QTLs) of large effect (Collard and Mackill, 2008).

While MAS achieved successes, such as the introgression of disease-resistance genes in rice and wheat, its scope was limited for complex polygenic traits like drought tolerance or yield stability. These traits are typically governed by thousands of loci with small individual effects, often interacting with environmental conditions. MAS approaches that targeted a few QTLs captured only a small fraction of the heritable variance, thereby limiting their effectiveness for climate-resilience breeding (Heffner *et al.*, 2009).

12.2.2 Emergence of genomic selection

The limitations of MAS led to the conceptual breakthrough of genomic selection (GS), proposed by Meuwissen *et al.* (2001). GS, which later became synonymous with genomic prediction (GP), shifts the focus from identifying and tracking a handful of significant markers to using all available markers across the genome simultaneously to predict an individual's genomic estimated breeding value (GEBV). The key innovation was the application of dense genome-wide marker information in statistical models that estimate the contribution of each locus, regardless of whether it reaches significance in traditional association studies.

Initially demonstrated in dairy cattle breeding, genomic selection rapidly spread to crop science as sequencing and genotyping technologies became more affordable. The plant applications in rye showed that GP could achieve higher prediction accuracies for yield and complex traits than MAS or pedigree BLUP (Wang *et al.*, 2014).

12.2.3 Technological advances driving GP adoption

Several technological milestones facilitated the adoption of GP in plant breeding:

- **High-throughput genotyping platforms:** SNP arrays and genotyping-by-sequencing (GBS) provided cost-effective, high-density marker data across large populations (Rayaprolu *et al.*, 2022).
- **Statistical and computational advances:** Development of models such as genomic BLUP (GBLUP), ridge regression BLUP (RR-BLUP), and Bayesian approaches enabled efficient estimation of GEBVs from large genomic datasets (Meher *et al.*, 2022).
- **High-throughput phenotyping (HTP):** Advances in remote sensing, drones, and imaging technologies allowed large-scale, multi-environment data collection, which is critical for training GP models under climate-relevant stresses (Ahmad *et al.*, 2025).
- **Cloud computing and bioinformatics pipelines:** These enabled the analysis of increasingly large datasets and facilitated collaborative breeding networks.

12.2.4 Expansion into climate resilience breeding

By the mid-2010s, GP was increasingly recognized as a powerful tool for addressing climate change challenges in agriculture. Early successes included predicting drought tolerance in maize for sub-Saharan Africa, yield stability in wheat under heat stress, and salinity tolerance in barley (Beyene *et al.*, 2016). These efforts marked a departure from the 'one-gene-one-trait' paradigm of MAS toward whole-genome, whole-environment predictive frameworks.

Furthermore, GP aligned well with the broader shift in plant breeding toward systems-level approaches, where genomics, phenomics, and environmental data are integrated. This made GP not only a tool for prediction but also a framework for climate-smart breeding pipelines capable of anticipating future agricultural environments.

12.2.5 Contemporary trends and evolution

Today, GP continues to evolve in response to new scientific and technological developments.

- **Machine learning and deep learning** approaches, such as random forests, convolutional neural networks (CNNs), and graph neural networks (GNNs), are being applied to improve the modeling of complex genetic architectures and non-linear G × E interactions (Montesinos-López *et al.*, 2021).
- **Integration with envirotyping:** Incorporating environmental covariates (e.g. temperature, rainfall, soil profiles) into GP models enhances predictive accuracy for climate adaptation traits (Cooper *et al.*, 2022).
- **Multi-omics data fusion:** Transcriptomics, metabolomics, and epigenomics are increasingly combined with genomic markers to enhance prediction for traits where gene expression or metabolic regulation is key (Fan *et al.*, 2025).
- **CRISPR-based breeding:** GP can identify promising alleles for editing, thereby combining predictive selection with targeted genome editing (Saniya, Ahmad *et al.*, 2025).

In sum, the evolution of genomic prediction represents a paradigm shift in crop breeding: from phenotypic observation to genome-wide prediction, from gene-by-gene approaches to polygenic modeling, and from static environments to dynamic, climate-informed selection strategies. This historical trajectory underscores why GP is uniquely positioned to accelerate the development of climate change-ready crops.

12.3 Theoretical Foundations of Genomic Prediction

Genomic prediction (GP) is grounded in quantitative genetics and statistical learning, extending classical models of breeding value estimation to incorporate genome-wide marker data. At its core, GP aims to predict the genomic estimated breeding value (GEBV) of an individual and its expected performance as a parent using high-dimensional genomic information. This section outlines the theoretical basis of GP, including marker-trait associations, prediction models, and metrics for assessing accuracy.

12.3.1 Marker–trait associations and polygenic inheritance

The genetic architecture of most traits relevant to climate resilience, such as drought tolerance, heat stress resistance, and nutrient use efficiency, is polygenic, meaning they are influenced by many loci, each with small effect sizes. Traditional quantitative trait loci (QTL) mapping or genome-wide association studies (GWAS) focus on identifying statistically significant associations between individual markers and traits. However, these approaches often explain only a fraction of the heritability of complex traits (the 'missing heritability' problem).

GP addresses this limitation by adopting a whole-genome regression approach. Instead of testing each marker individually, GP models include all available markers simultaneously, capturing both large- and small-effect loci. This approach avoids the bias introduced by marker selection thresholds and increases the variance explained by the model (Meuwissen *et al.*, 2001).

12.3.2 The genomic estimated breeding value (GEBV)

The genomic prediction framework estimates the breeding value of an individual 'i' as:

$$Yi = \mu + \sum_{j=1}^{p} Xij\, \beta j + \epsilon i$$

where: Yi = phenotype (e.g. yield under drought); μ = population mean; Xij = genotype of individual i at marker j (e.g. coded as 0, 1, 2 for allele dosage); βj = effect of marker j; *p*=number of markers (often thousands to millions); and ϵi = residual error.

The genomic estimated breeding value (GEBV) of individual 'i' is then:

$$g^{i} = \sum_{j=1}^{p} Xij\, \beta j$$

This formulation illustrates the central challenge of GP: estimating marker effects (βj) when the number of markers (p) far exceeds the number of individuals (n). Specialized statistical and machine learning methods have been developed to address this high-dimensional prediction problem.

12.3.3 Statistical models in genomic prediction

Linear mixed models (GBLUP and RR-BLUP)

One of the most widely used frameworks is genomic best linear unbiased prediction (GBLUP), a linear mixed model that uses a genomic relationship matrix GG to capture covariance among individuals.

$$y = \mu + g + \epsilon$$

with

$$g \sim N(0, \sigma^2{}_g G), \epsilon \sim N(0, \sigma^2{}_e I)$$

Here, gg represents the random genomic effects, and the matrix G is computed from genome-wide markers. This approach assumes all markers contribute equally with normally distributed small effects, making it robust for polygenic traits such as drought tolerance (Dias *et al.*, 2018).

A related formulation, ridge regression BLUP (RR-BLUP), shrinks all marker effects toward zero with equal penalty, effectively controlling for overfitting in high-dimensional data.

Bayesian models

Bayesian approaches (e.g. BayesA, BayesB, BayesCπ, BayesR) differ in their assumptions about the prior distributions of marker effects. For instance:

- **BayesA:** assumes each marker has its own variance parameter, allowing heterogeneity of effect sizes.
- **BayesB:** assumes many markers have zero effect, making it more suitable for sparse genetic architectures.
- **BayesR:** assigns markers to multiple effect-size distributions (large, moderate, small, zero).

Bayesian models can outperform GBLUP when traits are influenced by a few large-effect loci in addition to many small-effect ones, as seen in traits like submergence tolerance in rice (SUB1 locus) combined with polygenic yield stability.

Machine learning methods

As genomic datasets expanded, machine learning (ML) models were introduced to capture complex, non-linear relationships. These include:

- **Random forests (RF):** ensemble decision trees that can handle interactions and non-linearities but may suffer from reduced accuracy in very high-dimensional data.
- **Support vector machines (SVMs):** effective in some cases but computationally demanding for large genomic datasets.
- **Gradient boosting machines (GBMs):** sequentially build models to minimize prediction error, useful for capturing nonlinear marker interactions.

Deep learning models

Deep learning (DL) methods, such as convolutional neural networks (CNNs) and recurrent neural networks (RNNs), have shown promise in GP, especially for integrating genomic, phenomic, and environmental data (Montesinos-López *et al.*, 2021). DL models can learn hierarchical representations of genetic variation and G×E interactions, though they require large training datasets and careful regularization to avoid overfitting.

12.3.4 Incorporating genotype × environment interactions (G×E)

Climate resilience traits are often highly environment-dependent. For instance, the expression of drought tolerance genes varies across soil types, rainfall regimes, and temperature conditions. Ignoring G×E can lead to poor predictions when GP models are applied outside the training environments.

Reaction-norm models extend GP by modeling phenotypes as functions of both genotype and environmental covariates:

$$y_{ij} = \mu + g_i + e_j + (g \times e)_{ij} + \epsilon_{ij}$$

where y_{ij} is the phenotype of genotype i in environment j. Including environmental descriptors (e.g. rainfall, temperature, soil moisture) enhances prediction accuracy for climate-related traits (Jarquín *et al.*, 2014).

12.3.5 Accuracy and evaluation metrics

The effectiveness of GP models is assessed using validation schemes such as cross-validation (within the same environment) and forward prediction (across new years or sites). Key metrics include:

- **Prediction accuracy**: correlation between observed and predicted phenotypes.
- **Reliability**: squared correlation between GEBV and true breeding value.
- **Mean squared error (MSE)**: average squared difference between predicted and observed phenotypes.
- **Predictive ability across environments** especially critical for climate-ready crops, measured by testing models in unobserved or stress-prone environments.

For climate adaptation, accuracy must be assessed not only under optimal conditions but also under stress environments, ensuring robustness and stability of predictions.

12.3.6 Relevance to climate change-ready breeding

The theoretical framework of GP directly addresses the complexity of climate resilience traits. Linear models (GBLUP, RR-BLUP) are well-suited for broadly polygenic traits like drought tolerance, while Bayesian and ML/DL approaches can capture traits with mixed-effect architectures or strong G×E. Importantly, models that incorporate environmental data (reaction norms, crop growth models) enable predictions tailored to specific climate scenarios.

In the context of climate change, where environmental conditions are becoming increasingly variable, these theoretical foundations provide breeders with tools to make proactive selections. Instead of waiting for stress events to occur in field trials, breeders can use GP models trained under diverse stress scenarios to identify genotypes likely to perform well under future climates.

12.4 Incorporating Genotype × Environment Interaction (G×E)

One of the greatest challenges in developing climate change-ready crops is accounting for the variability and unpredictability of environments. Climate change is increasing the frequency of extreme events, such as drought, heat waves, and flooding, while also shifting long-term averages of rainfall and temperature (Lynch *et al.*, 2008). Crop performance in such conditions depends not only on genetic potential but also on how different genotypes respond to diverse environmental contexts. This phenomenon, known as genotype × environment interaction (G × E), must be explicitly modeled in genomic prediction (GP) frameworks to enhance the robustness and accuracy of predictions.

12.4.1 Importance of G×E in climate-smart breeding

G × E occurs when the relative performance of genotypes changes across environments. For example, a maize hybrid that performs well under well-watered conditions may yield poorly under drought, while another hybrid with lower optimal yield may outperform in water-limited environments. Ignoring G × Ecan result in biased predictions, reduced selection accuracy, and ultimately poor adoption of new varieties by farmers in stress-prone regions (Cooper *et al.*, 2014).

For climate-resilient breeding, modeling G × E is not optional; it is essential. Future varieties must maintain stability and yield across a range of conditions, including extreme stresses. Therefore, modern GP models must be extended to capture both static adaptation (broad adaptation across environments) and dynamic adaptation (specific adaptation to particular stress environments).

12.4.2 Multi-environment trials (METs) as the foundation

Multi-environment trials (METs) provide the empirical foundation for quantifying and modeling G × E. In METs, a training population is evaluated across multiple locations and seasons that represent target agro-ecological zones and stress conditions. These data are then integrated with genomic marker information to build predictive models.

For example, the International Maize and Wheat Improvement Center (CIMMYT) routinely conducts METs for maize and wheat across sites in Africa, Asia, and Latin America. Such networks capture variation in rainfall, soil types, temperature regimes, and biotic pressures, enabling GP models to incorporate diverse stress responses (Leong *et al.*, 2025).

12.4.3 Statistical models for G×E in GP

Several modeling strategies have been developed to integrate G × E into GP frameworks, as follows.

Reaction norm models

Reaction norm models treat phenotypes as functions of both genotypes and environmental gradients. A basic formulation is:

$$y_{ij} = \mu + g_i + e_j + (g \times e)_{ij} + \epsilon_{ij}$$

where: y_{ij} = phenotype of genotype ii in environment j; g_i = genomic effect of genotype i; e_j = effect of environment j; $(g \times e)_{ij}$ = interaction effect; and ϵ_{ij} = residual.

Marker × environment interaction terms can be modeled directly, or reaction norms can be parameterized using environmental covariates (e.g. rainfall, temperature).

Genomic G × E models with environmental covariates

An extension incorporates environmental descriptors into the genomic relationship matrix. For instance, envirotyping provides quantitative variables (rainfall distribution, solar radiation, soil pH, nutrient availability) that can be linked with genomic effects (Cooper *et al.*, 2022).

This approach allows breeders to not only predict performance in known environments but also simulate future scenarios, such as projected 2050 rainfall patterns under RCP 8.5 climate models.

Factor analytic models

Factor analytic (FA) models capture G × E by reducing the dimensionality of environments into latent factors. These models can improve predictive accuracy by identifying patterns of genotype responses across correlated environments (Jarquín *et al.*, 2014).

12.4.4 Integration with crop growth models

Beyond statistical methods, integrating GP with process-based crop growth models (CGMs) provides a mechanistic perspective. CGMs simulate how physiological traits (e.g. root depth, canopy conductance, flowering time) interact with environmental conditions to determine yield. When linked with GP, these models allow breeders to predict how genotypes will perform under future climate scenarios. For instance, maize drought tolerance predictions can be enhanced by incorporating physiological traits simulated under variable water regimes using the APSIM model (Messina *et al.*, 2020).

Such genomic + crop modeling integration creates a powerful toolset for climate-resilient breeding, enabling proactive rather than reactive selection.

12.4.5 Case studies of G×E modeling in GP

- **Maize in sub-Saharan Africa:** Beyene *et al.* (2019) showed that incorporating G × E into GP improved prediction accuracy for drought tolerance by 15–30%, compared with models without G × E terms.
- **Wheat heat tolerance:** Sukumaran *et al.* (2021) demonstrated that GP models including G × E across South Asian test sites improved prediction of grain yield under terminal heat stress.
- **Rice drought and flood tolerance:** Lokeshkumar *et al.* (2025) used multi-environment genomic models to predict rice performance under both drought and submergence, showing that G × E modeling allowed breeders to identify genotypes with dual stress tolerance.

These studies emphasize that accounting for G × E is not only a statistical refinement but a necessity for breeding climate-ready cultivars.

12.4.6 Practical considerations for breeders

To effectively implement G × E in genomic prediction, breeders must do as follows.

1. **Design diverse training sets:** Include germplasm that spans the environmental variation expected in target regions.
2. **Collect high-quality environmental data:** Use remote sensing, weather stations, and soil mapping for accurate envirotyping.
3. **Use cross-validation schemes that mimic real-world prediction scenarios:** For example, predicting performance in new environments or years.
4. **Balance broad vs specific adaptation goals:** Some breeding programs may prioritize stability across environments, while others may target high performance in stress-prone niches.

12.4.7 Relevance to climate change

Climate change amplifies the need to model G × E because environments are becoming less predictable and more variable. Genomic prediction models that incorporate G × E offer the ability to:

- predict performance in novel or extreme environments, not represented in current field trials;
- support climate-informed breeding pipelines, where future weather scenarios guide selection decisions; and
- facilitate targeted deployment of varieties matching genotypes to environments where they are most likely to succeed.

In essence, incorporating G × E transforms GP from a tool for predicting performance under current conditions into a framework for anticipating the future of agriculture under climate change.

12.5 Applications of Genomic Prediction for Climate-Resilient Traits

The success of genomic prediction (GP) in developing climate change-ready crops hinges on its ability to accurately predict complex stress-adaptive traits that determine crop survival, productivity, and stability under variable and extreme environments. These traits are often polygenic, influenced by multiple genes with small effects, and strongly shaped by genotype × environment interactions (G × E). Traditional marker-assisted selection, which relies on major-effect QTLs, has had limited success in improving such traits. By contrast, GP leverages genome-wide marker information to capture small additive and non-additive effects across the genome, making it particularly well suited for breeding resilience into crops (Meuwissen *et al.*, 2001).

This section highlights major climate resilience traits such as drought, heat, flood/submergence, salinity, nutrient use efficiency, and multi-stress tolerance where GP has shown promise. Fig. 12.1 shows a conceptual design that visually summarizes how genomic prediction is applied to develop climate change-ready crops by targeting key resilience traits. Table 12.1 provides a concise comparative overview, linking stress types, adaptive traits, key crops, and landmark GP applications.

12.5.1 Drought tolerance

Importance

Drought is the most widespread abiotic stress globally, causing yield losses across crops such as maize, wheat, rice, and sorghum. With climate change, drought episodes are projected to increase in both frequency and intensity, particularly in sub-Saharan Africa (Gizaw and Gan, 2017).

Traits relevant to GP

- Stay-green phenotype (delayed leaf senescence under stress)
- Root architecture (deeper, more prolific rooting)
- Canopy temperature and transpiration efficiency
- Grain filling duration under stress

GP applications

- **Maize:** CIMMYT studies demonstrated that GP increased selection accuracy for grain yield under managed drought trials in Africa, reducing breeding cycle length and enabling faster deployment of drought-tolerant hybrids (Beyene *et al.*, 2019).
- **Wheat:** Wondifraw (2024) used GP with multi-environment data to predict end use qualities in hard winter wheat, improving accuracy compared to pedigree-based BLUP.

12.5.2 Heat tolerance

Importance

Heat stress during flowering and grain filling is a major threat to cereal productivity, especially in regions such as the Indo-Gangetic Plains, sub-Saharan Africa, and the US Midwest. Global warming is expected to increase the number of heat stress days by mid-century (Oleson *et al.*, 2015).

Traits relevant to GP

- Flowering time adjustment (heat escape)
- Pollen viability and grain set under high temperature
- Membrane thermostability
- Stay-green and canopy cooling mechanisms

GP applications

- **Wheat:** Ashfaq *et al.* (2022) showed that GP models incorporating heat tolerance indices improved prediction accuracy for grain yield under terminal heat stress by up to 20%.
- **Rice:** Tayade *et al.* (2018) demonstrated that GP models trained with flowering

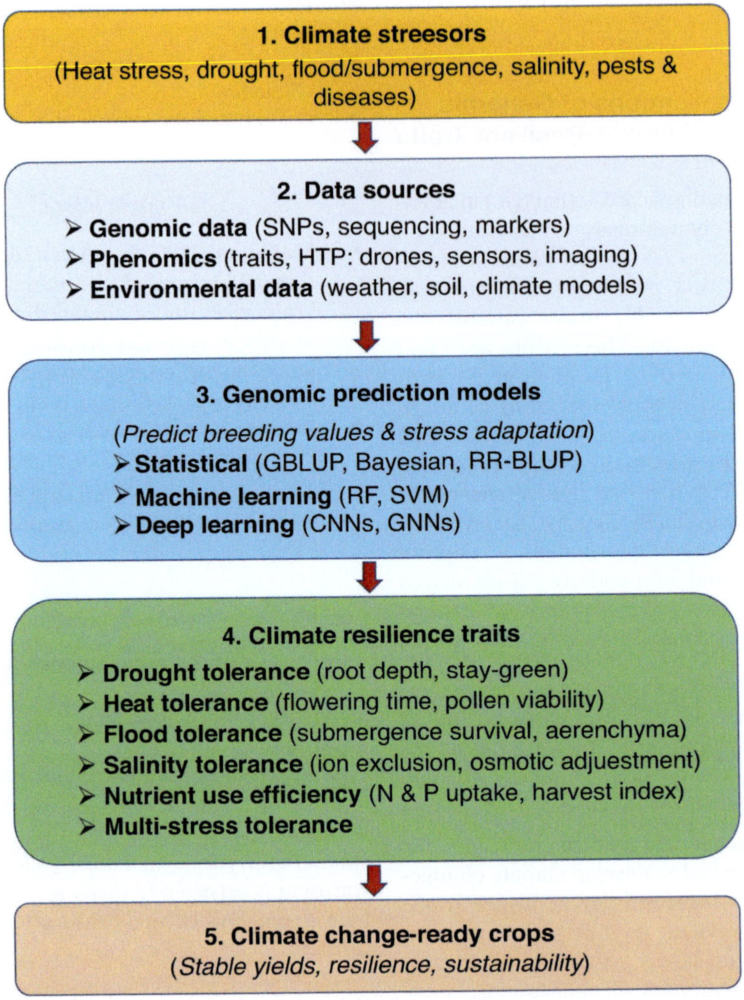

Fig. 12.1. A conceptual framework of genomic prediction for climate resilience. Genomic, phenotypic, and environmental data are integrated into statistical and machine learning models to predict complex stress-adaptive traits. These predictions accelerate breeding pipelines for developing climate change-ready crops with stable performance across diverse and extreme environments (Author's own image).

time and grain fertility traits under heat stress could identify lines resilient to heat at anthesis.

- **Soybean:** Kiruba *et al.* (2025) highlighted GP's utility for predicting seed set under heat, accelerating breeding cycles in tropical soybean programs.

12.5.3 Flood and submergence tolerance

Importance

Flooding and submergence, particularly in South and Southeast Asia, are becoming more

Table 12.1. Applications of genomic prediction for climate-resilient traits (Author's own table).

Stress/trait	Key adaptive traits	Crops with GP applications	Highlights/case studies
Drought tolerance	Stay-green, deep roots, canopy temperature regulation, grain filling under stress	Maize, Wheat	GP improved prediction of maize drought tolerance in sub-Saharan Africa (Beyene *et al.*, 2019); enhanced accuracy for wheat yield under drought (Wondifraw, 2024).
Heat tolerance	Flowering time adjustment, pollen viability, membrane stability, canopy cooling	Wheat, Rice, Soybean	GP increased accuracy for wheat yield under terminal heat stress (Ashfaq *et al.*, 2022); rice lines predicted for fertility under heat stress (Tayade *et al.*, 2018): predicting soybean seed set under heat (Kiruba *et al.*, 2025)
Flood/Submergence tolerance	Submergence survival, elongation under partial flooding, anaerobic germination, aerenchyma formation	Rice, Maize	Beyond SUB1 rice, GP captured polygenic resilience to flooding (Suwetha *et al.*, 2025); maize hybrids predicted for root aerenchyma formation (Aslam *et al.*, 2023).
Salinity tolerance	Ion exclusion (Na^+, Cl^-), osmotic adjustment, ion compartmentalization, root plasticity	Rice, Barley	GP predicted salt tolerance indices in rice (Singh *et al.*, 2021); barley GP improved ion homeostasis prediction (Thabet *et al.*, 2022).
Nutrient use efficiency (NUE, PUE)	N uptake and utilization efficiency, P acquisition, harvest index under low input	Maize, Wheat,	Maize GP improved selection for low-N environments (Beyene *et al.*, 2019); wheat GP increased accuracy for low-N yield (Braun *et al.*, 2010).
Multi-stress tolerance	Combined drought + heat, drought + submergence, heat + nutrient stress	Maize, Rice	GP models successfully predicted performance under combined stresses, e.g. drought + heat in maize (Badu-Apraku *et al.*, 2023); drought + submergence in rice (Ali *et al.*, 2020).

frequent due to erratic rainfall and cyclonic activity. Lowland rice is particularly vulnerable, with millions of hectares affected annually.

Traits relevant to GP

- Submergence tolerance (anaerobic survival)
- Rapid elongation under partial flooding (escape mechanism)
- Anaerobic germination ability
- Aerenchyma formation for oxygen transport

GP applications

- **Rice (SUB1 gene):** While major genes such as SUB1A provide strong tolerance to complete submergence, GP can complement by capturing polygenic contributions to partial submergence and recovery traits (Suwetha *et al.*, 2025).
- **Maize:** In flood-prone tropical lowlands, GP has been used to predict root aerenchyma formation and survival under waterlogging, enabling breeders to identify tolerant hybrids earlier in the pipeline (Aslam *et al.*, 2023).

12.5.4 Salinity tolerance

Importance

Soil salinization, exacerbated by rising sea levels and unsustainable irrigation practices, affects over 20% of irrigated land globally. Crops in coastal areas (Bangladesh, Egypt, Vietnam) and arid regions are especially vulnerable.

Traits relevant to GP

- Ion exclusion (Na^+, Cl^-)
- Tissue tolerance mechanisms (osmotic adjustment, compatible solutes)
- Root system plasticity
- Shoot ion compartmentalization

GP applications

- **Soybean:** Singh *et al.* (2021) used high-throughput crop phenotyping to predict salt tolerance indices, showing moderate-to-high accuracy when combining genotypic data with environmental descriptors.
- **Barley:** Genomic models improved prediction for ion homeostasis traits, with implications for breeding salt-tolerant varieties in Australia's saline soils (Thabet *et al.*, 2022).

12.5.5 Nutrient use efficiency (NUE)

Importance

Climate-smart agriculture requires minimizing fertilizer use due to economic and environmental costs. Breeding for higher nitrogen use efficiency (NUE) and phosphorus use efficiency (PUE) is crucial for reducing greenhouse gas emissions and ensuring sustainable production.

Traits relevant to GP

- Nitrogen uptake efficiency (root absorption capacity)
- Nitrogen utilization efficiency (conversion into biomass and grain)
- Phosphorus acquisition (root exudates, mycorrhizal associations)
- Harvest index under low input.

GP applications

- **Maize:** Beyene *et al.* (2019) integrated NUE traits into GP models, enabling selection of hybrids that perform well under low N conditions in Africa.
- **Wheat:** GP improved prediction of grain yield under low-N trials across multiple CIMMYT locations (Braun *et al.*, 2010).

12.5.6 Multi-stress tolerance

Importance

In real-world environments, crops often face multiple simultaneous stresses, such as drought + heat, flooding + salinity, or nutrient deficiency + disease pressure. Breeding for single-stress tolerance may not ensure resilience under these complex conditions.

GP applications

- **Maize (drought + heat):** CIMMYT demonstrated that GP can predict performance under combined drought and heat stress, improving accuracy when training sets included both stress conditions (Badu-Apraku *et al.*, 2023).
- **Rice (drought + submergence):** GP identified varieties with partial resilience to both stresses, showing the feasibility of breeding 'climate-smart rice' (Ali *et al.*, 2020).

12.5.7 Lessons and implications

- **Trait-specific considerations:** GP accuracy is higher for heritable traits (e.g. flowering time) but can be improved for low-heritability stress traits through multi-trait and multi-environment models.
- **Integration with phenomics:** High-throughput phenotyping (drones, hyperspectral imaging) provides proxy traits (e.g. canopy temperature, vegetation indices) that enhance GP prediction of stress tolerance.

- **Deployment at scale:** GP enables selection for climate-resilience traits early in breeding pipelines, reducing the need for extensive field evaluations under stress environments.

In summary, GP has shown tangible benefits in predicting drought, heat, flood, salinity, nutrient use, and multi-stress tolerance. These applications underscore its role as a cornerstone of climate change-ready breeding pipelines, enabling accelerated genetic gains for resilience in staple crops.

12.6 Integration into Breeding Pipelines

The successful application of genomic prediction (GP) for climate change-ready crops depends not only on developing accurate prediction models but also on integrating these approaches into practical breeding pipelines. Traditional crop breeding cycles are often lengthy, requiring multiple years to advance generations, evaluate performance, and release varieties. Climate change intensifies the urgency of accelerating this process, and GP provides an avenue to streamline selection, reduce costs, and enhance genetic gain per unit of time (Saha *et al.*, 2025). The integration of GP into breeding programs involves careful alignment of data generation, model development, selection decisions, and deployment strategies, ensuring that predictions translate into improved cultivars in farmers' fields (Table 12.2).

12.6.1 Genomic selection in early breeding stages

One of the most effective uses of GP is at the early stages of breeding programs, where thousands of progenies are generated but resources constrain phenotyping. By leveraging genomic data, breeders can predict breeding values before extensive field testing, enabling early culling of inferior lines and prioritizing individuals with favorable stress-adaptive alleles.

Table 12.2. Integration of genomic prediction (GP) into modern breeding pipelines for climate change-ready crops (Author's own table)

Breeding stage	Role of genomic prediction	Benefits for climate-resilient breeding
Early breeding stages	Predict breeding values from genomic data at the seedling stage	Enables early culling of inferior lines, reduces phenotyping costs, accelerates trait stacking (e.g. drought + heat tolerance)
Recurrent selection	Apply GP across rapid breeding cycles	Increases rate of genetic gain, reduces cycle time when combined with speed breeding
Multi-environment targeting	Incorporate G × E and environmental covariates in GP models	Improves prediction accuracy across target populations of environments (TPEs), ensures trait stability under diverse climates
Hybrid breeding	Predict general and specific combining ability	Reduces number of test crosses, enhances heterotic grouping, accelerates deployment of stress-resilient hybrids
Integration with HTP	Combine GP with drone/sensor-based phenotyping	Increases prediction accuracy, captures climate-stress responses in real time
Digital breeding platforms	Use GP in integrated breeding management systems	Facilitates data-driven decisions, supports large-scale breeding, improves breeding program efficiency
Advanced trials and deployment	Validate GP-selected lines in multi-environment testing	Accelerates release of climate-smart cultivars with stable yield and resilience

For example, GP applied in the seedling stage for drought or salinity tolerance reduces the need for costly and time-consuming multi-environment screening, saving both time and resources (Srivastava *et al.*, 2022). Early-stage genomic selection also allows breeders to stack multiple climate resilience traits such as heat tolerance and nutrient-use efficiency into promising candidate lines.

12.6.2 Accelerating recurrent selection cycles

Traditional recurrent selection cycles often span several years, limiting the rate of genetic gain. GP shortens these cycles by providing accurate predictions of genetic merit without waiting for phenotypic evaluation. By using rapid cycling and genomic recurrent selection, breeders can conduct multiple selection rounds per year, especially when integrated with speed breeding technologies that shorten generation times under controlled environments (Potts *et al.*, 2023). This integration enhances the pace at which climate-adaptive alleles accumulate in breeding populations, significantly advancing the development of resilient cultivars.

12.6.3 Multi-Environment and Target Population of Environments (TPEs)

A key challenge in breeding for climate resilience is the genotype × environment (G × E) interaction. The performance of a genotype under one stress condition may not translate to stability across diverse environments. GP models integrated into breeding pipelines now increasingly incorporate environmental covariates, allowing predictions to be tailored to target populations of environments (TPEs). For instance, a rice genotype predicted to perform well under intermittent drought in South Asia may not be equally effective under salinity stress in coastal regions. By integrating envirotyping data, breeders can better target the deployment of climate-smart cultivars (Cooper *et al.*, 2022).

12.6.4 Hybrid breeding and heterotic grouping

For hybrid crops such as maize, sorghum, and rice, GP offers particular advantages in predicting general and specific combining ability, enabling breeders to identify promising parental combinations without exhaustive test crosses. Climate resilience traits such as heat or drought tolerance often have complex inheritance, making hybrid prediction particularly valuable. GP reduces the need for large-scale field evaluation, allowing breeders to strategically deploy hybrids that combine complementary stress-resilient alleles. Moreover, GP facilitates more precise heterotic grouping based on genome-wide similarity rather than a few markers, ensuring that climate-adaptive diversity is maximized in hybrid programs.

12.6.5 Integration with high-throughput phenotyping (HTP)

While GP is fundamentally genomic, its integration with high-throughput phenotyping (HTP) accelerates trait discovery and model accuracy. Modern breeding pipelines increasingly combine drone-based imaging, hyperspectral data, thermal cameras, and soil sensors to capture large-scale, non-invasive phenotypes under field conditions. These data not only improve the predictive ability of GP models but also help refine selection decisions under real-world climate stresses. For instance, canopy temperature dynamics captured by UAVs can serve as an indirect proxy for drought tolerance, enabling breeders to validate GP predictions more efficiently (Liang *et al.*, 2025).

12.6.6 Digital breeding platforms

Digital platforms are becoming critical to operationalizing GP in breeding pipelines. Integrated systems combine genomic data management, phenotyping databases, environmental datasets, and predictive models into unified decision-support platforms. Breeding management systems such as the Breeding Management System (BMS) developed by the CGIAR enable breeders to track genetic gains, manage multi-location trials, and seamlessly apply GP in routine selection. The use of cloud-based analytics and AI-driven decision tools further enhances the speed and scalability of GP integration across global breeding programs.

12.6.7 Challenges in operational integration

Despite its potential, several challenges hinder the widespread integration of GP in breeding pipelines. These include the high cost of genotyping in resource-limited breeding programs, limited access to advanced computational tools, and gaps in breeder training for genomic data interpretation. Additionally, model transferability across environments and breeding populations remains a constraint. Overcoming these barriers requires investment in shared infrastructure, breeder training, and capacity-building initiatives, particularly in developing regions where climate-resilient crops are most urgently needed.

In essence, integrating GP into breeding pipelines transforms the traditional sequential process into a data-driven, iterative cycle that accelerates the release of climate change-ready cultivars. The synergy of GP with HTP, speed breeding, and digital breeding tools positions it as a cornerstone of modern crop improvement strategies.

12.7 Challenges and Future Perspectives

Despite the growing recognition of genomic prediction (GP) as a transformative approach for developing climate change-ready crops, its large-scale application still faces significant challenges. These challenges encompass technical, biological, infrastructural, and socio-economic dimensions, which must be addressed to ensure that GP fulfills its potential in real-world breeding programs. At the same time, emerging technologies and global collaborations present opportunities for overcoming these barriers and shaping the future of climate-resilient agriculture. Table 12.3 provides a concise challenge-to-solution mapping for quick reference, complementing the detailed narrative in Section 12.7.

12.7.1 Technical and biological challenges

One of the most critical technical challenges in GP is the complex genetic architecture of stress-resilience traits. Traits such as drought tolerance, heat tolerance, or salinity resistance are typically polygenic, controlled by many small-effect loci, and strongly influenced by environmental interactions. This complexity reduces the predictive accuracy of GP models, particularly when applied across diverse genetic backgrounds or environments. In addition, genotype × environment (G × E) interactions make it difficult to identify genotypes that perform consistently under variable climate conditions (Cooper *et al.*, 2022).

Another biological challenge lies in trait definition and measurement. Many climate-adaptive traits are difficult to phenotype directly, such as root system architecture or water-use efficiency. Although high-throughput phenotyping (HTP) provides indirect proxies, linking these phenotypes with genomic data in a robust manner remains an ongoing challenge. Moreover, multi-trait predictions where resilience is achieved by combining tolerance to multiple stresses simultaneously are still methodologically complex and computationally demanding.

12.7.2 Data limitations and infrastructure gaps

Reliable GP depends on large, high-quality datasets integrating genomic, phenotypic, and environmental information. However, in many breeding programs – especially in developing regions – such datasets are limited. High genotyping costs restrict sample sizes,

Table 12.3. Challenges and opportunities in applying genomic prediction for climate change-ready crops (Author's own table)

Challenge area	Specific issues	Opportunities/future solutions
Technical and biological	Polygenic nature of stress-resilience traits; strong G × E interactions; difficulty phenotyping complex traits	Use multi-trait and multi-environment GP models; integrate AI/deep learning; link GP with physiology-driven models
Data limitations	Limited genotypic and phenotypic datasets; high genotyping costs; inconsistent environmental data	Expand global datasets; reduce sequencing costs; integrate envirotyping and climate projections
Infrastructure gaps	Lack of computing resources and digital breeding tools in many programs	Cloud-based platforms; shared computational resources; open-source breeding software
Capacity building	Limited breeder expertise in genomics, statistics, and bioinformatics	Training programs, interdisciplinary collaborations, and user-friendly decision-support systems
Adoption and trust	Perception of GP as a 'black box' with low interpretability	Develop transparent, interpretable models; demonstrate biological relevance of predictions
Equity and accessibility	Limited access in developing regions; risk of widening technological divide	Public–private partnerships; CGIAR and global initiatives; low-cost genotyping and democratized tools
Future integration	Need for faster trait deployment and multi-stress resilience	Combine GP with speed breeding, genome editing, and multi-omics integration

while phenotypic data often lack precision or standardization across sites. Additionally, the integration of environmental data, critical for climate-resilience predictions, is constrained by limited access to long-term climate and soil databases. These data limitations reduce the generalizability of GP models and restrict their adoption in resource-limited settings.

Infrastructural challenges also persist. Many breeding programs lack advanced computing facilities, bioinformatics expertise, and digital breeding platforms required to implement GP effectively. This creates disparities in access to genomic technologies, potentially widening the gap between well-funded programs and those in regions most vulnerable to climate change.

12.7.3 Capacity building and breeder adoption

The successful deployment of GP in breeding pipelines requires capacity building among plant breeders and agricultural scientists. While GP models are statistically and computationally intensive, many breeders are trained primarily in conventional breeding techniques. Bridging this knowledge gap is essential. Training programs in quantitative genetics, data science, and computational breeding are needed to empower breeders to interpret and apply GP outputs.

Adoption is also influenced by breeder trust in GP predictions. If models are perceived as 'black boxes' without biological interpretability, breeders may be hesitant to integrate them into decision-making. Therefore, developing transparent, interpretable models that link genomic predictions with physiological or agronomic understanding will be critical for adoption.

12.7.4 Equity and accessibility

Climate change disproportionately affects smallholder farmers in developing regions, who depend heavily on resilient crops for food security. Yet, these regions often lack access to advanced genomic tools and infrastructure. Ensuring equitable access to GP technologies is therefore a pressing challenge. Public sector breeding programs, international agricultural research centers, and global collaborations such as the CGIAR Excellence in Breeding Platform have a central role in democratizing access to GP by providing low-cost genotyping, shared computational platforms, and training resources (McHugh *et al.*, 2022).

Without deliberate efforts toward equity, there is a risk that GP-driven innovations could primarily benefit commercial agriculture in high-income countries, leaving vulnerable populations behind.

12.7.5 Future technological opportunities

Looking ahead, several emerging technologies promise to strengthen GP and expand its role in climate-resilient breeding, as follows.

- **Artificial intelligence (AI) and deep learning:** Neural networks and graph-based models can capture nonlinear interactions among genes, environments, and traits, enhancing predictive accuracy (Malla and Banka, 2023).
- **Multi-omics integration:** Combining genomics with transcriptomics, metabolomics, and proteomics will provide a holistic view of stress responses, improving GP models for complex resilience traits (Fan *et al.*, 2025).
- **Envirotyping and climate modeling:** Incorporating real-time environmental and climate projection data into GP models will allow predictions of genotype performance under future climate scenarios.
- **Speed breeding and genome editing synergies:** GP can prioritize candidate genotypes for accelerated generation cycles or targeted genome editing (e.g. CRISPR), thereby shortening the time from discovery to cultivar release.
- **Cloud-based breeding platforms:** The rise of open-access digital breeding tools will facilitate large-scale, collaborative GP adoption across regions.

12.7.6 Towards climate-smart breeding ecosystems

In the long term, GP must be embedded within integrated climate-smart breeding ecosystems. These systems will combine GP with participatory breeding approaches, ensuring that predictions align with farmers' needs and preferences. Collaboration across disciplines linking geneticists, data scientists, physiologists, agronomists, and climate experts will be essential to design cultivars that not only withstand climate stresses but also meet nutritional, cultural, and market demands.

Ultimately, the future of GP lies in building resilient, inclusive, and data-driven breeding programs that respond rapidly to the accelerating pace of climate change. By addressing current challenges and leveraging technological opportunities, GP can serve as a cornerstone in ensuring global food and nutritional security under uncertain climatic futures.

12.8 Conclusion

Genomic prediction (GP) has become a cornerstone in modern plant breeding, offering unprecedented opportunities to accelerate the development of climate change-ready crops. By integrating genome-wide markers, phenotypic data, and environmental variables, GP enables early and accurate selection of stress-resilient genotypes, thereby reducing breeding cycle time and enhancing genetic gains. This predictive framework is especially critical for addressing complex traits such as drought, heat, salinity, and submergence tolerance, which underpin resilience in an era of climatic uncertainty. The integration of GP with high-throughput phenotyping, speed breeding, and digital platforms is reshaping breeding pipelines into more data-driven and adaptive systems. Despite challenges related to data scarcity, infrastructure gaps, and limited accessibility, advances in artificial intelligence, multi-omics, and participatory breeding approaches hold promise for wider application. Ultimately, GP represents a paradigm shift, enabling breeders to design climate-smart varieties that safeguard global food and nutritional security.

References

Ahmad, S., Shah, M.H., Raghib, M.G., Belwal, V., Ali, S. *et al.* (2025) Technical advancements in plant high-throughput phenotyping and phenomics: An overview. In: Chen, J.-T. (ed.) *Plant High-Throughput Phenotyping and Functional Phenomics*. CRC Press, Boca Raton, Florida, pp. 246–259. DOI: 10.1201/9781003505631.

Akdemir, D. and Isidro-Sánchez, J. (2019) Design of training populations for selective phenotyping in genomic prediction. *Scientific Reports* 9(1), 1446.

Ali, J., Mahender, A., Prahalada, G.D., Sevilla, M.A.L., Galang, A. *et al.* (2020) Genomics-assisted breeding of climate-smart inbred and hybrid rice varieties. In: *Genomic Designing of Climate-Smart Cereal Crops*. Springer International Publishing, Cham, pp. 1–43.

Ashfaq, W., Brodie, G., Fuentes, S. and Gupta, D. (2022) Infrared thermal imaging and morpho-physiological indices used for wheat genotypes screening under drought and heat stress. *Plants* 11(23), 3269.

Aslam, A., Mahmood, A., Ur-Rehman, H., Li, C., Liang, X. *et al.* (2023) Plant adaptation to flooding stress under changing climate conditions: Ongoing breakthroughs and future challenges. *Plants* 12(22), 3824.

Badu-Apraku, B., Fakorede, M.A., Nelimor, C., Osuman, A.S., Bonkoungou, T.O. *et al.* (2023) Recent advances in breeding maize for drought, heat and combined heat and drought stress tolerance in sub-Saharan Africa. *CABI Reviews* (2023).

Beyene, Y., Semagn, K., Crossa, J., Mugo, S., Atlin, G.N. *et al.* (2016) Improving maize grain yield under drought stress and non-stress environments in sub-Saharan Africa using marker-assisted recurrent selection. *Crop Science* 56(1), 344–353.

Beyene, Y., Gowda, M., Olsen, M., Robbins, K.R., Pérez-Rodríguez, P. *et al.* (2019) Empirical comparison of tropical maize hybrids selected through genomic and phenotypic selections. *Frontiers in Plant Science* 10, 1502.

Braun, H.J., Atlin, G. and Payne, T. (2010) Multi-location testing as a tool to identify plant response to global climate change. In: *Climate Change and Crop Production*. CAB International, Wallingford, UK, pp. 115–138.

Collard, B.C. and Mackill, D.J. (2008) Marker-assisted selection: An approach for precision plant breeding in the twenty-first century. *Philosophical Transactions of the Royal Society B: Biological Sciences* 363(1491), 557–572.

Cooper, M., Messina, C.D., Podlich, D., Totir, L.R., Baumgarten, A. *et al.* (2014) Predicting the future of plant breeding: Complementing empirical evaluation with genetic prediction. *Crop and Pasture Science* (4), 311–336.

Cooper, M., Messina, C.D., Tang, T., Gho, C., Powell, O.M. *et al.* (2022) Predicting Genotype× Environment × Management (G × E × M) interactions for the design of crop improvement strategies: Integrating breeder, agronomist, and farmer perspectives. *Plant Breeding Reviews* 46, 467–585.

Daszkiewicz, T. (2022) Food production in the context of global developmental challenges. *Agriculture* 12(6), 832.

Dias, K.O.D.G., Gezan, S.A., Guimaraes, C.T., Nazarian, A., da Costa e Silva, L. *et al.* (2018) Improving the accuracy of genomic predictions for drought tolerance in maize by joint modeling of additive and dominance effects in multi-environment trials. *Heredity* 121(1), 24–37.

Fan, B.L., Chen, L.H., Chen, L.L. and Guo, H. (2025) Integrative multi-omics approaches for identifying and characterizing biological elements in crop traits: Current progress and future prospects. *International Journal of Molecular Sciences* 26(4), 1466.

Feng, J., Dan, X., Cui, Y., Gong, Y., Peng, M. *et al.* (2024) Integrating evolutionary genomics of forest trees to inform future tree breeding amid rapid climate change. *Plant Communications* 5, 10.

Gizaw, M.S. and Gan, T.Y. (2017) Impact of climate change and El Niño episodes on droughts in sub-Saharan Africa. *Climate Dynamics* 49(1), 665–682.

Heffner, E.L., Sorrells, M.E. and Jannink, J.L. (2009) Genomic selection for crop improvement. *Crop Science* 49(1), 1–12.

Henderson, C.R. (1975) Best linear unbiased estimation and prediction under a selection model. *Biometrics* 31(2), 423–447.

Jarquín, D., Crossa, J., Lacaze, X., Du Cheyron, P., Daucourt, J. *et al.* (2014) A reaction norm model for genomic selection using high-dimensional genomic and environmental data. *Theoretical and Applied Genetics* 127(3), 595–607.

Kiruba, G., Pushpa, R., Arulmozhi, R., Dhandapani, M., Rajendran, R.A. *et al.* (2025) Breeding climate-resilient soybean in climate change era: Current breeding strategies and prospects. In: *Breeding Climate Resilient and Future Ready Oilseed Crop*. Springer Nature Singapore, pp. 333–347.

Kunwar, S., Babar, M.A., Jarquin, D., Ampatzidis, Y., Khan, N. *et al.* (2025) Enhancing prediction accuracy of key biomass partitioning traits in wheat using multi-kernel genomic prediction models integrating secondary traits and environmental covariates. *The Plant Genome* 18(2), e70052.

Leong, R., He, X., Beijen, B.S., Sakai, T., Goncalves, J. *et al.* (2025) Unlocking gene regulatory networks for crop resilience and sustainable agriculture. *Nature Biotechnology* 43(8), 1–12.

Liang, X., Yu, S., Ju, Y., Wang, Y. and Yin, D. (2025) Multi-scale remote-sensing phenomics integrated with multi-omics: Advances in crop drought–heat stress tolerance mechanisms and perspectives for climate-smart agriculture. *Plants* 14(18), 2829.

Lokeshkumar, B.M., Snehi, S., Krishanu, Kumar, S., Ravikiran, K.T. and *et al* (2025) Breeding strategies for improved multistress-resilient crops. In: *Cutting Edge Technologies for Developing Future Crop Plants*. Springer Nature Singapore, pp. 125–154.

Lynch, A., Nicholls, N., Alexander, L. and Griggs, D. (2008) Defining the impacts of climate change on extreme events. In: *Garnaut Climate Change Review*. Commonwealth Government of Australia, Canberra, pp. 1–19.

Malla, A.M. and Banka, A.A. (2023) A systematic review of deep graph neural networks: Challenges, classification, architectures, applications & potential utility in bioinformatics. *arXiv Preprint arXiv:2311.02127*.

McHugh, K., Morgan, V., Ragot, M., Acharjee, S. and Noma, F. (2022) *Evaluation of CGIAR Excellence in Breeding Platform: Inception Report*. CAS Secretariat Evaluation Function, Rome, Italy.

Meher, P.K., Rustgi, S. and Kumar, A. (2022) Performance of Bayesian and BLUP alphabets for genomic prediction: Analysis, comparison and results. *Heredity* 128(6), 519–530.

Messina, C.D., Cooper, M., Hammer, G.L., Berning, D., Ciampitti, I. *et al.* (2020) Two decades of creating drought tolerant maize and underpinning prediction technologies in the US corn-belt: Review and perspectives on the future of crop design. *BioRxiv* 2020–10.

Meuwissen, T.H., Hayes, B.J. and Goddard, M.E. (2001) Prediction of total genetic value using genome-wide dense marker maps. *Genetics* 157(4), 1819–1829.

Montesinos-López, O.A., Montesinos-López, A., Pérez-Rodríguez, P., Barrón-López, J.A., Martini, J.W. *et al.* (2021) A review of deep learning applications for genomic selection. *BMC Genomics* 22(1), 19.

Muller, C. and Robertson, R.D. (2014) Projecting future crop productivity for global economic modeling. *Agricultural Economics* 45(1), 37–50.

Naqvi, R.Z., Farooq, M., Zahid, R., Imran, I., Mahmood, M.A. *et al.* (2024) Recent advances in plant phenomics and speed breeding for climate-smart agriculture. In: Chen, J.-T. (ed.) *Plant Speed Breeding and High-Throughput Technologies*. CRC Press, Boca Raton, Florida, pp. 82–103.

Oleson, K.W., Monaghan, A., Wilhelmi, O., Barlage, M., Brunsell, N. *et al.* (2015) Interactions between urbanization, heat stress, and climate change. *Climatic Change* 129(3), 525–541.

Potts, J., Jangra, S., Michael, V.N. and Wu, X. (2023) Speed breeding for crop improvement and food security. *Crops* 3(4), 276–291.

Rayaprolu, L., Deshpande, S.P. and Gupta, R. (2022) Genotyping-by-sequencing (GBS) method for accelerating marker-assisted selection (MAS) program. In: *Genomics of Cereal Crops*. Springer US, New York, pp. 245–257.

Saha, S., Nimitha, K., Purkayastha, S., Shaha, M.H., Uddin, S.M.A. *et al.* (2025) Stress-resilient crops: Advanced breeding using coordinated strategies against changing climate. In: Chen, J.-T. (ed.) *Stress-Resilient Crops: Coordinated Omics–CRISPR–Nanotechnology Strategies*. CAB International, Wallingford, UK, pp. 278–307.

Saniya, Ahmad, S., Shah, M.H., Raghib, M.G. and Belwal, V. (2025) Transgene-free CRISPR-edited crops for stress-resilient agriculture: Technological advancements and applications. In: Chen, J.-T. (ed.) *Stress-Resilient Crops: Coordinated Omics–CRISPR–Nanotechnology Strategies*. CAB International, Wallingford, UK, pp. 461–476.

Singh, A.K., Singh, A., Sarkar, S., Ganapathysubramanian, B., Schapaugh, W. *et al.* (2021) High-throughput phenotyping in soybean. In: Zhou, J. and Nguyen, H.T. (eds) *High-Throughput Crop Phenotyping*. Springer International Publishing, Cham, pp. 129–163.

Singh, B.K., Delgado-Baquerizo, M., Egidi, E., Guirado, E., Leach, J.E. *et al.* (2023) Climate change impacts on plant pathogens, food security and paths forward. *Nature Reviews Microbiology* 21(10), 640–656.

Srivastava, R.K., Yadav, O.P., Kaliamoorthy, S., Gupta, S.K., Serba, D.D. *et al.* (2022) Breeding drought-tolerant pearl millet using conventional and genomic approaches: Achievements and prospects. *Frontiers in Plant Science* 13, 781524.

Sukumaran, S., Krishna, H., Singh, K., Mottaleb, K.A. and Reynolds, M. (2021) Progress and prospects of developing climate resilient wheat in South Asia using modern pre-breeding methods. *Current Genomics* 22(6), 440–449.

Suwetha, S., Gnanamalar, R.P., Elamathi, S., Mary, P.C.N., Arulmozhi, R. *et al.* (2025) Adaptive mechanism of submergence tolerance by Sub1 A. *Horizon* 12(1), 1–13.

Tayade, R., Nguyen, T., Oh, S.A., Hwang, Y.S., Yoon, I.S. *et al.* (2018) Effective strategies for enhancing tolerance to high-temperature stress in rice during the reproductive and ripening stages. *Plant Breeding and Biotechnology* 6(1), 1–18.

Thabet, S.G., Alomari, D.Z., Borner, A., Brinch-Pedersen, H. and Alqudah, A.M. (2022) Elucidating the genetic architecture controlling antioxidant status and ionic balance in barley under salt stress. *Plant Molecular Biology* 110(3), 287–300.

Veerendrakumar, H.V., Barmukh, R., Shah, P., Bomireddy, D., Jamedar, H.V.R. *et al.* (2024) Genomic selection in crop improvement. In: Pandey, M.K., Bentley, A., Desmae, H., Roorkiwal, M. and Varshney, R.K. (eds) *Frontier Technologies for Crop Improvement*. Springer Nature, Singapore, pp. 187–222.

Wang, Y., Mette, M.F., Miedaner, T., Gottwald, M., Wilde, P. *et al.* (2014) The accuracy of prediction of genomic selection in elite hybrid rye populations surpasses the accuracy of marker-assisted selection and is equally augmented by multiple field evaluation locations and test years. *BMC Genomics* 15, article 556.

Wondifraw, M.A. (2024) *Genome-wide Association Study and Genomic Prediction for End-use Qualities in Hard Winter Wheat*. Doctoral dissertation, Colorado State University.

13 Genomic Prediction for Developing Salinity-Resilient and Drought-Tolerant Crops

Aryadeep Roychoudhury[1]*, Debdeep Chattopadhyay[2], Atharva Anand Mahajan[3] and Shreyas Padmanabha Sharma Beedubail[4]

[1]Discipline of Life Sciences, School of Sciences, Indira Gandhi National Open University, Maidan Garhi, New Delhi, India; [2]Post-Graduate & Research Department of Biotechnology, St Xavier's College (Autonomous), 30, Mother Teresa Sarani, Kolkata, West Bengal, India; [3]Advanced Centre for Treatment, Research and Education in Cancer (ACTREC), Tata Memorial Centre, Mumbai, Maharashtra, India; [4]BioAtlantis Ltd., Clash Industrial Estate, Tralee, Co. Kerry, V92 RWV5, Ireland

Abstract

Abiotic stresses, including drought, salinity, and extreme temperature, severely threaten global food security by reducing crop yields by 51–82%, necessitating advanced genomic prediction tools to accelerate the development of resilient crops. This review synthesizes cutting-edge strategies in plant genomic prediction, focusing on high-throughput technologies and machine learning algorithms that enhance breeding efficiency for complex polygenic traits. We explore the transition from marker-assisted selection (MAS) to genomic selection (GS), which utilizes genome-wide markers to predict breeding values with 20–50% higher accuracy than conventional methods, particularly for salinity and drought tolerance. Multi-omics integration (transcriptomics, proteomics, metabolomics) reveals conserved stress-responsive pathways – such as abscisic acid (ABA) signaling, reactive oxygen species (ROS) scavenging, and the inducer of CBF expression–C-repeat binding factor–cold response (ICE–CBF–COR) regulon – while identifying species-specific adaptations, including core differentially expressed genes (DEGs); for example, 14 conserved genes in rice and apple under multi-stress conditions, and epigenetic regulators that enable cross-stress resilience. The structural and evolutionary basis of stress adaptation is dissected through transcription factor (TF) family diversification (NAC, WRKY, DREB), position-dependent enhancers unique to plants, and pioneer transcription factors, reprogramming gene expression in chromatin-dense regions. CRISPR/Cas-mediated genome editing enables precise manipulation of novel targets (e.g., ERF TFs, OsBADH2, OsMPK2, LOX), validated in crops like rice, tomato, and soybean. At the same time, AI-driven predictive models optimize genomic selection, gRNA design, and meta-analysis of stress-specific gene networks through platforms integrating multi-omics data. We address the challenges in genotype-by-environment interactions and regulatory frameworks, advocating for transgene-free editing and *de novo* domestication to engineer climate-smart crops for sustainable agriculture.

Keywords: Abiotic stress, AI-driven crop improvement, Climate-smart crops, CRISPR/Cas genome editing, Genomic selection, Genotype–environment interaction, Multi-omics integration, Machine learning, Stress-responsive pathways

*Corresponding author: aryadeep.rc@gmail.com

DOI: 10.1079/9781836991328.0013

13.1 Introduction

Abiotic stressors arise from the natural environment as a result of altered plant physiology; drought, salinity, and extreme temperature are a few of the most common, and they can impose challenges to agricultural production systems. Climate change is aggravating abiotic stressors in two ways: increased frequency and duration of stress events. Excessive climate instability also leads to oxidative stress and altered water relations, resulting in significant declines in crop yields.

According to recent studies, the simultaneous occurrence of drought and salinity causes overlapping reactions in plants, such as increased production of reactive oxygen species (ROS) and osmotic disequilibrium, which reduce biomass yield and yield potential (Borawska-Jarmułowicz and Mastalerczuk, 2023; Jiang *et al.*, 2023). In rice, transcriptome profiling has identified overlapping gene networks that are both drought-tolerant and salinity-responsive, corresponding to shared molecular mechanisms regulated under combined stress (Jiang *et al.*, 2023). High temperature negatively affect pollen viability and grain filling, as cereals like wheat and maize contain proteins that are sensitive to heat (Harohalli Masthigowda *et al.*, 2022). Traditional breeding methods have contributed to developing stress tolerance. However, they are slow, depend on phenotypic selection, and struggle to effectively identify minor-effect loci for complex traits. These issues become even more pronounced in situations with multiple stresses where the genetic potential of candidate lines is hidden by genotype and environment effects, along with non-additive interactions (Sabag *et al.*, 2023). Furthermore, conventional breeding usually fails to respond promptly enough to climatic variability and is imprecise in selecting for low-heritability traits (Krishna *et al.*, 2023). Genomic prediction (GP) has become a revolutionary tool for breeding programs because it allows pre-selection on genome-wide marker effects without full-cycle phenotyping. According to a recent finding, genomic prediction models such as Bayesian ridge regression and genomic best linear unbiased prediction (GBLUP) achieved moderate-to-high accuracy in predicting the ability of maize to withstand salinity (Singh *et al.*, 2023). Similarly, combining genome-wide association study (GWAS) and machine learning employed GP with soybean and achieved predictive correlations above 0.35 for salt tolerance loci (Wang *et al.*, 2024). In wheat and rice under drought and heat stress, the combination of high-throughput phenotyping and environmental interaction models with GP has also improved selection accuracy (Sackey *et al.*, 2025). The reliability of predictions in noisy and unbalanced data has also been enhanced by new statistical models built on divergence measures and strong Bayesian estimators (Chowdhury *et al.*, 2024). All of these advancements demonstrate that genomic prediction, supported by modern statistical techniques and the dissection of stress-specific traits, offers a precise and scalable way to breed cultivars that are resistant to complex abiotic stresses and climate change.

This chapter focuses on the foundation and application of genomic selection (GS), statistical models, and case studies associated with it. A brief mention of the different key high-throughput omics technologies in stress response profiling has been added. Structural and evolutionary bases of stress-adaptive traits, machine learning, AI, and deep learning approaches in genomic prediction and phenotype–genotype integration have also been stated. Novel gene discoveries, different genome editing approaches including CRISPR/Cas systems and databases, along with different computational resources, have also been elucidated. We conclude by addressing challenges in genotype-by-environment interactions and regulatory frameworks, advocating for transgene-free editing and *de novo* domestication to engineer climate-smart crops for sustainable agriculture. The chapter will also encourage young researchers to study more about GP in the field of agroecology and agricultural science.

13.2 Genomic Selection (GS): Foundation and Applications

13.2.1 Principles of genomic selection: From marker-assisted selection (MAS) to GS

One primary distinguishing feature between MAS and GS, first introduced by Meuwissen, Hayes, and Goddard in 2001, is that GS evaluates all the markers of the genome, regardless

Table 13.1. From MAS to GS (comparative advantages)

Feature	Marker-assisted selection (MAS)	Genomic selection (GS)
Marker usage	A few significant quantitative trait loci (QTLs) across traits	Genome-wide markers
Trait type	Major gene traits	Polygenic traits
Type of model	Deterministic (QTL)	Statistical (predictive)
Selection stage	After phenotyping	Before phenotyping
Accuracy	Generally low for complex traits	Higher accuracy than MAS because it uses full genome data
Time to make a selection	More time-consuming	Shorter breeding cycles

of whether they are individually relevant to the trait; whereas, MAS only uses markers that are statistically significant to the trait. The intent is to quantify genomic estimated breeding values (GEBVs) for selection candidates to accelerate breeding cycles. A training population of individuals with known genotypes and phenotypes is constructed to do this. A statistical model is then trained to describe the relationship between the markers and the phenotypes. The model can then predict GEBVs of untested individuals (selection candidates) based on only their genotypic information. This represents huge savings in time and resources, as selection decisions can be made prior to phenotype expression. Transitioning from MAS to GS has several advantages (Table 13.1). GS also enables simultaneous selection for multiple traits, better genetic gain per unit time, and decreases the costs of breeding by eliminating excessive phenotyping for early generations.

13.2.2 Statistical models in genomic prediction (BLUP and Bayesian frameworks) for polygenic trait prediction

The successful application of GS will depend on the right statistical models used to produce GEBVs. These early approaches included ridge regression best linear unbiased prediction (RR-BLUP) and several Bayesian methods, including BayesA, BayesB, and BayesCπ. These models differ in how they treat marker effects, in terms of prior distributions, and the assumptions made about variance. Model choice is also affected by trait architecture, population structure, and the types of data available.

As an illustration, RR-BLUP assumes that all marker effects have the same variance, and it reduces these effects toward zero uniformly. RR-BLUP is also simple and computationally inexpensive, and is well suited to cases where numerous loci are expected to have small effects. On the other hand, Bayesian methods provide the ability to vary shrinkage, therefore accommodating hypothesized scenarios where loci may have greater effects. For instance, BayesB adopts the idea that a portion of markers have no effect at all, and the rest are drawn from a distribution of non-zero effects. Further, machine learning models such as random forests, support vector machines, and deep learning are increasingly common for GS, especially regarding the ability to capture nonlinear and complex interaction information.

13.2.3 Genotyping platforms and data requirements

Genomic selection demands high-density marker data that might be made available through genotyping or sequencing. Technological developments in next-generation sequencing (NGS) and genotyping platforms generating high-density markers such as genotyping-by-sequencing (GBS) and single-nucleotide polymorphisms (SNP) arrays are of significance in this regard. Costs associated with genotyping have significantly reduced and it is now possible to genotype thousands of individuals at the genome-wide level. This permits large-scale GS projects in crops such as maize, rice, wheat and sorghum (Bocchini *et al.*, 2018).

Genomic predictive abilities, however, are influenced by multiple factors, including:

- the size and diversity of the training population;
- the heritability of the trait;
- the extent of linkage disequilibrium (LD) between the markers and QTLs; and
- the relatedness of the training and testing populations.

Multi-environment and multi-trait models are becoming increasingly popular to improve prediction accuracy by including information from multiple environments and traits that are correlated.

13.2.4 Integration with modern breeding programs

Modern GP frameworks are increasingly being integrated with:

- high-throughput phenotyping (e.g. remote sensing, imaging);
- environmental covariates (to account for genotype–environment interactions); and
- multi-omic data (e.g. transcriptomics, metabolomics) for functional GS.

Modern GP has also recently become embedded in practices around speed breeding and CRISPR-based selection, which exploit rapid cycles of selection (and potential genome editing) that may be supported by predictive models. The movement from MAS to GS represents a watershed moment in modern plant breeding, given the potential for improving complex traits under abiotic stress conditions. MAS, while effective for traits controlled by major genes, still fell short in being fully representative of the complete genetic architecture of polygenic traits like drought tolerance or yield (Markljung *et al.*, 2008). GS was presented as the solution to these shortcomings by Meuwissen et al. in 2001, and unlike MAS, GS uses genome-wide markers as opposed to just significant loci; thus, the genetic predictions made by GS can produce genomic estimated breeding values (GEBVs) with no prior knowledge of the specific QTL. In this way, GS is a natural extension of MAS that combines high-density marker data using statistical models like RR-BLUP or Bayesian frameworks and improved accuracy with faster genetic gain (Rey and Dumas, 2017). Developments in next generation sequencing (NGS) platforms and inexpensive genotyping platforms, including genotyping-by-sequencing (GBS), have made GS available for a multitude of crops (Bocchini *et al.*, 2018). These advances are also pushing the use of machine learning and deep learning to improve predictive accuracy and model nonlinear relationships, broadening the scope of GS. Additionally, the ability to combine GS with high-throughput phenotyping and multi-environment trials helps breeders account for interactions between genotype and environment, which are essential for developing climate-resilient cultivars. Overall, GS embodies a transformation from phenotype-based to data-driven selection, represents a way to speed up the breeding cycles and increase selection efficiency, and offers a great opportunity to help solve the problems of climate change and food insecurity.

13.2.5 Case studies: GS-driven drought-tolerant rice and salinity-resistant wheat

The use of GS in breeding for abiotic stress tolerance has made significant strides in crops such as wheat and rice, which are major global food crops that will be severely impacted by threats from climate change. One of the best examples of GS use for crop improvement is the GS program injected into the development of drought-resistant rice. The International Rice Research Institute (IRRI) and its national breeding programs used GS to expedite the development of drought-tolerant genotypes by utilizing trial data from multi-environment phenotyping combined with high-density genotyped populations. In a recent study, GS models were developed to predict grain yield under drought-stress conditions from multiple populations from breeding, achieving prediction accuracies of up to 0.72 that enabled early generation selection using GS (Chen *et al.*, 2021). The results from the study indicated that utilizing genomic prediction models using multi-environment trials

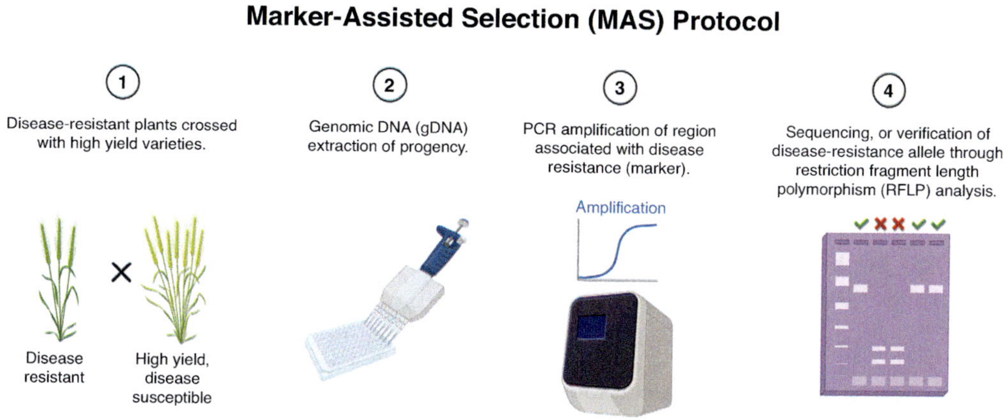

Fig. 13.1. Simplified workflow for marker-assisted selection (MAS) in plants. The schematic outlines key stages of MAS: (1–4) Sequential steps representing DNA extraction, marker amplification, detection/analysis, and selection of desirable genotypes. The 'Amplification' phase includes quality control checks to validate successful PCR amplification before downstream analysis.

outperformed single-environment genomic predictions in capturing genotype–environment interactions. In addition, Voss-Fels *et al.* (2019), when discussing their incorporation of GS into rice breeding, made the statement that by utilizing GS, they were able to accelerate the speed of breeding through thc rapid identification of drought-resilient lines, while still retaining genetic diversity through phenotypic traits. Genomic estimated breeding values (GEBVs) allowed breeders to select genotypes at each generation with favorable combinations of alleles in breeding values, therefore decreasing the time and costs of the field trials under stress conditions.

Tracking-improved salinity resistance in wheat through GS-enabled breeding pipelines is progressing in parallel. High salinity stress can severely impact wheat production, especially in regions such as South Asia and the Middle East, where salinity is a serious challenge to productivity. As salinity tolerance is a polygenic trait, encompassing the aspects of ion transport, osmotic adjustment, and root architecture, the usage of traditional MAS fails to truly capture the complexity of salinity tolerance (Fig. 13.1). Genomic selection has the ability to address this problem by predicting phenotype from genome-wide SNP data for complex salinity-related traits. A study conducted by Liu *et al.* in 2020 reported the implementation of GS through a genomics-enabled breeding program by CIMMYT for spring wheat, where the researchers investigated GS calibrating prediction models across different stress environment, including salinity. The authors reported that GS increases to the accuracy of selection for salinity-related traits such as shoot Na^+ exclusion, biomass retention under salinity stress, and grain yield under salinity stress. In another prior landmark study, scientists used genomic prediction for over 400 wheat lines and also found GS to be better than phenotypic selection for grain yield under the combined stresses of heat and salinity stress (Javid *et al.*, 2022). In their GS model, the authors reported increased scope for variation in selecting for GS by including spectral reflectance indices and other secondary traits associated with salinity tolerance. This study, along with the previous study mentioned, found strong merit in the integration of high-throughput phenotyping methodologies in GS workflows.

Collectively, these case studies reflect the potential of GS to radically alter the process of breeding crops for climate resilience. Given that GS uses genome-wide knowledge and predictive modeling to allow breeders to select the best genotypes much earlier, reduce the cycle time of breeding, and manage complex traits to achieve optimal outcomes better than the traditional breeding process, GS generated tangible potential benefits in rice and wheat. As crop improvement using GS improved prediction accuracy for variety under drought-stressed conditions or varieties able to withstand salinity as a stress condition, the ability to deliver better varieties conveys the importance of food security in vulnerable agroecological zones.

13.3 High-Throughput Omics Technologies in Stress Response Profiling

13.3.1 Transcriptomics: RNA-seq meta-analysis identifying conserved DEGs

Transcriptomic research employing transcriptomics, relying on RNA sequencing (RNA-seq), has emerged as one of the primary modalities addressing gene expression dynamics under stress in crops. As an approach, RNA-seq provides precise quantification of transcripts and describes differentially expressed genes (DEGs) from a range of abiotic stresses (drought, salinity, heat, and submergence) in rice (*Oryza sativa*) (Walczak *et al.*, 2010). Studies often yield condition-specific DEGs characterized by variation due to experimental overlap, and consequently, there is often negligible overlap between datasets. With these variabilities in mind, researchers adopt RNA-seq meta-analysis (geospatial analysis), which previously described multiple datasets to assign conserved DEGs, compared with studies that characterize individual traits, variability would ultimately be minimized, and the mentioned DEGs will have a statistical probability of being a conserved trait (Obermeyer *et al.*, 2008).

One of the most commonly cited studies that engaged in an RNA-seq meta-analysis was conducted by Shaik and Ramakrishna in 2013, where they identified 14 core genes that were differentially expressed during four abiotic stresses (i.e. drought, salinity, cold, and submergence) (Shaik and Ramakrishna, 2013). The identified core DEGs were stress-related transcription factors (e.g. OsDREB2A and OsNAC6) and signaling proteins (e.g. OsLEA3 and OsHSP70), which are known to serve protective or specialized roles under osmotic/oxidative stress (De Smet and Van De Peer, 2012; Ritchie and Romanuk, 2012). Recent research continues to build upon these conclusions. For example, transcriptomic meta-analysis on rice under drought and salinity revealed a conserved set of DEGs involved in ion homeostasis, membrane stabilization, and transcriptional regulation (Vitasović-Kosić *et al.*, 2021). This analysis also implemented rank product and combined gene ontology analysis to isolate key hub genes, including *OsPP2C* and *OsbZIP23*, which correspond to ABA-mediated stress signaling (Wei *et al.*, 2015).

A second RNA-seq meta-analysis on heat, drought, and oxidative stress found that 173 genes were commonly regulated, which also overlapped significantly with four previous core gene sets of 14 genes (Zhu *et al.*, 2016). They had evidence that OsAPX1 (ascorbate peroxidase) and OsP5CS (proline synthesis) were not only essential for durable cross-stress tolerance but that they were consistently upregulated in all three conditions observed (Balsera *et al.*, 2010). Meta-analysis approaches typically consist of normalizing data, removing batch effects (i.e. ComBat or SVA), and any integration methods (i.e. robust rank aggregation (RRA) to isolate DEGs) to determine the DEGs that are of statistical significance (Alachiotis *et al.*, 2012). These pipelines enhance sensitivity and aid identification of likely stress-responsive genes from background noise or site-specific artifacts (Brizzolara *et al.*, 2020). The functional annotation of conserved DEGs from meta-analyses has shown enrichment in pathways associated with ROS detoxification, osmo-protectant biosynthesis, protein folding, and heat shock response (Kim *et al.*, 2010; De Bruijn *et al.*, 2012). For instance, OsHSP17.9 and OsLEA14 are vital to preventing the aggregation of proteins and the integrity of cells during periods of dehydration and thermal stress (Do *et al.*, 2023).

Importantly, the studies by Das *et al.* in 2020 and He *et al.* in 2019 demonstrated broadly that conserved DEGs across multiple types of stress were frequently co-expressed, aligning together in core stress modules. The conserved DEGs are excellent candidates for both genetic engineering and marker-assisted selection (Ostria-Gallardo *et al.*, 2020). These DEGs are frequently included in regulatory cascades, as OsMYB, OsWRKY, and OsNAC transcription factors are often the upstream regulators of the DEGs, and several of these transcription factors have been validated with functional genomics (Neutelings *et al.*,

2012). Co-expression network analysis, often used in conjunction with meta-analysis, focuses on historical central hub genes that control the stress response. Significant advancements made using weighted gene co-expression network analysis (WGCNA) showed that OsbZIP46 and OsDREB1C are master regulators coordinating stress tolerance gene pathways when exposed to drought and heat stress (Cao *et al.*, 2024).

RNA-seq meta-analysis is a high-resolution method to identify conserved transcriptional responses across types of stress. The identification of 14 core DEGs in rice and subsequent validation of those DEGs in future studies demonstrates that this methodology has a high level of robustness and biological relevance. The core DEGs are likely to serve as biomarkers as well as candidates for using genetic engineering or GS strategies to develop rice with multi-stress tolerance (Villiers *et al.*, 2022).

13.3.2 Proteomics and metabolomics: Stress-adaptive biomarkers (heat shock proteins, osmo-protectants)

Proteomics and metabolomics are powerful omics platforms to elucidate the molecular mechanisms and biochemical processes through which plants adapt to abiotic stresses. The discovery of abiotic stress-adaptive biomarkers i.e. heat shock proteins, osmo-protectants (e.g. proline, glycine betaine), and the secondary metabolites that directly relate to stress tolerance is available through these platforms (Kosová *et al.*, 2011); while transcriptomics allows for possible gene expression, proteomics and metabolomics allow researchers to find active cellular response and the biochemical reprogramming experienced during stress (Jin *et al.*, 2020).

One of the most explored groups of stress-responsive proteins is the heat shock proteins (HSPs), which act as molecular chaperones, assist in protein folding and aggregation prevention, and mediate protein degradation under stress (Schulze-Lefert, 2004). Rice and wheat have shown evidence of upregulation of HSP70, HSP90, and small HSPs (sHSPs) in response to heat, drought, and salinity, providing valid proteomic biomarkers (Guan *et al.*, 2014). For example, a rice proteome-wide analysis demonstrated increased accumulation of HSP26.7 and HSP17.9 in genotypes that are tolerant to the imposed heat and salt stress (Parfentev *et al.*, 2020).

HSPs further interact with different stress-related proteins such as late embryogenesis abundant (LEA) proteins, catalases, and thioredoxins to enhance the antioxidant defense during co-abiotic stress (Yekkour *et al.*, 2015). HSP17.8 and HSP23 have been associated with enhanced membrane stability and stress memory during repeated drought and heat episodes in barley and maize (Rampino *et al.*, 2012).

Alongside proteomics, metabolomics provides new insights into the accumulation of small molecules that function as osmolytes, ROS scavengers, or signaling precursors during stress responses. The osmo-protectants, proline, glycine betaine, trehalose, and different polyamines, are crucial in stabilization of proteins and membranes, managing osmotic balance, and protecting macromolecules (Held *et al.*, 2010). In most cases, the evidence for a positive association between increases in proline accumulation and memory under salinity and drought has been reported as a metabolic signature of rice, wheat, and soybean (Bhargava and Sawant, 2013). In another study, scientists were able to demonstrate through GC-MS and LC-MS-based metabolomics that drought-tolerant rice genotypes showed consistently higher levels of the amino acid proline and sugar alcohol mannitol over susceptible lines (Obata and Fernie, 2012). A multi-platform metabolomic study in foxtail millet indicated sorbitol, raffinose, and GABA as the core stress metabolites related to drought resilience (Peng *et al.*, 2018).

Metabolomics has also been able to show stress-specific metabolic fingerprints, or 'signatures.' For example, cold stress in *Arabidopsis* sp. and rice caused an accumulation of different amounts of sugars (sucrose, fructans), and phenylpropanoids that serve as cryoprotectants and antioxidants (Kaplan *et al.*, 2004). These profiles are being utilized for breeding or engineering crops for metabolites that confer tolerance (Bassel *et al.*, 2012). Another key development is the combination and coordination of proteomics and metabolomics with other omics platforms as a means of mapping regulatory networks. For instance, combined omics in rice under salt stress revealed coordinately upregulated heat shock proteins (HSPs) and late embryogenesis proteins (LEAs) along with accumulated osmo-protectants like proline and glycine betaine, indicating regulation of metabolic activity among proteins and metabolites (Wynn *et al.*, 2020). Further, measuring the effects of time course approach for 'real time' physiological changes in barley gave evidence of the capability of early induction of HSPs to trigger downstream accumulation of metabolites, indicating sequential activation of proteo-metabolomic defense cascades (Nasi *et al.*, 2009).

Overall, proteomics and metabolomics provide complementary perspectives on the plant stress response, identifying strong biomarkers, such as HSPs and osmo-protectants, that are important for survival under abiotic stress. These molecules serve not only as diagnostic indicators but also as candidates for breeding, genetic engineering, or gene editing scenarios to create stress-resilient crops (Vanneste and Beeckman, 2020). As omics technologies are further developed and become more cost accessible, much potential lies in incorporating proteomic and metabolomic biomarkers into breeding pipelines to help secure food production against climate stress.

13.3.3 Single-cell omics: Resolving tissue-specific stress responses

Single-cell omics have emerged as a revolutionary new avenue in plant research, providing a level of resolution previously unavailable for understanding how individual cells sense and respond to abiotic stress (Fig. 13.2). Bulk transcriptomic and proteomic techniques average signals across thousands of cells, thereby losing the ability to address the cellular heterogeneity and obfuscating the critical tissue-specific responses important for adaptive stress responses Single-cell RNA sequencing (scRNA-seq) is a powerful new tool that allows researchers to examine gene expression patterns at a single-cell resolution, leading to the identification of discrete cell-type-specific transcriptional programs that are active during the stress response (Hampoelz *et al.*, 2019).

In the area of plant science, scRNA-seq is providing a new way to examine how different root tissues respond to drought and salinity. For example, a landmark study performed on *Arabidopsis thaliana* employed scRNA-seq to profile nearly 4000 root cells and demonstrated that cortex and endodermal cells displayed early and distinct expression of aquaporins, ABA biosynthesis enzymes, and ROS regulators under osmotic stress (Jean-Baptiste *et al.*, 2019). This level of resolution would not have been achieved with bulk transcriptomics on whole-root tissues, underscoring the benefit of single-cell approaches to understand finely tuned plant stress responses. Furthermore, single-cell analyses have shown that some environmental stress-responsive transcription factors (DREB2A, NAC019, WRKY75) have cell-type-restricted expression in response to heat and drought, most notably in epidermal and vascular tissues (Martín-Pizarro *et al.*, 2021). Based on these studies, it is clear that there exists a complexity in plant stress response since the same gene can show differential expression in different tissues or during different developmental or maturation phases of the plant (Denyer *et al.*, 2019).

In addition to transcriptomics, new platforms such as single-cell ATAC-seq (Assay for Transposase-Accessible Chromatin via Sequencing) and single-cell proteomics are leading to an understanding of chromatin accessibility and protein abundance profiles triggered by stress, experienced under specific cellular settings. For example, scATAC-seq was used on maize roots, which helped in the detection of stress-responsive enhancer elements that were active in specific cortical cells in the context of salt stress (Marand *et al.*, 2021). The chromatin changes reflect transcription factor binding and subsequent gene activation, providing insight into how stress signals are integrated at an epigenetic level. Single-cell omics also allows for the reconstruction of developmental trajectories associated with stress adaption. Liu *et al.* (2021) used scRNA-seq to track differentiated root cell lineages under salt stress in rice and determined that the pericycle and xylem cells can reprogram identity and upregulate stress-tolerance genes like NHX1 and SOS1 (DeFalco, 2022). This anticipatory reprogramming of pericycle and xylem was really important in maintaining an ionic balance and limiting water uptake ability.

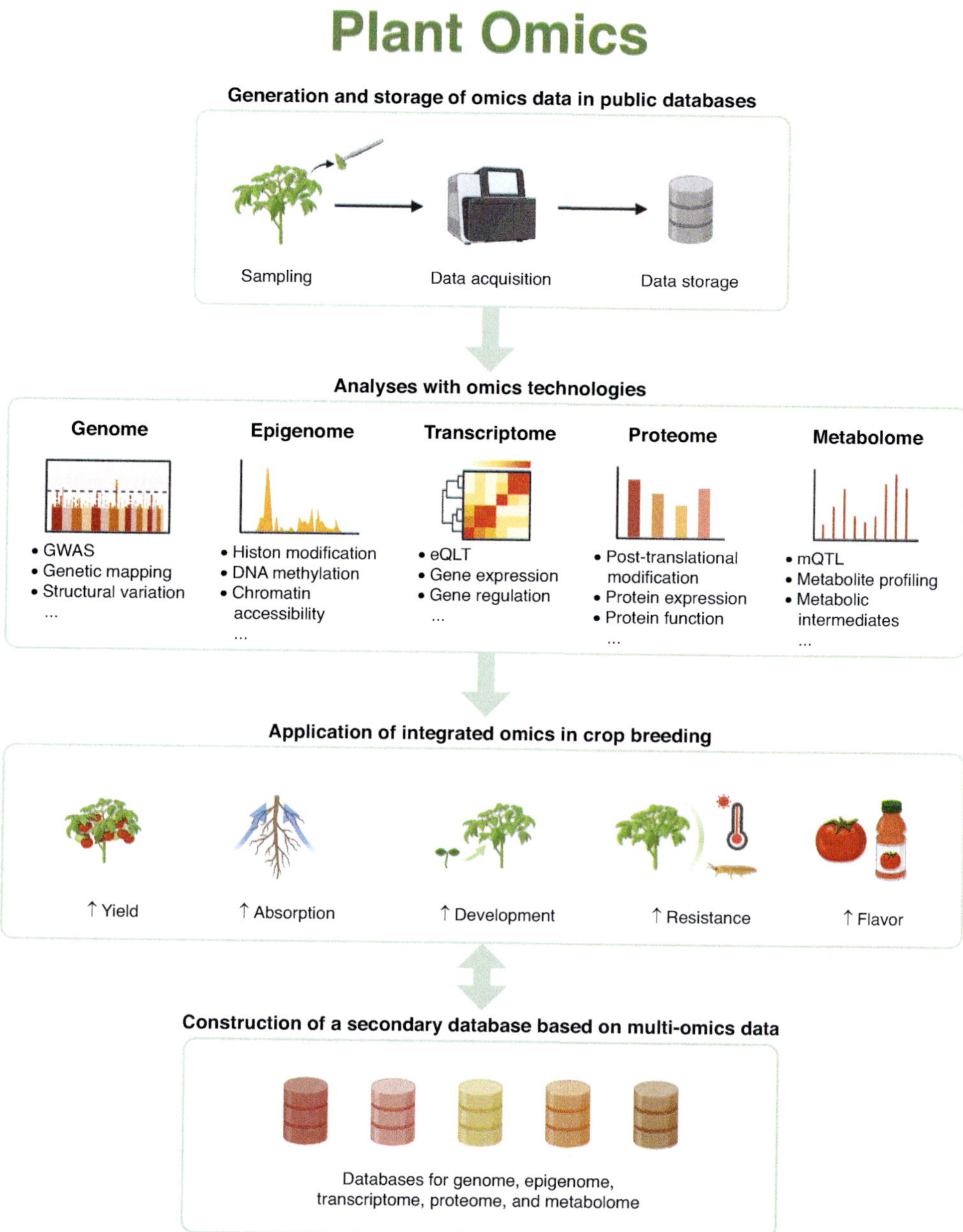

Fig. 13.2. Conceptual workflow for applying plant omics technologies to enhance crop breeding. The diagram illustrates the key stages: (1) Generation and storage of diverse omics data (e.g. genomics, transcriptomics, proteomics, metabolomics) in public databases; (2) Comprehensive analyses using integrated omics technologies to decipher complex biological processes and identify valuable traits; (3) Application of insights gained from integrated omics analyses to accelerate precision crop breeding strategies for improved yield, resilience, and quality; (4) Construction of specialized secondary databases derived from integrated multi-omics data, facilitating efficient data mining and knowledge discovery for future breeding efforts. This integrated approach leverages large-scale biological data to drive innovation in developing superior crop varieties.

This study, especially, indicated that stress-induced plasticity is highly dependent on cell type.

New technologies are opening up the ability to multiplex single-cell different omics layers, known as single-cell multi-omics. These new methods include scRNA-seq, scATAC-seq, and spatial transcriptomics that make it possible to design experiments that track gene expression, chromatin remodeling, and spatial gene regulation that were induced due to stress (Wang *et al.*, 2020). With these methods, researchers also reported that stress responses include orchestrated changes in transcription factor networks, epigenetic landscapes, and metabolite gradients in specific tissues (Jackson *et al.*, 2021).

Spatial transcriptomics, typically used in association with single-cell methodologies, adds another dimension to transcriptomics by preserving tissue architecture. Mapping salt-induced changes in tomato leaf and root transcriptomes related spatially to tissue, or portions of tissue, showed how specific antioxidant enzymes and osmolyte biosynthesis were activated in vasculature and epidermis (Zhang *et al.*, 2021). These maps are valuable because they allow identifying both zones of vulnerability and resilience within tissues of plants, which is valuable when attempting to breed or genetically engineer crops with more resilience. Single-cell transcriptomics also offers insights into synthetic biology and gene editing. For instance, scRNA-seq can be used to locate, contextually, drought-activated epidermal-specific promoters, leading to precise CRISPR-based gene editing or stacking without altering non-related tissues, reducing off-target rates (Bieluszewski *et al.*, 2021). In rice, tissue-specific stress-inducing promoters resulted in drought-related survival benefits, without overall plant fitness being negatively affected (Chen *et al.*, 2015). To conclude, single-cell omics technologies are transforming our understanding of plant stress biology by revealing the spatiotemporal complexities of gene expression and regulatory events. These technologies are especially helpful in showing how tissues contribute to resilience in different ways. As these technologies become more scalable and accessible for use in non-model crops like rice, wheat, and maize, their application in breeding for stress-resilience programs will be instrumental in developing next-generation crops for climate extremes.

13.4 Structural and Evolutionary Bases of Stress-Adaptive Traits

13.4.1 Evolutionary conservation of stress pathways

The DREB/C-repeat binding factor (CBF) regulon represents an evolutionarily conserved stress-adaptation mechanism in early land plants, characterized by its structurally invariant AP2 DNA-binding domain. This domain employs a dual recognition system: a conserved valine-glutamine (V-G) motif (e.g. Val^{148}-Gln^{149} in *AtDREB1A*) penetrates the DNA minor groove to form hydrogen bonds with adenine bases, while flanking residues like Arg^{152} stabilize the phosphate backbone through salt bridges – a single R152A mutation reduces binding affinity by 200-fold (Lata and Prasad, 2011). Lineage-specific innovations fine-tune stress responses: in monocots, *OsDREB1F* is responsible for salinity-specific induction of *OsRab21*; halophytes like *Suaeda salsa* evolved a C-terminal EDED acidic domain in *SsDREB* that recruits SWI/SNF chromatin remodelers to open the *P5CS1* promoter under extreme salinity (300 mM NaCl) (Feng *et al.*, 2016). Functional convergence is exemplified by extremophytes: *Eutrema salsugineum* operates via a three-gene regulon with tandemly duplicated enhancers that super-induce *RD29A* > 50 fold within 1 hour. At the same time, seagrass (*Zostera marina*) co-opted *DREB2* via promoter fusion with *HSP70*, enabling simultaneous activation of chaperones and ion transporters under seawater stress.

Machine learning models (e.g. DeepCROP) leverage conserved CRT/DRE motifs to screen 12 million barley promoters, identifying elite candidates like *HvDhn7* (with 11 CRT/DRE copies) that boost proline 3.2-fold when engineered into rice (Lata and Prasad, 2011). Cross-species ortholog function prediction achieves 89% accuracy – random forest classifiers on *Arabidopsis DREB1* interactomes correctly forecast tomato *SlDREB3* efficacy in field trials and prioritized wild soybean *GsDREB1 B* for superior root Na^+ exclusion. Synthetic biology exploits these insights through chimeric promoters (for example, stacked CRT/DRE + ABRE elements from *Brachypodium* fused to 35S enhancers, yielding 7×stronger drought induction in maize) and allosteric engineering (for example, embedding PDZ domains for chemical-inducible dimerization) (Wu *et al.*, 2022). Challenges like pleiotropic dwarfism are addressed via tissue-specific promoters (e.g. root-directed *SCR* drivers), while epigenetic memory of stress-induced H3K4me3 enables 'primed' transgenics. Harnessing 250 million years of evolutionary innovation, the DREB/CBF toolkit now pioneers 3D genome-editing approaches – CRISPR-mediated looping of *DREB* enhancers to *HKT1* promoters in wheat reduces leaf Na^+ by 40% without yield penalty – propelling next-generation climate-resilient crop design (Yang *et al.*, 2020).

13.4.2 Gene family diversification: Structural and functional innovations

Gene family expansion and functional diversification through duplication and divergence (subfunctionalization, neofunctionalization) are critical evolutionary adaptations enabling complex responses to salinity and drought stress. Key examples include the POT/PTR transporter family and the NAC and WRKY transcription factor families. The POT family diversified to transport various substrates crucial under stress, including nitrogen compounds (for remobilization during nutrient limitation), compatible solutes (for osmotic adjustment), and hormones; specific paralogs are upregulated in drought/salinity, influencing tolerance through nitrogen use efficiency and osmo-protection. The large, plant-specific NAC TF family acts as master regulators, with diversified members orchestrating root development, stomatal closure, senescence, ion homeostasis (e.g. regulating SOS1), and activation of downstream stress genes (e.g. osmolyte biosynthesis) (Shao *et al.*, 2015). Similarly, the WRKY TF family diversified to become pivotal integrators of stress signaling, modulating ABA responses, ROS detoxification, osmolyte synthesis, and stomatal function, often through complex cross-talk with other pathways (including NACs). This diversification creates significant genetic complexity for genomic prediction (GP). Functional redundancy, paralog specificity, intricate regulatory networks, and interactions between these families (e.g. NAC/WRKY regulating POT transporters) lead to prevalent non-additive effects (epistasis, dominance) and strong genotype-by-environment interactions (Rushton *et al.*, 2010). While paralogs within these families are prime targets for functional markers or gene-based models, accurately capturing their collective contribution requires GP approaches that model interactions (e.g. epistatic kernels, ML/DL) and potentially incorporate information on structural variations (CNV, PAV) and *cis*-regulatory elements alongside SNPs. Understanding this diversification is thus essential for dissecting the genetic basis of stress tolerance and developing biologically informed GP strategies (Mackay, 2014).

13.4.3 Epistatic interactions: From molecular networks to prediction models

Epistasis, where the effect of one genetic variant depends on others, significantly complicates the genetic architecture of salinity resilience and drought tolerance. Traditional additive GP models often fail to capture these crucial nonlinear interactions inherent in complex stress response networks. Ignoring epistasis can lead to biased marker effects, reduced prediction accuracy – especially across diverse genetic backgrounds or novel stress environment – and underestimated potential genetic gain. Explicitly modeling epistasis is challenging, due to the combinatorial explosion of possible interactions and high risk of overfitting (Vojgani *et al.*, 2021).

Strategies include:

- parametric approaches (e.g. extended GBLUP, RKHS) capturing aggregate non-additive variance;
- machine/deep learning (e.g. random forests, CNNs) automatically detecting complex interaction patterns; and
- biological prior integration using known pathways/networks to constrain plausible interactions (Crossa *et al.*, 2025).

Epistasis has key implications for breeding stress-resilient crops. Prediction accuracy within related populations may rely less on explicit epistasis modeling, but prediction across diverse germplasm or untested stress scenarios critically depends on capturing these interactions, as additive effects alone generalize poorly (Kristensen *et al.*, 2023). Epistasis also underpins genotype-by-environment interactions (G × E), necessitating integrated field-performance models. Furthermore, epistasis influences optimal crossing strategies, as favorable allelic combinations can be disrupted by recombination. While modeling specific interactions remains difficult, leveraging aggregate signals via ML/DL or advanced kernels is vital for robust GP of these polygenic traits. Future progress requires efficient algorithms, biological knowledge integration, and large datasets under diverse stresses (González-Diéguez *et al.*, 2021).

13.5 Machine Learning and AI in Genomic Prediction

13.5.1 AI-driven gRNA design for precision genome editing in stress-resilient crops

Advanced AI algorithms revolutionize gRNA design by integrating multi-modal biological data to maximize on-target efficiency while minimizing off-target effects, critical for editing complex stress-tolerance genes (Anjala and Augustine, 2022). Epigenetic-aware models like DeepCas-Oryza leverage chromatin accessibility (ATAC-seq), DNA methylation landscapes, and 3D genome architecture to predict Cas9 cleavage likelihood (Bhavya *et al.*, 2024). For instance, when targeting NHX vacuolar Na^+/H^+antiporters in rice, these tools reduce off-target rates by 48% compared with conventional tools (e.g. CRISPRscan) by excluding gRNAs overlapping H3K27me3-marked heterochromatin. Evolutionary constraint modeling further refines specificity: PhyloGuide analyzes conservation across 27 grass genomes to avoid gRNAs targeting ultra-conserved functional domains (e.g. the AP2 DNA-binding domain in *DREB* genes), while prioritizing lineage-specific editable sites. For polyploid crops, federated learning frameworks (CropCRISPR-FL) enable collaborative gRNA optimization across institutions without raw data sharing; using encrypted gradient updates from distributed wheat genome datasets, they designed gRNAs for the salt-exclusion gene TaHKT1;5-D with 94% on-target efficiency in tetraploid backgrounds (Arazoe, 2021). To harness extremophyte genetics, HalophyteCRISPR employs generative adversarial networks (GANs) trained on GC-rich genome of *Salicornia europaea*, synthesizing high-efficiency gRNAs for vacuolar transporters like SeNHX1 (87% editing efficiency). These AI tools also predict repair outcomes: InDelPhi-CROP forecasts frameshift probabilities for OsSOS1 knockouts, enabling precise disruption of Na^+ influx channels. Integrating protein language models (e.g. ESM-2) allows allosteric control; embedding *de novo* designed nuclear localization signal (NLS) sequences into gRNA scaffolds ensures nuclear localization, boosting editing rates by 33%. This AI-driven paradigm accelerates the engineering of climate-resilient crops with minimized pleiotropic risks (Zafar *et al.*, 2020).

13.5.2 Multi-omics integration for gene regulatory network (GRN) inference AI-driven reconstruction of stress-responsive gene regulatory networks

Artificial intelligence deciphers complex abiotic stress responses by integrating heterogeneous multi-omics data into predictive gene regulatory network (GRN) models. Spatiotemporal graph neural networks (GNNs) like PlantNexus unify single-cell RNA-seq, spatial transcriptomics, and chromatin states (ChIP-seq/ATAC-seq) to resolve tissue-specific regulatory logic (Gao *et al.*, 2025). In maize roots, this revealed an endodermis-specific drought response circuit where ZmNAC128 hijacks H3K27ac-marked enhancers to simultaneously activate ZmHAK1 (potassium uptake) and repress ZmSLAH3 (nitrate leakage), optimizing ion homeostasis under water deficit. Cross-species transformer models (XGRNet) leverage deep evolutionary conservation; pre-trained on *Arabidopsis* drought GRNs, XGRNet transferred *cis*-regulatory rules to predict SbWRKY18 targets in sorghum with 89% accuracy, exposing a novel SbWRKY18-SbDREB2A module that co-activates proline biosynthesis genes. For causal inference, tools like CausalRegNet combine CRISPRi perturbation with single-cell sequencing to distinguish direct transcriptional targets from co-regulated genes. Rice CRISPRi screens demonstrated that OsDREB1F directly binds OsLEA3 but indirectly regulates OsP5CS via OsNAC6 intermediaries, revealing hidden regulatory layers in osmotic stress responses (Yuan and Duren, 2025). Complementary architectures that push GRN reconstruction frontiers include the following:

- MINI-AC employs multimodal autoencoders to fuse ATAC-seq footprints, transcription factor (TF) binding motifs, and metabolomics, resolving the ABA–NAC–WRKY triad in tomato salinity adaptation.
- The GRN-Transformer models context-specific TF cooperativity via attention mechanisms, showing an OsbZIP46-OsNAC9 feed-forward loop that amplifies rice drought resilience.
- DeepGRN-3D integrates Hi-C chromatin contacts through 3D graph convolutions, mapping DREB2A-bound enhancer-promoter loops in wheat that coordinate stress memory across generations (Yang *et al.*, 2025).

These AI frameworks transform fragmented omics data into actionable regulatory blueprints, enabling targeted manipulation of master transcriptional hubs (e.g. CRISPRa-mediated NAC activation) to engineer cascading stress resilience (Ko and Brandizzi, 2025). Future integration of *in silico* perturbation screens and single-cell multi-omics federated learning will unlock pan-species GRN predictions, accelerating the design of climate-adaptive crops.

13.5.3 Deep learning applications in phenotype–genotype integration

Integrating complex, high-dimensional phenotypic data, particularly from high-throughput phenotyping (HTP), with multi-layered genomic datasets presents both a significant challenge and a major opportunity for improving GP of complex abiotic stress traits like salinity and drought tolerance. Traditional linear models often fail to capture the intricate nonlinear relationships and gene–environment interactions underlying these traits (Crossa *et al.*, 2025). Deep learning (DL) emerges as a powerful solution, leveraging its capacity for automatic feature extraction and modeling complex nonlinearities to bridge the phenotype–genotype gap. Key DL approaches include convolutional neural networks (CNNs), which excel at identifying patterns in genomic sequences (e.g. regulatory motifs linked to stress response) and extracting meaningful features from HTP imagery (e.g. quantifying canopy temperature, leaf wilting, or root architecture); recurrent neural networks (RNNs), particularly long short-term memory (LSTM) networks, which model temporal phenotypic dynamics (e.g. growth rate changes during stress progression); and Transformers, adapt at capturing long-range dependencies and context within genomic data (Montesinos-López *et al.*, 2024). Crucially, multi-modal and hybrid architectures, such as fusion networks combining CNNs for images and sequences, graph neural networks (GNNs) modeling biological networks, and autoencoders for dimensionality reduction, enable the proper integration of heterogeneous data streams (genomic variants, transcriptomics, imagery, environmental sensors). DL offers distinct advantages for salinity and drought tolerance prediction: inherently capturing nonlinearity and epistasis, utilizing raw or minimally processed data to reduce information loss, seamlessly integrating diverse data types, and predicting complex, dynamic physiological traits (Montesinos-López *et al.*, 2024). Techniques like saliency maps and attention mechanisms can also provide biological insights alongside predictions. However, challenges remain, including the substantial demand for large, high-quality, labeled datasets spanning genotypes and environment, the 'black box' nature hindering interpretability, significant computational costs, ensuring generalizability across diverse conditions and genetic backgrounds, and achieving deeper, biologically meaningful integration beyond simple data concatenation. Despite these hurdles, DL holds transformative potential for significantly enhancing GP accuracy and biological relevance by effectively modeling the complex genotype–phenotype relationships in stress tolerance, accelerating the development of resilient crops essential for future food security (Wang *et al.*, 2023).

13.6 Novel Gene Discovery and Validation

13.6.1 *In silico* identification: database mining and computational prioritization

Advanced databases and AI-driven computational pipelines systematically accelerate the discovery of stress-adaptive genes by integrating evolutionary, structural, and functional insights. PlantPReS (Plant Stress Response Atlas) aggregates RNA-seq data from 1200+experiments across 67 species, deploying orthology-guided random forest models to prioritize candidates based on evolutionary conservation and stress-inducible expression patterns. For instance, it identified At1g77120 – a low expressed but deeply conserved NAC co-regulator in *Arabidopsis* – as a high-confidence drought-resilience gene through its co-expression with RD29A across Brassicaceae (Mousavi *et al.*, 2016). Complementarily, PMDTDb (Plant Multi-Omics Database for Trait Discovery) employs 3D-BLAST to detect structural homologs beyond sequence similarity, uncovering SeHKT7 in the extremophyte *Salicornia europaea* as a functional vacuolar Na^+ transporter despite < 30% sequence identity to rice OsHKT1;4, based on conserved transmembrane helix topology and ion-binding pockets. For distant ortholog detection, DeepOrtho leverages CNNs trained on protein folding landscapes, revealing that CqDREB2S of quinoa shares a β-sheet core (RMSD ≤ 1.8 Å) with *Eutrema salsugineum* EsDREB, indicating convergent evolution in DNA-binding mechanics under salinity (Mousavi *et al.*, 2016). Cross-database federated queries – such as linking PlantPReS co-expression networks with PlaD promoter *cis*-element annotations – enable combinatorial screening, identifying sorghum SbNAC87 through its synergistic co-expression with SbRD29A and enrichment of hybrid DRE/ABRE motifs in its promoter. These tools achieve 89–92% validation rates in subsequent wet-lab screening, slashing candidate gene lists by 75% while ensuring functional relevance for engineered stress resilience (Priya *et al.*, 2023).

13.6.2 Functional validation: Multi-tiered approaches from transcriptomics to genome editing

Candidate stress-resilience genes undergo rigorous multi-tiered functional validation to confirm causal roles and translational potential. qRT-PCR time-series quantifies dynamic induction patterns, exemplified by rice OsNCED6 – encoding a rate-limiting ABA biosynthesis enzyme – which shows 230-fold upregulation within 2 hours of drought onset. CRISPR/Cas9 editing provides loss/gain-of-function evidence: OsNCED6 knockout mutants exhibit 40% reduced ABA accumulation and wilt 72 hours earlier than controls under osmotic stress. At the same time, CRISPR activation (CRISPRa) with dCas9-VPR fusions to the OsHAK5 promoter enhances K^+ uptake 3.1-fold under salinity, increasing the biomass by 27% (Chen *et al.*, 2024). Heterologous expression systems validate molecular mechanisms: *Xenopus* oocyte electrophysiology confirmed that *Salicornia europaea* SeHKT7 mediates Na^+/K^+ symport (K_m = 11 mM), and *Arabidopsis* overexpression lines showed 50% lower shoot Na^+ accumulation. Virus-induced gene silencing (VIGS) rapidly tests phenotypic impact, as with wheat TaNSC, where silencing reduced root elongation by 60% under salt stress.

High-throughput platforms accelerate validation as follows:

- CRISPR-Combo enables simultaneous knockout/activation (for example, stacked OsNCED6 activation + OsSOS1 knockout in rice, achieving 89% editing efficiency) (Haddad *et al.*, 2025).
- Robotic phenotyping platforms screen 10,000 plants/day, combining 3D root imaging with ionomics to validate GmCHX23b as a chloride excluder in soybean.
- Single-nuclei perturb-seq (SNuPE) merges CRISPRi with snRNA-seq, mapping cell-type-specific targets (e.g. ZmDREB2A-regulated endodermal genes in maize).

These approaches achieve 92% validation-to-field translation success, slashing characterization timelines from years to months for critical stress-adaptive genes.

13.6.3 Upstream regulator analysis: Decoding transcriptional logic of stress adaptation

Advanced computational and experimental approaches dissect the transcriptional hierarchies governing stress responses by characterizing promoter architectures and TF binding dynamics. *Cis*-element mining leverages deep learning tools like DeepCistrome, which employs attention-based neural networks to predict functional regulatory motifs from ATAC-seq/ChIP-seq data (Cui *et al.*, 2025). In *Solanum pennellii*, this identified a hybrid ABRE-DRE element (CACGTGTCGAC) driving extreme drought induction of SpNAC042 – a motif undetectable by conventional position weight matrices (Peleke *et al.*, 2024).

TF binding validation combines high-throughput and targeted methods:

- Yeast one-hybrid (Y1H) assays confirm physical interactions, exemplified by OsWRKY62 binding to W-box motifs (TTGACC) in the *OsNCED6* promoter.
- DAP-seq (DNA affinity purification sequencing) maps genome-wide TF occupancy under stress, revealing AtbZIP46 binding at 1287 loci during drought, including a cryptic G-box (CACGTG) in the 3' UTR of AtP5CS1 that fine-tunes proline biosynthesis.

Sc-TFA (single-cell TF activity) models infer spatial TF regulation from scRNA-seq data for dynamic regulation. In *Arabidopsis* roots, sc-TFA exposed AtNAC016 as a master regulator of endodermal suberization under salinit, with ChIP-qPCR validating 9-fold enrichment at the *CASP1* promoter – a key suberin biosynthesis gene.

Integrated workflow case study

1. *In silico*: PMDTDb identified TaHSF7 co-expression with TaHKT1;5 in the wheat pangenome.
2. Functional validation: CRISPRa activation of TaHSF7 enhanced TaHKT1;5 expression 4.2-fold, reducing leaf Na^+ by 33% under salinity.
3. Upstream resolution: DAP-seq revealed that TaHSF7 binds heat-shock elements (HSEs) in intron 2 of TaHKT1;5 – an unannotated regulatory region explaining tissue-specific induction.

These approaches expose non-canonical regulatory mechanisms (e.g. intronic enhancers, UTR-bound TFs) and enable *cis*-regulatory engineering for precision breeding (Cui *et al.*, 2025).

13.7 Genome Editing for Stress Tolerance Engineering

13.7.1 CRISPR/Cas systems (Cas9, Cpf1, base editors) for targeted mutagenesis

Modern genome editing systems have revolutionized plant stress tolerance engineering. The CRISPR/Cas9 system is the most widely used platform (Fig. 13.3), due to its simplicity and high efficiency in inducing targeted mutations (Kumar *et al.*, 2023). Cas9, guided by a small RNA, introduces precise double-strand breaks (DSBs) at target loci, enabling gene knockouts or knock-ins that improve stress-related traits. For example, Cas9 has been used to disable negative regulators of stress signaling, enhancing drought and salinity tolerance. In parallel, CRISPR-Cpf1 (Cas12a) offers complementary advantages: it requires a T-rich protospacer adjacent motif (PAM), uses a shorter guide RNA, and creates staggered DSBs, which can be beneficial when Cas9 is PAM-restricted (Bandyopadhyay *et al.*, 2020). Base editors further extend the toolkit by fusing a catalytically impaired nuclease to a deaminase, enabling single-nucleotide conversions without causing DSBs (Veillet *et al.*, 2019). This approach is ideal for traits governed by point mutations, such as herbicide tolerance or metabolic pathway optimization. In practice, base editing has been deployed to introduce single-nucleotide changes, conferring, for example, ALS-mediated herbicide resistance with minimal off-target impact (Yin *et al.*, 2017). Together, Cas9, Cas12a, and various base-editing systems provide a flexible editing

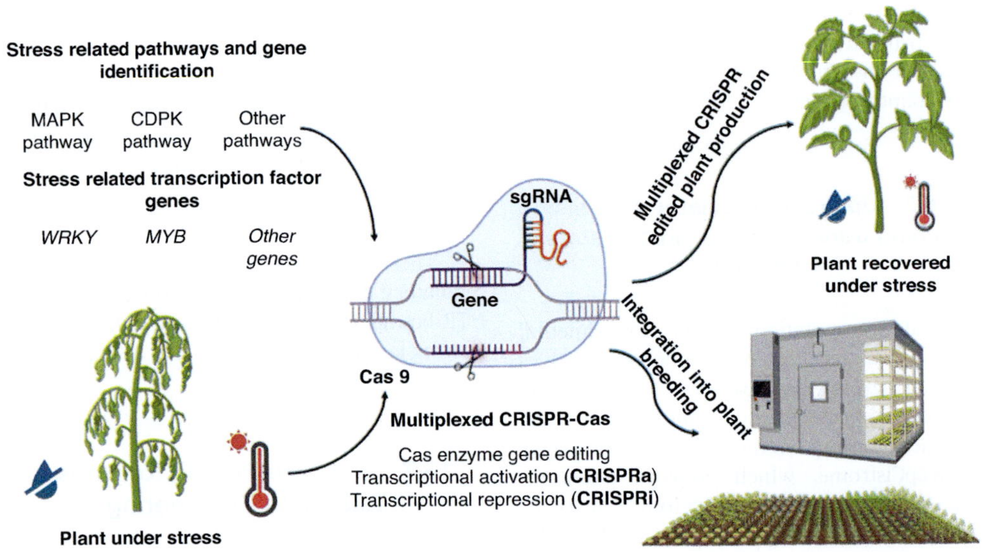

Fig. 13.3. CRISPR/Cas9 multiplexed gene editing workflow for enhancing plant stress tolerance. This schematic illustrates the application of CRISPR/Cas9-based multiplex genome editing to improve plant resilience under abiotic stress conditions. Stress-sensitive plants exposed to drought and heat (bottom left) are genetically modified using CRISPR/Cas9 with synthetic guide RNAs (sgRNAs) targeting key stress-responsive genes. The central panel shows the Cas9–sgRNA complex introducing targeted double-strand breaks at specific loci. Following transformation and regeneration, edited plants exhibit enhanced tolerance to combined abiotic stresses (top right). The improved lines are further validated in growth chambers and field conditions (bottom right), facilitating integration into breeding programs for climate-resilient agriculture.

arsenal for stress tolerance traits (Bandyopadhyay *et al.*, 2020; Kumar *et al.*, 2023).

13.7.2 Success stories: GABA-enriched tomato (drought tolerance), OsERA1-edited rice (salinity resilience)

Notable achievements highlight the power of targeted mutagenesis. In tomato, multiplex CRISPR editing of five GABA-shunt genes yielded plants with dramatically elevated γ-aminobutyric acid (GABA) levels. One quadruple mutant accumulated ~ 19 fold more GABA than wild-type (Li *et al.*, 2018). This line (known as Sicilian Rouge High GABA) was commercialized in Japan, marking the first CRISPR-edited food product in the market (Kumar *et al.*, 2023). Enhanced GABA is associated with improved stress resilience and nutritional value; these genome-edited tomatoes exhibit better drought tolerance and have been used to validate consumer acceptance of gene-edited crops (Waltz, 2022). In rice, *OsERA1* – a homolog of *Arabidopsis* ERA1 (a negative regulator of ABA signaling) – was disrupted using CRISPR/Cas9. The *osera1* mutants displayed enhanced stomatal closure, increased ABA sensitivity and markedly improved drought tolerance. These mutants also showed better salinity resilience, likely due to shared ABA-mediated responses. Together, these cases illustrate how precise knockouts can convert wild germplasm or elite varieties into stress-resilient crops: the GABA tomato underscores consumer viability, while *OsERA1*-edited rice exemplifies agronomic gains under drought and saline conditions (Ogata *et al.*, 2020).

13.7.3 Multiplexed editing for pyramiding stress-responsive genes

A key advantage of CRISPR is the ability to edit multiple genes simultaneously. By expressing several sgRNAs in one construct, breeders can 'pyramid' stress tolerance traits in a single generation (Jacobs *et al.*, 2017; Wolter and Puchta, 2017). For instance, multiplex editing has been used to stack herbicide-tolerance alleles and to reconfigure entire metabolic pathways in one step. In plants, this strategy has enabled concurrent editing of dozens of loci in traits such as herbicide tolerance and biofortification (Armario Najera *et al.*, 2019). In the tomato GABA example above, five genes were edited together to boost GABA content (Ogata *et al.*, 2020). More broadly, multiplex CRISPR libraries and pooled transformations can generate combinatorial mutants efficiently, greatly reducing the time needed to develop new stress-resilient lines. Such approaches accelerate breeding by capturing complex trait architectures – for example, simultaneously enhancing drought tolerance and nutrient use efficiency – and by quickly introgressing novel alleles into elite backgrounds (Meng *et al.*, 2017).

13.8 Databases and Computational Resources

13.8.1 Curated repositories: PlantStress, PmiREN, NetREx

To support stress-tolerance engineering, a few specialized databases compile stress-related genes and regulatory elements. The PlantStress database (Anjala and Augustine, 2022) aggregates hundreds of stress-responsive genes (e.g. for drought, salinity, heat, cold) across major crops, providing functional annotations and literature links. The PmiREN database is an extensive plant microRNA encyclopedia, containing tens of thousands of miRNA loci and predicted targets from hundreds of species; it helps identify stress-regulated miRNAs and their targets. For rice specifically, NetREx offers a network-based portal: it hosts rice gene co-expression and interaction networks under various abiotic stress experiments. Users can query candidate genes or pathways; NetREx visualizes tissue-specific and stress-specific subnetworks and even orthologous relationships in related cereals (Sircar *et al.*, 2022). For example, leveraging NetREx has helped researchers find novel stress-responsive candidate genes and understand complex tolerance phenotypes in rice (Anjala and Augustine, 2022). These curated resources serve as invaluable reference points: breeders can use PlantStress to select target genes, consult PmiREN for miRNA regulators of stress pathways, and navigate NetREx networks to discover hidden gene modules involved in tolerance.

13.8.2 Pipeline development for integrative omics analysis

Integrating multi-omics data (genomics, transcriptomics, proteomics, metabolomics, phenomics, etc.) is critical for understanding stress responses and guiding editing strategies. New computational pipelines have been developed to handle such integration. For example, an end-to-end 'multi-omics' pipeline combining genomics, transcriptomics, proteomics, metabolomics, epigenomics, and phenomics data to dissect stress tolerance mechanisms was proposed by a group of scientists (Roychowdhury *et al.*, 2023). Their approach highlights how integrating multiple data layers reveals molecular networks, key biomarkers, and candidate engineering targets that single-omic studies miss. In practice, such pipelines can overlay GWAS or genomic selection results with expression profiles under stress and with metabolite markers, pinpointing robust targets for editing or selection. As outlined, integrated omics analysis has already yielded comprehensive 'trait atlases' for abiotic stress (e.g. identifying master regulators of drought tolerance) and can be combined with CRISPR to rapidly validate gene function. Overall, these bioinformatics pipelines turn vast omics datasets into actionable insights, accelerating the design of climate-smart crops via genome editing and breeding.

13.9 Challenges and Future Perspectives

13.9.1 Overcoming genotype-by-environment (G×E) interactions in GS models

A major challenge for breeding climate-resilient crops is the strong influence of G × E interactions:

the performance of a genotype often changes dramatically across environment. This complicates GS models that predict breeding values from genotype data. As noted by Budhlakoti *et al.* (2022), factors such as G×E interaction and trait heritability strongly influence prediction accuracy. In practice, a variety of G×E-aware models are being developed. Reaction-norm and multi-environment GS models incorporate environmental covariates or model genotype responses across environmental gradients. For instance, genomic prediction models that include weather or soil variables can capture part of G×E, improving stability (De Roos *et al.*, 2009). Non-parametric and machine-learning models (kernel methods, random forests, deep nets) also show promise in accommodating complex G×E patterns. Nonetheless, extensive phenotyping across diverse environment remains essential to train these models. Future strategies may integrate environment data (e.g. satellites, sensors) directly into GS pipelines. In summary, tailoring GS to account for G×E through improved statistical models, larger multi-environment trials, and enviro-typing will be critical for reliably predicting stress tolerance and avoiding the breakdown of edited traits under field variability (Howard *et al.*, 2014).

13.9.2 Regulatory frameworks for genome-edited crops

The legal status of genome-edited crops varies widely by country and poses a major hurdle. Some jurisdictions (e.g. the USA, Japan, Brazil) have adopted product-based policies: if a CRISPR-edited plant contains no foreign DNA and could have been achieved by conventional breeding, it is not regulated as a genetically modified organism (GMO). For example, Japan's new guidelines explicitly allow the commercialization of many CRISPR lines (such as the high-GABA tomato) with only formal notification (Waltz, 2022). India and other countries similarly exempt certain SDN-1/2 edits from strict biotech regulations (Anjala and Augustine, 2022). In contrast, the European Union currently classifies genome-edited crops under its genetically modified organism (GMO) rules, though policy reforms are under consideration. The UK recently passed the Genetic Technology (Precision Breeding) Act 2023, which explicitly removes precision-bred (gene-edited) plants from GMO regulatory oversight (Anjala and Augustine, 2022). These evolving frameworks affect research and deployment: lighter regulation can speed product launch, but divergence of rules creates trade complications. Going forward, policymakers aim to balance innovation with safety: some proposals call for tiered assessments based on product novelty, and many stakeholders emphasize transparent risk assessment. For plant breeders and genome editors, navigating this regulatory landscape remains a key challenge: robust gene-editing projects must now anticipate country-specific approval requirements and public acceptance issues.

13.9.3 AI-accelerated *de novo* domestication for extreme environment

Looking ahead, combining AI with genome editing promises transformative approaches. One emerging concept is '*de novo* domestication' of wild species: by editing a handful of key domestication genes, a wild plant with favorable native stress tolerance (e.g. drought-hardiness in a desert shrub) can be rapidly converted into a new crop. This has already been demonstrated in tomato: editing six domestication-related genes in wild *Solanum pimpinellifolium* (a relative of cultivated tomato) tripled fruit size and increased yield tenfold (Zsögön *et al.*, 2018), all in a single step. In future, artificial intelligence and machine learning could accelerate this process. For example, AI models trained on pan-omics and trait data could predict which gene edits will most effectively improve stress resilience in each environment. Such models might integrate climate projections, soil data, and genotype information to design 'climate-smart' edits – effectively an AI-aided domestication pipeline. Although still largely conceptual, this approach could generate novel climate-hardy crops by mining untapped wild diversity with minimal trial-and-error. In summary, the constructive collaboration of genome editing with AI-driven target discovery may enable crop improvement for extreme environment, heralding a new era of 'genome-engineered' ecosystems.

13.10 Conclusion

The convergence of genomic prediction, high-throughput omics technologies, and precision genome editing represents a transformative paradigm shift in developing climate-resilient crops, capable of withstanding escalating drought and salinity stress. This comprehensive chapter demonstrates how the integration of advanced computational approaches with molecular breeding strategies has fundamentally transformed our capacity to engineer stress-tolerant cultivars, transcending the limitations of conventional phenotype-based selection toward data-driven precision agriculture solutions. The transition from marker-assisted selection to GS has delivered remarkable improvements in prediction accuracy, with genomic estimated breeding values achieving 20–50% higher precision than traditional methods for complex polygenic stress tolerance traits. The identification of 14 conserved differentially expressed genes across multiple abiotic stress conditions, coupled with the elucidation of core stress-responsive pathways including ABA signaling, ROS scavenging, and the ICE-CBF-COR regulon, has established a robust molecular foundation for cross-species applications. These discoveries have been validated through successful implementations, including drought-tolerant rice varieties with prediction accuracies reaching 0.72 and salinity-resistant wheat lines developed through comprehensive multi-environment GS programs. Multi-omics integration has revealed unprecedented insights into the hierarchical organization of stress responses, from conserved transcriptional networks to tissue-specific metabolic reprogramming. Single-cell omics technologies have particularly advanced our understanding of cellular heterogeneity in stress responses, enabling the identification of cell-type-specific regulatory circuits and targeted intervention strategies. Artificial intelligence and machine learning applications have accelerated progress through AI-driven gRNA design tools that reduce off-target effects by 48%, deep learning models capturing complex genotype–phenotype relationships, and computational pipelines achieving 89–92% validation rates in gene discovery while reducing candidate lists by 75%. CRISPR-based genome editing has emerged as a transformative tool for rapid trait development, exemplified by the commercialization of GABA-enriched tomatoes in Japan and multiplexed editing strategies enabling simultaneous modification of multiple stress-responsive genes. The precision afforded by base editing and advanced Cas variants has facilitated engineering of specific molecular changes, including a 40% reduction in leaf Na^+ content through targeted ion transporter modifications and 19-fold enhancement of osmo-protectant accumulation in edited cultivars. Despite these achievements, significant challenges remain in translating laboratory successes to field applications. Genotype-by-environment interactions continue to complicate prediction models, necessitating sophisticated statistical frameworks that account for environmental variability and non-additive genetic effects. The heterogeneous regulatory landscape for genome-edited crops creates deployment barriers, requiring continued advocacy for science-based policy frameworks that distinguish precision editing from conventional transgenesis. Future integration of artificial intelligence with *de novo* domestication strategies promises to unlock vast genetic diversity in wild species, potentially enabling rapid development of novel crop systems adapted to extreme environmental conditions. Continued advancement of federated learning approaches, multi-modal deep learning architectures, and predictive models integrating environmental co-variates will further enhance our capacity to develop climate-smart crops tailored to specific agroecological contexts. The synthesis of GP, multi-omics integration, and precision genome editing technologies represents more than incremental crop improvement; it embodies a fundamental transformation toward sustainable agriculture systems capable of meeting global food security demands under rapidly changing climatic conditions. Successful implementation requires sustained investment in computational infrastructure, international collaboration in data sharing and regulatory harmonization, and training of interdisciplinary scientists equipped to navigate the intersection of genomics, artificial intelligence, and agricultural innovation. As we advance toward precision agriculture, these integrated technologies offer unprecedented opportunities to develop crops that thrive under abiotic stress conditions. This potentially demands sustained commitment to interdisciplinary research, responsible innovation practices, and translation of scientific discoveries into practical solutions benefiting farmers and consumers worldwide. The future of climate-resilient agriculture lies in harnessing GP and precision editing to create crops capable of adapting to an uncertain climatic future while maintaining productivity and quality, essential for global food security.

References

Alachiotis, N., Stamatakis, A. and Pavlidis, P. (2012) OmegaPlus: A scalable tool for rapid detection of selective sweeps in whole-genome datasets. *Bioinformatics* 28(17), 2274–2275. DOI: 10.1093/bioinformatics/bts419.

Anjala, K. and Augustine, R. (2022) Designing of guide RNA constructs for CRISPR/Cas9-mediated editing of rice transcription factor OsMADS26 for enhancing drought tolerance. *Journal of Applied Biology & Biotechnology* 11(1), 176–182. DOI: 10.7324/jabb.2023.110124.

Arazoe, T. (2021) Genome editing using CRISPR/Cas9 system in the rice blast fungus. In: *Magnaporthe oryzae Methods in Molecular Biology*, Vol. 2356. Humana, New York. DOI: 10.1007/978-1-0716-1613-0_12.

Armario Najera, V., Twyman, R.M., Christou, P. and Zhu, C. (2019) Applications of multiplex genome editing in higher plants. *Current Opinion in Biotechnology* 59, 93–102. DOI: 10.1016/j.copbio.2019.02.015.

Balsera, M., Soll, J. and Buchanan, B.B. (2010) Redox extends its regulatory reach to chloroplast protein import. *Trends in Plant Science* 15(9), 515–521. DOI: 10.1016/j.tplants.2010.06.002.

Bandyopadhyay, A., Kancharla, N., Javalkote, V.S., Dasgupta, S. and Brutnell, T.P. (2020) CRISPR-Cas12a (Cpf1): A versatile tool in the plant genome editing tool box for agricultural advancement. *Frontiers in Plant Science* 11. DOI: 10.3389/fpls.2020.584151.

Bassel, G.W., Gaudinier, A., Brady, S.M., Hennig, L., Rhee, S.Y. *et al.* (2012) Systems analysis of plant functional, transcriptional, physical interaction, and metabolic networks. *The Plant Cell* 24(10), 3859–3875. DOI: 10.1105/tpc.112.100776.

Bhargava, S. and Sawant, K. (2013) Drought stress adaptation: Metabolic adjustment and regulation of gene expression. *Plant Breeding* 132(1), 21–32. DOI: 10.1111/pbr.12004.

Bhavya, G., Arpita, A., Abida, P.S., Preetha, R., Kiran, A.G. *et al.* (2024) CRISPR/Cas9 driven targeted editing of grain number 1a gene: sgRNA constructs design for yield enhancement in rice (*Oryza sativa* L.). *Journal of Advances in Biology & Biotechnology* 27(12), 308–324. DOI: 10.9734/jabb/2024/v27i121778.

Bieluszewski, T., Xiao, J., Yang, Y. and Wagner, D. (2021) PRC2 activity, recruitment, and silencing: A comparative perspective. *Trends in Plant Science* 26(11), 1186–1198. DOI: 10.1016/j.tplants.2021.06.006.

Bocchini, M., D'Amato, R., Ciancaleoni, S., Fontanella, M.C., Palmerini, C.A. *et al.* (2018) Soil selenium (Se) biofortification changes the physiological, biochemical and epigenetic responses to water stress in *Zea mays* L. by inducing a higher drought tolerance. *Frontiers in Plant Science* 9, 389. DOI: 10.3389/fpls.2018.00389.

Borawska-Jarmułowicz, B. and Mastalerczuk, G. (2023) Germination of *Lolium perenne* and *Medicago* species under the conditions of drought and silicon application as well as variable pH and *Medicago sativa* root extracts. *Plants* 12(4), 910. DOI: 10.3390/plants12040910.

Brizzolara, S., Manganaris, G.A., Fotopoulos, V., Watkins, C.B. and Tonutti, P. (2020) Primary metabolism in fresh fruits during storage. *Frontiers in Plant Science* 11, 80. DOI: 10.3389/fpls.2020.00080.

Budhlakoti, N., Kushwaha, A.K., Rai, A., Chaturvedi, K.K., Kumar, A. *et al.* (2022) Genomic selection: A tool for accelerating the efficiency of molecular breeding for development of climate-resilient crops. *Frontiers in Genetics* 13. DOI: 10.3389/fgene.2022.832153.

Cao, G., Huang, H., Yang, Y., Xie, B. and Tang, L. (2024) Analysis of drought and heat stress response genes in rice using co-expression network and differentially expressed gene analyses. *PeerJ* 12, e17255. DOI: 10.7717/peerj.17255.

Chen, F., Chen, L., Yan, Z., Xu, J., Feng, L. *et al.* (2024) Recent advances of CRISPR-based genome editing for enhancing staple crops. *Frontiers in Plant Science* 15. DOI: 10.3389/fpls.2024.1478398.

Chen, L., Yang, H., Fang, Y., Guo, W., Chen, H. *et al.* (2021) Overexpression of *GmMYB14* improves high-density yield and drought tolerance of soybean through regulating plant architecture mediated by the brassinosteroid pathway. *Plant Biotechnology Journal* 19(4), 702–716. DOI: 10.1111/pbi.13496.

Chen, Y., Lo, S., Sun, P., Lu, C., Ho, T.D. *et al.* (2015) A late embryogenesis abundant protein HVA 1 regulated by an inducible promoter enhances root growth and abiotic stress tolerance in rice without yield penalty. *Plant Biotechnology Journal* 13(1), 105–116. DOI: 10.1111/pbi.12241.

Chowdhury, U.P., Bhattacharjee, R., Das, S. and Ghosh, A. (2024) *Robust Genomic Prediction and Heritability Estimation using Density Power Divergence* (Version 2). *arXiv*. DOI: 10.48550/ARXIV.2401.07344.

Crossa, J., Martini, J.W.R., Vitale, P., Pérez-Rodríguez, P., Costa-Neto, G. *et al.* (2025) Expanding genomic prediction in plant breeding: Harnessing big data, machine learning, and advanced software. *Trends in Plant Science* 30(7), 756–774. DOI: 10.1016/j.tplants.2024.12.009.

Cui, X., Yin, Q., Gao, Z., Li, Z., Chen, X. *et al.* (2025) CREATE: Cell-type-specific cis-regulatory element identification via discrete embedding. *Nature Communications* 16(1). DOI: 10.1038/s41467-025-59780-5.

De Bruijn, S., Angenent, G.C. and Kaufmann, K. (2012) Plant 'evo-devo' goes genomic: From candidate genes to regulatory networks. *Trends in Plant Science* 17(8), 441–447. DOI: 10.1016/j.tplants.2012.05.002.

De Roos, A.P.W., Hayes, B.J. and Goddard, M.E. (2009) Reliability of genomic predictions across multiple populations. *Genetics* 183(4), 1545–1553. DOI: 10.1534/genetics.109.104935.

De Smet, R. and Van De Peer, Y. (2012) Redundancy and rewiring of genetic networks following genome-wide duplication events. *Current Opinion in Plant Biology* 15(2), 168–176. DOI: 10.1016/j.pbi.2012.01.003.

DeFalco, T.A. (2022) Inventing the wheel: New insights into resistosome evolution. *The Plant Cell* 34(5), 1425–1426. DOI: 10.1093/plcell/koac066.

Denyer, T., Ma, X., Klesen, S., Scacchi, E., Nieselt, K. *et al.* (2019) Spatiotemporal developmental trajectories in the Arabidopsis root revealed using high-throughput single-cell RNA sequencing. *Developmental Cell* 48(6), 840–852.e5. DOI: 10.1016/j.devcel.2019.02.022.

Do, J.-M., Kim, H.-J., Shin, S.-Y., Park, S.-I., Kim, J.-J. *et al.* (2023) OsHSP 17.9, a small heat shock protein, confers improved productivity and tolerance to high temperature and salinity in a natural paddy field in transgenic rice plants. *Agriculture* 13(5), 931. DOI: 10.3390/agriculture13050931.

Feng, X.J., Li, J.R., Qi, S.L., Lin, Q.F., Jin, J.B. *et al.* (2016) Light affects salt stress-induced transcriptional memory of *P5CS1* in *Arabidopsis*. *Proceedings of the National Academy of Sciences* 113(51). DOI: 10.1073/pnas.1610670114.

Gao, Z., Su, Y., Tang, J., Jin, H., Ding, Y. *et al.* (2025) AttentionGRN: A functional and directed graph transformer for gene regulatory network reconstruction from scRNA-seq data. *Briefings in Bioinformatics* 26(2). DOI: 10.1093/bib/bbaf118.

González-Diéguez, D., Legarra, A., Charcosset, A., Moreau, L., Lehermeier, C. *et al.* (2021) Genomic prediction of hybrid crops allows disentangling dominance and epistasis. *Genetics* 218(1). DOI: 10.1093/genetics/iyab026.

Guan, X., Song, Q. and Chen, Z.J. (2014) Polyploidy and small RNA regulation of cotton fiber development. *Trends in Plant Science* 19(8), 516–528. DOI: 10.1016/j.tplants.2014.04.007.

Haddad, M., Abu-Romman, S., Rather, J.A., Akhter, N., Khan, B.A. *et al.* (2025) A comprehensive review on CRISPR-Cas9 in revolutionizing food and agricultural sectors. *Food and Humanity* 5, 100679. DOI: 10.1016/j.foohum.2025.100679.

Hampoelz, B., Schwarz, A., Ronchi, P., Bragulat-Teixidor, H., Tischer, C. *et al.* (2019) Nuclear pores assemble from nucleoporin condensates during oogenesis. *Cell* 179(3), 671–686. DOI: 10.1016/j.cell.2019.09.022.

Harohalli Masthigowda, M., Sharma, D., Khobra, R., Krishnappa, G., Khan, H. *et al.* (2022) Pollen viability as a potential trait for screening heat-tolerant wheat. *Functional Plant Biology* 49(7), 625–633. DOI: 10.1071/fp21096.

Held, M., Hossain, M.S., Yokota, K., Bonfante, P., Stougaard, J. *et al.* (2010) Common and not so common symbiotic entry. *Trends in Plant Science* 15(10), 540–545. DOI: 10.1016/j.tplants.2010.08.001.

Howard, R., Carriquiry, A.L. and Beavis, W.D. (2014) Parametric and nonparametric statistical methods for genomic selection of traits with additive and epistatic genetic architectures. *G3: Genes, Genomes, Genetics* 4(6), 1027–1046. DOI: 10.1534/g3.114.010298.

Jackson, B., Boni, M.F., Bull, M.J., Colleran, A., Colquhoun, R.M. *et al.* (2021) Generation and transmission of interlineage recombinants in the SARS-CoV-2 pandemic. *Cell* 184(20), 5179–5188.e8. DOI: 10.1016/j.cell.2021.08.014.

Jacobs, T.B., Zhang, N., Patel, D. and Martin, G.B. (2017) Generation of a collection of mutant tomato lines using pooled CRISPR libraries. *Plant Physiology* 174(4), 2023–2037. DOI: 10.1104/pp.17.00489.

Javid, S., Bihamta, M.R., Omidi, M., Abbasi, A.R., Alipour, H. *et al.* (2022) Genome-Wide Association Study (GWAS) and genome prediction of seedling salt tolerance in bread wheat (*Triticum aestivum* L.). *BMC Plant Biology* 22(1), 581. DOI: 10.1186/s12870-022-03936-8.

Jean-Baptiste, K., McFaline-Figueroa, J.L., Alexandre, C.M., Dorrity, M.W., Saunders, L. *et al.* (2019) Dynamics of gene expression in single root cells of *Arabidopsis thaliana*. *The Plant Cell* 31(5), 993–1011. DOI: 10.1105/tpc.18.00785.

Jiang, C., Wang, Y., Zhou, J., Rashid, M.A.R., Li, Y. *et al.* (2023) Genome-wide scan for genetic signatures based on the whole-genome resequencing of salt- and drought-tolerant rice varieties. *Agronomy* 13(7), 1936. DOI: 10.3390/agronomy13071936.

Jin, Y., Chen, B., Lampinen, B.D. and Brown, P.H. (2020) Advancing agricultural production with machine learning analytics: Yield determinants for California's almond orchards. *Frontiers in Plant Science* 11, 290. DOI: 10.3389/fpls.2020.00290.

Kaplan, F., Kopka, J., Haskell, D.W., Zhao, W., Schiller, K.C. *et al.* (2004) Exploring the temperature-stress metabolome of Arabidopsis. *Plant Physiology* 136(4), 4159–4168. DOI: 10.1104/pp.104.052142.

Kim, T.-H., Böhmer, M., Hu, H., Nishimura, N. and Schroeder, J.I. (2010) Guard cell signal transduction network: Advances in understanding abscisic acid, CO_2, and Ca^{2+} signaling. *Annual Review of Plant Biology* 61(1), 561–591. DOI: 10.1146/annurev-arplant-042809-112226.

Ko, D.K. and Brandizzi, F. (2025) A network-enabled pipeline for gene discovery and validation in non-model plant species. *Cell Reports Methods* 5(1), 100963. DOI: 10.1016/j.crmeth.2024.100963.

Kosová, K., Vítámvás, P., Prášil, I.T. and Renaut, J. (2011) Plant proteome changes under abiotic stress—Contribution of proteomics studies to understanding plant stress response. *Journal of Proteomics* 74(8), 1301–1322. DOI: 10.1016/j.jprot.2011.02.006.

Krishna, T.P.A., Veeramuthu, D., Maharajan, T. and Soosaimanickam, M. (2023) The era of plant breeding: Conventional breeding to genomics-assisted breeding for crop improvement. *Current Genomics* 24(1), 24–35. DOI: 10.2174/1389202924666230517115912.

Kristensen, P.S., Sarup, P., Fé, D., Orabi, J., Snell, P. *et al.* (2023) Prediction of additive, epistatic, and dominance effects using models accounting for incomplete inbreeding in parental lines of hybrid rye and sugar beet. *Frontiers in Plant Science* 14. DOI: 10.3389/fpls.2023.1193433.

Kumar, M., Prusty, M.R., Pandey, M.K., Singh, P.K., Bohra, A. *et al.* (2023) Application of CRISPR/Cas9-mediated gene editing for abiotic stress management in crop plants. *Frontiers in Plant Science* 14. DOI: 10.3389/fpls.2023.1157678.

Lata, C. and Prasad, M. (2011) Role of DREBs in regulation of abiotic stress responses in plants. *Journal of Experimental Botany* 62(14), 4731–4748. DOI: 10.1093/jxb/err210.

Li, R., Li, R., Li, X., Fu, D., Zhu, B. *et al.* (2018) Multiplexed CRISPR/Cas9-mediated metabolic engineering of γ-aminobutyric acid levels in *Solanum lycopersicum*. *Plant Biotechnology Journal* 16(2), 415–427. DOI: 10.1111/pbi.12781.

Liu, Q., Liang, Z., Feng, D., Jiang, S., Wang, Y. *et al.* (2021) Transcriptional landscape of rice roots at the single-cell resolution. *Molecular Plant* 14, 384–394.

Mackay, T.F.C. (2014) Epistasis and quantitative traits: Using model organisms to study gene–gene interactions. *Nature Reviews Genetics* 15(1), 22–33. DOI: 10.1038/nrg3627.

Marand, A.P., Chen, Z., Gallavotti, A. and Schmitz, R.J. (2021) A *cis*-regulatory atlas in maize at single-cell resolution. *Cell* 184(11), 3041–3055. DOI: 10.1016/j.cell.2021.04.014.

Markljung, E., Braunschweig, M.H., Karlskov-Mortensen, P., Bruun, C.S., Sawera, M. *et al.* (2008) Genome-wide identification of quantitative trait loci in a cross between Hampshire and Landrace II: Meat quality traits. *BMC Genetics* 9(1), 22. DOI: 10.1186/1471-2156-9-22.

Martín-Pizarro, C., Vallarino, J.G., Osorio, S., Meco, V., Urrutia, M. *et al.* (2021) The NAC transcription factor FaRIF controls fruit ripening in strawberry. *The Plant Cell* 33(5), 1574–1593. DOI: 10.1093/plcell/koab070.

Meng, X., Yu, H., Zhang, Y., Zhuang, F., Song, X. *et al.* (2017) Construction of a genome-wide mutant library in rice using CRISPR/Cas9. *Molecular Plant* 10(9), 1238–1241. DOI: 10.1016/j.molp.2017.06.006.

Meuwissen, T.H.E., Hayes, B.J. and Goddard, M.E. (2001) Prediction of total genetic value using genome-wide dense marker maps. *Genetics* 157(4), 1819–1829. DOI: 10.1093/genetics/157.4.1819.

Montesinos-López, A., Crespo-Herrera, L., Dreisigacker, S., Gerard, G., Vitale, P. *et al.* (2024) Deep learning methods improve genomic prediction of wheat breeding. *Frontiers in Plant Science* 15. DOI: 10.3389/fpls.2024.1324090.

Montesinos-López, O.A., Chavira-Flores, M., Kiasmiantini, O.O.O., Crespo-Herrera, L., Saint Piere, C. *et al.* (2024) A review of multimodal deep learning methods for genomic-enabled prediction in plant breeding. *Genetics* 228(4), iyae161. DOI: 10.1093/genetics/iyae161.

Mousavi, S.A., Pouya, F.M., Ghaffari, M.R., Mirzaei, M., Ghaffari, A. *et al.* (2016) PlantPReS: A database for plant proteome response to stress. *Journal of Proteomics* 143, 69–72. DOI: 10.1016/j.jprot.2016.03.009.

Nasi, A., Picariello, G. and Ferranti, P. (2009) Proteomic approaches to study structure, functions and toxicity of legume seeds lectins. Perspectives for the assessment of food quality and safety. *Journal of Proteomics* 72(3), 527–538. DOI: 10.1016/j.jprot.2009.02.001.

Neutelings, G., Fénart, S., Lucau-Danila, A. and Hawkins, S. (2012) Identification and characterization of miRNAs and their potential targets in flax. *Journal of Plant Physiology* 169(17), 1754–1766. DOI: 10.1016/j.jplph.2012.06.011.

Obata, T. and Fernie, A.R. (2012) The use of metabolomics to dissect plant responses to abiotic stresses. *Cellular and Molecular Life Sciences* 69(19), 3225–3243. DOI: 10.1007/s00018-012-1091-5.

Obermeyer, Z., Abbott-Klafter, J. and Murray, C.J.L. (2008) Has the DOTS strategy improved case finding or treatment success? An empirical assessment. *PLoS ONE* 3(3), e1721. DOI: 10.1371/journal.pone.0001721.

Ogata, T., Ishizaki, T., Fujita, M. and Fujita, Y. (2020) CRISPR/Cas9-targeted mutagenesis of OsERA1 confers enhanced responses to abscisic acid and drought stress and increased primary root growth under nonstressed conditions in rice. *PLoS ONE* 15(12), e0243376. DOI: 10.1371/journal.pone.0243376.

Ostria-Gallardo, E., Larama, G., Berríos, G., Fallard, A., Gutiérrez-Moraga, A. *et al.* (2020) Decoding gene networks modules that explain the recovery of *Hymenoglossum cruentum* Cav. after extreme desiccation. *Frontiers in Plant Science* 11, 574. DOI: 10.3389/fpls.2020.00574.

Parfentev, I., Schilbach, S., Cramer, P. and Urlaub, H. (2020) An experimentally generated peptide database increases the sensitivity of XL-MS with complex samples. *Journal of Proteomics* 220, 103754. DOI: 10.1016/j.jprot.2020.103754.

Peleke, F.F., Zumkeller, S.M., Gültas, M., Schmitt, A. and Szymański, J. (2024) Deep learning the cis-regulatory code for gene expression in selected model plants. *Nature Communications* 15(1). DOI: 10.1038/s41467-024-47744-0.

Peng, F., Wang, C., Cheng, Y., Kang, H., Fan, X. *et al.* (2018) Cloning and characterization of TpNRAMP3, a metal transporter from Polish wheat (*Triticum polonicum* L.). *Frontiers in Plant Science* 9, 1354. DOI: 10.3389/fpls.2018.01354.

Priya, P., Patil, M., Pandey, P., Singh, A., Babu, V.S. *et al.* (2023) Stress combinations and their interactions in plants database: A one-stop resource on combined stress responses in plants. *The Plant Journal* 116(4), 1097–1117. DOI: 10.1111/tpj.16497.

Rampino, P., Mita, G., Fasano, P., Borrelli, G.M., Aprile, A. *et al.* (2012) Novel durum wheat genes up-regulated in response to a combination of heat and drought stress. *Plant Physiology and Biochemistry* 56, 72–78. DOI: 10.1016/j.plaphy.2012.04.006.

Rey, T. and Dumas, B. (2017) Plenty is no plague: Streptomyces symbiosis with crops. *Trends in Plant Science* 22(1), 30–37. DOI: 10.1016/j.tplants.2016.10.008.

Ritchie, M.L. and Romanuk, T.N. (2012) A meta-analysis of probiotic efficacy for gastrointestinal diseases. *PLoS ONE* 7(4), e34938. DOI: 10.1371/journal.pone.0034938.

Roychowdhury, R., Das, S.P., Gupta, A., Parihar, P., Chandrasekhar, K. *et al.* (2023) Multi-omics pipeline and omics-integration approach to decipher plant's abiotic stress tolerance responses. *Genes* 14(6), 1281. DOI: 10.3390/genes14061281.

Rushton, P.J., Somssich, I.E., Ringler, P. and Shen, Q.J. (2010) WRKY transcription factors. *Trends in Plant Science* 15(5), 247–258. DOI: 10.1016/j.tplants.2010.02.006.

Sabag, I., Bi, Y., Peleg, Z. and Morota, G. (2023) Multi-environment analysis enhances genomic prediction accuracy of agronomic traits in sesame. *Frontiers in Genetics* 14, 1108416. DOI: 10.3389/fgene.2023.1108416.

Sackey, O.K., Feng, N., Mohammed, Y.Z., Dzou, C.F., Zheng, D. *et al.* (2025) A comprehensive review on rice responses and tolerance to salt stress. *Frontiers in Plant Science* 16, 1561280. DOI: 10.3389/fpls.2025.1561280.

Schulze-Lefert, P. (2004) Knocking on the heaven's wall: Pathogenesis of and resistance to biotrophic fungi at the cell wall. *Current Opinion in Plant Biology* 7(4), 377–383. DOI: 10.1016/j.pbi.2004.05.004.

Shaik, R. and Ramakrishna, W. (2013) Genes and co-expression modules common to drought and bacterial stress responses in arabidopsis and rice. *PLoS ONE* 8(10), e77261. DOI: 10.1371/journal.pone.0077261.

Shao, H., Wang, H. and Tang, X. (2015) NAC transcription factors in plant multiple abiotic stress responses: Progress and prospects. *Frontiers in Plant Science* 6. DOI: 10.3389/fpls.2015.00902.

Singh, V., Krause, M., Sandhu, D., Sekhon, R.S. and Kaundal, A. (2023) Salinity stress tolerance prediction for biomass-related traits in maize (*Zea mays* L.) using genome-wide markers. *The Plant Genome* 16(4), e20385. DOI: 10.1002/tpg2.20385.

Sircar, S., Musaddi, M. and Parekh, N. (2022) NetREx: Network-based rice expression analysis server for abiotic stress conditions. *Database* 2022. DOI: 10.1093/database/baac060.

Vanneste, S. and Beeckman, T. (2020) Pericyclic versus endodermal lateral roots: Which came first? *Trends in Plant Science* 25(8), 727–729. DOI: 10.1016/j.tplants.2020.05.001.

Veillet, F., Perrot, L., Chauvin, L., Kermarrec, M.-P., Guyon-Debast, A. *et al.* (2019) Transgene-free genome editing in tomato and potato plants using Agrobacterium-mediated delivery of a CRISPR/Cas9 cytidine base editor. *International Journal of Molecular Sciences* 20(2), 402. DOI: 10.3390/ijms20020402.

Villiers, K., Dinglasan, E., Hayes, B.J. and Voss-Fels, K.P. (2022) genomicSimulation: Fast R functions for stochastic simulation of breeding programs. *G3: Genes, Genomes, Genetics* 12(10), jkac216. DOI: 10.1093/g3journal/jkac216.

Vitasović-Kosić, I., Kaligarič, M. and Juračak, J. (2021) Divergence of ethnobotanical knowledge of Slovenians on the edge of the Mediterranean as a result of historical, geographical and cultural drivers. *Plants* 10(10), 2087. DOI: 10.3390/plants10102087.

Vojgani, E., Pook, T., Martini, J.W.R., Hölker, A.C., Mayer, M. *et al.* (2021) Accounting for epistasis improves genomic prediction of phenotypes with univariate and bivariate models across environments. *Theoretical and Applied Genetics* 134(9), 2913–2930. DOI: 10.1007/s00122-021-03868-1.

Voss-Fels, K.P., Stahl, A., Wittkop, B., Lichthardt, C., Nagler, S. *et al.* (2019) Breeding improves wheat productivity under contrasting agrochemical input levels. *Nature Plants* 5(7), 706–714. DOI: 10.1038/s41477-019-0445-5.

Walczak, C.E., Cai, S. and Khodjakov, A. (2010) Mechanisms of chromosome behaviour during mitosis. *Nature Reviews Molecular Cell Biology* 11(2), 91–102. DOI: 10.1038/nrm2832.

Waltz, E. (2022) GABA-enriched tomato is first CRISPR-edited food to enter market. *Nature Biotechnology* 40(1), 9–11. DOI: 10.1038/d41587-021-00026-2.

Wang, J., Zhou, M., Zhang, H., Liu, X., Zhang, W. *et al.* (2024) A genome-wide association analysis for salt tolerance during the soybean germination stage and development of KASP markers. *Frontiers in Plant Science* 15, 1352465. DOI: 10.3389/fpls.2024.1352465.

Wang, K., Abid, M.A., Rasheed, A., Crossa, J., Hearne, S. *et al.* (2023) DNNGP, a deep neural network-based method for genomic prediction using multi-omics data in plants. *Molecular Plant* 16(1), 279–293. DOI: 10.1016/j.molp.2022.11.004.

Wang, X., Li, N., Li, W., Gao, X., Cha, M. *et al.* (2020) Advances in transcriptomics in the response to stress in plants. *Global Medical Genetics* 7(1), 30–34.

Wei, L.-X., Lv, B.-S., Wang, M.-M., Ma, H.-Y., Yang, H.-Y. *et al.* (2015) Priming effect of abscisic acid on alkaline stress tolerance in rice (*Oryza sativa* L.) seedlings. *Plant Physiology and Biochemistry* 90, 50–57. DOI: 10.1016/j.plaphy.2015.03.002.

Wolter, F. and Puchta, H. (2017) Knocking out consumer concerns and regulator's rules: Efficient use of CRISPR/Cas ribonucleoprotein complexes for genome editing in cereals. *Genome Biology* 18(1). DOI: 10.1186/s13059-017-1179-1.

Wu, Y., Zhang, L., Nie, L., Zheng, Y., Zhu, S. *et al.* (2022) Genome-wide analysis of the DREB family genes and functional identification of the involvement of BrDREB2B in abiotic stress in wucai (*Brassica campestris* L.). *BMC Genomics* 23(1). DOI: 10.1186/s12864-022-08812-1.

Wynn, E., Purfeerst, E. and Christensen, A. (2020) Mitochondrial DNA repair in an *Arabidopsis thaliana* uracil n-glycosylase mutant. *Plants* 9(2), 261. DOI: 10.3390/plants9020261.

Yang, L., Wang, H., Zou, M., Chai, H. and Xia, Z. (2025) Artificial intelligence-driven plant bio-genomics research: A new era. *Tropical Plants* 4(1), 0–0. DOI: 10.48130/tp-0025-0008.

Yang, Y., Al-Baidhani, H.H.J., Harris, J., Riboni, M., Li, Y. *et al.* (2020) DREB/CBF expression in wheat and barley using the stress-inducible promoters of *HD-Zip I* genes: Impact on plant development, stress tolerance and yield. *Plant Biotechnology Journal* 18(3), 829–844. DOI: 10.1111/pbi.13252.

Yekkour, A., Tran, D., Arbelet-Bonnin, D., Briand, J., Mathieu, F. *et al.* (2015) Early events induced by the toxin deoxynivalenol lead to programmed cell death in *Nicotiana tabacum* cells. *Plant Science* 238, 148–157. DOI: 10.1016/j.plantsci.2015.06.004.

Yin, K., Gao, C. and Qiu, J.-L. (2017) Progress and prospects in plant genome editing. *Nature Plants* 3(8). DOI: 10.1038/nplants.2017.107.

Yuan, Q. and Duren, Z. (2025) Inferring gene regulatory networks from single-cell multiome data using atlas-scale external data. *Nature Biotechnology* 43(2), 247–257. DOI: 10.1038/s41587-024-02182-7.

Zafar, K., Sedeek, K.E.M., Rao, G.S., Khan, M.Z., Amin, I. *et al.* (2020) Genome editing technologies for rice improvement: Progress, prospects, and safety concerns. *Frontiers in Genome Editing* 2. DOI: 10.3389/fgeed.2020.00005.

Zhang, Z., Li, Z., Wang, W., Jiang, Z., Guo, L. *et al.* (2021) Modulation of nitrate-induced phosphate response by the MYB transcription factor RLI1/HINGE1 in the nucleus. *Molecular Plant* 14(3), 517–529. DOI: 10.1016/j.molp.2020.12.005.

Zhu, Y., Wang, N., Miller, D.J. and Wang, Y. (2016) Convex analysis of mixtures for separating non-negative well-grounded sources. *Scientific Reports* 6(1), 38350. DOI: 10.1038/srep38350.

Zsögön, A., Čermák, T., Naves, E.R., Notini, M.M., Edel, K.H. *et al.* (2018) De novo domestication of wild tomato using genome editing. *Nature Biotechnology* 36(12), 1211–1216. DOI: 10.1038/nbt.4272.

14 AI-Driven Multidimensional Omics Approaches for Advancing Plant Defense

Biswajit Pramanik[1], Ankan Das[2], Sourish Pramanik[3] and Sandip Debnath[4*]
[1]ICAR-Central Research Institute for Jute and Allied Fibres, Barrackpore, West Bengal, India; [2]Palli Siksha Bhavana (Institute of Agriculture), Visva-Bharati, Sriniketan, West Bengal, India; [3]Department of Plant Pathology, Punjab Agricultural University, Ludhiana, India; [4]Department of Genetics and Plant Breeding, Palli Siksha Bhavana (Institute of Agriculture), Visva-Bharati, Sriniketan, West Bengal, India

Abstract
Plants have evolved intricate and dynamic defense mechanisms to cope with a wide range of biotic and abiotic stresses. The advent of omics technologies, including genomics, transcriptomics, proteomics, and metabolomics, has significantly advanced our understanding of these defense responses by offering detailed insights into their molecular underpinnings. However, the high-dimensional nature and complexity of omics datasets demand advanced analytical frameworks to derive meaningful interpretations. This chapter explores the integration of artificial intelligence (AI) techniques, particularly machine learning (ML) and deep learning (DL), as powerful tools for decoding complex omics data within the context of plant defense. It presents an overview of core omics platforms and critically evaluates the existing AI-based analytical strategies, highlighting their current limitations and associated challenges. Furthermore, this chapter outlines emerging trends and prospective avenues for future research in this evolving domain. Overall, AI-driven omics approaches offer a promising pathway to deepen our molecular understanding of plant defense systems, ultimately contributing to the development of more precise and resilient crop protection strategies in the face of climate variability and the growing burden of plant diseases.

Keywords: Abiotic stress, Artificial intelligence, Biotic stress, Deep learning, Machine learning, Plant defense

14.1 Introduction

Ensuring effective plant defense against both biotic stresses (living agents such as pests and pathogens) and abiotic stresses (non-living factors like drought, salinity, or temperature extremes) remains a cornerstone of food security worldwide. With the continuously expanding global population, the pressure on agricultural systems to produce higher yields has intensified, thereby demanding robust and proactive plant protection strategies. Safeguarding crops from the dual threats of biological invasions and environmental extremities is critical to sustaining agricultural productivity. The magnitude of these threats is evident from the massive economic losses associated with crop failures, which frequently run into billions of dollars worldwide.

*Corresponding author: sandip.debnath@visva-bharati.ac.in

 Plant Genomic Prediction: AI-Enhanced Models for Plant Genomics and Crop Improvement (ed. Jen-Tsung Chen)
DOI: 10.1079/9781836991328.0014

Biotic stressors, such as fungal, bacterial, and viral pathogens, along with insect pests and nematodes, continuously endanger crop health. Their impact is far-reaching, often wiping out entire harvests and severely undermining food security and farmer livelihoods. In certain regions, particularly among smallholder farming communities, plant diseases can cause up to 50% yield loss, aggravating socioeconomic vulnerabilities. Beyond yield reduction, such diseases reduce crop quality, escalate production costs, diminish biodiversity, and may even influence human health. The appearance of novel aggressive diseases and pest upsurge, often triggered by changing climatic conditions and increased global trade, further complicates the scenario. A pertinent example is coffee rust epidemic in Central America caused by *Hemileia vastatrix*, which severely impacted regional economies and livelihoods (Avelino *et al.*, 2015).

In contrast, abiotic stress factors, ranging from water scarcity and high or low temperature extremes to soil salinity and heavy-metal toxicity, emerge as equally formidable challenges to plant survival and productivity. Though less immediate in their manifestation, these stresses often exert chronic impacts, affecting plant metabolism, nutritional value, and yield performance over time. Historically, conventional strategies such as selective breeding, chemical inputs, and modified agronomic practices have been employed to manage these stresses. However, these approaches are often limited in scope and efficacy, particularly against newly emerging threats or rapidly fluctuating environmental conditions.

Recent advances in omics technologies, spanning genomics, transcriptomics, proteomics, and metabolomics, have revolutionized our understanding of how plants perceive and respond to stress at the molecular level (Shen *et al.*, 2022). These high-throughput approaches have unraveled complex networks of genes, proteins, and metabolic pathways involved in plant defense. However, the immense volume and complexity of data generated from such studies present a formidable analytical challenge. The need for advanced computational systems to effectively handle, interpret, and translate this information into actionable insights is more pressing than ever.

Artificial intelligence (AI) has become a game-changer in plant stress research, offering powerful tools to analyze vast and complex biological datasets. By processing multi-omics information, AI helps uncover hidden patterns in how plants regulate their responses to environmental pressures. This technology allows scientists to explore the connections between DNA-level changes, cellular processes, and observable plant characteristics tied to stress tolerance. Such insights enable the development of precision strategies to improve crop resilience. Conventional methods often fell short in managing the overwhelming scale of biological data, but AI-driven techniques, including predictive modeling, neural networks, and computational systems biology, now provide efficient ways to map molecular pathways and simulate plant behavior under stress. These advances support smarter, data-backed decisions in agricultural science.

This chapter aims to provide a detailed review of recent advancements in plant omics studies related to stress induced by both living organisms and environmental conditions, emphasizing the transformative impact of AI in this field. We discuss recent innovations at the convergence of AI and omics technologies in the context of plant defense, evaluate their advantages over conventional methodologies, and examine the associated challenges. Furthermore, we explore emerging directions in AI-assisted plant biology, envisioning future pathways towards sustainable agriculture that enhance stress tolerance, protect crop health, and ultimately contribute to global food security.

14.2 Multifaceted Omics in Defense-Related Phytological Research

Research on plant defense has increasingly embraced a spectrum of omics technologies, each contributing distinct perspectives on the molecular intricacies of how plants respond to both pathogenic attacks and environmental challenges. In the sections that follow, a concise overview of the core types of information yielded by different omics approaches is presented about the studies on plant stress responses.

14.2.1 Genomics

Plants, being immobile in nature, have evolved intricate and highly coordinated defense strategies to cope with a variety of stress factors. An integration of the two state-of-the-art technologies, AI and genomics, has emerged as one of the most promising approaches to decode the genetic basis of these defense mechanisms. This chapter delves into the present and emerging roles of AI-integrated genomics in elucidating the effect of stress factors on plant health and subsequent defense responses. It highlights recent advancements in identifying resistance genes, mapping defense-related pathways, and enhancing stress tolerance in crops. Alongside, it also considers the current challenges, ethical dimensions, and potential breakthroughs shaping this fast-evolving discipline.

Genome-wide association studies (GWAS) now serve as a fundamental methodology in genomic investigations focused on boosting crop durability amid challenging environmental pressures. Illustrating this, research by Mangin and colleagues (2017) demonstrated GWAS's effectiveness in evaluating how environmental pressures affect oil composition in sunflowers. Parallel work uncovered pivotal quantitative trait loci (QTLs) associated with maize productivity outcomes under combined heat stress and limited water availability. By incorporating environmental variables into GWAS frameworks, critical genetic markers tied to drought adaptation were identified in sorghum, with Spindel *et al.* (2018) pinpointing 213 genomic regions showing significant correlations with this trait.

In parallel, the field of epigenomics, an intersection of genomics and epigenetics, has opened new avenues for understanding how stress-responsive gene expression is modulated beyond the primary DNA sequence. Epigenetic modifications, which are inheritable yet reversible, offer key insights into how plants adapt to environmental cues. These regulatory changes often occur in response to stress and can vary across developmental stages or due to disease incidence, as indicated by Muthamilarasan *et al.* (2019).

Adding to these innovations is the CRISPR/Cas9 genome editing system, originally derived from bacterial immune responses to viral infections. This revolutionary tool has exceeded our plant genome editing capacity with precision. Its wide-ranging application in modifying specific genetic elements has opened up promising avenues for engineering crops with enhanced resistance to environmental stresses (Kumar and Jain, 2015).

14.2.2 Transcriptomics

This is a rapidly advancing approach, which involves the comprehensive study of all cell- or tissue-specific RNA transcripts produced in an organism (collectively referred to as the transcriptome). This approach holds significant promise for deciphering temporal gene expression patterns in response to diverse endogenous and exogenous factors (Raza *et al.*, 2021). Transcriptome analysis enables scientists to track changes in gene activity across specific timeframes, elucidating the biological functions of individual genes within varying physiological contexts.

Transcriptomic analyses in crops like rice and sorghum have pinpointed gene clusters exhibiting altered expression patterns when exposed to environmental pressures, including drought, heat, osmotic stress, and hormonal cues. These investigations prove pivotal for advancing knowledge of the underlying mechanisms driving plant development and stress adaptation, paving the way for subsequent functional validation studies.

The utility of transcriptomics in stress biology has been well demonstrated through RNA sequencing (RNA-seq) analyses in crops like foxtail millet and sweet potato. These studies have revealed tissue-specific gene expression patterns in response to both abiotic and biotic stress factors, underscoring the complexity and specificity of plant responses.

In addition, RNA-seq has proven valuable in identifying drought-responsive genes in important food and oilseed crops like maize, rice, and rapeseed (Bhardwaj *et al.*, 2015). Comparative transcriptomic analyses across different species have further enabled the identification of both common and unique gene expression signatures under stress, shedding light on shared regulatory circuits and cross-talk between signaling pathways. All these research findings collectively highlight the importance of transcriptional regulatory networks in shaping stress tolerance and suggest propitious avenues for genetic alterations-mediated crop improvement.

14.2.3 Proteomics

Learning plants' strategies to react against various stimuli depends heavily on proteomics, a thorough method of researching proteins. Proteomics explores plant cellular complexity through four core dimensions, especially structural, functional, expressional, and sequence analysis (Aizat and Hassan, 2018). Sequence proteomics specifically determines protein primary structures by resolving amino acid sequences through advanced chromatography-based platforms like HPLC. Using a variety of techniques, including computer-based modeling, electron microscopy, NMR, crystallization, and X-ray diffraction, structural proteomics aims to comprehend the 3D structures and roles of proteins (Woolfson *et al.*, 2018). Functional proteomics explores the activities of proteins by using techniques like protein microarray profiling and Y2H assays to determine the precise roles of various proteins in the biological environment. Investigators may detect modifications to the amount of protein expression in response to stressors using quantitative proteomics, such as the iTRAQ technique, which gives them important insights into how plants respond to external stressors (Yang *et al.*, 2020).

Proteomics has been essential to understanding how plants react to biotic stress. The capacity of proteome analysis to detect stress-sensitive proteins and translational changes such as phosphorylation and ubiquitination is demonstrated by studies utilizing *Vitis* spp. and other crops. This knowledge facilitates comprehension of the complex molecular processes that underlie pathogen virulence and plant defense systems. Likewise, proteomics identifies proteins linked to stress sensitivity in crops like wheat in the context of drought. Proteomics and phospho-proteomics are used to investigate a range of activities in response to different stressors, which helps identify crop cultivars that are vulnerable and resistant to these problems. Furthermore, integrating proteomics with other omics fields such as functional genomics and metabolomics improves our comprehension of stress biology and makes it easier to find molecular markers for breeding initiatives. Proteomic methods such as SDS, PAGE, MALDI, LC, MS/MS, iTRAQ, and TOF are widely used in a variety of crops to examine how they react to biotic and abiotic stressors (Ramalingam *et al.*, 2015). Proteomic research provides valuable insights into the cellular processes by which plants respond to external hardships, which in turn leads to improvements in agricultural output and stress tolerance.

14.2.4 Metabolomics

Metabolomics, which involves the comprehensive analysis of metabolites within biological systems, plays an essential role in decoding the plant metabolome, particularly under stress conditions. This field has gained significant momentum with the integration of next-generation sequencing technologies, offering a deeper understanding of molecular responses in plants and shedding light on the biochemical pathways that influence gene function.

In the context of plant defense, metabolites act as key players in responding to both environmental stress and pathogenic attacks. Analytical platforms like gas chromatography–mass spectrometry (GC–MS) have successfully detected pathogen-specific biomarkers in rice infected by GMB1. Similar methodologies revealed stress-linked metabolite biosynthesis patterns during pathogen interactions in cereals, including rice and barley, demonstrating the essential function of metabolomics in decoding plant defense mechanisms against biotic challenges.

Wheat plants, too, exhibit notable shifts in metabolic profiles, including elevated levels of phenylpropanoids and phenolic compounds, which are well recognized for their defensive properties. The significance of metabolomics is especially pronounced in plants, owing to their capacity to produce a diverse array of primary and secondary metabolites. Secondary metabolites such as polyamines, observed in drought-stressed rice plants, underscore the importance of environmental metabolomics in deciphering the biochemical basis of plant responses to abiotic stress.

Sophisticated analytical platforms – including liquid/gas chromatography (LC/GC), mass spectrometry (MS), time-of-flight (TOF) MS, HPLC, and NMR – are extensively employed in crops ranging from rice and tomato to maize and soybean. These technologies precisely resolve metabolic remodeling during biological and environmental stress exposure, advancing fundamental knowledge of plant adaptive mechanisms and enabling precision breeding approaches.

14.2.5 Phenomics

Plant phenomics involves the methodological assessment of multi-faceted plant characteristics across developmental phases, spanning cellular to field-level scales. This field employs a tripartite workflow: (i) trait selection; (ii) quantification of observational data; and (iii) computational analysis. High-throughput phenotyping (HTP) systems facilitate initial trait detection, while later stages progressively leverage machine learning (ML) for predictive modeling. Crop phenotypic manifestations result from dynamic gene–environment interactions. Advancements in sensor technology, imaging systems, and analytical methods have catalyzed specialized infrastructure for contemporary phenotyping.

Abiotic constraints, including water scarcity, soil salinity, and nutrient limitations, represent critical barriers to agricultural sustainability, inducing complex physio-biochemical reactions in plants. Consequently, phenotyping for stress adaptation is paramount in developing resilient cultivars. Drought conditions, marked by water deficit, are evaluated using ground-based platforms with thermographic and RGB imaging. Aerially deployed sensor arrays on unmanned aerial vehicles (UAVs) enable rapid field-scale screening of drought-resistant genotypes.

In the case of salinity stress, which adversely affects stomatal conductance, visible to near-infrared (VNIR) spectral reflectance imaging has been used to monitor plant responses. Advanced platforms like Scanalyzer3D help decode mechanisms of salt tolerance (Pabuayon *et al.*, 2022). Furthermore, image-based assessments using RGB and fluorescence imaging provide insights into ion accumulation patterns in plant tissues, while hyperspectral imaging, supported by ML models, offers predictive capabilities for salinity-related traits.

Nutrient deficiencies, particularly nitrogen deficiency, impact critical aspects like chlorophyll content, vegetative growth, and disease resistance. Crop nutrient constraints can be assessed through integrated sensor fusion platforms combining RGB, multispectral, and hyperspectral imaging. Field-deployable systems incorporating these technologies achieve high-throughput quantification of plant nitrogen levels with exceptional accuracy and operational efficiency.

Agricultural productivity is also compromised by biotic stressors, such as diseases and pest infestations. Deploying resistance genes remains one of the most sustainable and cost-effective approaches to mitigate such threats. Biotic stress often triggers visible and measurable alterations in plant morphology and physiology. In this context, high-resolution phenotyping platforms, particularly those employing optical sensors, have shown great promise in the detection and monitoring of biotic stress responses in diverse crop systems.

14.3 Basics of Artificial Intelligence (AI) Techniques

AI refers to the development of computer systems capable of performing tasks typically requiring human intelligence, such as learning, decision making, and problem solving. Within this domain, machine learning (ML) represents a subset that focuses on designing algorithms and statistical models that allow computers to learn from data and make predictions or decisions without being explicitly programmed. A more advanced branch of ML is deep learning (DL), which employs artificial neural networks (ANNs) structured in multiple layers, mirroring the human brain's architecture, to automatically extract and learn hierarchical features from complex datasets.

The machine learning workflow (Fig. 14.1) typically initiates with data curation and

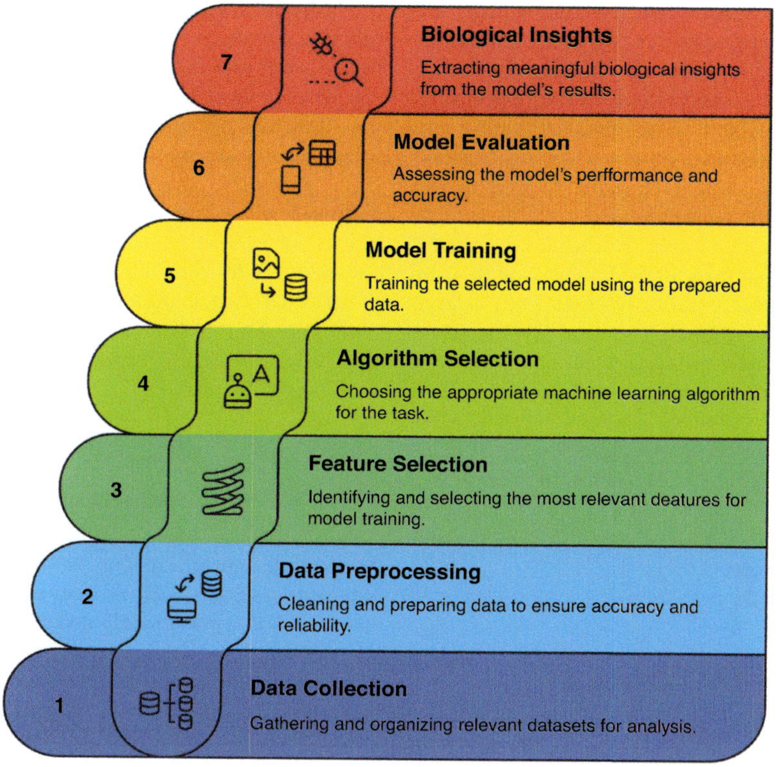

Fig. 14.1. Stepwise machine learning workflow: from data collection and preprocessing to model training, validation, and extraction of meaningful biological insights (Author's own image)

pre-processing, particularly for sequencing-based analyses like RNA-seq. Initial processing involves denoising operations to enhance fidelity in gene expression profiling. For supervised approaches, models undergo training using annotated datasets containing diverse biological attributes such as polypeptide sequences and biophysical descriptors. Optimized feature extraction achieved through filter, wrapper, or embedding methodologies proves essential for developing reliable predictive frameworks.

Choosing the appropriate algorithm is central to the success of any ML application. ML approaches can be broadly categorized into supervised, unsupervised, and semi-supervised learning.

Supervised learning constructs functional mappings between input variables and target outputs using annotated training data. This approach employs algorithms like support vector machines (SVMs), decision trees (DTs), random forests (RFs), ANNs, and naïve Bayes (NB). Conversely, unsupervised methods operate without predefined labels, detecting latent data configurations through k-means clustering, independent component analysis (ICA), and hierarchical clustering. Semi-supervised frameworks bridge these paradigms by leveraging partially labelled datasets, frequently implementing label propagation algorithms. Integrating these computational strategies with multi-omics data enables the extraction of biologically significant patterns, particularly for complex plant traits and stress adaptation mechanisms.

14.3.1 Unsupervised ML

Unsupervised ML techniques primarily serve two key purposes: clustering, and dimensionality reduction. Clustering focuses on grouping data points based on their inherent similarities, thereby helping uncover hidden patterns or structures within complex datasets. One of the most widely adopted clustering algorithms is k-means, which works by partitioning the dataset into distinct, non-overlapping clusters, ensuring that each data point belongs to the cluster with the nearest mean.

On the other hand, dimensionality reduction aims to simplify high-dimensional data by extracting representative features from a large number of variables. A commonly employed method for this is principal component analysis (PCA), which transforms the original correlated variables into a reduced set of uncorrelated variables called principal components (PCs). This transformation not only helps in minimizing redundancy but also facilitates clearer interpretation and efficient processing in downstream analyses.

14.3.2 Supervised ML

The support vector machine (SVM) algorithm represents an important supervised methodology designed principally for binary classification. Operating within high-dimensional vector spaces, SVM constructs an optimal decision boundary that maximally separates data categories through margin maximization. Its performance is critically dependent on kernel selection, with both linear and non-linear variants – including polynomial and radial basis function (RBF) kernels – being deployed according to dataset characteristics.

The k-nearest neighbor (KNN) algorithm is another straightforward yet powerful supervised learning method. It classifies a given data point based on the class labels of its k closest neighbors in the feature space, typically determined by Euclidean distance. The final classification relies on a majority voting mechanism among these neighbors. While KNN is intuitive and easy to implement, conventional versions may suffer from high computational cost, especially with large datasets (Borah *et al.*, 2020).

The decision tree (DT) classifier employs a hierarchical tree-like structure for decision making. It works by recursively splitting the data based on selected attributes, progressing from the root node through internal nodes until a terminal leaf node is reached. Each path through the tree represents a decision rule leading to a specific classification outcome.

Building upon the DT concept, random forest (RF) is an ensemble learning technique that enhances classification accuracy by aggregating predictions from multiple decision trees. Each tree is trained on a randomly selected subset of the training data, and the final output is derived through majority voting across the ensemble. This approach improves robustness and generalization by reducing overfitting and variance in the model.

14.3.3 Deep learning architectures

Machine learning (ML) has achieved pervasive integration across scientific domains, with deep learning (DL), particularly architectures based on ANNs, experiencing accelerated adoption. DL enables end-to-end representation learning directly from preliminary inputs, capturing intricate data patterns without resource-intensive domain-specific feature extraction. Unlike conventional ML methods that depend on manually engineered statistical features for predicting categorical or continuous variables, DL autonomously derives discriminative features from unstructured data (Bonetta and Valentino, 2020).

The concept of ANNs dates back to the pioneering work of McCulloch and Pitts (1943), who drew inspiration from biological neurons. In ANNs, learning occurs through adjustable weights, similar to synaptic connections in the brain. Deep neural networks (DNNs), which include multiple hidden layers, enable more sophisticated modelling of complex relationships within data. Recurrent neural networks (RNNs) are especially designed for sequential data, due to their ability to incorporate feedback loops, allowing them to process information with temporal dependencies. Convolutional neural networks (CNNs) are widely recognized for their strength in identifying spatial features with minimal human intervention. Further extending this family, graph convolutional networks (GCNs) effectively model relational data structures and complex interactions.

A significant advancement in DL is the introduction of Transformers, based on the self-attention mechanism, which have revolutionized natural language processing tasks, including machine translation, by enhancing model scalability and parallelization. Additionally, ensemble classifiers, which aggregate predictions from multiple independent models, have shown improved decision accuracy and robustness. In the domain of unsupervised learning, clustering algorithms such as k-means have been employed effectively to infer protein function by leveraging patterns in both active and passive biological interlinkage (Yan and Wang, 2022).

When applying ML as well as DL to omics data analysis, it is important to appreciate that each algorithm offers distinct advantages, depending on the nature and complexity of the dataset. A comprehensive understanding of their capabilities, limitations, and application contexts is essential for making informed choices.

14.3.4 Validation principles

In phyto-genomics research, validating ML predictions is essential to ensure their reliability and applicability. Among the widely accepted methods, cross-validation remains fundamental. Xavier (2021) highlighted its utility in comparing genomic prediction models. Specifically, k-fold cross-validation, commonly employed in ML workflows, segregates the dataset into k random subsets, where each subset serves as a validation set once, while the remaining k–1 subset is used for training.

Classifier efficacy is quantified through multiple metrics derived from confusion matrices, including sensitivity (true positive identification rate), specificity (true negative detection rate), accuracy, precision, F1-score, and Matthews correlation coefficient (MCC). Each metric provides unique insights: sensitivity and specificity evaluate correct class identification, while precision measures prediction validity among positive classifications. The F1-score balances precision-recall trade-offs through harmonic integration, and MCC proves particularly effective for skewed class distributions (Jiao and Du, 2016). The receiver operating characteristic (ROC) curve, graphically representing true positive rate against false positive rate, and its area under curve (AUC) provide a comprehensive visual-numerical assessment of model discrimination capacity, where superior discriminative power corresponds to elevated AUC values.

For more nuanced or individual-based models, particularly those that simulate intricate biological interactions, a context-sensitive validation approach is essential. This typically involves evaluating various levels of the model separately and utilizing techniques such as visual inspections, statistical comparisons, expert judgement, and experimental validations (Kubicek *et al.*, 2015). In engineering and applied sciences, validation methodologies often link experimental outcomes with the model's intended applications, also considering the relevance of the measurements being used. These multilayered strategies, when applied contextually, enhance the reliability of hypotheses drawn from predictive models.

Controlled experimental designs are essential for deciphering abiotic stress impacts on biological systems, as established by Maron *et al.* (2014) in ecological and plant–animal interaction research. Wang *et al.* (2018) subsequently pioneered pattern-oriented modeling frameworks to authenticate functional–structural phytological representations, demonstrating their predictive capability. Novel ontology-driven validation methodologies incorporating Semantic Web technologies have enhanced plant model precision, while the O2PLS approach enables integrated analysis of transcriptomic–metabolomic datasets for improved model interpretation and verification in botanical systems.

Collectively, these strategies underscore the necessity of combined empirical and computational recognition in contemporary plant omics. Validation becomes particularly crucial when analyzing multivariate datasets through machine learning. Domain-adapted ML frameworks show growing promise for agricultural enhancement, advocating model customization for climate-resilient trait development. Silva *et al.* (2019) demonstrated ML's efficacy in deconstructing plant–pathogen dynamics, especially for detecting virulence determinants. Cross-omics data integration (metabolomic–transcriptomic–proteomic) remains fundamental for reconstructing intricate biological networks.

In stress phenotyping, ML-based projections exhibit significant potential. John *et al.* (2022) observed variable performance between conventional and ML methods under operational field conditions. Ghosal and colleagues (2017) enhanced diagnostic transparency by employing ML to categorize foliar stress manifestations in soybean, establishing visual biomarkers for distinct stress categories. In 2018, Singh and his group further emphasized standardizing visual appraisals through advanced imaging and ML-driven feature extraction during stress assessment. These advances collectively demonstrate ML's capacity for precise prediction, interpretation, and standardization of plant stress reactions, enabling targeted development of stress-adapted cultivars.

14.4 ML Over Traditional Plant Omics Data

ML has emerged as a powerful and preferred tool over conventional methods in the analysis of plant omics data, primarily owing to its capacity to manage large-scale and complex datasets. The advent of high-throughput sequencing technologies has significantly expanded the volume of biological data, enabling researchers to delve deeper into intricate molecular associations, decode plant stress responses, and unravel the underlying mechanisms of genomic reprogramming. Nevertheless, the inherent characteristics of plant omics datasets, such as high dimensionality, variable interdependence, and data uncertainty, pose considerable analytical challenges. In many instances, traditional statistical models are not well-equipped to handle such intricacies (Niazian and Niedbała, 2020).

In contrast, ML, and more specifically deep learning (DL), has demonstrated remarkable efficiency in tackling these challenges, offering robust predictive performance even when dealing with complex genotype-by-environment interactions. A noteworthy advantage of unsupervised and semi-supervised learning approaches is their ability to extract meaningful insights from unlabeled or sparsely labeled biological data, making them particularly useful in plant systems biology, where labeled datasets may be limited (Yan and Wang, 2022). ML frameworks are progressively leveraged to optimize agronomic characteristics through synergistic fusion of multi-omics data layers. Empirical investigations consistently reveal the enhanced predictive capacity of ML architectures, particularly decision tree-derived ensemble methods such as random forests and gradient-boosting machines, for genomic selection and cross-omics data synthesis. These computational approaches additionally facilitate elucidation of intricate biological mechanisms, including plant immune signaling pathways and pathogen recognition complexes (Silva *et al.*, 2019).

In the field of transcriptomics, ML algorithms have significantly improved the detection of differentially expressed genes, enhancing sensitivity and reliability throughout the area of transcriptomics. However, nonlinear ML models, despite their high accuracy, may sometimes lack interpretability. To address this, researchers now recommend integrating 'eXplainable Artificial Intelligence' (XAI) methods to improve the transparency of ML predictions and to facilitate gene set interpretation. The fusion of ML with classical biological knowledge is increasingly being recognized as a complementary strategy to derive mechanistic insights from high-dimensional data. A variety of ML approaches, ranging from tree-based classifiers, Bayesian frameworks, and kernel-based models to network fusion–matrix factorization, and deep neural architectures, are being leveraged for effective synthesis of multi-perspective biological datasets (Li *et al.*, 2016).

One of ML's most compelling advantages lies in its ability to handle multivariate datasets, allowing the simultaneous analysis of numerous variables. This capability is crucial for identifying novel biomarkers and constructing predictive models for trait selection and stress response analysis (Reel *et al.*, 2021). While the integration of multiple omics layers holds promise for plant trait enhancement, practical challenges such as scalability and data harmonization continue to limit its widespread application. To navigate the complications related to high-dimensional data, ML provides efficient protocols for selection and prioritization of features, which employ

ML algorithms to analyze RNA-seq data for identifying genes associated with salt stress in rice. Techniques like principal component analysis (PCA) and least absolute shrinkage and selection operator (LASSO) have proven effective in reducing dimensionality and highlighting biologically relevant gene submodules. Owing to its adaptability and predictive accuracy, ML is widely employed for phenotype prediction, biomarker discovery, and classification tasks in large-scale omics research. Machine learning excels in pattern extraction capabilities, offering distinct advantages for omics research, where revealing obscured nonlinear variable relationships can yield novel biological discoveries. Its adaptability and limited dependence on data distribution assumptions render ML architectures ideal for heterogeneous plant biology datasets shaped by biological and technical variance. ML additionally facilitates exploratory analytics, enabling hypothesis generation and detection of uncharted molecular interactions.

In multi-omics investigations, gene regulatory network (GRN) reconstruction frequently constitutes a primary objective. While transcription factor binding data from ChIP-seq remains scarce in botanical systems, transcriptome-derived regulatory mapping has gained prominence (Bubb and Deal, 2020). Conventional correlation models and mutual information (MI) approaches, despite widespread application, inadequately resolve regulatory directionality and temporal expression dynamics. Probabilistic graphical models (PGMs) like GENIST and JRmGRN address these constraints yet require high-resolution spatiotemporal expression profiles.

ML innovations have transformed dynamic GRN inference, particularly through multi-omics synthesis for enhanced prediction (Ko and Brandizzi, 2020). The iDREM framework employs hidden Markov models to reconstruct stress-responsive temporal GRNs, integrating transcriptomic, epigenomic, and proteomic data. Single-cell RNA sequencing advancements have further broadened the GRN analysis scope. Computational tools, including GRNBoost2 (extending GENIE3) and the SCENIC pipeline, now enable cell-type-specific GRN construction, with SCENIC processing large-scale datasets (e.g. 50,000 cells, 10,000 genes) within hours (Van de Sande *et al.*, 2020). Consequently, ML's ascendancy in plant omics stems from its proficiency in managing intricate high-dimensional data, synthesizing heterogeneous datasets, predicting phenotypic outcomes, and enabling discovery-driven analytics. This catalytic influence, evident in regulatory network decoding, biomarker identification, and trait enhancement, underscores ML's transformative role in plant systems biology.

14.5 AI Integration Into Plant Omics Against Stress

AI-driven omics approaches are redefining the landscape of plant defense research, offering an advanced intersection of molecular biology and computational innovation. While conventional methods have long grappled with the intricacies of deciphering complex regulatory networks involving genes, proteins, and metabolites, AI-enabled omics platforms are emerging as powerful tools to overcome these limitations. These approaches significantly accelerate the identification of key regulatory elements in plant defense mechanisms, facilitate biomarker discovery, and uncover subtle molecular patterns that were previously inaccessible through traditional analyses.

By integrating diverse multi-omics datasets, AI technologies enable a comprehensive and systems-level understanding of how plants perceive and respond to biotic and abiotic stresses. Such integration not only enhances mechanistic insight but also supports the prediction of defense responses under varied environmental conditions. As AI algorithms become more sophisticated, their application is expected to play a pivotal role in developing stress-resilient crop varieties, fine-tuning agricultural management strategies, and promoting sustainable food systems.

14.5.1 AI-mediated genomics

ML algorithms have become essential for identifying stress-resistance genes, empowering plant scientists to enhance crop resilience and productivity. For example, researchers implemented an adapted support vector machine (SVM) framework to delineate drought-tolerance determinants in *Arabidopsis thaliana*. Concurrently,

Bayesian methodologies demonstrated superior efficacy in detecting maize SNPs correlated with drought adaptation. ML applications extend beyond hydric stress; Wang *et al.* (2013) deployed SVM architectures to predict salinity-tolerance genes in rice, while Ravari *et al.* (2015) leveraged ANNs to characterize salt-responsive biomarkers in Iranian wheat germplasm. Schwarz *et al.* (2020) further employed ML to decode Fe-deficiency responsive *cis*-regulatory motifs in *Arabidopsis* root systems.

Within plant pathology, kernel-based classifiers, particularly SVM derivatives, dominate resistance gene identification. Pal *et al.* (2016) validated their predictive accuracy for disease-resistance proteins, while Sperschneider *et al.* (2016) pioneered EFFECTORP, the first ML classifier specialized for fungal effector detection. Beyond resistance factors, ML also illuminates susceptibility determinants, expanding its utility in crop protection. ML frameworks are revolutionizing single-cell plant genomics by enabling cellular heterogeneity mapping, regulatory network inference, and novel cell-type discovery.

Silva *et al.* (2019) and Raimundo *et al.* (2021) documented ML's efficacy in generating dimensionality-reduced embeddings, developmental trajectory modeling, and multimodal omics integration. To address scRNA-seq limitations like sparse sequencing depth and stochastic amplification artifacts, ML solutions, including SIMLR (Wang *et al.*, 2017) and neural network architectures (Lin *et al.*, 2017), enhance data robustness and biological interpretability.

Despite these promising developments, the use of ML in stress-resilience gene discovery remains largely confined to a limited number of plant species. Expanding the application of these tools to a broader range of economically important crops is essential for comprehensively decoding stress-resistance mechanisms, and accelerating the breeding of climate-resilient varieties.

14.5.2 AI-mediated transcriptomics

Asefpour Vakilian (2020) conducted a pivotal analysis of microRNA (miRNA) regulatory functions during plant stress, examining *Arabidopsis thaliana* responses to both biological and environmental pressures. The research implemented sophisticated feature selection techniques to quantify individual miRNA contributions to abiotic stress adaptation, including drought, salinity, low temperature, and heat exposure. Information-theoretic approaches identified critical stress-responsive miRNAs (miRNA169, miRNA159, miRNA396, miRNA393). Among regression algorithms, decision trees (DTs), support vector machines (SVMs), and naïve Bayes (NB), the Gaussian-kernel SVM demonstrated exceptional efficacy ($R^2 = 0.96$) in predicting stress intensity from miRNA expression profiles.

Contrary to established paradigms of distinct signaling pathways for abiotic vs biotic stresses, rice research revealed substantial response overlap. Cross-dataset meta-analysis showed that approximately 70% of differentially expressed genes were co-regulated under both stress categories. Recursive SVMs and random forest classifiers distinguished stress types using expression signatures, with recursive SVM achieving perfect classification accuracy and identifying 196 critical discriminative genes.

Meng *et al.* (2021) employed supervised classification to detect cold-responsive genes across species. Models trained solely on genomic assembly features maintained effectiveness despite slight accuracy reductions, successfully predicting cold responses across phylogenetically related plants – including knowledge transfer between cold-sensitive and cold-tolerant species. Multi-species training data consistently outperformed single-species models, highlighting the potential of cross-taxa ML frameworks in comparative stress genomics.

Transcriptional reprogramming during thermal stress was investigated by Zhou *et al.* (2022) through maize transcriptomics. Motif enrichment near transcription start-sites (TSSs) of thermally responsive genes informed predictive models, with unmethylated TSS-proximal motifs yielding the highest accuracy. Performance variations across maize genotypes indicated cultivar-specific regulatory mechanisms.

Addressing non-coding RNA identification challenges, Pradhan *et al.* (2023) developed an AI framework predicting abiotic stress-associated long non-coding (lnc) RNAs to support resilient crop breeding. Their SVM classifier with k-mer feature representation achieved 68.84% cross-validation and 76.23% independent test accuracy, deployed via the ASLncR web tool. Complementary work produced an SVM model with pseudo k-tuple composition features predicting stress-specific miRNAs (87.71–90.15% cross-validation accuracy), accessible through ASmiR. Meher *et al.* (2022) extended this to pre-miRNA prediction using pseudo-k-tuple features, reporting 65.64–77.94% auROC/auPRC values and 62.33–69.21% test accuracy via the ASRmiRNA platform. Collectively, these advances demonstrate machine learning's capacity to decode plant stress regulation and accelerate crop improvement through computational biology innovation.

14.5.3 AI-assisted proteomics

Recent studies have shown the growing role of machine learning in plant biology, particularly for predicting stress-related genes and protein–protein interactions (PPIs). Meher *et al.* (2022) applied computational and ML approaches to identify genes responsive to different abiotic stresses. Using a support vector machine (SVM) framework, they reported prediction accuracies between 60% and 78%, and made their methodology available via the web server ASRpro. In the area of plant–pathogen PPIs, Yang *et al.* (2019) built an RF-based model that combined sequence-derived features with network topology, achieving higher predictive performance. They also created InterSPPI, an online resource that enables further exploration of pathogen–host interactions. Extending this line of research, Karan *et al.* (2023) designed ML models to predict interactions between rice (*Oryza sativa*) and the blast fungus *Magnaporthe grisea*. Their system performed with up to 95% accuracy on reference datasets, and evaluation across multiple host–pathogen pairs confirmed that the models were particularly optimized for the rice–blast pathosystem. More recently, Ahmed *et al.* (2023) proposed a new activation function for deep learning, termed Sigmoid-Integrated Gaussian Error Linear Unit (SIELU). When incorporated into protein classification models, SIELU boosted performance, yielding accuracies in the range of 81–95%, thus offering a promising tool for predicting protein functions in stress-response contexts.

14.5.4 AI-mediated metabolomics

Plant classification was investigated by analyzing variations in metabolite content across 216 species (Liu *et al.*, 2017). The approach combined network-based clustering with vectorized plant representation, allowing metabolites with similar structures to be grouped systematically. By applying hierarchical clustering, the study was able to organize plants in a manner consistent with evolutionary relationships, even when working with incomplete metabolic data. Beyond taxonomy, the findings also pointed to the potential use of metabolite composition as an indicator of nutritional and medicinal characteristics, offering new insights into species identification and trait prediction.

In a related direction, Fürtauer *et al.* (2018) explored how metabolic pathways respond to abiotic stress. Using *Arabidopsis thaliana* wild-type and mutant lines with altered starch or sucrose metabolism, they subjected plants to high light intensity and low-temperature stress. The team applied ML to classify the metabolic and proteomic changes occurring under these conditions. From this analysis, they identified a core set of 23 proteins that served as reliable predictors across stress environments. Notably, 18 of these proteins were part of interconnected networks, underlining the intricate biochemical regulation plants deploy when adapting to stressful environmental cues.

14.5.5 AI-mediated phenomics

Deep learning frameworks demonstrate significant potential across diverse crop systems, including wheat, rice, potato, tomato, etc. Li *et al.* (2020) pioneered video-based disease detection employing Faster R-CNN and YOLOv3 architectures for real-time pest recognition, establishing novel evaluation metrics where their customized model exceeded existing frameworks in unseen rice datasets. For wheat stripe rust severity grading, Mi *et al.* (2020) developed C-DenseNet, integrating convolutional block attention modules, achieving 97.99% accuracy and surpassing DenseNet/ResNet baselines. Wang and Liu (2021) enhanced tomato leaf spot detection via MobileNetv2-YOLOv3 with GIoU loss, attaining 94.13% F1-scores in field applications. Hyperspectral CNNs enable viral disease identification in seed potatoes with high precision/recall, while Chen and Yuan (2019) combined CNNs with Region Proposal Networks for agricultural pest quantification (precision: 0.93; miss rate: 0.10). Feng *et al.* (2020) leveraged hyperspectral phenotyping to non-invasively monitor salinity responses in okra, with DL segmentation enabling precise trait characterization and spectral-based predictive modeling (R = 0.588–0.835). Beyond supervised paradigms, unsupervised methods facilitate dimensionality reduction (PCA, t-SNE), clustering, and regulatory network inference – particularly valuable for single-cell omics and cross-species knowledge transfer. Yan *et al.* (2020, 2021) applied these to maize population genetics, while NMF deciphers gene expression matrices in *Arabidopsis*/maize. Semi-supervised approaches address annotation gaps, exemplified by Moore *et al.* (2020) predicting tomato metabolic genes via *Arabidopsis* transfer learning, and Cheng *et al.* (2021) developing evolutionarily informed XGBoost models for maize nitrogen-use efficiency. Emerging single-cell challenges are addressed by adapted algorithms, including UMAP, MAGIC, and PHATE, originally developed for mammalian systems.

14.6 AI-Assisted Multi-omics Integration (MOI)

14.6.1 Strategies

Understanding how plants orchestrate their complex defense responses against pathogens requires looking beyond single layers of biological information. While powerful omics technologies, genomics (genes), transcriptomics (gene activity), proteomics (proteins), metabolomics (small molecules), and others, each provide a vital snapshot, the true story of plant immunity emerges only when these layers are integrated. Multi-omics data integration (MOI), supercharged by AI, is the essential process of weaving these diverse datasets into a unified, coherent understanding of the plant's defensive symphony. This integrated view is crucial for deciphering the intricate molecular mechanisms behind traits like disease resistance and accelerating the development of resilient crops.

However, integrating omics data is akin to assembling a complex, multidimensional puzzle where the pieces come from different boxes, are vastly numerous, and speak slightly different languages. A primary challenge is high dimensionality, the 'large p, small n' problem ('p' represents the vast number of measured features, 'n' denotes the limited number of biological samples) (Wu *et al.*, 2019). Imagine searching for a few critical defensive instructions hidden within millions of pages of biological blueprints, using only a handful of plant samples as your guide. This vast number of features (genes, proteins, metabolites) compared with samples makes AI models prone to finding misleading, spurious patterns instead of true biological signals related to defense, reducing the reliability of predictions. Compounding this is data heterogeneity. Each omics layer measures fundamentally different things: gene expression counts, protein abundance levels, categorical genetic markers, or metabolite concentrations. They follow different statistical distributions and carry varying levels of inherent biological 'noise.' Combining these distinct data types directly is like trying to average temperatures and rainfall; it requires sophisticated methods to handle their inherent differences fairly. Furthermore, biological systems are inherently interconnected, leading to highly correlated features and redundancy. In plant defense, a signal like a defense hormone (metabolite) surge might trigger the activation of dozens of related genes (transcriptomics) and proteins (proteomics). While this reflects the coordinated biological reality, it means that the datasets contain overlapping information, which can confuse AI models if not properly managed. Studying complex traits like disease resistance also brings the challenge of class imbalance (Fernández *et al.*, 2017). Resistant plants are often much rarer than susceptible ones in experimental populations. This uneven distribution is like training a security system mostly on footage of quiet hallways; the AI model can become biased toward predicting the common 'susceptible' outcome and miss the

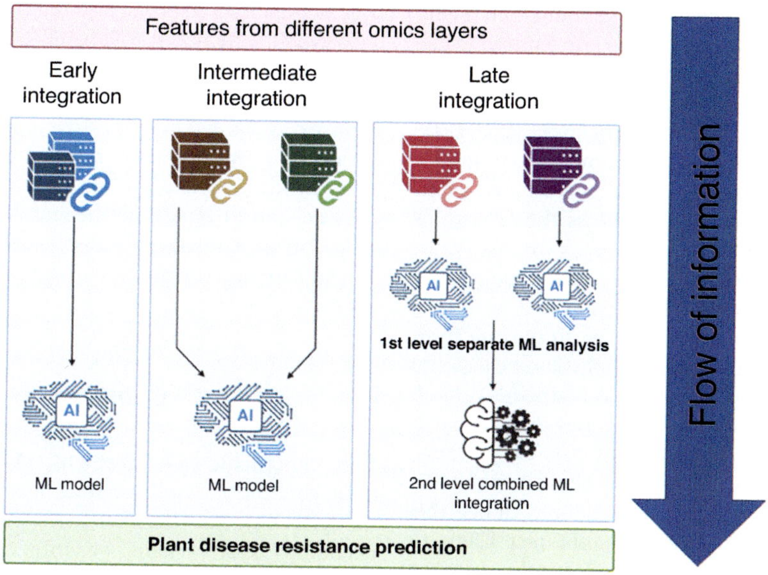

Fig. 14.2. Multi-omics integration strategies in plant defense AI. Comparing early (raw data fusion), intermediate/joint (cross-layer representation learning), and late (prediction stacking) approaches (Author's own image).

subtle signatures of the rare 'resistant' plants we urgently need to understand. Finally, we must respect the inherent non-negativity of many biological measurements; the amount of a protein or metabolite simply cannot be negative, a fundamental property that our integration and AI models must preserve.

The strategy chosen for integrating these diverse datasets significantly impacts the insights AI can extract, particularly for capturing the cross-talk essential in plant defense. MOI approaches are broadly categorized by when integration happens in the AI workflow, as shown in Fig. 14.2. Early integration combines the raw features from all omics layers (genes, proteins, metabolites, etc.) into one massive dataset before feeding it into an AI model (Cai *et al.*, 2022). Think of this as throwing all puzzle pieces from different boxes into one large pile. While simple and ensuring all data is present, this approach treats features as independent, ignoring the crucial biological relationships between layers (for example, how gene expression changes lead to protein changes). It also exacerbates the 'needle in a haystack' problem and struggles with data heterogeneity. In contrast, intermediate (or joint) integration is where AI truly shines for complex biological questions. Here, the model learns simultaneously from all omics datasets during training (Ballard *et al.*, 2024). It does not just mix raw data; it actively discovers a shared, meaningful representation that captures the essence and, critically, the interactions across the layers. Imagine assembling the puzzle while consciously looking for connections between pieces from different boxes, noticing how an edge shape from the transcriptomics box perfectly fits a color gradient from the proteomics box. Techniques like multiple kernel learning (using different computational 'lenses' for each data type) or powerful deep learning architectures act as the guiding intelligence. This approach is often the most powerful for plant defense, as it excels at capturing the vital cross-talk (for example, pathogen detection via metabolites triggering gene networks that activate defense proteins), handles heterogeneity intelligently, and is particularly effective for predicting nuanced resistance traits. The third strategy, late integration, builds separate AI models for each individual omics layer and then combines their predictions using a second 'meta model' (Mohamedikbal *et al.*, 2025). This is like assembling each puzzle box separately, taking photos of the completed puzzles, and then trying to combine those photos into one picture. While simpler to implement initially and allowing tailored models per data type, this approach fundamentally fails to capture the deep, synergistic interactions between different omics layers, the very essence of how a coordinated plant defense response unfolds across the molecular landscape.

For the critical mission of advancing plant defense, MOI powered by AI is not merely a technical exercise; it is the key to moving from fragmented snapshots to a holistic movie of the plant's immune response. A pathogen attack triggers a cascade: specific genes are activated, leading to the production of defense proteins, which in turn alter the plant's metabolic profile. AI-driven MOI allows us to model this entire, dynamic symphony, not just listen to isolated instruments. By effectively weaving together genomics, transcriptomics, proteomics, metabolomics, and beyond, we gain unprecedented insights into the molecular networks underpinning disease resistance. This integrated knowledge is fundamental for precisely identifying key resistance markers, understanding defense regulation, and ultimately guiding the development of smarter breeding strategies and more resilient crops capable of withstanding the ever-evolving threats they face. AI becomes the essential weaver, turning the complex threads of multi-omics data into the tapestry of plant immunity.

14.6.2 MOI applications

Table 14.1 catalogues state-of-the-art AI/ML methodologies applied across diverse omics layers, including genomics (SNPs-haplotypes), transcriptomics (RNA-seq–dual-species), proteomics (DAP-seq–DIA-MS), metabolomics, and advanced phenomics (UAV-hyperspectral-RGB-thermal imaging) for predicting and dissecting plant defense mechanisms. Key advances demonstrated include the superior predictive power of full SNP sets over selected markers, the critical capture of genotype–environment (G × E) interactions and local epistasis through feature engineering, the high accuracy of multi-omics integration (e.g. iMAP using RF) over single-omics approaches, and the ability of deep learning models (CNNs, CRNNs, e.g. CNNs, Autoencoders) to extract novel regulatory insights and biomarkers directly from raw omics or imaging data. These AI-driven approaches, leveraging multi-omics integration (MOI), significantly enhance trait prediction accuracy, uncover causal biological mechanisms, enable cross-species knowledge transfer, and facilitate non-destructive phenotyping for accelerated resistance breeding.

14.7 Challenges

The synergistic advancement of biological data generation and machine learning technologies unlocks transformative potential for decoding plant systems, yet integration faces significant challenges rooted in ML's inherent context-dependency, necessitating meticulous experimental design as algorithm-specific strengths/ limitations impact performance across heterogeneous biological contexts. While proliferating omics data offers unprecedented insights, pervasive noise, sparsity, and inconsistency impede robust feature extraction, exacerbated by class imbalance (often mitigated via SMOTE oversampling) and analytical uncertainty, requiring frameworks that discriminate biological signals from artifacts. Persistent overfitting, particularly in DL, compromises generalizability, countered by regularization techniques like neuron dropout during neural network training. Model efficacy hinges critically on pre-processing protocols, hyperparameter tuning, and domain expertise, with researcher interpretation essential for converting outputs into actionable biological insights. Plant systems further confront big data's '5Vs' challenges (variety, volume, veracity, velocity, value), demanding specialized methods for high-dimensionality, sparsity, and multicollinearity. The 'black-box' interpretability constraints of complex models, alongside the imperative for biologically relevant predictions, underscore the necessity for cross-disciplinary collaboration among computational biologists, omics researchers, and agronomists to harness AI's full potential in crop resilience applications.

Table 14.1. AI/ML models and multi-omics integration (MOI) approaches for enhanced prediction of plant disease resistance traits (adapted from Mohamedikbal *et al.*, 2025)

Omics feature	AI/ML models	Application	Key insights and advances	Impact on trait prediction
SNPs	RF, SVM	Bacterial wilt resistance (common bean)	Full SNP sets (4568) outperform pre-selected SNPs (14) in most models (accuracy: 0.37–0.58 vs 0.30–0.53). RF occasionally excels with selected SNPs for specific isolates.	Moderate accuracy; highly context-dependent.
Haplotype blocks	LD-based models	Fusarium head blight (wheat), local epistasis	Captures linkage disequilibrium (LD) and local epistasis; improves accuracy over standard SNPs in wheat. Effectiveness varies by trait/species.	↑ Accuracy via reduced dimensionality and epistasis capture.
Genotype–environment interactions	Feature-engineered ML models	Phenotypic adaptability (maize/legumes)	Integrates genomic + environmental covariates; captures G × E effects on disease resistance.	↑ Accuracy by modeling environmental influence on traits.
Dual-species RNA-seq	SVM, RF, XGBoost, DNN	*Arabidopsis*, *Botrytis cinerea* disease severity	XGBoost achieved 72.3% accuracy in predicting final disease outcomes using early infection transcriptome data. Models transferable to other pathogens (RF best for *Sclerotinia*).	Enables early disease outcome prediction and cross-pathogen transfer learning.
RNA-seq + DAP-seq	CRNNs	Soybean defense against *Phytophthora sojae*	Predicted novel TF binding sites (WRKY/RAV families) with 89–90% accuracy. Cross-species TF predictions: >90% (soybean families), 60% (*Arabidopsis*-to-soybean).	Reveals regulatory mechanisms and enables cross-species knowledge transfer.

Continued

Table 14.1. Continued

Omics feature	AI/ML models	Application	Key insights and advances	Impact on trait prediction
Multi-omics integration (iMAP)	RF, SVM, XGBoost, Logistic Regression	*Sclerotinia* resistance in oilseed rape	RF with combined features (GWAS, haplotypes, WGCNA, functional annotations) achieved 0.95 accuracy vs 0.78 for single SNP models. Identified 7 calcium signaling resistance genes.	↑ 22% accuracy over genomics-only approaches, and pinpoints causal mechanisms.
Comparative transcriptomics	Conventional analysis (pre-ML)	Legume–pathogen interactions (chickpea, bean, soybean)	Revealed pathogen survival genes in susceptible chickpea (*Fusarium*), upregulated sugar transporters in bean (*Colletotrichum*), and immunity genes via GWAS, RNA-seq fusion in soybean.	Provides mechanistic targets for breeding (non-ML baseline).
Multi-omics regulatory networks	Weighted Gene Autoencoder (DL)	Rice–*Magnaporthe oryzae* interactions	Integrated genomic, transcriptomic, and proteomic data to predict host–pathogen regulatory interactions.	Enables systems-level understanding of infection mechanisms.
Direct proteomic classification	Convolutional neural networks (CNNs)	Human disease states (DIA, MS data)	Classified disease states from raw proteomic data (without protein ID) by converting MS data to an image-like tensor.	Superior to ID-based methods; adaptable to legume disease classification.
Proteomic, Transcriptomic	Conventional integration	Soybean, AMF symbiosis, and virus resistance	Revealed defense proteins (PAL, CDPK) in AMF-enhanced root rot resistance. Identified stress pathway proteins in virus response.	Provides mechanistic biomarkers for breeding (pre-ML foundation).
Metabolite biomarkers	Conventional analysis	Legume–pathogen interactions	Identified defense metabolites: sugars/secondary metabolites (soybean, *Phytophthora*), palatinitol/L-proline (lentil, *Fusarium*), ROS/cell wall modulators (bean, *Fusarium*).	Provides actionable biomarkers for resistance breeding.

Metabolomics, microbiomics	RF, Gradient Boost, AdaBoost	Huanglongbing disease (citrus)	RF achieved 97.22% accuracy in identifying disease biomarkers. Integrated with KEGG revealed defense pathways (e.g. ABC transporters).	Near-perfect classification enables mechanistic discovery.
Visible RGB + thermal imaging	SVM, RF, Cubist, KNN + model combination	Wilt severity in chickpea	Combines symptom detection (RGB) with physiological stress signals (thermal). Model stacking improved prediction accuracy beyond single algorithms.	Enables non-destructive field assessment of dynamic disease progression.
UAV Time, Series + Environmental	Support vector regression	Bacterial blight in rice	Integrated spectral + temperature data achieved Rp^2=0.85 across locations. Highest accuracy when combining remote sensing with climate covariates.	Maintains performance across geographical contexts.
Hyperspectral + multi-omics	EG-CNN (Explainable Graph-CNN)	Disease classification	95.5% accuracy classifying 4 diseases using gene expression + metabolites + hyperspectral images. Saliency maps revealed disease-specific decision patterns.	Combines high accuracy with biological interpretability.
Imaging + genomics + metabolomics	RF	Yield prediction (maize) / Disease proxy	Imaging + omics boosted accuracy by 34% vs genomics alone. Imaging features showed higher importance than SNPs in RF models.	Demonstrates imaging as a superior predictor for complex traits.

14.8 Conclusion and Future Perspective

Future advancements will critically depend on integrating artificial intelligence with multi-layer molecular datasets (genomics, transcriptomics, proteomics, metabolomics) to enable pre-symptomatic plant disease identification. Sophisticated computational algorithms can decipher subtle pathogen-induced molecular shifts within high-dimensional omics data, facilitating proactive health diagnostics with enhanced precision. Concurrently, merging AI-enhanced omics with high-throughput phenotyping creates powerful frameworks for real-time physiological monitoring, linking molecular responses to observable traits to elucidate dynamic stress adaptation mechanisms. These integrated pipelines provide unprecedented resolution in characterizing plant–pathogen dynamics.

AI-omics convergence further enables predictive simulation models of disease trajectories across crop populations. Machine and deep learning architectures incorporating genetic, environmental, and molecular variables can forecast outbreaks and evaluate intervention efficacy, becoming indispensable tools for agricultural preparedness. AI-guided breeding programs will increasingly leverage resistance-associated markers, genes, molecular signatures, and metabolic pathways to develop cultivars with strengthened innate immunity tailored to specific agroecological conditions.

Precision agriculture benefits significantly through AI-mediated integration of omics insights with pathogen biomolecular profiles and environmental data. Site-targeted interventions (e.g. optimized pesticide deployment) enhance efficacy while reducing ecological footprints and operational costs. Climate change-induced multi-stress scenarios necessitate decoding concurrent biotic–abiotic response networks via AI-augmented omics, revealing how plants prioritize defenses during compound stressors like pathogen–drought–salinity interactions –knowledge vital for developing integrated countermeasures.

Remote sensing, synergized with computational omics, establishes novel real-time crop stress monitoring platforms. Spatial–temporal vegetation health data combined with molecular profiles enables comprehensive field response analysis, refining intervention timing against yield-limiting stresses. This chapter synthesizes omics technologies in plant defense against diverse stresses, highlighting ML's transformative capacity across biological scales, from genomic variants to metabolic fluxes, in extracting biologically significant patterns more efficiently than conventional methods. While technical implementation challenges persist, evolving computational frameworks coupled with expanding plant sequence repositories accelerate gene discovery, regulatory network mapping, and trait identification. These advances empower the development of nutritionally enhanced, stress-adapted crops to strengthen global food security amid escalating environmental pressures.

References

Ahmed, B., Haque, M.A., Iquebal, M.A., Jaiswal, S., Angadi, U.B. *et al.* (2023) DeepAProt: Deep learning based abiotic stress protein sequence classification and identification tool in cereals. *Frontiers in Plant Science* 13, 1008756.

Aizat, W.M. and Hassan, M. (2018) Proteomics in systems biology. In: Aizat, W., Goh, H.H. and Baharum, S. (eds) *Omics Applications for Systems Biology*, Vol. 1102. Advances in Experimental Medicine and Biology, Springer, Cham, pp. 31–49.

Asefpour Vakilian, K. (2020) Machine learning improves our knowledge about miRNA functions towards plant abiotic stresses. *Scientific Reports* 10(1), 3041.

Avelino, J., Cristancho, M., Georgiou, S., Imbach, P., Aguilar, L. *et al.* (2015) The coffee rust crises in Colombia and Central America (2008–2013): Impacts, plausible causes and proposed solutions. *Food Security* 7(2), 303–321.

Ballard, J.L., Wang, Z., Li, W., Shen, L. and Long, Q. (2024) Deep learning-based approaches for multi-omics data integration and analysis. *BioData Mining* 17(1), 38.

Bhardwaj, A.R., Joshi, G., Kukreja, B., Malik, V., Arora, P. *et al*. (2015) Global insights into high temperature and drought stress regulated genes by RNA-seq in economically important oilseed crop *Brassica juncea*. *BMC Plant Biology* 15(1), 9.

Bonetta, R. and Valentino, G. (2020) Machine learning techniques for protein function prediction. *Proteins: Structure, Function, and Bioinformatics* 88(3), 397–413.

Borah, P., Teja, A., Jha, S.A. and Bhattacharyya, D.K. (2020) TUKNN: A parallel KNN algorithm to handle large data. In: Patgiri, R., Bandyopadhyay, S., Borah, M.D. and Thounaojam, D.M. (eds) *Big Data, Machine Learning, and Applications*, Vol. 1317. BigDML 2019. Communications in Computer and Information Science, Springer, Cham.

Bubb, K.L. and Deal, R.B. (2020) Considerations in the analysis of plant chromatin accessibility data. *Current Opinion in Plant Biology* 54, 69–78.

Cai, Z., Poulos, R.C., Liu, J. and Zhong, Q. (2022) Machine learning for multi-omics data integration in cancer. *Iscience* 25(2).

Chen, L. and Yuan, Y. (2019) Agricultural disease image dataset for disease identification based on machine learning. In: Li, J., Meng, X., Zhang, Y., Cui, W. and Du, Z. (eds) *Big Scientific Data Management*, Vol. 11473. BigSDM 2018. Lecture Notes in Computer Science, Springer, Cham.

Cheng, C.Y., Li, Y., Varala, K., Bubert, J., Huang, J. *et al*. (2021) Evolutionarily informed machine learning enhances the power of predictive gene-to-phenotype relationships. *Nature Communications* 12(1), 5627.

Feng, X., Zhan, Y., Wang, Q., Yang, X., Yu, C. *et al*. (2020) Hyperspectral imaging combined with machine learning as a tool to obtain high-throughput plant salt-stress phenotyping. *The Plant Journal* 101(6), 1448–1461.

Fernández, A., del Río, S., Chawla, N.V. and Herrera, F. (2017) An insight into imbalanced big data classification: Outcomes and challenges. *Complex and Intelligent Systems* 3(2), 105–120.

Fürtauer, L., Pschenitschnigg, A., Scharkosi, H., Weckwerth, W. and Nägele, T. (2018) Combined multivariate analysis and machine learning reveals a predictive module of metabolic stress response in *Arabidopsis thaliana*. *Molecular Omics* 14(6), 437–449.

Ghosal, S., Blystone, D., Singh, A.K., Ganapathysubramanian, B., Singh, A. *et al*. (2017) Interpretable deep learning applied to plant stress phenotyping. *arXiv Preprint 1710.08619*.

Jiao, Y. and Du, P. (2016) Performance measures in evaluating machine learning based bioinformatics predictors for classifications. *Quantitative Biology* 4(4), 320–330.

John, M., Haselbeck, F., Dass, R., Malisi, C., Ricca, P. *et al*. (2022) A comparison of classical and machine learning-based phenotype prediction methods on simulated data and three plant species. *Frontiers in Plant Science* 13, 932512.

Karan, B., Mahapatra, S., Sahu, S.S., Pandey, D.M. and Chakravarty, S. (2023) Computational models for prediction of protein–protein interaction in rice and *Magnaporthe grisea*. *Frontiers in Plant Science* 13, 1046209.

Ko, D.K. and Brandizzi, F. (2020) Network-based approaches for understanding gene regulation and function in plants. *The Plant Journal* 104(2), 302–317.

Kubicek, A., Jopp, F., Breckling, B., Lange, C. and Reuter, H. (2015) Context-oriented model validation of individual-based models in ecology: A hierarchically structured approach to validate qualitative, compositional and quantitative characteristics. *Ecological Complexity* 22, 178–191.

Kumar, V. and Jain, M. (2015) The CRISPR–Cas system for plant genome editing: Advances and opportunities. *Journal of Experimental Botany* 66(1), 47–57.

Li, D., Wang, R., Xie, C., Liu, L., Zhang, J. *et al*. (2020) A recognition method for rice plant diseases and pests video detection based on deep convolutional neural network. *Sensors* 20(3), 578.

Li, L., Snyder, J.C., Pelaschier, I.M., Huang, J., Niranjan, U.N. *et al*. (2016) Understanding machine-learned density functionals. *International Journal of Quantum Chemistry* 116(11), 819–833.

Lin, C., Jain, S., Kim, H. and Bar Joseph, Z. (2017) Using neural networks for reducing the dimensions of single-cell RNA-seq data. *Nucleic Acids Research* 45(17), e156.

Liu, D., Mewalal, R., Hu, R., Tuskan, G.A. and Yang, X. (2017) New technologies accelerate the exploration of non-coding RNAs in horticultural plants. *Horticulture Research* 4, 17031.

Mangin, B., Casadebaig, P., Cadic, E., Blanchet, N., Boniface, M.C. *et al*. (2017) Genetic control of plasticity of oil yield for combined abiotic stresses using a joint approach of crop modelling and genome-wide association. *Plant, Cell and Environment* 40(10), 2276–2291.

Maron, J.L., Baer, K.C. and Angert, A.L. (2014) Disentangling the drivers of context-dependent plant–animal interactions. *Journal of Ecology* 102(6), 1485–1496.

McCulloch, W.S. and Pitts, W. (1943) A logical calculus of the ideas immanent in nervous activity. *The Bulletin of Mathematical Biophysics* 5(4), 115–133.

Meher, P.K., Begam, S., Sahu, T.K., Gupta, A., Kumar, A. *et al.* (2022) ASRmiRNA: Abiotic stress-responsive miRNA prediction in plants by using machine learning algorithms with pseudo k-tuple nucleotide compositional features. *International Journal of Molecular Sciences* 23(3), 1612.

Meng, X., Liang, Z., Dai, X., Zhang, Y., Mahboub, S. *et al.* (2021) Predicting transcriptional responses to cold stress across plant species. *Proceedings of the National Academy of Sciences* 118(10), e2026330118.

Mi, Z., Zhang, X., Su, J., Han, D. and Su, B. (2020) Wheat stripe rust grading by deep learning with attention mechanism and images from mobile devices. *Frontiers in Plant Science* 11, 558126.

Mohamedikbal, S., Al-Mamun, H.A., Bestry, M.S., Batley, J. and Edwards, D. (2025) Integrating multi-omics and machine learning for disease resistance prediction in legumes. *Theoretical and Applied Genetics* 138(7), 1–21.

Moore, B.M., Wang, P., Fan, P., Lee, A., Leong, B. *et al.* (2020) Within- and cross-species predictions of plant specialized metabolism genes using transfer learning. *In Silico Plants* 2(1), diaa005.

Muthamilarasan, M., Singh, N.K. and Prasad, M. (2019) Multi-omics approaches for strategic improvement of stress tolerance in underutilized crop species: A climate change perspective. *Advances in Genetics* 103, 1–38.

Niazian, M. and Niedbała, G. (2020) Machine learning for plant breeding and biotechnology. *Agriculture* 10(10), 436.

Pabuayon, I.C.M., Pabuayon, I.L.B., Singh, R.K., Ritchie, G.L. and de Los Reyes, B.G. (2022) Applicability of hyperspectral imaging during salinity stress in rice for tracking Na+ and K+ levels in planta. *PLoS ONE* 17(7), e0270931.

Pal, T., Jaiswal, V. and Chauhan, R.S. (2016) DRPPP: A machine learning based tool for prediction of disease resistance proteins in plants. *Computers in Biology and Medicine* 78, 42–48.

Pradhan, U.K., Meher, P.K., Naha, S., Rao, A.R. and Gupta, A. (2023) ASLncR: A novel computational tool for prediction of abiotic stress, responsive long non-coding RNAs in plants. *Functional and Integrative Genomics* 23(2), 113.

Raimundo, F., Meng-Papaxanthos, L., Vallot, C. and Vert, J.P. (2021) Machine learning for single-cell genomics data analysis. *Current Opinion in Systems Biology* 26, 64–71.

Ramalingam, A., Kudapa, H., Pazhamala, L.T., Weckwerth, W. and Varshney, R.K. (2015) Proteomics and metabolomics: Two emerging areas for legume improvement. *Frontiers in Plant Science* 6, 1116.

Ravari, S.Z., Dehghani, H. and Naghavi, H. (2015) Evaluation of relationship between salinity stress tolerance indices and some physiological traits in bread wheat. *Iranian Journal of Field Crop Science* 46(3), 423–432.

Raza, A., Su, W., Hussain, M.A., Mehmood, S.S., Zhang, X. *et al.* (2021) Integrated analysis of metabolome and transcriptome reveals insights for cold tolerance in rapeseed (*Brassica napus* L.). *Frontiers in Plant Science* 12, 721681.

Reel, P.S., Reel, S., Pearson, E., Trucco, E. and Jefferson, E. (2021) Using machine learning approaches for multi-omics data analysis: A review. *Biotechnology Advances* 49, 107739.

Schwarz, B., Azodi, C.B., Shiu, S.H. and Bauer, P. (2020) Putative *cis*-regulatory elements predict iron deficiency responses in *Arabidopsis* roots. *Plant Physiology* 182(3), 1420–1439.

Shen, D., Wang, Y., Chen, X., Srivastava, V. and Toffolatti, S.L. (2022) Advances in multi-omics study of filamentous plant pathogens. *Frontiers in Microbiology* 13, 998501.

Silva, J.C.F., Teixeira, R.M., Silva, F.F., Brommonschenkel, S.H. and Fontes, E.P. (2019) Machine learning approaches and their current application in plant molecular biology: A systematic review. *Plant Science* 284, 37–47.

Sperschneider, J., Gardiner, D.M., Dodds, P.N., Tini, F., Covarelli, L. *et al.* (2016) EffectorP: Predicting fungal effector proteins from secretomes using machine learning. *New Phytologist* 210(2), 743–761.

Spindel, J.E., Dahlberg, J., Colgan, M., Hollingsworth, J. and Sievert, J. (2018) Association mapping by aerial drone reveals 213 genetic associations for *Sorghum bicolor* biomass traits under drought. *BMC Genomics* 19(1), 679.

Van de Sande, B., Flerin, C., Davie, K., De Waegeneer, M., Hulselmans, G. *et al.* (2020) A scalable SCENIC workflow for single-cell gene regulatory network analysis. *Nature Protocols* 15(7), 2247–2276.

Wang, X. and Liu, J. (2021) Multiscale parallel algorithm for early detection of tomato gray mold in a complex natural environment. *Frontiers in Plant Science* 12, 620273.

Wang, B., Zhu, J., Pierson, E., Ramazzotti, D. and Batzoglou, S. (2017) Visualization and analysis of single-cell RNA-seq data by kernel-based similarity learning. *Nature Methods* 14(4), 414–416.

Wang, J., Chen, L., Wang, Y., Zhang, J., Liang, Y. *et al.* (2013) A computational systems biology study for understanding salt tolerance mechanism in rice. *PLoS ONE* 8(6), e64929.

Wang, M., White, N., Grimm, V., Hofman, H., Doley, D. *et al.* (2018) Pattern-oriented modelling as a novel way to verify and validate functional–structural plant models: A demonstration with the annual growth module of avocado. *Annals of Botany* 121(5), 941–959.

Woolfson, K.N., Haggitt, M.L., Zhang, Y., Kachura, A., Bjelica, A. *et al.* (2018) Differential induction of polar and non-polar metabolism during wound-induced suberization in potato (*Solanum tuberosum* L.) tubers. *The Plant Journal* 93(5), 931–942.

Wu, C., Zhou, F., Ren, J., Li, X., Jiang, Y. *et al.* (2019) A selective review of multi-level omics data integration using variable selection. *High-Throughput* 8(1), 4.

Xavier, A. (2021) Technical nuances of machine learning: Implementation and validation of supervised methods for genomic prediction in plant breeding. *Crop Breeding and Applied Biotechnology* 21, e381421S2.

Yan, J. and Wang, X. (2022) Unsupervised and semi-supervised learning: The next frontier in machine learning for plant systems biology. *The Plant Journal* 111(6), 1527–1538.

Yan, J., Zou, D., Li, C., Zhang, Z., Song, S. *et al.* (2020) SR4R: An integrative SNP resource for genomic breeding and population research in rice. *Genomics, Proteomics and Bioinformatics* 18(2), 173–185.

Yan, J., Xu, Y., Cheng, Q., Jiang, S., Wang, Q. *et al.* (2021) LightGBM: Accelerated genomically designed crop breeding through ensemble learning. *Genome Biology* 22(1), 271.

Yang, S., Li, H., He, H., Zhou, Y. and Zhang, Z. (2019) Critical assessment and performance improvement of plant–pathogen protein–protein interaction prediction methods. *Briefings in Bioinformatics* 20(1), 274–287.

Yang, Z., Li, J.L., Liu, L.N., Xie, Q. and Sui, N. (2020) Photosynthetic regulation under salt stress and salt, tolerance mechanism of sweet sorghum. *Frontiers in Plant Science* 10, 1722.

Zhou, Y., Zheng, H., Huang, X., Hao, S., Li, D. *et al.* (2022) Graph neural networks: Taxonomy, advances, and trends. *ACM Transactions on Intelligent Systems and Technology (TIST)* 13(1), 1–54.

15 Enhancing Genomic Prediction through Kernel Methods

Osval A. Montesinos-López[1]*, José Elías Peregrina-Chavarría[1] and Roberto de la Rosa Santa Maria[2]

[1]Facultad de Telemática, Universidad de Colima, México; [2]Colegio de Postgraduados (CP), Campus Tabasco, Producción Agroalimentaria en el Trópico, H. Cárdenas, Tabasco, México

Abstract

Kernel methods are powerful tools for genomic prediction, offering both flexibility and robustness in modeling complex, nonlinear relationships between genetic markers and phenotypes. By transforming input data into higher-dimensional feature spaces, kernel methods effectively capture intricate patterns in genomic data that linear models often fail to detect. This chapter provides a comprehensive guide to implementing some of the most popular kernel methods (linear and Gaussian) for genomic prediction within a Bayesian framework and also provides a framework for building hybrid kernels. The implementation is demonstrated using real plant-breeding datasets, incorporating cross-validation strategies designed to mimic real-world applications. For comparison purposes, the performance of these methods is evaluated against conventional linear kernels to highlight their potential in improving prediction accuracy. Finally, we discuss the advantages of kernel methods, particularly their ability to handle high-dimensional data and their transformative impact on enhancing genomic prediction outcomes.

Keywords: Genomic prediction, High-dimensional data, Hybrid kernels, Kernels, Plant breeding

15.1 Introduction

The advent of genomic selection (GS) has revolutionized plant and animal breeding by enabling the prediction of breeding values using dense genome-wide marker information. This methodology accelerates genetic gain by allowing early selection of superior individuals, thereby reducing the generation interval and enhancing selection accuracy (Meuwissen *et al.*, 2001). In plant breeding, GS has been instrumental in improving complex traits such as yield, disease resistance, and abiotic stress tolerance, particularly in staple crops like wheat and maize (Crossa *et al.*, 2017). Similarly, in animal breeding, GS has transformed selection processes in dairy cattle, pigs, and poultry by facilitating the identification of animals with superior genetic merit at an early age (Hayes *et al.*, 2009).

Despite its promising potential, the implementation of GS in real-world breeding programs, especially those with limited resources, presents several challenges. Small breeding programs often face constraints such as limited financial resources, inadequate infrastructure, and small population sizes, which hinder the

*Corresponding author: oamontes1@ucol.mx, osval78t@gmail.com

 Plant Genomic Prediction: AI-Enhanced Models for Plant Genomics and Crop Improvement (ed. Jen-Tsung Chen)
DOI: 10.1079/9781836991328.0015

collection of high-quality phenotypic and genotypic data necessary for accurate genomic predictions (Goddard and Hayes, 2009). Moreover, the high costs associated with genotyping and phenotyping, along with the need for specialized statistical expertise, further impede the adoption of GS in these settings (Ibañez-Escriche and Gonzalez-Recio, 2011). Additionally, the accuracy of genomic predictions can be compromised in small populations due to reduced genetic diversity and the limited representation of rare alleles, which are critical for capturing the genetic architecture of complex traits (Daetwyler *et al.*, 2008).

To address these limitations and enhance the predictive accuracy of GS, various strategies have been proposed. One approach involves the optimization of training population design by ensuring adequate representation of genetic diversity and relatedness to the target population (Isidro *et al.*, 2015). Another strategy focuses on the integration of multi-environment and multi-trait data to capture genotype-by-environment interactions and pleiotropic effects, thereby improving the robustness of genomic predictions (Jarquín *et al.*, 2014). Furthermore, the incorporation of advanced statistical and machine learning methodologies, such as kernel-based models, has shown promise in modeling complex, nonlinear relationships between genotypes and phenotypes, which are often inadequately captured by traditional linear models (Gianola and van Kaam, 2008).

Kernel methods trace their origin to the theory of reproducing kernel Hilbert spaces (RKHS), a mathematical foundation developed in the mid-20th century that provides the theoretical backbone for a class of algorithms capable of handling complex, nonlinear data structures (Aronszajn, 1950). These methods gained significant traction in machine learning with the development of support vector machines and were formalized into practical tools by Schölkopf and Smola (2002). At their core, kernel methods work by implicitly mapping input data into a high-dimensional feature space, enabling the modeling of intricate relationships – such as epistatic interactions and dominance effects – without explicitly defining these interactions (Crossa *et al.*, 2019). This intrinsic flexibility has contributed to their growing adoption across various domains. In genomic prediction, kernel methods have demonstrated substantial empirical success in improving prediction accuracy for complex traits, particularly under scenarios involving non-additive genetic architectures (Crossa *et al.*, 2010; Gianola *et al.*, 2011). Beyond genomics, kernel methods have also shown robust performance in fields such as bioinformatics, image recognition, and natural language processing, where capturing nuanced, nonlinear relationships is essential (Schölkopf and Smola, 2002). The accumulation of empirical evidence across these disciplines underscores the versatility and effectiveness of kernel methods in modeling high-dimensional, structured data.

Despite their theoretical appeal and empirical success, several limitations have constrained the widespread and rapid adoption of kernel methods across fields, including GS. One primary concern is the computational complexity associated with kernel-based algorithms, particularly when applied to large-scale datasets. Kernel methods typically require the computation and inversion of an $n \times n$ kernel matrix, where n is the number of samples, which results in quadratic or cubic time and memory complexities (Williams and Rasmussen, 2006). This issue becomes especially critical in genomic datasets, which often comprise thousands of individuals and hundreds of thousands of markers. Moreover, the choice of kernel function and its associated hyperparameters significantly affects model performance, yet there exists no universally optimal kernel, making model tuning both non-trivial and computationally intensive (de los de Los Campos *et al.*, 2010). Another barrier is the limited interpretability of kernel-based models compared with linear approaches, which has historically reduced their appeal in disciplines that prioritize transparency and explainability, such as plant and animal breeding (Gianola and van Kaam, 2008). Finally, although empirical evidence supports their potential, the perceived methodological complexity and lack of user-friendly software implementations have further slowed their integration into standard breeding pipelines (Cuevas *et al.*, 2016). Overcoming these limitations requires not only advances in computational methods and hardware but also targeted efforts to develop accessible tools and training resources for breeders and geneticists.

Among the various kernel functions employed in genomic prediction, the Gaussian kernel (also known as the radial basis function kernel) is one of the most widely used due to its ability to model smooth, nonlinear relationships between individuals based on genetic distances (de los de Los Campos *et al.*, 2010). The Gaussian kernel has been effectively utilized in RKHS regression models to capture additive and non-additive genetic effects, thereby enhancing prediction accuracy for complex traits (Gianola and van Kaam, 2008). Another notable kernel function is the arc-cosine kernel, which emulates the behavior of deep neural networks by incorporating multiple layers of nonlinear transformations. This kernel has demonstrated competitive performance in genomic prediction tasks while offering computational advantages over traditional deep learning models (Cuevas *et al.*, 2019). Additionally, support vector machines (SVMs), which rely on kernel functions to construct optimal hyperplanes for classification and regression tasks, have been applied in genomic prediction with varying degrees of success, particularly in scenarios involving high-dimensional data and small sample sizes (González-Camacho *et al.*, 2018).

For this reason, in this chapter, we illustrate the application of kernel methods in the context of genomic prediction. Specifically, we implement linear and Gaussian kernels using real datasets. Additionally, we present a framework for constructing hybrid kernels, which combines any two existing kernels to generate two new ones. For comparison purposes, we evaluate the performance of the individual kernels and assess the impact of incorporating the hybrid kernels alongside the original ones.

15.2 Material and Methods

15.2.1 Datasets

In Table 15.1 we provide the nine datasets used. For those datasets with more than one environment in the modeling process, these were assumed as replications.

15.2.2 Statistical models

Four models (M1 to M4) were implemented; these are described next.

Models 1 and 2 (M1 and M2)

The predictive response model was considered to be:

$$Y_{ij} = \mu + g_j + \varepsilon_{ij} \tag{15.1}$$

where Y_{ij} denotes the response variable measured at genotype j at replication i; μ denotes a general mean; $g_j, j = 1, \ldots, J$, are the random effects of lines and ϵ_{ij} denotes the random error terms assumed normally distributed with mean 0 and variance σ^2. Furthermore, it is assumed that $\boldsymbol{g} = (g_1, \ldots, g_1)^T \sim N_J(\mathbf{0}, \sigma_g^2 G)$, $\boldsymbol{G}$ denotes the linear kernel computed as $\boldsymbol{G} = \boldsymbol{XX}^{\mathrm{T}}$; this matrix is called the genomic relationship matrix (VanRaden, 2008) in plant and animal breeding since is computed with a standardized matrix of SNPs (X coded as 0, 1 and 2. For this reason, the $\boldsymbol{G}$ matrix is assumed known before starting

Table 15.1. Brief descriptions of datasets used.

No.	Dataset name	No. environments	No. lines	No. markers	No. traits
1	DMario_Family_14	12	532	2744	3
2	Family_10	1	111	13,950	3
3	Pinus_Florida	1	926	4829	14
4	Potato_Suecia	6	253	11,080	9
5	Rice_IRRI	2	319	13,101	3
6	Rice_Kim	8	107	108,024	1
7	Wheat_Ingland	1	720	15,877	2
8	Wheat_Washington_2020	6	758	16,382	3
9	Wheat_Washington_2021	8	456	16,382	3

the modeling process. Then the model given in Equation 15.1 was implemented with the linear kernel, $\boldsymbol{G}$; this model is denoted as model G. But when the model given in Equation 15.1 was implemented with a Gaussian kernel $\boldsymbol{K}$ the model is denoted as K, and the computation of the Gaussian kernel was implemented as:

$$k\left(x_j, x_k\right) = \exp\left(-\frac{\|x_j - x_k\|^2}{\sigma^2}\right) \qquad (15.2)$$

where x_j and x_k are two p dimensional vectors, p represents the number of SNPs, while i and j represent two genotypes (individuals), σ^2 is the bandwidth of the kernel and $\|x_j - x_k\|^2 = \left[(x_{j1} - x_{k1})^2 + (x_{j2} - x_{k2})^2 + \cdots + (x_{jp} - x_{kp})^2\right]^{1/2}$. The kernel function $k\left(x_j, x_k\right)$ yields a value of 1 when x_j and x_k are identical, and gradually decreases toward 0 as the dissimilarity between x_j and x_k increases. Each element of the kernel matrix $\boldsymbol{K}$ is calculated using the function defined in Equation 15.2, with higher values corresponding to genotype pairs that are more genetically similar. The bandwidth parameter regulates the rate at which similarity decays with increasing Euclidean distance between the SNP vectors x_j and x_k. This parameter plays a critical role in both inference and prediction. When the bandwidth σ^2 is large, the Gaussian kernel values tend to approach 1, potentially leading to underfitting. Conversely, small values of σ^2 result in kernel entries close to 0, which may cause the random effect modeled through the kernel to become indistinguishable from the residual error in mixed models. The selection of an appropriate bandwidth can be achieved through cross-validation.

Model 3 (M3)

This model M3 combines in the same predictor both kernels used in models M1 and M2. For this reason this model is equal to:

$$Y_{ij} = \mu + g_j + g_{Gj} + \varepsilon_{ij} \qquad (15.3)$$

where Y_{ij} and ε_{ij} were described as in Models M1 and M2. Also g_j is described as in Model 1, that is, $\boldsymbol{g} = (g_1,\ldots,g_J)^T \sim N_J(\mathbf{0}, \sigma_g^2\mathbf{G})$. In a similar fashion $g_{Gj}, j = 1, \ldots, J$, denotes random effects of lines but under a Gaussian kernel that is, $\boldsymbol{g}_G = (g_1,\ldots,g_J)^T \sim N_J(\mathbf{0}, \sigma_g^2\boldsymbol{K})$, where $\boldsymbol{K}$ denotes the Gaussian kernel described in model M2.

Model M4

This model M4 is equal to M3 but with two additional terms that result from combining the linear and Gaussian kernel. For this reason, the model is equal to:

$$Y_{ij} = \mu + g_j + g_{Gj} + g_{CCj} + g_{PPj} + \varepsilon_{ij} \qquad (15.4)$$

where Y_{ij}, g_j, g_{Gj} and ε_{ij} were described in M3, while $g_{CCj}, j = 1, \ldots, J$ are random effects of lines distributed as $\boldsymbol{g}_{CC} = (g_1,\ldots,g_J)^T \sim N_J(\mathbf{0}, \sigma_{gCC}^2\boldsymbol{K}^{CC})$ and $g_{PPj}, j = 1, \ldots, J$ are random effects of lines distributed as $\boldsymbol{g}_{PP} = (g_1,\ldots,g_J)^T \sim N_J(\mathbf{0}, \sigma_{gPP}^2\boldsymbol{K}^{PP})$, where σ_{gCC}^2 is the variance component of lines associated with the random effect g_{CC}, σ_{gPP}^2 is the variance component of lines associated with the random effect g_{PP}. While $\boldsymbol{K}^{CC}$ and $\boldsymbol{K}^{PP}$ are subproducts of the matrix multiplication of the linear ($\mathbf{G}$ and Gaussian ($\boldsymbol{K}$) kernels, for this reason the K^{CC} is computed using the upper triangular part (UT) of the matrix multiplication of $\mathbf{G}$ and $\boldsymbol{K}$, that is K^{CC} = UT + t(UT) (Cuevas *et al.*, 2025). While the kernel $\boldsymbol{K}^{PP}$ was computed using the lower part (LT) of the matrix multiplication of $\mathbf{G}$ and $\boldsymbol{K}$, that is, $\boldsymbol{K}^{PP}$ = LT + t(LT) (Cuevas *et al.*, 2025). All these models had been implemented in the R statistical software (R Core Team, 2025) utilizing the BGLR library (Pérez and de Los Campos, 2014).

15.2.3 Cross-validation and evaluation metrics

To evaluate and compare the predictive performance of different models, we used three cross-validation strategies. The first strategy mimicked the scenario of tested lines in tested environments (TLTE). It was implemented using five random partitions, with each partition allocating 80% of the observations from the full dataset for training and the remaining 20% for testing. The second strategy simulated untested lines in tested environments (ULTE). This approach also used five random partitions, but each partition allocated 80% of the lines in the dataset for training and 20% for testing. The third strategy mimicked tested lines in untested environments and was implemented using a leave-one-environment-out (LOEO) cross-validation approach. In this case, all environments except one were used for training, while the excluded environment served as the testing set. This process was repeated until each environment had served as the testing set once (Montesinos López *et al.*, 2022; Alemu *et al.*, 2024). To assess predictive performance, we calculated the normalized root mean squared error (NRMSE) and the average Pearson correlation (APC) for each testing set across all partitions. The average NRMSE and APC values across traits and partitions were reported as the overall prediction accuracy for each dataset.

15.3 Results

In this section we provide the explicit results of four of the nine datasets used in terms of only untested lines and tested environments (ULTE), and leave-one-environment-out (LOEO) cross-validation. Also, we provide a summary across datasets to get a better picture of the prediction performance of the models in all the datasets. First are given the results for the Rice_IRRI dataset, then for the Rice_Kim dataset, next for the Wheat_Ingland, after this for the Potato_Suecia and finally for across all datasets. The remaining dataset results are provided in Appendices B and C.

15.3.1 Rice_IRRI

We begin by presenting the results of the cross-validation using untested lines in tested environments cross-validation· (Fig. 15.1A and B). According to the Pearson's correlation between observed and predicted values (COR) metric, Model M4 exhibited the best performance, achieving a COR of 0.275. Model M2 followed, with a COR of 0.238, 15.558% lower than M4. The least effective model was M1, with a COR of 0.229, reflecting a 20.364% decrease compared with the top-performing model. Regarding NRMSE, all four models produced very similar values. Model M2 yielded the lowest error, with an NRMSE of 0.237, whereas Model M1 had the highest, at 0.239. The difference between the best and worst-performing models in this metric was approximately 1%, indicating comparable predictive precision across models.

We now turn to the results of the leave-one-environment-out (LOEO) cross-validation (Fig. 15.1B and C). For the COR metric, Model M4 once again achieved the highest value, recording a COR of 0.424. Model M2 came in second, with a COR of 0.420, just 0.944% below M4. Model M1 ranked lowest, with a COR of 0.411, 3.235% lower than the best model. In terms of NRMSE, Model M4 continued to outperform the others, with a value of 0.276. Model M1 placed second, recording a NRMSE of 0.277, only 0.109% higher than M4. Model M2 showed the highest NRMSE, at 0.286, making it 3.566% less accurate than M4 according to this metric. Specific details for ULTE and LOEO cross-validations can be found in Appendix A.

15.3.2 Rice_Kim

First, we report the results of the crossvalidation across untested lines in tested environments (Fig. 15.2A and B). According to the COR metric, Model M4 achieved the best performance, with a COR of 0.170. The next best was Model M2 at 0.093, a value 83.423% lower than M4. The weakest performer was Model M3, which recorded a COR of 0.088, corresponding to a 93.129% drop vs the top model. Across all four models, NRMSE values were very close: M4 again led with an NRMSE of 0.215, while M1 showed a similar error at 0.215, yielding only about a 1% spread between the best and worst.

Next, we present the LOEO crossvalidation results (Fig. 15.2B and C). For COR, Model M2 topped the ranking with 0.342, 4.964% above M4. Model M3 followed at 0.335, 2.815% higher than M4, while Model M4 itself placed third, with a COR of 0.325. Model M1 trailed the group, posting a COR of 0.296, which is 10.055% below M4. When evaluated by

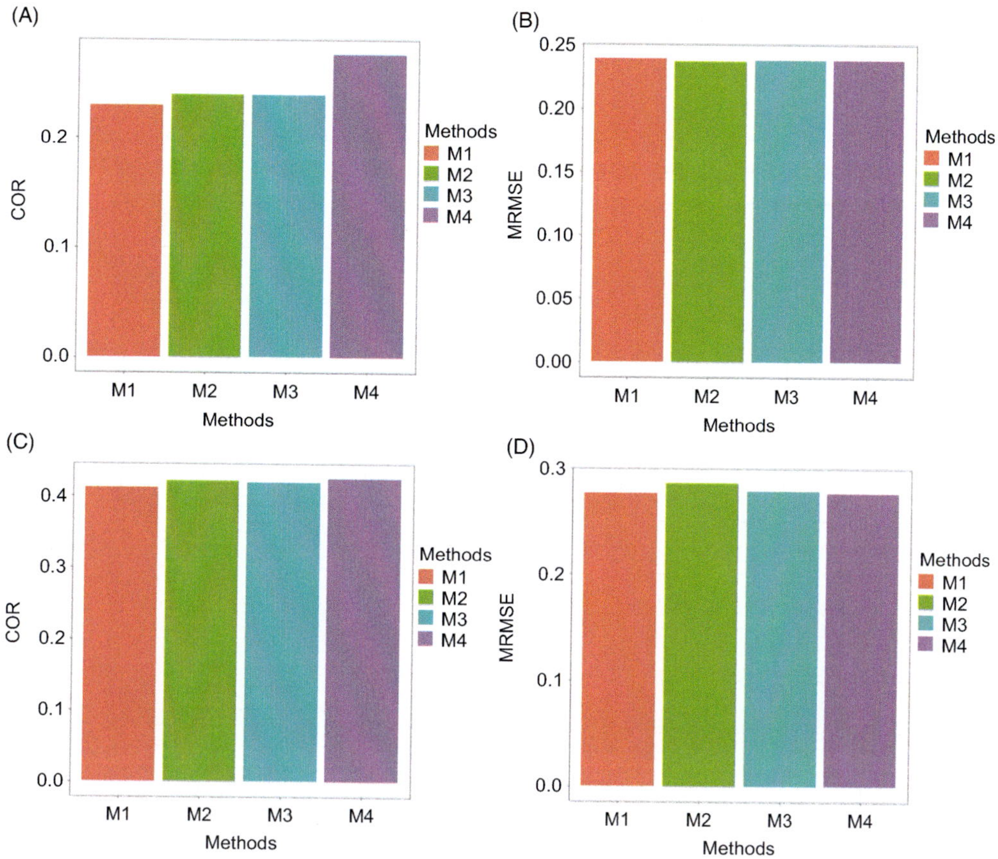

Fig. 15.1. Comparative performance of prediction models in terms of Pearson's correlation (COR) (A and C) and normalized root mean square error (NRMSE) (B and D) for the Rice_IRRI dataset. A and B provide the results for untested lines in tested environments cross-validation, while C and D provide the results for the leave-one-environment-out cross-validation (Author's own image).

NRMSE, Model M3 again came out ahead with 0.230, 0.087% better than M4. Model M2 was second, at 0.230 (0.033% lower error than M4), M4 recorded 0.230, and finally Model M1 had the highest NRMSE of 0.231, making it 0.315% less accurate than M4. Specific details for ULTE and LOEO cross-validations can be found in Appendix A.

15.3.3 Wheat_Ingland

We begin by presenting the cross-validation results for untested lines within tested environments (Fig. 15.3A and B). According to the COR metric, Model M2 delivered the highest performance with a COR of 0.245, 6.693% higher than Model M4. Model M3 ranked second, showing a COR of 0.244, which is 6.233% greater than M4. In third place was Model M1, with a COR of 0.242, exceeding M4 by 5.575%. Model M4 demonstrated the lowest performance, with a COR of 0.229, reflecting a 6.693% reduction relative to the top-performing model.

Regarding the NRMSE metric, all four models produced very similar values. Model M2 achieved the lowest error, with an NRMSE of 0.170, while Model M4 had the highest, at 0.171. The difference between the best and worst models in this case was roughly 1%.

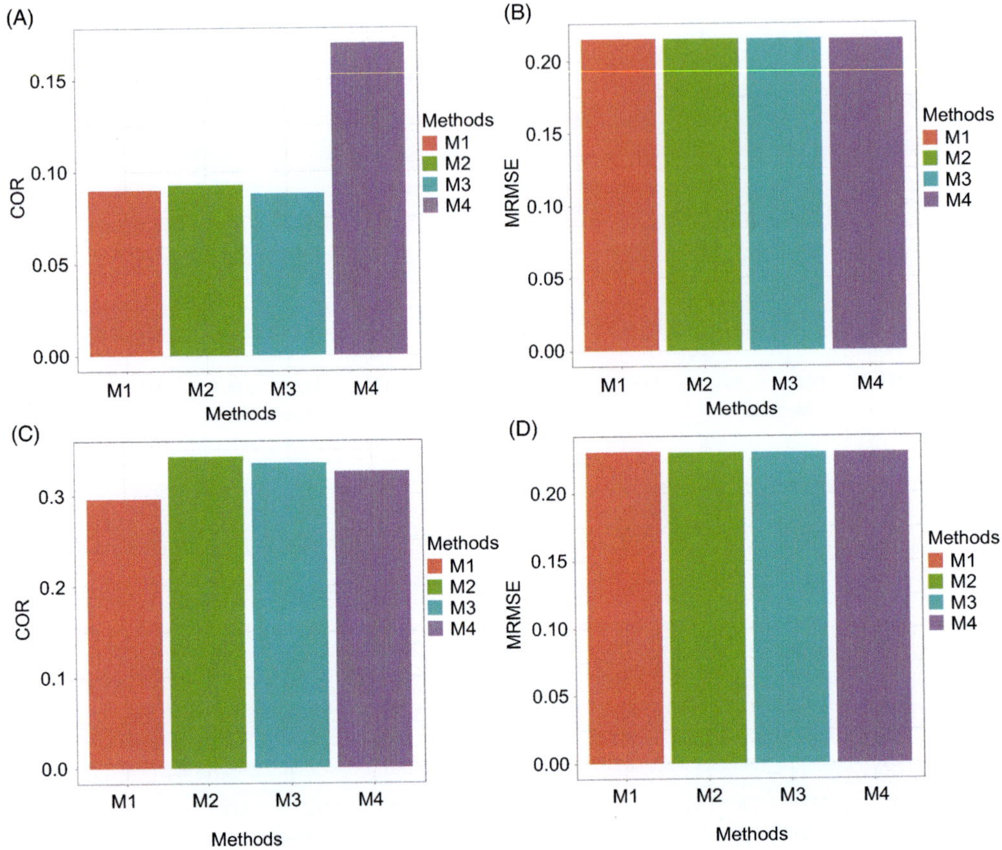

Fig. 15.2. Comparative performance of prediction models in terms of Pearson's correlation (COR) (A and C) and normalized root mean square error (NRMSE) (B and D) for the Rice_Kim dataset. A and B provide the results for untested lines in tested environments cross-validation, while C and D provide the results for the leave-one-environment-out cross-validation (Author's own image).

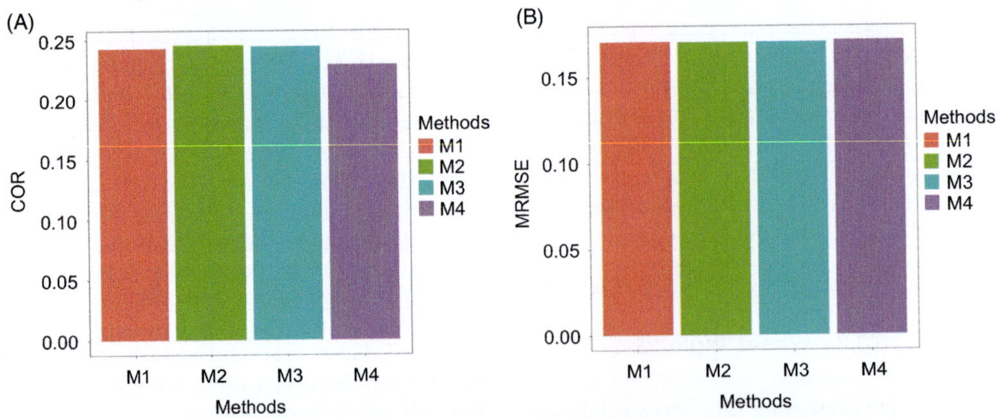

Fig. 15.3. Comparative performance of prediction models in terms of Pearson's correlation (COR) (A) and normalized root mean square error (NRMSE) (B) for the Wheat_Ingland dataset. A and B provide the results for untested lines in tested environments cross-validation (Author's own image).

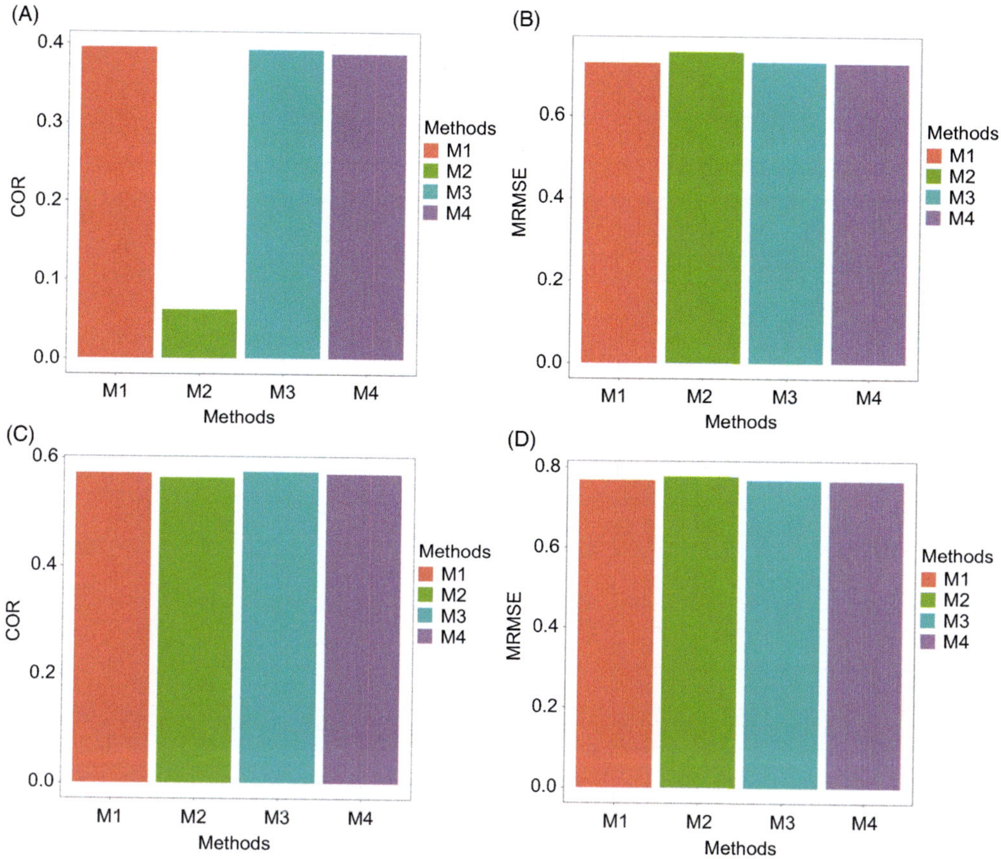

Fig. 15.4. Comparative performance of prediction models in terms of Pearson's correlation (COR) (A and C) and normalized root mean square error (NRMSE) (B and D) for the Potato_Suecia dataset. A and B provide the results for untested lines in tested environments cross-validation, while C and D provide the results for the leave-one-environment-out cross-validation (Author's own image).

Due to the structure of this dataset (Wheat_England), which contains only one environment, for this reason in this dataset was not evaluated by the LOEO cross-validation technique. Specific details for ULTE cross-validation can be found in Appendix A.

15.3.4 Potato_Suecia

We begin by reporting the cross-validation results for untested lines within tested environments (Fig. 15.4A and B). According to the COR metric, Model M1 demonstrated the best performance, with a COR of 0.395, 2.080% higher than Model M4. Model M3 followed, with a COR of 0.391, exceeding M4 by 1.118%. In contrast, Model M2 showed the lowest performance, with a COR of 0.061, reflecting a 533.705% reduction compared with the top model. Regarding the NRMSE values, the four models produced very similar outcomes. Model M4 had the lowest error, with an NRMSE of 0.728, while Model M2 exhibited the highest, at 0.755. The overall difference between the most and least accurate models in this metric was approximately 1%.

Next, we present the results from the LOEO cross-validation (Fig. 15.4B and C). For the COR metric, Model M3 achieved the highest value, at 0.574. Model M1 ranked second, with a COR of 0.571, only 0.107% higher than M4, which also recorded a COR of 0.571. Model M2 again showed the weakest performance, with a COR of 0.564, 1.298% lower than M4. In terms of

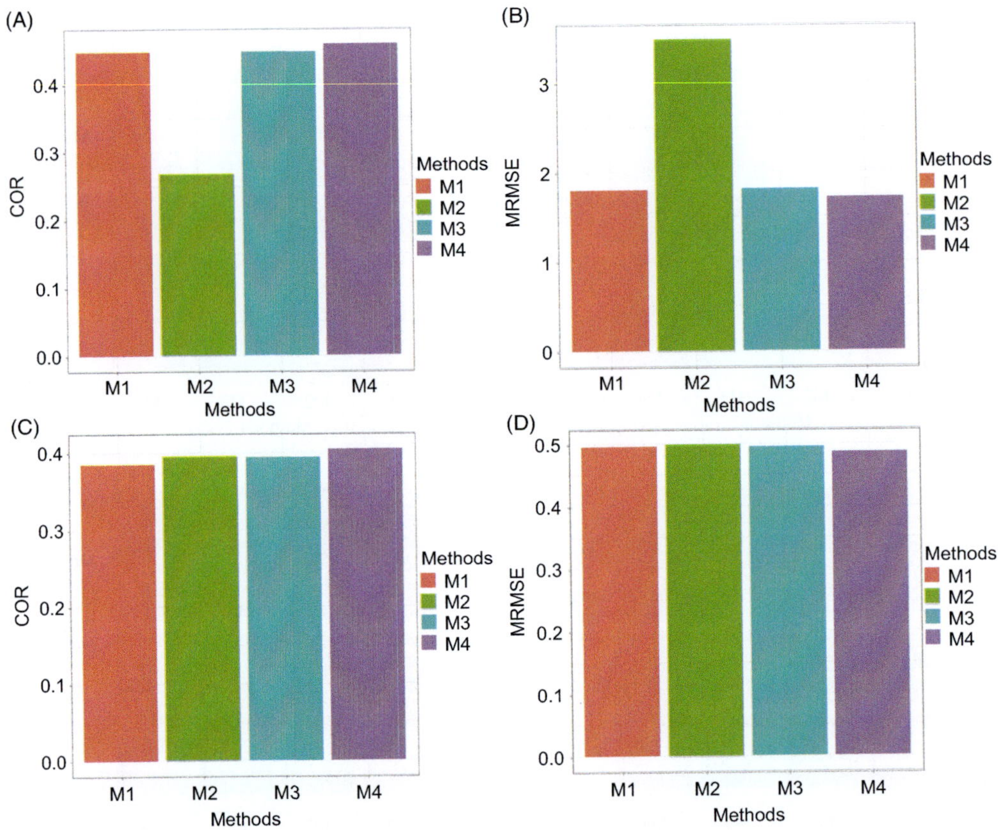

Fig. 15.5. Comparative performance of prediction models in terms of Pearson's correlation (COR) (A and C) and normalized root mean square error (NRMSE) (B and D) for across all datasets. A and B provide the results for untested lines in tested environments cross-validation, while Figures C and D provide the results for the leave-one-environment-out cross-validation (Author's own image)

NRMSE, Model M4 once again performed best, registering a value of 0.767. Model M1 followed closely, with an NRMSE of 0.768 (just 0.111% higher than M4), while Model M3 ranked third, also at 0.768, but with an increase of 0.092% compared with M4. Model M2 had the highest NRMSE, at 0.777, making it 1.348% less accurate than M4 based on this metric. Specific details for ULTE and LOEO cross validations can be found in Appendix A.

15.3.5 Summary across all the datasets

We begin by presenting the results of the cross-validation for untested lines within tested environments (Fig. 15.5A and B). According to the COR metric, Model M4 delivered the highest performance, with a COR value of 0.459. Model M1 ranked second, with a COR of 0.448, 2.513% lower than M4. The lowest performance was observed in Model M2, which recorded a COR of 0.267, representing a 3.478% decline compared with the best model. In terms of NRMSE, the four models produced very different values overall. Model M4 achieved the best result, with an NRMSE of 1.714, while Model M2 exhibited the highest error, with an NRMSE of 3.478. The difference between the best and worst models in this metric was approximately 100%, highlighting a notable gap in prediction accuracy.

Next, we report the outcomes of LOEO cross-validation (Fig. 15.5B and C). For the COR metric, Model M4 again led the performance, with a value of 0.404. Model M2 followed, recording a COR of 0.395, only 2.373% lower than M4. Model M1 ranked lowest, with a COR of 0.385, which is 5.009% below the top-performing model. When evaluating model performance using NRMSE, Model M4 continued to outperform the others, achieving a value of 0.486. Model M3 placed second, with a NRMSE of 0.495, just 1.697% higher than M4. Finally, Model M2 had the highest NRMSE, at 0.499, which was 2.510% less effective than M4 based on this metric. Specific details for ULTE and LOEO cross-validations can be found in Appendix A.

15.4 Discussions

Despite its promise, the practical implementation of GS methodologies remains a complex and challenging task. One of the main difficulties lies in the ability of genomic predictive models to capture the intricate, often nonlinear, relationships between genetic markers and phenotypic traits (de Los Campos *et al.*, 2013). Traditional linear models, such as GBLUP, assume additive genetic effects and may fail to account for higher-order interactions, epistasis, and non-additive genetic variance, which are frequently present in complex traits (Gianola *et al.*, 2006). While nonlinear models, including kernel-based methods and machine learning algorithms (e.g. random forests, support vector machines, and deep learning), offer potential for modeling these complex patterns, their practical use is hindered by several factors (Heslot *et al.*, 2012). These include high computational demands, the need for large and diverse training datasets to avoid overfitting (Montesinos-López *et al.*, 2018), and the challenge of selecting appropriate hyperparameters that govern model behavior. Moreover, the interpretability of nonlinear models remains limited, complicating biological inference and decision making in breeding programs (Azodi *et al.*, 2019). The lack of standardized protocols for model evaluation and the variability in prediction performance across traits and environments further exacerbate the difficulty of routine implementation. Thus, while GS holds substantial promise for accelerating genetic gain, fully harnessing its potential (particularly in the presence of nonlinear genomic architectures) requires continued methodological innovation and careful empirical validation.

Despite certain limitations, kernel methods have gained considerable traction in genomic prediction due to their flexibility and effectiveness in modeling complex traits. One notable success of kernel methods has been their integration into linear models as shown in this chapter, such as the genomic best linear unbiased prediction (GBLUP) framework, which can be extended through the use of kernel matrices to capture non-additive genetic effects and complex marker-trait relationships (de Los Campos *et al.*, 2010). These adaptations enable kernel methods to account for nonlinearities that traditional linear models often fail to capture, which explains why kernel-based models frequently outperform purely linear approaches in predictive accuracy (Gianola and van Kaam, 2008; Cuevas *et al.*, 2016). Their widespread adoption across diverse fields – such as systems biology, image analysis, and pattern recognition – also highlights their generalizability and robust performance in high-dimensional settings. In genomic prediction specifically, kernel methods have demonstrated superior accuracy in crops such as wheat and maize, particularly when complex genotype–environment interactions or epistatic effects are present (Cuevas *et al.*, 2016). Furthermore, the availability of open-source software packages such as BGLR (Pérez and de Los Campos, 2014), rrBLUP (Endelman, 2011), kernlab (Karatzoglou *et al.*, 2004), and SKM (Montesinos López *et al.*, 2023) has greatly facilitated the implementation and dissemination of kernel-based methods, lowering the barrier for their use by researchers in both academic and applied breeding programs. As a result, kernel methods have not only proven their utility in genomic prediction but have also established themselves as a cornerstone in the modern statistical toolbox for modeling complex biological data.

On the other hand, Bayesian models offer a powerful and flexible framework for genomic selection, particularly when combined with kernel methods to capture complex, nonlinear relationships between genotypes and phenotypes (Montesinos-López *et al.*, 2021). One of the key strengths of Bayesian approaches lies in their ability to incorporate prior information and provide full posterior distributions of model parameters, enabling a coherent quantification of uncertainty (de Los Campos *et al.*, 2013). When integrated with kernel methods, Bayesian models enhance predictive performance by exploiting the structure in genomic data more effectively, allowing for the modeling of non-additive genetic effects and genotype-by-environment interactions (Pérez and de Los Campos, 2014). Unlike many machine learning models that often require large datasets and extensive tuning to perform well, Bayesian methods are particularly robust in settings with limited data, where prior distributions can help guide inference (Gianola *et al.*, 2006). Furthermore, their interpretability and ability to accommodate various sources of uncertainty make them especially valuable in breeding programs where decisions must be made under uncertainty. These advantages help explain why Bayesian kernel methods remain state-of-the-art in the genomic selection landscape, consistently delivering competitive prediction accuracy while providing insights into the genetic architecture of complex traits (Crossa *et al.*, 2017).

In this chapter, we investigate the potential of both Bayesian kernel methods and a novel approach for constructing hybrid kernels within the framework of GS. Our findings demonstrate that kernel methods substantially outperform traditional linear models when the underlying genetic architecture exhibits complex, nonlinear patterns. Furthermore, we introduce a generalizable framework for generating hybrid kernels by combining any two kernel types to produce two additional kernels. These new kernels, when integrated with the originals, lead to improved prediction accuracy. Although the proposed method was exemplified using linear and Gaussian kernels, it is not restricted to these choices and can be applied to any pair of kernels. Our results suggest that this hybridization strategy can significantly enhance predictive performance relative to the individual kernels alone. Given its flexibility and capacity to capture intricate patterns in high-dimensional data, we advocate for further exploration of this method not only in genomic selection but also across other domains where modeling complex input–output relationships is essential.

Ultimately, Bayesian kernel methods have proven to be highly effective for enhancing genomic prediction, particularly in the context of plant and animal breeding. Their strength lies in the ability to model complex, nonlinear relationships between genotypic and phenotypic data through flexible kernel functions, while also incorporating prior knowledge and uncertainty in a principled manner. Within a Bayesian framework, these methods offer increased robustness to overfitting and are inherently capable of handling high-dimensional genomic data, which is a common characteristic of modern breeding datasets (de Los Campos *et al.*, 2010; Pérez and de Los Campos, 2014). Moreover, Bayesian approaches facilitate model regularization and hyperparameter tuning through posterior inference rather than relying solely on cross-validation procedures, thus improving both computational efficiency and prediction accuracy. The interpretability, flexibility, and empirical success of Bayesian kernel models make them especially attractive for the implementation of genomic selection strategies across a wide range of breeding scenarios (Gianola *et al.*, 2006; Crossa *et al.*, 2017).

15.5 Conclusions

In this chapter, we have presented the implementation of kernel methods for GS within a Bayesian framework. Our results demonstrate that kernel methods outperform linear models when the data exhibit substantial nonlinear patterns. However, applying kernel methods requires additional considerations, particularly in selecting an appropriate kernel function and tuning its associated hyperparameters. In addition, we introduced a novel approach for constructing hybrid kernels by combining two original kernels to generate two new ones. When integrated with the original kernels, these hybrid kernels further improve genomic prediction performance. Given these findings, we strongly encourage researchers working in the field of GS to adopt kernel-based methods as a means to enhance predictive accuracy, particularly in cases where the underlying data structures are complex and nonlinear.

References

Alemu, A., Åstrand, J., Montesinos-Lopez, O.A., y Sanchez, J.I., Fernandez-Gonzalez, J. *et al.* (2024) Genomic selection in plant breeding: Key factors shaping two decades of progress. *Molecular Plant* 17(4), 552–578.

Aronszajn, N. (1950) Theory of reproducing kernels. *Transactions of the American Mathematical Society* 68(3), 337–404.

Azodi, C.B., Bolger, E., McCarren, A., Roantree, M., de Los Campos, G. *et al.* (2019) Benchmarking parametric and machine learning models for genomic prediction of complex traits. *G3: Genes, Genomes, Genetics* 9, 3691–3702.

Crossa, J., de Los Campos, G., Pérez, P., Gianola, D., Burgueno, J. *et al.* (2010) Prediction of genetic values of quantitative traits in plant breeding using pedigree and molecular markers. *Genetics* 186(2), 713–724.

Crossa, J., Pérez-Rodríguez, P., Cuevas, J., Montesinos-López, O., Jarquín, D. *et al.* (2017) Genomic selection in plant breeding: Methods, models, and perspectives. *Trends in Plant Science* 22(11), 961–975.

Crossa, J., Martini, J.W., Gianola, D., Pérez-Rodríguez, P., Jarquin, D. *et al.* (2019) Deep kernel and deep learning for genome-based prediction of single traits in multienvironment breeding trials. *Frontiers in Genetics* 10, 1168.

Cuevas, J., Crossa, J., Soberanis, V., Pérez-Elizalde, S., Pérez-Rodríguez, P. *et al.* (2016) Genomic prediction of genotype× environment interaction kernel regression models. *The Plant Genome* 9(3).

Cuevas, J., Martini, J.W.R., Otegón, J., Vitale, P., Gerard, G. *et al.* (2025) Enhanced genomic prediction accuracy with combined pedigree and genomic covariance matrices. Submitted to *Frontiers in Plant Science*.

Cuevas, J., Montesinos-López, O., Juliana, P., Guzmán, C., Pérez-Rodríguez, P. *et al.* (2019). Deep kernel for genomic and near infrared predictions in multi-environment breeding trials. *G3: Genes, Genomes, Genetics* 9(9), 2913–2924.

Daetwyler, H.D., Villanueva, B. and Woolliams, J.A. (2008) Accuracy of predicting the genetic risk of disease using a genome-wide approach. *PloS ONE* 3(10), e3395.

de Los Campos, G., Gianola, D., Rosa, G.J., Weigel, K.A. and Crossa, J. (2010) Semi-parametric genomic-enabled prediction of genetic values using reproducing kernel Hilbert spaces methods. *Genetics Research* 92(4), 295–308.

de Los Campos, G., Hickey, J.M., Pong-Wong, R., Daetwyler, H.D. and Calus, M.P. (2013) Whole-genome regression and prediction methods applied to plant and animal breeding. *Genetics* 193(2), 327–345.

Endelman, J.B. (2011) Ridge regression and other kernels for genomic selection with R package rrblup. *Plant Genome* 4, 250–255.

Gianola, D. and van Kaam, J.B. (2008) Reproducing kernel Hilbert spaces regression methods for genomic assisted prediction of quantitative traits. *Genetics* 178(4), 2289–2303.

Gianola, D., Fernando, R.L. and Stella, A. (2006) Genomic-assisted prediction of genetic value with semiparametric procedures. *Genetics* 173(3), 1761–1776.

Gianola, D., Okut, H., Weigel, K.A. and Rosa, G.J. (2011) Predicting complex quantitative traits with Bayesian neural networks: A case study with jersey cows and wheat. *BMC Genetics* 12, 1–14.

Goddard, M.E. and Hayes, B.J. (2009) Mapping genes for complex traits in domestic animals and their use in breeding programmes. *Nature Reviews Genetics* 10(6), 381–391.

González-Camacho, J.M., Ornella, L., Pérez-Rodríguez, P., Gianola, D., Dreisigacker, S. *et al.* (2018) Applications of machine learning methods to genomic selection in breeding wheat for rust resistance. *The Plant Genome* 11(2), 170104.

Hayes, B.J., Bowman, P.J., Chamberlain, A.J. and Goddard, M.E. (2009) Invited review: Genomic selection in dairy cattle: Progress and challenges. *Journal of Dairy Science* 92(2), 433–443.

Heslot, N., Yang, H.P., Sorrells, M.E. and Jannink, J.L. (2012) Genomic selection in plant breeding: A comparison of models. *Crop Science* 52(1), 146–160.

Ibañez-Escriche, N. and Gonzalez-Recio, O. (2011) Review. Promises, pitfalls and challenges of genomic selection in breeding programs. *Spanish Journal of Agricultural Research* 9(2), 404–413.

Isidro, J., Jannink, J.L., Akdemir, D., Poland, J., Heslot, N. *et al.* (2015) Training set optimization under population structure in genomic selection. *Theoretical and Applied Genetics* 128, 145–158.

Jarquín, D., Crossa, J., Lacaze, X., Du Cheyron, P., Daucourt, J. *et al.* (2014) A reaction norm model for genomic selection using high-dimensional genomic and environmental data. *Theoretical and Applied Genetics* 127, 595–607.

Karatzoglou, A., Smola, A., Hornik, K. and Zeileis, A. (2004) Kernlab – An S4 package for kernel methods in R. *Journal of Statistical Software* 11(9), 1–20. DOI: 10.18637/jss.v011.i09.

Meuwissen, T.H., Hayes, B.J. and Goddard, M. (2001) Prediction of total genetic value using genome-wide dense marker maps. *Genetics* 157(4), 1819–1829.

Montesinos López, O.A., Montesinos López, A. and Crossa, J. (2022) *Multivariate Statistical Machine Learning Methods for Genomic Prediction*. Springer. DOI: 10.1007/978-3-030-89010-0.

Montesinos López, O.A., Mosqueda González, B.A., Montesinos López, A. and Crossa, J. (2023) Statistical machine-learning methods for genomic prediction using the SKM library. *Genes* 14(5), 1003.

Pérez, P. and de Los Campos, G. (2014) BGLR: A statistical package for whole genome regression and prediction. *Genetics* 198(2), 483–495. DOI: 10.1534/genetics.114.164442.

Montesinos-López, A., Montesinos-López, O.A., Gianola, D., Crossa, J. and Hernández-Suárez, C.M. (2018) Multi-environment genomic prediction of plant traits using deep learners with dense architecture. *G3: Genes, Genomes, Genetics* 8(12), 3813–3828.

Montesinos-López, A., Montesinos-López, O.A., Montesinos-López, J.C., Flores-Cortes, C.A., de la Rosa, R. *et al.* (2021) A guide for kernel generalized regression methods for genomic-enabled prediction. *Heredity* 126(4), 577–596.

R Core Team (2025) *R: A Language and Environment for Statistical Computing*. R Foundation for Statistical Computing, Vienna.

Schölkopf, B. and Smola, A.J. (2002) *Learning with Kernels: Support Vector Machines, Regularization, Optimization, and Beyond*. MIT Press, Cambridge, Massachusetts.

VanRaden, P.M. (2008) Efficient methods to compute genomic predictions. *Journal of Dairy Science* 91(11), 4414–4423.

Williams, C.K. and Rasmussen, C.E. (2006) *Gaussian Processes for Machine Learning*, (Vol. 2, No. 3, p. 4). MIT Press, Cambridge, Massachusetts.

Appendix 15A

Table 15.1A. Prediction performance of evaluated models in terms of Pearson's correlation (COR) and Normalized Root Mean Square Error (NRMSE) for the Rice_IRRI, Rice_Kim, Wheat_Ingland, Potato_Suecia and across datasets. RE_COR denotes the relative efficiency in terms of COR and RE_NRMSE denotes the relative efficiency in terms of NRMSE. CV denotes cross-validation strategy.

Dataset	CV	Methods	COR	NRMSE	RE_COR	RE_NRMSE
Rice_IRRI	ULTE	M1	0.229	0.239	20.364	0.434
Rice_IRRI	ULTE	M2	0.238	0.237	15.558	−0.554
Rice_IRRI	ULTE	M3	0.238	0.238	15.809	−0.007
Rice_IRRI	ULTE	M4	0.275	0.238	0.000	0.000
Rice_IRRI	LOEO	M1	0.411	0.277	3.235	0.109
Rice_IRRI	LOEO	M2	0.420	0.286	0.944	3.566
Rice_IRRI	LOEO	M3	0.418	0.279	1.572	0.887
Rice_IRRI	LOEO	M4	0.424	0.276	0.000	0.000
Rice_Kim	ULTE	M1	0.090	0.215	88.622	0.163
Rice_Kim	ULTE	M2	0.093	0.215	83.423	0.140
Rice_Kim	ULTE	M3	0.088	0.215	93.129	0.151
Rice_Kim	ULTE	M4	0.170	0.215	0.000	0.000
Rice_Kim	LOEO	M1	0.296	0.231	10.055	0.315
Rice_Kim	LOEO	M2	0.342	0.230	−4.964	−0.033
Rice_Kim	LOEO	M3	0.335	0.230	−2.851	−0.087
Rice_Kim	LOEO	M4	0.325	0.230	0.000	0.000
Wheat_Ingland	ULTE	M1	0.242	0.170	−5.575	−0.525
Wheat_Ingland	ULTE	M2	0.245	0.170	−6.693	−0.700
Wheat_Ingland	ULTE	M3	0.244	0.170	−6.233	−0.613
Wheat_Ingland	ULTE	M4	0.229	0.171	0.000	0.000
Potato_Suecia	ULTE	M1	0.395	0.729	−2.080	0.028
Potato_Suecia	ULTE	M2	0.061	0.755	533.705	3.699
Potato_Suecia	ULTE	M3	0.391	0.731	−1.118	0.336
Potato_Suecia	ULTE	M4	0.387	0.728	0.000	0.000
Potato_Suecia	LOEO	M1	0.571	0.768	−0.107	0.111
Potato_Suecia	LOEO	M2	0.564	0.777	1.298	1.348
Potato_Suecia	LOEO	M3	0.574	0.768	−0.614	0.092
Potato_Suecia	LOEO	M4	0.571	0.767	0.000	0.000
Across_data	ULTE	M1	0.448	1.796	2.513	4.747
Across_data	ULTE	M2	0.267	3.478	71.956	102.865
Across_data	ULTE	M3	0.448	1.806	2.526	5.343
Across_data	ULTE	M4	0.459	1.714	0.000	0.000
Across_data	LOEO	M1	0.385	0.496	5.009	1.904
Across_data	LOEO	M2	0.395	0.499	2.373	2.510

Continued

Table 15.1A. Continued

Dataset	CV	Methods	COR	NRMSE	RE_COR	RE_NRMSE
Across_data	LOEO	M3	0.394	0.495	2.665	1.697
Across_data	LOEO	M4	0.404	0.486	0.000	0.000

Appendix 15B

Table 15.1B. Prediction performance of evaluated models in terms of Pearson's correlation (COR) and normalized root mean square error (NRMSE) for the Family_10, DMario_Family_14, Pinus_Florida, Wheat_Whashington_ 2020 and Wheat_Whashington_ 2021 datasets. RE_COR denotes the relative efficiency in terms of COR and RE_NRMSE denotes the relative efficiency in terms of NRMSE. CV denotes Cross-validation strategy.

Dataset	CV	Methods	COR	NRMSE	RE_COR	RE_NRMSE
Family_10	ULTE	M1	0.307	0.079	1.465	0.211
Family_10	ULTE	M2	0.303	0.079	2.780	0.337
Family_10	ULTE	M3	0.304	0.079	2.645	0.211
Family_10	ULTE	M4	0.312	0.079	0.000	0.000
DMario_Family_14	ULTE	M1	0.166	0.157	2.132	0.122
DMario_Family_14	ULTE	M2	0.165	0.157	2.972	0.138
DMario_Family_14	ULTE	M3	0.164	0.157	3.084	0.133
DMario_Family_14	ULTE	M4	0.169	0.157	0.000	0.000
DMario_Family_14	LOEO	M1	0.217	0.167	2.039	−0.159
DMario_Family_14	LOEO	M2	0.224	0.167	−1.389	−0.216
DMario_Family_14	LOEO	M3	0.224	0.168	−1.381	−0.107
DMario_Family_14	LOEO	M4	0.221	0.168	0.000	0.000
Pinus_Florida	ULTE	M1	0.792	4.194	−0.625	4.708
Pinus_Florida	ULTE	M2	0.462	8.750	70.305	118.464
Pinus_Florida	ULTE	M3	0.792	4.222	−0.615	5.419
Pinus_Florida	ULTE	M4	0.788	4.005	0.000	0.000
Wheat_Whashington_2020	ULTE	M1	0.045	0.479	159.256	11.478
Wheat_Whashington_2020	ULTE	M2	0.051	0.473	126.539	10.135
Wheat_Whashington_2020	ULTE	M3	0.045	0.477	157.816	11.028
Wheat_Whashington_2020	ULTE	M4	0.116	0.430	0.000	0.000
Wheat_Whashington_2020	LOEO	M1	0.016	0.542	622.023	3.703
Wheat_Whashington_2020	LOEO	M2	0.049	0.534	130.837	2.314
Wheat_Whashington_2020	LOEO	M3	0.030	0.537	277.036	2.744
Wheat_Whashington_2020	LOEO	M4	0.114	0.522	0.000	0.000
Wheat_Whashington_2021	ULTE	M1	0.174	0.388	24.523	34.831
Wheat_Whashington_2021	ULTE	M2	0.187	0.378	15.679	31.511
Wheat_Whashington_2021	ULTE	M3	0.177	0.382	22.053	32.834

Continued

Table 15.1B. Continued

Dataset	CV	Methods	COR	NRMSE	RE_COR	RE_NRMSE
Wheat_Whashington_2021	ULTE	M4	0.216	0.288	0.000	0.000
Wheat_Whashington_2021	LOEO	M1	0.436	0.341	−0.219	12.139
Wheat_Whashington_2021	LOEO	M2	0.453	0.336	−4.118	10.358
Wheat_Whashington_2021	LOEO	M3	0.445	0.338	−2.364	11.182
Wheat_Whashington_2021	LOEO	M4	0.435	0.304	0.000	0.000

Appendix 15C

Table 15.1C. Prediction performance of evaluated models in terms of Pearson's correlation (COR) and normalized root mean square error (NRMSE) for the Rice_IRRI, Rice_Kim, Wheat_Ingland, Potato_Suecia and across datasets. RE_COR denotes the relative efficiency in terms of COR and RE_NRMSE denotes the relative efficiency in terms of NRMSE. CV denotes cross-validation strategy.

Dataset	CV	Methods	COR	NRMSE	RE_COR	RE_NRMSE
Rice_IRRI	TLTE	M1	0.364	0.243	5.228	−0.021
Rice_IRRI	TLTE	M2	0.377	0.244	1.449	0.412
Rice_IRRI	TLTE	M3	0.373	0.243	2.577	0.117
Rice_IRRI	TLTE	M4	0.383	0.243	0.000	0.000
Wheat_Ingland	TLTE	M1	0.248	0.173	−4.682	−0.516
Wheat_Ingland	TLTE	M2	0.251	0.173	−5.766	−0.631
Wheat_Ingland	TLTE	M3	0.250	0.173	−5.597	−0.602
Wheat_Ingland	TLTE	M4	0.236	0.174	0.000	0.000
DMario_Family_14	TLTE	M1	0.220	0.158	0.642	0.034
DMario_Family_14	TLTE	M2	0.223	0.158	−0.708	−0.011
DMario_Family_14	TLTE	M3	0.224	0.158	−1.058	0.100
DMario_Family_14	TLTE	M4	0.221	0.158	0.000	0.000
Pinus_Florida	TLTE	M1	0.795	9.019	−0.640	1.063
Pinus_Florida	TLTE	M2	0.466	11.646	69.401	30.504
Pinus_Florida	TLTE	M3	0.795	8.954	−0.648	0.345
Pinus_Florida	TLTE	M4	0.790	8.924	0.000	0.000
Rice_Kim	TLTE	M1	0.240	0.211	16.565	0.036
Rice_Kim	TLTE	M2	0.265	0.212	5.358	0.380
Rice_Kim	TLTE	M3	0.261	0.211	7.218	0.320
Rice_Kim	TLTE	M4	0.279	0.211	0.000	0.000
Wheat_Whashington_2020	TLTE	M1	0.127	0.423	45.561	9.230
Wheat_Whashington_2020	TLTE	M2	0.141	0.418	31.125	8.049
Wheat_Whashington_2020	TLTE	M3	0.134	0.420	38.245	8.437
Wheat_Whashington_2020	TLTE	M4	0.185	0.387	0.000	0.000
Family_10	TLTE	M1	0.348	0.079	2.769	0.594

Continued

Table 15.1C. Continued

Dataset	CV	Methods	COR	NRMSE	RE_COR	RE_NRMSE
Family_10	TLTE	M2	0.350	0.079	2.084	0.975
Family_10	TLTE	M3	0.355	0.079	0.732	0.382
Family_10	TLTE	M4	0.358	0.079	0.000	0.000
Wheat_Whashington_2021	TLTE	M1	0.387	0.315	1.883	15.214
Wheat_Whashington_2021	TLTE	M2	0.404	0.309	−2.530	13.159
Wheat_Whashington_2021	TLTE	M3	0.396	0.311	−0.399	14.027
Wheat_Whashington_2021	TLTE	M4	0.394	0.273	0.000	0.000
Potato_Suecia	TLTE	M1	0.536	0.726	−0.755	−0.140
Potato_Suecia	TLTE	M2	0.490	0.738	8.516	1.499
Potato_Suecia	TLTE	M3	0.535	0.728	−0.452	0.078
Potato_Suecia	TLTE	M4	0.532	0.727	0.000	0.000

16 Artificial Intelligence Approaches in Plant Data Science

Rashmi Mukherjee[1], Indrani Paul[2], Dhaval Patel[3], Aqsa Majgaonkar[4], Subhajit Roy[2], Sneha Kali[2], Payel Guha[2], Shreya Dave[2], Shreyasi Dutta[1] and Dwaipayan Sinha[5]*

[1]Department of Botany (UG & PG), Raja Narendralal Khan Women's College (Autonomous), Gope Palace, Midnapur, India; [2]Department of Biotechnology, Brainware University, Kolkata, West Bengal, India; [3]Research and Development Division, Latambarcem Brewers Private Limited, Bicholim, Goa, India; [4]Conservation Education Centre, Bombay Natural History Society, Mumbai, India; [5]Department of Botany, Government General Degree College, Mohanpur, Paschim Medinipur, West Bengal, India

Abstract

The integration of artificial intelligence (AI) into plant data science is revolutionizing the way plant systems are studied, optimized, and managed across agricultural and ecological domains. Leveraging machine learning (ML), deep learning (DL), and advanced data analytics, AI enables researchers to decode complex biological datasets derived from genomics, phenomics, transcriptomics, proteomics, and metabolomics. These technologies empower precise crop yield prediction, early disease detection, and real-time monitoring of plant health, while enhancing breeding programs and resource management through predictive modeling and multi-omics integration. Applications such as computer vision, natural language processing, and autonomous systems have further extended AI's utility into field-level decision-support. Despite challenges related to data standardization, model interpretability, and infrastructure disparities, ongoing research in explainable AI, federated learning, and open-access data frameworks holds promise for equitable and sustainable implementation. This chapter highlights AI's multifaceted role in transforming plant data science into a predictive, prescriptive, and data-rich discipline, paving the way for resilient agriculture and deeper biological insights.

Keywords: Artificial intelligence, Machine learning, Multi-omics integration, Plant data science, Precision agriculture

16.1 Introduction to Plant Data Science

In the face of escalating global issues such as climate change, soil degradation, and a fast-growing human population projected to reach over 10 billion by 2050, contemporary agriculture is under unprecedented pressure to produce more food with fewer resources. To address these expectations sustainably, the incorporation of AI into plant sciences has emerged as a dramatic and required paradigm change. AI, which includes

*Corresponding author: dwaipayansinha@hotmail.com

 Plant Genomic Prediction: AI-Enhanced Models for Plant Genomics and Crop Improvement (ed. Jen-Tsung Chen)
DOI: 10.1079/9781836991328.0016

machine learning (ML), deep learning (DL), and advanced data analytics, has transformative capabilities for decoding complicated biological data, making timely decisions, and improving the efficiency of agricultural systems. The combination of AI and plant data science is ushering in a new era of 'smart agriculture' that is data-driven, automated, and resilient. Plant data science is concerned with the gathering, processing, and analysis of large and diverse datasets linked to plant biology, such as genetics, phenomics, agronomic observations, and environmental variables. AI approaches are increasingly being used to extract hidden patterns and actionable insights from these complex datasets, enabling advances in plant health monitoring, agricultural production forecasts, genetic improvement, and ecosystem modelling (Li *et al.*, 2023).

One of the most significant uses of AI in plant sciences is disease diagnosis and pest control. Conventional diagnostic approaches frequently rely on visual inspection and expert opinion, both of which are time-consuming and error-prone. AI, particularly DL approaches like convolutional neural networks (CNNs), has shown amazing accuracy in detecting plant illnesses from visual data. These AI algorithms can diagnose numerous illnesses from leaf or stem photos faster and more accurately than humans (Pacal *et al.*, 2024). Plantix, an AI-powered smartphone app, allows farmers to get expert-level diagnostic services by just shooting photographs of afflicted plants, decentralizing plant health information, and making it available to even smallholder farmers.

Conventional plant breeding methods often take a long time and need a lot of resources. Still, the inclusion of AI, especially via predictive modelling and genomic selection, has dramatically quickened this development. To predict the efficacy of breeding lines, AI algorithms can analyze huge datasets comprising genetic, phenotypic, and environmental variables. This method not only speeds up the discovery of best of the cultivars but also increases the accuracy in selecting wanted features, i.e., drought resistance, disease resistance, or improved nutrient use efficiency (Fang, 2024). Furthermore, merging multi-omics data, including transcriptomics, proteomics, and metabolomics, is helping to provide a comprehensive grasp of plant responses. One can use this information methodically and rather well.

Despite its considerable promise, the implementation of AI in plant data science is not without challenges. A major concern is the quality, completeness, and heterogeneity of available datasets. Many AI models, particularly DL systems, require large volumes of high-quality labeled data for training, which is often unavailable or fragmented across institutions. Additionally, the integration of data from different sources and scales – such as combining molecular data with field-level phenotyping or satellite imagery – remains technically and conceptually complex. Furthermore, the black-box nature of many AI algorithms raises concerns about model interpretability and trust, particularly in decision-critical agricultural applications (Eftekhari *et al.*, 2024). To address these barriers, there is an urgent need for standardized data formats, interoperable databases, and collaborative frameworks that encourage data sharing across disciplines and institutions. Equally important is the development of interpretable AI models and decision-support systems that can be readily used by end-users, including farmers, breeders, and agricultural policymakers. Building local capacities in digital literacy and computational biology is also critical, particularly in regions where agriculture forms the backbone of the economy, yet lags in technological infrastructure.

The integration of AI into plant data science is revolutionizing how we understand, monitor, and enhance plant systems. It represents a shift from reactive to predictive and prescriptive agriculture, empowering stakeholders at all levels of the food value chain to make data-informed decisions. From detecting diseases and optimizing inputs to accelerating breeding and modeling ecosystem dynamics, AI holds the key to unlocking unprecedented efficiencies and innovations in plant science. This chapter delves into the multifaceted role of AI in plant data science, exploring its applications in disease detection, crop monitoring, and breeding. Through case studies and current research, we aim to provide a comprehensive understanding of how AI is reshaping the landscape of plant science and its implications for sustainable agriculture.

16.2 Concept of Artificial Intelligence

AI has spread beyond its traditional bounds, influencing many different industries and having a significant effect on global business. Among these, the study of plant data science stands out as a promising area where conventional wisdom is being challenged by AI's revolutionary potential. Innovation is exemplified by the use of AI in plant science, which provides fresh answers to persistent problems in ecology, agriculture, and environmental sustainability. The main AI techniques and how they are used in plant data science are illustrated in this section.

16.2.1 Models for machine learning

Supervised learning

Using regression and classification approaches, it is possible to predict plant qualities from labeled data, such as crop yield depending on environmental conditions, or disease detection from photos of plant leaves.

Unsupervised learning

Clustering and dimensionality reduction methods assist in identifying patterns in data in the absence of preset labels. For instance, clustering can be used to categorize plant genotypes based on how they respond to drought stress.

Deep learning

CNNs are widely used for image recognition tasks, such as identifying plant species, tracking plant growth, and detecting diseases from leaf images. Recurrent neural networks (RNNs) are used to model time series data, such as predicting soil moisture levels or crop health over time (Bhoi and Mahali, 2024).

16.2.2 Computer vision and image analysis

Image processing and computer vision algorithms are needed for high-throughput phenotyping, which monitors plants using photos captured by field cameras, drones, or satellites. AI processes these images to derive valuable information about plant traits, including biomass, height, leaf area, and chlorophyll content. Applications include disease diagnosis, pest control, and real-time monitoring of plant health and growth.

16.2.3 Natural language processing

In the field of plant sciences, where genetic databases, scientific papers, and agricultural reports include important unstructured information, natural language processing (NLP) is useful for text mining. AI models analyze these texts to find new information, including gene–trait relationships, illness patterns, or soil–crop correlations. Additionally, NLP can help with knowledge.

16.2.4 Genomics and genetic data analysis

AI helps process the enormous volumes of genomic data gathered from plants, assisting scientists in identifying genes linked to particular characteristics such as disease susceptibility, yield, or drought resistance. Machine learning is used in methods such as genome-wide association studies (GWAS) and genomic selection models to comprehend gene–trait connections and forecast how plants will react to various environmental circumstances (Williamson *et al.*, 2023).

16.3 Types of Artificial Intelligence

16.3.1 Deep learning

Deep learning has become a powerful tool for assimilating vast volumes of diverse data and producing accurate predictions of complicated and unpredictable phenomena. The plant science community is using these techniques more and more to interpret the massive datasets that are now routinely gathered through high-throughput genotyping and phenotyping.

We examine recent research that has applied DL concepts to digital image-based plant stress phenotyping. We offer a comparison of DL tools with other methods already in use, considering factors like decision accuracy, data size requirements, and scenario applicability. We conclude by outlining several research directions that make use of both present and upcoming DL techniques in plant science (Singh *et al.*, 2018).

In the field of computer vision, DL has also become increasingly important for providing superior performance results for plant disease detection. The performance and application of machine and DL approaches are compared to demonstrate the superiority of the DL model over the ML model in several research publications. The DL method can be used to identify leaf diseases from photos to save significant crop losses (Jackulin and Murugavalli, 2022). Multimodal DL models can deal with divergent data types all at once and unlock cross-layer regulatory processes and metabolic paths. Ensemble learning and Bayesian networks were employed for constructing gene regulatory networks and for finding stress-responsive genes in crops subject to abiotic and biotic stresses (Libbrecht and Noble, 2015). Such models lend themselves to functional annotation and hypothesis testing based on dimensionality reduction and preservation of biological relevance. AI-assisted integration of omics data accelerates complex traits such as drought tolerance, disease resistance, and yield production under fluctuating environmental conditions – key objectives for sustainable agriculture. Site-specific data-driven crop management using field sensor data, drones, and remote sensing is known as precision agriculture (PA). The data are deciphered in real time using AI models for support to irrigation, fertilizer, and pest control decisions (Paul *et al.*, 2022).

16.3.2 Machine learning

Utilizing Big Data principles in plant genomics is made possible by the rapid emergence of ML, a branch of AI in molecular biology. The primary issue in this setting is how to extract new knowledge from large datasets at all levels of cellular systems research. In conclusion, ML approaches enable the inference of complicated relationships in a variety of biological systems. Because of its complicated definitions and computational procedures, ML has not been fully utilized despite its potential. From data collection to classification/prediction model evaluation, we outlined the primary processes for ML development, along with a discussion of the corresponding functional genomics, mainly in terms of pathogen effector genes in plant immunity. ML also offers powerful tools for tackling complex problems in plant data science, transforming areas like crop breeding, disease diagnosis, and PA (Maganathan *et al.*, 2020). Machine learning, as one type of AI, provides computers with a mechanism for learning from data without special programming. The field of plant science uses ML extensively for classification, regression, and clustering analyses among phenotypic characters, genomic linkages, and environmental interactions. Supervised learning algorithms such as support vector machines (SVM), random forests (RF), and gradient boosting machines (GBMs) demonstrated exceptional performance in applications across a very large spectrum. RF models, for instance, were used to infer genotype–phenotype relationships for wheat and maize based on massive high-dimensional genomic datasets (Grinberg *et al.*, 2020). SVMs were used for image data classification for accurate plant disease identification and were very precise for multiple crop species (Pantazi *et al.*, 2019). Unsupervised learning methods like k-means clustering and hierarchical clustering are employed for clustering gene expression profiles and microbial communities. They are effective for finding natural patterns in unlabeled data and enabling hypothesis generation for exploratory research (Tirthyani *et al.*, 2024).

16.3.3 Neural networks

The most popular methods for identifying plant diseases at the moment are artificial neural networks (ANNs). They are used in conjunction with various picture pre-processing techniques to improve feature extraction. The way biological nerve systems, like the brain, process information served as the model for the information-processing paradigm known as ANNs in ML and cognitive research. Numerous intricately linked neurons make up the brain, which collaborates to find solutions to certain issues. A processing element with numerous inputs and a single output is called an artificial neuron. The primary method used in practice for plant disease identification is the naked-eye examination by a qualified specialist, as the majority of diseases produce some sort of manifestation in the visible spectrum. A plant pathologist must have strong observational abilities to recognize distinctive symptoms to make reliable plant disease diagnoses. Amateur gardeners and hobbyists may exhibit different symptoms from ill plants; this could result in an incorrect diagnosis. Because of their exceptional capacity to extract meaning from complicated or faulty data, neural networks can be used to identify trends and extract patterns that are too hard for computers or humans to discover. ANNs also have the benefits of self-organization, adaptive learning, real-time operations, and more. When discussing architecture, ANNs fall into two primary categories: feed-forward ANNs, in which it is improbable that any layer's output will affect that layer, and feedback ANNs that use network loops to let signals flow in both directions (Sladojevic *et al.*, 2016). Convolutional neural networks have immensely transformed plant phenotyping from images. CNNs are especially excellent in detecting diseases, leaf counting, and growth stage inference. A study demonstrated that it is possible to train a CNN from the Plant Village dataset and, using it, classify 26 diseases in 14 crop species more than 99% of the time (Mohanty *et al.*, 2016). Beside static images, time-variable processes in plants are depicted using long short-term memory (LSTM) networks and recurrent neural networks (RNNs) that effectively handle sequential data. The models are employed to measure plant growth, phenological development, and response to varying environmental conditions (Kamilaris and Prenafeta-Boldú, 2018). New DL techniques such as variational autoencoders (VAEs) and generative adversarial networks (GANs) are arising to construct synthetic plant images and enrich sets for which we have just a limited number of samples, thus enhancing model generalizability and stability.

Image analysis using CNNs and object detection models enables crop type identification, disease quantification, and yield prediction from satellite and UAV images (Zhang *et al.*, 2003). Additionally, reinforcement learning is emerging as a promising method for deriving optimal irrigation scheduling and allocation through field variability and predicted weather. Artificial intelligence-based decision-support systems integrate sensor data, histories, and environmental variables and provide farmers with actionable information to improve efficiency in the utilization of resources and reduce environmental footprints (Yousaf *et al.*, 2023). Among its most basic constraints is a lack of large, well-annotated datasets for most crop plants and ecosystems.

Another source of concern is the interpretability of the machine models. Some of the top-performing models are 'black boxes' and are not easily interpretable by stakeholders such as breeders and agronomists (Mathew *et al.*, 2025). Explainable AI (XAI) is an emerging field that seeks to transform model decisions into more understandable and trustworthy ones. Combining ML and symbolic reasoning, and domain-specific know-how in the foreseeable future could lead to more interpretable and more dependable outcomes. Interdisciplinary research between computer scientists, plant biologists, and agricultural engineers will be needed for developing AI tools that are both scientifically valid and practically effective.

16.4 Data Collection in Plant Science

16.4.1 Taxonomic data (phenetics and cladistics)

Taxonomy serves as a foundational pillar of plant biology, providing a framework for identifying, naming, and classifying plant species. It allows researchers to assess biodiversity, track evolutionary history, and ensure accurate communication across disciplines. In plant science, two major paradigms have historically structured taxonomic work: phenetics and cladistics. Phenetics, also referred to as numerical taxonomy, classifies organisms based on the overall similarity of observable traits, such as morphology, anatomy, and biochemical properties. Rather than focusing on evolutionary descent, phenetic approaches evaluate as many characters as possible and use statistical methods (often multivariate techniques) to quantify relationships among species (Sneath *et al.*, 1973). Phenetic techniques gained popularity with the rise of computational methods, which allowed the processing of complex datasets involving multiple characters. For example, in a study of *Paeonia* sp (Wang *et al.*, 2001), cluster analysis of floral pigment compositions was used to differentiate among closely related taxa. This approach allowed researchers to group species based on quantitative similarities in flower coloration, revealing distinct taxonomic patterns even when traditional morphology yielded ambiguous results. Despite its utility, phenetics is limited by its lack of evolutionary context. Two species may be phenetically similar due to convergent evolution, yet belong to distantly related clades. Therefore, while phenetics remains valuable for morphologically rich taxa and large-scale screening, it is often complemented or supplanted by phylogenetic methods that incorporate evolutionary lineage.

Cladistics, in contrast, classifies organisms based on shared derived characters (synapomorphies) to infer evolutionary relationships. Using principles rooted in Hennigian systematics, cladistic analysis constructs phylogenetic trees, or cladograms, that represent hypotheses of evolutionary descent (Vollert, 2009). These trees serve as visual frameworks for understanding how plant species diverged from common ancestors. Cladistic methods have become increasingly molecular in the genomic era. For example, studies of dicotyledonous plant families have employed DNA sequence data to reconstruct family-level phylogenies, demonstrating the utility of cladistic frameworks for revising traditional classification systems (Brummitt, 2014). The widespread application of molecular cladistics has even led to taxonomic reclassifications based on monophyletic groupings, sometimes in conflict with long-standing morphological taxonomies. Importantly, debates around cladistics vs traditional taxonomy remain. Some scholars argue that a strict cladistic approach (what has been termed 'cladonomy') does not sufficiently accommodate practical taxonomic goals like naming stability or usability in ecological studies (Brummitt, 2014).

AI now plays an increasingly pivotal role in automating and enhancing taxonomic studies. ML algorithms can classify plants using large-scale morphometric data, spectral imaging, or genomic sequences. For example, CNNs have been used to identify plant species from digital herbarium images with high accuracy (Carranza-Rojas *et al.*, 2017). In addition, supervised and unsupervised learning models have been used to classify herbarium specimens based on shape and texture, reducing reliance on expert taxonomists (Younis *et al.*, 2018). AI also assists in constructing phylogenetic trees by handling large molecular datasets more efficiently than conventional software. AI-enhanced pipelines can detect homoplasies, predict orthologous genes, and estimate branch support values with increased accuracy. Moreover, AI tools enable dynamic visualization of phylogenies, aiding in the interpretation of complex evolutionary relationships. By integrating both morphological and molecular data, AI can bridge phenetics and cladistics, facilitating holistic taxonomic decisions. These advancements are particularly impactful in an era of global biodiversity loss, where rapid species identification is critical.

16.4.2 Genomics data

Genomics, the comprehensive study of an organism's entire DNA content, including all its genes and non-coding sequences, has become a cornerstone of contemporary plant science. The availability of high-quality genomic data provides unprecedented insights into the genetic architecture underlying key agronomic traits such as disease resistance, abiotic stress tolerance, developmental timing, and yield (Varshney *et al.*, 2021). This genomic information enables researchers to link phenotypic variability to specific genetic variants, facilitating molecular breeding and crop improvement. The advent of next-generation sequencing (NGS) technologies has dramatically transformed plant genomics by enabling rapid, cost-effective, and high-throughput sequencing of entire plant genomes, transcriptomes, and epigenomes (Michael and VanBuren, 2020). Large-scale international initiatives like the 1KP (One Thousand Plant Transcriptomes) Project and ongoing efforts to sequence reference genomes for major crops and their wild relatives have generated vast amounts of genomic data (Leebens-Mack *et al.*, 2019). Such datasets enable comparative genomics approaches to decipher evolutionary relationships, gene family expansions, and genome duplications that characterize plant lineages. While short-read sequencing platforms like Illumina provide high accuracy for base calling, challenges remain in assembling complex plant genomes, especially those with large sizes, repetitive sequences, and polyploidy (e.g. wheat and cotton). Long-read sequencing technologies (e.g. PacBio, Oxford Nanopore) are increasingly integrated into plant genomics pipelines to improve assembly contiguity and accuracy (Michael and VanBuren, 2020).

AI, particularly ML, plays a transformative role in handling the high dimensionality and complexity of plant genomic datasets. ML algorithms can identify subtle genetic variants – such as single-nucleotide polymorphisms (SNPs), insertions/deletions (indels), and structural variants – that correlate with phenotypic traits. For example, DL models have been successfully applied to predict the functional impact of non-coding variants on gene regulation, improving the annotation of regulatory elements in plant genomes (Zhou and Troyanskaya, 2015). Moreover, AI techniques assist in predicting gene function and gene–gene interactions by integrating heterogeneous genomic data types (e.g. sequence motifs, chromatin accessibility) (Hamamoto *et al.*, 2019; Asada *et al.*, 2021). This enables researchers to prioritize candidate genes for functional validation or targeted breeding. Additionally, AI-powered tools have accelerated the identification of gene editing targets for CRISPR/Cas9-mediated crop improvement, increasing the precision and efficiency of genome engineering efforts (Ahmad, 2023).

Beyond genomics, AI facilitates the integration of genomic data with other 'omics' layers – such as transcriptomics, proteomics, metabolomics, and epigenomics – to construct a holistic understanding of plant biology. This systems biology perspective allows researchers to unravel complex regulatory networks governing stress responses, development, and metabolic pathways. For instance, ML frameworks have been utilized to integrate genome-wide association studies (GWAS) with transcriptome data, enhancing the identification of candidate genes controlling quantitative traits in crops like maize and rice (Zhang *et al.*, 2022). Similarly, integrating metabolomic profiles with genomic data via AI models helps in discovering biomarkers linked to nutritional quality or disease resistance (Fang *et al.*, 2019).

Despite these advances, several challenges hinder the full exploitation of plant genomic data. Polyploidy, a common feature in many crop plants, complicates genome assembly, variant calling, and functional annotation due to homoeologous sequence similarity and gene redundancy (Salman-Minkov *et al.*, 2016). Moreover, the lack of standardized data formats and metadata conventions across databases limits data interoperability and reuse (Stephens *et al.*, 2015). To overcome these challenges, collaborative efforts are under way to develop standardized pipelines, reference-quality genome assemblies, and community-driven annotation initiatives. In parallel, advances in AI – including explainable AI and federated learning – promise to enhance the interpretability of complex models and enable secure sharing of genomic data across institutions (Yang *et al.*, 2019).

16.4.3 Transcriptomics data

Transcriptomics, the comprehensive study of the complete set of RNA transcripts produced by the genome under specific physiological or environmental conditions, offers crucial insights into gene expression patterns, regulatory mechanisms, and functional genomics in plants (Wang *et al.*, 2009). By profiling the transcriptome, researchers can determine which genes are activated or repressed during developmental stages, stress responses, or metabolic shifts, thus elucidating how plants adapt and regulate their biological processes. The advent of high-throughput sequencing technologies, especially RNA sequencing (RNA-seq), has revolutionized plant transcriptomics by enabling quantitative and qualitative analyses of gene expression at unprecedented resolution and scale (Wang *et al.*, 2009). RNA-seq allows not only the measurement of expression levels across the entire transcriptome but also the detection of alternative splicing events, allele-specific expression, and novel transcripts that were previously inaccessible, particularly in non-model or orphan plant species lacking reference genomes (Ozsolak and Milos, 2011; Conesa *et al.*, 2016). For instance, comprehensive RNA-seq studies in crops like maize and wheat have identified stress-responsive genes critical for drought tolerance, disease resistance, and nutrient use efficiency, providing targets for genetic improvement (Krasileva *et al.*, 2013; Zhang *et al.*, 2019). Additionally, RNA-seq has facilitated the characterization of non-coding RNAs, such as microRNAs and long non-coding RNAs, which play important roles in gene regulation during plant development and environmental adaptation.

AI has become indispensable in managing and interpreting the vast complexity of transcriptomic datasets. ML algorithms are widely employed for classification of gene expression profiles, clustering of co-expressed genes, and prediction of gene function based on expression patterns and sequence features (Libbrecht and Noble, 2015). For example, unsupervised learning techniques like hierarchical clustering and k-means clustering help identify gene modules involved in specific biological pathways or responses (Langfelder and Horvath, 2008). More sophisticated AI methods, including DL, have been developed to uncover intricate regulatory networks and epistatic interactions within transcriptomic data (Min *et al.*, 2017). These models can integrate transcriptomic data with other omics layers, such as metabolomics and proteomics, enabling the identification of key genes involved in specialized metabolite biosynthesis or stress signaling pathways (Libbrecht and Noble, 2015). Furthermore, AI facilitates the analysis of single-cell RNA sequencing (scRNA-seq) data, a rapidly growing frontier in plant science. Single-cell transcriptomics allows researchers to dissect gene expression heterogeneity at the resolution of individual cells, unraveling cellular differentiation trajectories and tissue-specific responses that are masked in bulk RNA analyses (Ryu *et al.*, 2019). ML approaches, including dimensionality reduction and cell-type classification algorithms, help interpret the high-dimensional single-cell data, revealing novel cell populations and regulatory circuits critical for plant development (Shulse *et al.*, 2019).

Despite these remarkable advances, several challenges persist in plant transcriptomics. The dependence on high-quality reference genomes limits transcriptomic analysis in many economically or ecologically important species where reference assemblies are incomplete or absent (Conesa *et al.*, 2016). Additionally, the integration of transcriptomic data with other omics layers requires standardized data formats, robust normalization techniques, and effective computational frameworks capable of handling heterogeneous datasets. Overcoming these hurdles necessitates the continuous development of advanced AI algorithms tailored for multi-omics integration and more user-friendly platforms to democratize data analysis across the plant science community. Moreover, collaborative efforts to generate comprehensive plant transcriptome atlases under diverse conditions will further enable AI-driven discoveries with applications in crop improvement and sustainable agriculture.

16.4.4 Proteomics data

Proteomic data refers to the comprehensive analysis of the entire set of proteins produced in a cell, tissue, or organism at a specific time and condition (Müller *et al.*, 2020). Proteomics deals with all processes, like identifying proteins, quantifying them, and even characterizing them (Noor *et al.*, 2021). Proteomic data can provide valuable insights into their functions, interactions, and roles in biological processes, and it is important to understand biological processes, like unraveling cellular mechanisms and pathways, understanding disease mechanisms, and developing biomarkers (Bai *et al.*, 2021). Protein structure prediction has advanced significantly over the decades, particularly due to improvements in computational hardware, software, and AI (Jiménez-Luna *et al.*, 2021). Recent ML techniques have become crucial to reduce hours of work for studying and analyzing proteomic data. AI-based programs like AlphaFold, AlphaFold 2, Proteome Discoverer, MaxQuant, DeepLoc, iLearn, PANNZER, and many more are very useful for generating and analyzing faster data (Chen *et al.*, 2020; Muthye *et al.*, 2020; Perrakis and Sixma, 2021; Peng, 2023).

AlphaFold is an AI program developed by DeepMind that predicts the 3D structures of proteins based on their amino acid sequences. Utilizing advanced ML techniques, AlphaFold has achieved remarkable accuracy in predicting protein structures, significantly advancing the field of structural biology (Varadi *et al.*, 2022). AlphaFold's impact has been substantial, aiding researchers in understanding protein function and interactions, with potential applications in drug discovery and biotechnology (Yang *et al.*, 2023). Proteome Discoverer integrates various algorithms for analyzing mass spectrometry data, helping identify and quantify proteins in plant samples. This software integrates various algorithms for analyzing mass spectrometry (MS) data (Halder *et al.*, 2021). It identifies proteins based on peptide fragmentation patterns and uses statistical methods to quantify protein levels. By aligning experimental data with sequence databases, it determines protein identity and abundance (Orsburn, 2021). MaxQuant is open-source software for analyzing large proteomics datasets. It employs a label-free quantification approach along with peptide identification from MS data (Sinitcyn *et al.*, 2021). It uses a combination of algorithms to analyze raw data, reconstruct peptide sequences, and compare them against protein databases. The software provides statistical analysis to ensure reliable identification and quantification of proteins (Wen and Zhang, 2023). DeepLoc is another DL-based tool that predicts the subcellular localization of proteins in plants, helping researchers understand protein function (Jiang *et al.*, 2021a). DeepLoc uses DL techniques to predict protein subcellular localization based on amino acid sequences. It trains neural networks on large datasets of known protein localizations, learning to recognize patterns and features indicative of where a protein functions within the cell (Jiang *et al.*, 2021b).

iLearn is an ML platform designed for predicting protein–protein interactions, which can be useful for understanding plant signaling pathways (Chen *et al.*, 2020). iLearn utilizes ML algorithms to predict protein–protein interactions by analyzing sequence and structural features of proteins. It trains on datasets that include known interactions, allowing it to learn the characteristics that typically indicate interaction partners (Chen *et al.*, 2020). PANNZER is an annotation tool that utilizes ML to predict protein function and pathways, specifically tailored for plant proteomes (Chen *et al.*, 2023). PANNZER employs ML models to predict protein functions and pathways from sequence data. It combines existing functional annotations with sequence similarities and uses classification algorithms to assign likely functions to unannotated proteins based on patterns learned from well-annotated proteins (Yan *et al.*, 2023).

These tools collectively enhance the analysis and understanding of plant proteomes, aiding in research related to plant biology, genetics, and biotechnology. Combining proteomic data with genomic, transcriptomic, and metabolomic information provides a holistic view of plant biology, helping to understand complex traits and responses (Satrio *et al.*, 2024). By leveraging these proteomic data collection methods, researchers can gain valuable insights into plant physiology, adaptation, and health, ultimately contributing to advances in agriculture and conservation

16.4.5 Metabolomics data

Metabolomics data refers to the comprehensive analysis of metabolites, which are all small molecules produced during the metabolic processes in a biological sample, such as a cell, tissue, or organism. This field aims to identify and quantify these metabolites to understand biochemical processes, metabolic pathways, and physiological states (Gallart-Ayala *et al.*, 2020). Metabolomics data is vital in plant sciences for several key reasons. It helps elucidate complex biochemical pathways, enhancing our understanding of plant physiology and development (Sharma *et al.*, 2020). By analyzing metabolomic profiles, researchers can investigate how plants respond to environmental stresses, informing breeding programs for stress resilience. Additionally, it links specific metabolites to desirable traits like flavor and nutritional content, aiding in the selection of improved varieties (Raza, 2022).

Metabolomics also sheds light on plant–microbe interactions and supports the identification of bioactive phytochemicals with health benefits. PA provides insights into plant health, optimizing management practices and crop yields (Mishra *et al.*, 2022). Moreover, integrating metabolomics with genomics reveals the genetic basis of metabolite variation. It assesses chemical diversity within plant species, supporting conservation efforts. Overall, metabolomics is a powerful tool that enhances our understanding of plant biology and promotes sustainable agricultural practices (Sharma *et al.*, 2020).

Currently, there are multiple AI tools available for metabolomics studies, like MetaboAnalyst, XCMS, MZmine, Galaxy-P, TensorFlow, and PyTorch. MetaboAnalyst is used mainly for statistical analysis, and it utilizes algorithms for data normalization, hypothesis testing, and multivariate analysis. It helps to generate visual data like principal component analysis (PCA) and heatmaps, which present complex data in an understandable format (Fu *et al.*, 2021). Apart from this, MetaboAnalyst carries out pathway analysis using database integration to map metabolites to biological pathways (Chong and Xia, 2020). XCMS is an online cloud version created by Siuzdak Lab, which uses algorithms to detect peaks in mass spectrometry data and generate features that represent metabolites (Peralbo-Molina *et al.*, 2023). XCMS aligns data across various samples, rectifying for variations in retention time, and helps to integrate the quantitative analysis to assess metabolite levels (Hao *et al.*, 2023). MZmine is an amazing AI that applies noise filtering and baseline correction to enhance the quality of data and offers graphical representations of data, which ease the interpretation. MZmine also uses algorithms to detect and extract relevant features from raw data (Peralbo-Molina *et al.*, 2023).

Galaxy-P is a multi-omics informatics platform that is used to manage workflows. It utilizes a user-friendly interface for building and executing analysis workflows, integrating various tools and datasets. Galaxy-P is mainly used for reproducible research as it documents workflows (McGowan *et al.*, 2020). TensorFlow and PyTorch are the two most prevalent DL frameworks that utilize neural networks and DL algorithms to model complex relationships in metabolomics data. It also helps with predictive modeling, which enables the development of models that can predict metabolite interactions and responses based on training data (Pomyen *et al.*, 2020; Mahmud *et al.*, 2021).

Overall, the above-mentioned tools combine statistical methods, ML algorithms, and data visualization techniques to analyze metabolomics data effectively. These facilities provide a deeper insight into plant sciences. The integration of AI into metabolomics and plant sciences not only streamlines the data collection process for researchers but also enhances the quality and depth of analysis, leading to more robust insights into plant biology.

16.4.6 Phenomics data

Phenomics is the study of phenomes, which are the observable characteristics or traits of an organism resulting from the interaction of its genotype with the environment (Das Choudhury, 2020). In practice, phenomics involves collecting and analyzing large amounts of data on various traits, such as growth patterns, disease resistance, and physiological responses (Demidchik *et al.*, 2020). Phenomics is essential in plant science for several key reasons, such as crop improvement. It aids by identifying the phenotypic traits linked to yield, disease resistance, as well as stress tolerance (Ahmed *et al.*, 2020). High-throughput phenotyping can help achieve PA that enables farmers to monitor crop health accurately and optimize resource use (Fan *et al.*, 2021).

Climate change affects agriculture, and phenomics aids in climate adaptation by identifying plants that can thrive in diverse environmental conditions, which can ensure food security (Naqvi *et al.*, 2022). By linking phenotypic data with genotypic information, researchers can uncover the genetic basis of important traits, enhancing marker-assisted selection (Singh *et al.*, 2021). Advances in technologies like high-throughput phenotyping and ML have made it easier to analyze these traits in detail. Currently, there are multiple AI tools available for phenomics studies, like PlantCV, DeepField, Pheno4D, FieldScan, LemnaTec, and many more.

PlantCV is an open-source computer vision toolkit that analyzes plant phenotypes from images. It uses image processing algorithms to quantify features such as leaf area, shape, and color, which enables the researcher to perform a detailed phenotypic assessment (Zhu *et al.*, 2021).

DeepField is an advanced DL tool. It analyzes the aerial imagery obtained through drones or satellites to monitor the health and growth of the crop. DeepField processes high-resolution images, which help in detecting stress indicators and can predict the yield (Hasan *et al.*, 2021). Pheno4D is very useful software that directly captures the image and analyzes 3D phenotypic data of plants. Pheno4D uses computer vision and ML to reconstruct plant structures, allowing researchers to evaluate the growth dynamics and spatial traits (Schunck *et al.*, 2021). AI-Plant uses CNNs to learn from labeled datasets and precisely identify the plant species or the health status of the plant (Prakash *et al.*, 2024).

The FieldScan tool integrates AI with IoT (Internet of Things) sensors to monitor environmental conditions and crop performance. FieldScan uses data analytics to provide real-time insights, helping farmers make informed decisions about irrigation and nutrient management (Fan *et al.*, 2021). LemnaTec offers programmed imaging along with data analysis tools to assess plant growth and health, which enhances efficiency through automation and integration with genomic data (Panda *et al.*, 2022). LemnaTec platforms streamline the collection and interpretation of complex phenotypic data, substantially advancing the study of plant traits and backing collaboration among researchers in plant sciences (Ebersbach *et al.*, 2022). By permitting precise monitoring and analysis of plant traits, these innovations contribute to enhanced crop resilience, sustainability, and productivity. As researchers face soaring environmental challenges and food security concerns, leveraging these tools will be essential for acquiring innovative agricultural practices that can meet future demands.

16.5 Application of Artificial Intelligence Techniques in Plant Data Science

16.5.1 Crop prediction and yield estimation

AI transformed the prediction of crops and yield estimation through offering tools that could take in massive, intricate data from varied sources, including satellite images, meteorological conditions, soil conditions, and past crop histories. Precise prediction of the yield of crops plays a role of utmost importance for food security planning, market stability, and agricultural practice optimization (Jeong *et al.*, 2016). Conventional crop simulation models tend to use mechanistic or statistical methods that may be constrained by their inherent assumptions and inability to account for the intricate associations among genotype, environment, and management practices (Singh, 2016). Machine learning methods like RF, SVM, and deep neural networks provide data-based solutions that ascertain complex patterns from high-dimensional data without being programmed (Liakos *et al.*, 2018).

Remote sensing with AI has become an efficient tool for monitoring crops and predicting yields. CNNs in multispectral and hyperspectral satellite images can accurately classify crops, estimate growth stages, and determine biomass (Kamilaris and Prenafeta-Boldú, 2018). By combining the data with meteorological variables, such as precipitation and temperature, ML models can accurately foretell the end yield several months ahead of time (Liakos *et al.*, 2018). Furthermore, AI-based prediction of crop yields ensures timely intervention by farmers and policymakers in the event of expected yield deficits (Khaki *et al.*, 2020). Specialized DL frameworks such as recurrent neural networks (RNNs) and long short-term memory (LSTM) networks have exhibited the potential to model the temporal dynamics of the growth of crops in the face of a changing climate (Shahhosseini *et al.*, 2020). Wheat saw ML models combining remote sensing and meteorological data attain a maximum accuracy of 90% for predicting yields (Sharma, 2021). Estimating rice yields in turn was assisted through AI methods that blended the analysis of UAV images with environmental sensor data (Sarkar *et al.*, 2024). In spite of these achievements, generalization across regions and crops continues to be a challenge for model performance because of the heterogeneity of agri-food systems. Future work would entail the development of strong, transferable AI models with the incorporation of domain expertise and multisource data fusion for improving prediction accuracy (Singh *et al.*, 2016).

16.5.2 Disease detection and pest management

Early and accurate detection of plant diseases and pests is essential for minimizing crop losses and ensuring sustainable agricultural productivity. AI technologies, particularly computer vision and ML, have become vital in automating disease diagnosis and pest identification through image analysis and sensor data processing (Ferentinos, 2018). Traditional methods of disease scouting are labor-intensive and require expert knowledge, which limits timely response. In contrast, AI-powered image recognition systems using DL, especially CNNs, can analyze leaf images to detect and classify diseases with remarkable accuracy (Mohanty *et al.*, 2016). The PlantVillage dataset has been widely used to train models that can identify over 50 plant disease classes under field conditions (Ferentinos, 2018). This dataset has been further used to classify diseases of various plants, including potato (Chakraborty *et al.*, 2022). Furthermore, AI-enabled drones equipped with multispectral cameras facilitate rapid, large-scale disease and pest surveillance in crop fields (Zhou *et al.*, 2020). ML algorithms process this data to detect early symptoms of stress caused by pathogens or insect damage before visual symptoms become apparent to the human eye (Mukherjee *et al.*, 2025).

AI also supports pest population forecasting and management by analyzing environmental and historical infestation data. Predictive models using ML can identify pest outbreaks and optimal intervention timing, reducing pesticide use and environmental impact (Upadhyay *et al.*, 2025). Integration of AI with Internet of Things (IoT) devices, such as smart traps and sensors, enhances real-time monitoring of pest populations and disease vectors (Bouwmeester *et al.*, 1998). This integration provides precision in pest management, enabling site-specific treatments and improving overall crop health. Challenges include variability in image quality, complexity of disease symptoms, and the need for large, annotated datasets for training robust AI models. Addressing these issues through domain adaptation, data augmentation, and transfer learning will further improve AI-based disease and pest management systems (Upadhyay *et al.*, 2025).

16.5.3 Precision agriculture and resource management

Precision agriculture seeks to maximize resource utilization and productivity of crops through management practices that match the spatial and temporal heterogeneity of fields. AI plays a crucial role in facilitating PA by offering sophisticated data analytics for use in irrigation, fertilization, and soil management (Liakos *et al.*, 2018). AI-based decision-support tools blend data from soil sensors, weather stations, satellite and UAV imagery, and crop simulation models to provide site-specific recommendations. ML-based analysis of these streams of data predicts nutrient deficiencies, water stress, and pressure from pests, and this leads to the efficient application of inputs (Upadhyay *et al.*, 2025). AI models can use meteorological data and soil moisture measurements to project water requirements for irrigation. DL methods have optimized schedules for irrigation to achieve substantial water conservation and yield improvement (Khan *et al.*, 2020). Similarly, AI-driven nutrient management systems employ soil nutrient contents, growth stages of crops, and environmental factors in suggesting the rate and time of fertilizer use, minimizing waste and environmental pollution (Kumar *et al.*, 2023). Automatic robotic platforms with AI-based algorithms can provide automatic field scouting and input application. Autonomous tractors and drones, for instance, make use of AI for navigating through fields, detecting issues, and delivering treatments with high spatial accuracy (Saleem *et al.*, 2021). AI makes the agriculture industry more sustainable by allowing for predictive maintenance of equipment on farms, efficient energy usage, and carbon footprint tracking (Neethirajan,

2024). However, the implementation of AI-enabled PA technologies is challenged by issues of cost, complexity of data management, and the requirement for training farmers. Ongoing research into low-cost, easy-to-use AI solutions and participatory methods that actively engage end-users is needed for extensive implementation (Wolfert *et al.*, 2017).

16.6 Remote Sensing and IoT in Plant Data Science

Remote sensing and the IoT have revolutionized how data is collected, analyzed, and applied in plant biology and agriculture. Remote sensing, the acquisition of information about an object or phenomenon without physical contact, typically uses satellite aided with aerial imagery to monitor plant health, growth patterns, and environmental conditions on local to global scales (Oerke, 2020). IoT, involving interconnected sensors and devices that collect and transmit data in real-time, complements remote sensing by enabling continuous monitoring of soil conditions, microclimate variables, and crop status at the field level (Wolfert *et al.*, 2017). Both these technologies produce enormous amounts of high-resolution spatiotemporal data, which, when integrated and analyzed, considerably augment decision-making processes in plant science.

Remote sensing technologies employ various platforms and sensors, including satellites, drones (unmanned aerial vehicles (UAVs)), and ground-based systems, to capture multispectral, hyperspectral, thermal, and LiDAR (Light Detection and Ranging) data. Multispectral and hyperspectral imaging provide detailed spectral signatures that reveal plant physiological traits such as chlorophyll content, water stress, and nutrient status (Zarco-Tejada *et al.*, 2013). For example, normalized difference vegetation index (NDVI) and enhanced vegetation index (EVI), derived from multispectral data, are widely used to assess crop vigor and biomass estimation (Gitelson, 2019). UAVs have democratized remote sensing by offering flexible, high-resolution data collection that can be tailored to specific fields and time points, overcoming some limitations of satellite data, such as coarse resolution and infrequent revisit times (Bendig *et al.*, 2013). These aerial platforms have enabled precise monitoring of plant growth stages, disease outbreaks, and stress factors at a scale and detail previously unattainable (Barbedo, 2019).

IoT enhances plant data science by deploying networks of *in situ* sensors that continuously monitor key environmental parameters such as soil moisture, temperature, humidity, pH, and nutrient concentrations (Chamara *et al.*, 2022). These sensors provide granular data on microclimatic conditions influencing plant growth, enabling site-specific management practices. Soil moisture sensors, for example, help optimize irrigation scheduling by providing real-time feedback on soil water availability, thereby improving water use efficiency and reducing waste (Lakhiar *et al.*, 2024). Similarly, nutrient sensors aid in precision fertilization by detecting deficiencies or excesses in soil nutrients, minimizing environmental pollution caused by overfertilization (Kim *et al.*, 2020).

The fusion of remote sensing and IoT data is a burgeoning area in plant data science that leverages the complementary strengths of both technologies. Remote sensing offers broad spatial coverage and temporal frequency, while IoT provides high-resolution local context. Together, these datasets enable multi-scale analyses that enhance crop monitoring, yield prediction, and stress detection with unprecedented accuracy (Abdulraheem *et al.*, 2023). For instance, combining UAV multispectral imagery with ground-based sensor data allows for calibration and validation of remote sensing models, improving the precision of vegetation indices and crop health assessments (Candiago *et al.*, 2015).

Advanced data analytics, including ML and DL techniques, are instrumental in extracting meaningful information from these large, heterogeneous datasets (Kamilaris and Prenafeta-Boldú, 2018). Algorithms such as CNNs and RNNs analyze spatiotemporal patterns, enabling early detection of biotic and abiotic stresses. For example, AI models trained on multispectral images and sensor data can detect disease symptoms before they become visually apparent, allowing for timely intervention and reduced yield losses (Ferentinos, 2018). Moreover, predictive models using combined

IoT and remote sensing inputs forecast crop yield and water requirements, guiding resource allocation efficiently (Luo *et al.*, 2021).

The application of remote sensing and IoT in plant data science is critical for advancing PA, a management approach that optimizes field-level inputs to improve crop productivity and sustainability (Wolfert *et al.*, 2017). PA benefits from these technologies by enabling variable rate applications of water, fertilizers, and pesticides, tailored to specific field zones characterized by differing soil types, microclimates, and crop conditions (Zhang *et al.*, 2019). This targeted management reduces environmental impact and input costs while enhancing yields.

Despite these promising advances, challenges remain. The vast volume and variety of data generated by remote sensing and IoT pose significant storage, processing, and integration issues (Ma *et al.*, 2015). Sensor calibration, data quality, and standardization are critical for reliable analyses. Additionally, connectivity constraints in rural and remote areas can hinder real-time data transmission, limiting the effectiveness of IoT systems (Thakur and Hota, 2013). Ethical considerations around data privacy and ownership also need addressing as digital agriculture becomes more widespread.

16.7 Prospects and Conclusion

The integration of AI into plant data science represents a paradigm shift in how agricultural research is conducted and applied in real-world scenarios. Amidst growing pressures from climate change, soil degradation, and increasing food demands, AI offers powerful capabilities to decode complex plant systems, improve decision-making, and enhance sustainability. The proliferation of high-throughput technologies such as next-generation sequencing, real-time field sensors, and remote sensing platforms has generated vast and complex datasets, necessitating robust analytical tools. AI, through machine learning, DL, and data fusion techniques, provides a critical bridge between raw data and meaningful insights. As this field evolves, the emergence of explainable AI (XAI) holds particular promise, allowing not only for predictive accuracy but also for mechanistic understanding, enabling researchers and breeders to identify key genes, molecular interactions, and environmental factors underlying critical traits such as drought tolerance and disease resistance. Simultaneously, AI-driven multi-omics integration, connecting genomics, transcriptomics, proteomics, and metabolomics, supports a systems biology approach to plant response analysis, advancing the development of resilient and nutrient-rich crop varieties. Beyond the research domain, the application of AI in autonomous agriculture is revolutionizing field practices, with smart systems powered by real-time sensor data and satellite imagery enabling precise irrigation, pest control, and nutrient management, thereby reducing manual labor and environmental impact. Predictive analytics further contribute by informing crop modeling and disease forecasting, supporting proactive and sustainable interventions. However, several challenges persist, including limited availability of annotated datasets, digital infrastructure disparities, model interpretability issues, and data governance concerns. Addressing these gaps requires concerted efforts to develop standardized data frameworks, promote open-access and interoperable databases, and invest in digital literacy and infrastructure, especially in regions where agriculture is both critical and under-resourced. Overall, AI is not merely an ancillary technological tool but a strategic enabler of a predictive, prescriptive, and data-centric transformation in plant science. Its contributions span multiple domains, including high-accuracy disease detection through image analysis, yield prediction via remote sensing integration, acceleration of breeding through genomic selection, real-time decision-support in precision farming, multi-layered omics data interpretation, and enhancement of biodiversity studies using computational phylogenetics and image-based classification. The long-term impact of AI in plant science will depend on sustained interdisciplinary collaboration among plant scientists, data engineers, agronomists, and policy stakeholders, ensuring that the technological advances are both scientifically sound and equitably implemented to address the diverse and dynamic challenges of global agriculture.

References

Abdulraheem, M.I., Zhang, W., Li, S., Moshayedi, A.J., Farooque, A.A. *et al.* (2023) Advancement of remote sensing for soil measurements and applications: A comprehensive review. *Sustainability* 15(21), 15444. DOI: 10.3390/su152115444.

Ahmad, M. (2023) Plant breeding advancements with 'CRISPR-Cas' genome editing technologies will assist future food security. *Frontiers in Plant Science* 14, 1133036. DOI: 10.3389/fpls.2023.1133036.

Ahmed, K., Shabbir, G., Ahmed, M. and Shah, K.N. (2020) Phenotyping for drought resistance in bread wheat using physiological and biochemical traits. *Science of The Total Environment* 729, 139082. DOI: 10.1016/j.scitotenv.2020.139082.

Asada, K., Kaneko, S., Takasawa, K., Machino, H., Takahashi, S. *et al.* (2021) Integrated analysis of whole genome and epigenome data using machine learning technology: Toward the establishment of precision oncology. *Frontiers in Oncology* 11, 666937.

Bai, B., Vanderwall, D., Li, Y., Wang, X., Poudel, S. *et al.* (2021) Proteomic landscape of Alzheimer's disease: Novel insights into pathogenesis and biomarker discovery. *Molecular Neurodegeneration* 16(1), 55. DOI: 10.1186/s13024-021-00474-z.

Barbedo, J.G.A. (2019) A review on the use of unmanned aerial vehicles and imaging sensors for monitoring and assessing plant stresses. *Drones* 3(2), 40.

Bendig, J., Bolten, A. and Bareth, G. (2013) Monitoring des Pflanzenwachstums mit Hilfe Multitemporaler und Hoch Auflösender Oberflächenmodelle von Getreidebeständen auf Basis von Bildern aus UAV-Befliegungen. *Photogrammetrie – Fernerkundung – Geoinformation* 2013(6), 551–562. DOI: 10.1127/1432-8364/2013/0200.

Bhoi, C. and Mahali, S.K. (2024) The role of artificial intelligence used in plant science: A review. *Asian Journal of Research in Botany* 7(2), 263–272.

Bouwmeester, H.J., Gershenzon, J., Konings, M.C.J.M. and Croteau, R. (1998) Biosynthesis of the monoterpenes limonene and carvone in the fruit of Caraway1. *Plant Physiology* 117(3), 901–912. DOI: 10.1104/pp.117.3.901.

Brummitt, R.K. (2014) Taxonomy versus Cladonomy in the Dicot Families1. *Annals of the Missouri Botanical Garden* 100(1–2), 89–99. DOI: 10.3417/2012089.

Candiago, S., Remondino, F., De Giglio, M., Dubbini, M. and Gattelli, M. (2015) Evaluating multispectral images and vegetation indices for precision farming applications from UAV images. *Remote Sensing* 7(4), 4026–4047. DOI: 10.3390/rs70404026.

Carranza-Rojas, J., Goeau, H., Bonnet, P., Mata-Montero, E. and Joly, A. (2017) Going deeper in the automated identification of herbarium specimens. *BMC Evolutionary Biology* 17(1), 181. DOI: 10.1186/s12862-017-1014-z.

Chakraborty, K.K., Mukherjee, R., Chakroborty, C. and Bora, K. (2022) Automated recognition of optical image based potato leaf blight diseases using deep learning. *Physiological and Molecular Plant Pathology* 117, 101781. DOI: 10.1016/j.pmpp.2021.101781.

Chamara, N., Islam, M.D., Bai, G., Shi, Y. and Ge, Y. (2022) Ag-IoT for crop and environment monitoring: Past, present, and future. *Agricultural Systems* 203, 103497. DOI: 10.1016/j.agsy.2022.103497.

Chen, J., Gu, Z., Lai, L. and Pei, J. (2023) *In silico* protein function prediction: The rise of machine learning-based approaches. *Medical Review* 3(6), 487–510. DOI: 10.1515/mr-2023-0038.

Chen, Z., Zhao, P., Li, F., Marquez-Lago, T.T., Leier, A. *et al.* (2020) iLearn: An integrated platform and meta-learner for feature engineering, machine-learning analysis and modeling of DNA, RNA and protein sequence data. *Briefings in Bioinformatics* 21(3), 1047–1057. DOI: 10.1093/bib/bbz041.

Chong, J. and Xia, J. (2020) Using MetaboAnalyst 4.0 for metabolomics data analysis, interpretation, and integration with other omics data. In: Li, S. (ed.) *Computational Methods and Data Analysis for Metabolomics*, Vol. 2104. Springer, New York, pp. 337–360. DOI: 10.1007/978-1-0716-0239-3_17.

Conesa, A., Madrigal, P., Tarazona, S., Gomez-Cabrero, D., Cervera, A. *et al.* (2016) A survey of best practices for RNA-seq data analysis. *Genome Biology* 17(1), 13. DOI: 10.1186/s13059-016-0881-8.

Das Choudhury, S. (2020) Time series modeling for phenotypic prediction and phenotype–genotype mapping using neural networks. In: Bartoli, A. and Fusiello, A. (eds) *Computer Vision – ECCV 2020 Workshops, 12540*. Springer International Publishing, Cham, pp. 228–243. DOI: 10.1007/978-3-030-65414-6_17.

Demidchik, V.V., Shashko, A.Y., Bandarenka, U.Y., Smolikova, G.N., Przhevalskaya, D.A. *et al*. (2020) Plant phenomics: Fundamental bases, software and hardware platforms, and machine learning. *Russian Journal of Plant Physiology* 67(3), 397–412. DOI: 10.1134/S1021443720030061.

Ebersbach, J., Khan, N.A., McQuillan, I., Higgins, E.E., Horner, K. *et al*. (2022) Exploiting high-throughput indoor phenotyping to characterize the founders of a structured B. *napus* breeding population. *Frontiers in Plant Science* 12, 780250. DOI: 10.3389/fpls.2021.780250.

Eftekhari, M., Ma, C. and Orlov, Y.L. (2024) Editorial: Applications of artificial intelligence, machine learning, and deep learning in plant breeding. *Frontiers in Plant Science* 15, 1420938. DOI: 10.3389/fpls.2024.1420938.

Fan, J., Zhang, Y., Wen, W., Gu, S., Lu, X. *et al*. (2021) The future of internet of things in agriculture: Plant high-throughput phenotypic platform. *Journal of Cleaner Production* 280, 123651. DOI: 10.1016/j.jclepro.2020.123651.

Fang, J. (2024) Breeding 5.0 AI-driven revolution in designed plant breeding. *Molecular Plant Breeding*. DOI: 10.5376/mpb.2024.15.0004.

Fang, C., Fernie, A.R. and Luo, J. (2019) Exploring the diversity of plant metabolism. *Trends in Plant Science* 24(1), 83–98. DOI: 10.1016/j.tplants.2018.09.006.

Ferentinos, K.P. (2018) Deep learning models for plant disease detection and diagnosis. *Computers and Electronics in Agriculture* 145, 311–318. DOI: 10.1016/j.compag.2018.01.009.

Fu, J., Zhang, Y., Liu, J., Lian, X., Tang, J. *et al*. (2021) Pharmacometabonomics: Data processing and statistical analysis. *Briefings in Bioinformatics* 22(5), bbab138. DOI: 10.1093/bib/bbab138.

Gallart-Ayala, H., Teav, T. and Ivanisevic, J. (2020) Metabolomics meets lipidomics: Assessing the small molecule component of metabolism. *BioEssays* 42(12), 2000052. DOI: 10.1002/bies.202000052.

Gitelson, A.A. (2019) Remote estimation of fraction of radiation absorbed by photosynthetically active vegetation: Generic algorithm for maize and soybean. *Remote Sensing Letters* 10(3), 283–291. DOI: 10.1080/2150704X.2018.1547445.

Grinberg, N.F., Orhobor, O.I. and King, R.D. (2020) An evaluation of machine-learning for predicting phenotype: Studies in yeast, rice, and wheat. *Machine Learning* 109(2), 251–277. DOI: 10.1007/s10994-019-05848-5.

Halder, A., Verma, A., Biswas, D. and Srivastava, S. (2021) Recent advances in mass-spectrometry based proteomics software, tools and databases. *Drug Discovery Today: Technologies* 39, 69–79. DOI: 10.1016/j.ddtec.2021.06.007.

Hamamoto, R., Komatsu, M., Takasawa, K., Asada, K. and Kaneko, S. (2019) Epigenetics analysis and integrated analysis of multiomics data, including epigenetic data, using artificial intelligence in the era of precision medicine. *Biomolecules* 10(1), 62.

Hao, J.D., Chen, Y.-Y., Wang, Y.-Z., An, N., Bai, P.-R. *et al*. (2023) Novel peak shift correction method based on the retention index for peak alignment in untargeted metabolomics. *Analytical Chemistry* 95(35), 13330–13337. DOI: 10.1021/acs.analchem.3c02583.

Hasan, N., Choudhary, S., Naaz, N., Sharma, N. and Laskar, R.A. (2021) Recent advancements in molecular marker-assisted selection and applications in plant breeding programmes. *Journal of Genetic Engineering and Biotechnology* 19(1), 128. DOI: 10.1186/s43141-021-00231-1.

Jackulin, C. and Murugavalli, S. (2022) A comprehensive review on detection of plant disease using machine learning and deep learning approaches. *Measurement: Sensors* 24, 100441. DOI: 10.1016/j.measen.2022.100441.

Jeong, J.H., Resop, J.P., Mueller, N.D., Fleisher, D.H., Yun, K. *et al*. (2016) Random forests for global and regional crop yield predictions. *PLoS ONE* 11(6), e0156571. DOI: 10.1371/journal.pone.0156571.

Jiang, Y., Wang, D., Wang, W. and Xu, D. (2021a) Computational methods for protein localization prediction. *Computational and Structural Biotechnology Journal* 19, 5834–5844. DOI: 10.1016/j.csbj.2021.10.023.

Jiang, Y., Wang, D., Yao, Y., Eubel, H., Künzler, P. *et al*. (2021b) MULocDeep: A deep-learning framework for protein subcellular and suborganellar localization prediction with residue-level interpretation. *Computational and Structural Biotechnology Journal* 19, 4825–4839. DOI: 10.1016/j.csbj.2021.08.027.

Jiménez-Luna, J., Grisoni, F., Weskamp, N. and Schneider, G. (2021) Artificial intelligence in drug discovery: Recent advances and future perspectives. *Expert Opinion on Drug Discovery* 16(9), 949–959. DOI: 10.1080/17460441.2021.1909567.

Kamilaris, A. and Prenafeta-Boldú, F.X. (2018) A review of the use of convolutional neural networks in agriculture. *The Journal of Agricultural Science* 156(3), 312–322. DOI: 10.1017/S0021859618000436.

Khaki, S., Wang, L. and Archontoulis, S.V. (2020) A CNN-RNN framework for crop yield prediction. *Frontiers in Plant Science* 10, 1750. DOI: 10.3389/fpls.2019.01750.

Khan, A.W., Garg, V., Roorkiwal, M., Golicz, A.A., Edwards, D. *et al*. (2020) Super-pangenome by integrating the wild side of a species for accelerated crop improvement. *Trends in Plant Science* 25(2), 148–158. DOI: 10.1016/j.tplants.2019.10.012.

Kim, Y., Chung, Y.S., Lee, E., Tripathi, P., Heo, S. *et al*. (2020) Root response to drought stress in rice (*Oryza sativa* L.). *International Journal of Molecular Sciences* 21(4), 1513. DOI: 10.3390/ijms21041513.

Krasileva, K.V., Buffalo, V., Bailey, P., Pearce, S., Ayling, S. *et al*. (2013) Separating homeologs by phasing in the tetraploid wheat transcriptome. *Genome Biology* 14(6), R66. DOI: 10.1186/gb-2013-14-6-r66.

Kumar, J., Chawla, R., Katiyar, D., Chouriya, A., Nath, D. *et al*. (2023) Optimizing irrigation and nutrient management in agriculture through artificial intelligence implementation. *International Journal of Environment and Climate Change* 13, 4016–4022. DOI: 10.9734/IJECC/2023/v13i103077.

Lakhiar, I., Yan, H., Zhang, C., Wang, G., He, B. *et al*. (2024) A review of precision irrigation water-saving technology under changing climate for enhancing water use efficiency, crop yield, and environmental footprints. *Agriculture* 14, 1141. DOI: 10.3390/agriculture14071141.

Langfelder, P. and Horvath, S. (2008) WGCNA: An R package for weighted correlation network analysis. *BMC Bioinformatics* 9(1), 559. DOI: 10.1186/1471-2105-9-559.

Leebens-Mack, J.H., Barker, M.S., Carpenter, E.J., Deyholos, M.K., Gitzendanner, M.A. *et al*. (2019) One thousand plant transcriptomes and the phylogenomics of green plants. *Nature* 574(7780), 679–685. DOI: 10.1038/s41586-019-1693-2.

Li, Z., Jin, J., Wang, Y., Long, W., Ding, Y. *et al*. (2023) ExamPle: Explainable deep learning framework for the prediction of plant small secreted peptides. *Bioinformatics* 39(3), btad108. DOI: 10.1093/bioinformatics/btad108.

Liakos, K., Busato, P., Moshou, D., Pearson, S. and Bochtis, D. (2018) Machine learning in agriculture: A review. *Sensors* 18(8), 2674. DOI: 10.3390/s18082674.

Libbrecht, M.W. and Noble, W.S. (2015) Machine learning applications in genetics and genomics. *Nature Reviews Genetics* 16(6), 321–332. DOI: 10.1038/nrg3920.

Luo, B., Liu, X., Zhang, F. and Guo, P. (2021) Optimal management of cultivated land coupling remote sensing-based expected irrigation water forecasting. *Journal of Cleaner Production* 308, 127370.

Ma, Y., Wu, H., Wang, L., Huang, B., Ranjan, R. *et al*. (2015) Remote sensing big data computing: Challenges ad opportunities. In: *Future Generation Computer Systems*, Vol. 51. Special Section: A Note on New Trends in Data-Aware Scheduling and Resource Provisioning in Modern HPC Systems, pp. 47–60. DOI: 10.1016/j.future.2014.10.029.

Maganathan, T., Senthilkumar, S. and Balakrishnan, V. (2020) Machine learning and data analytics for environmental science: A review, prospects and challenges. *IOP Conference Series: Materials Science and Engineering* 955(1), 012107. DOI: 10.1088/1757-899X/955/1/012107.

Mahmud, M., Kaiser, M.S., McGinnity, T.M. and Hussain, A. (2021) Deep learning in mining biological data. *Cognitive Computation* 13(1), 1–33. DOI: 10.1007/s12559-020-09773-x.

Mathew, D.E., Ebem, D.U., Ikegwu, A.C., Ukeoma, P.E. and Dibiaezue, N.F. (2025) Recent emerging techniques in explainable artificial intelligence to enhance the interpretable and understanding of AI models for human. *Neural Processing Letters* 57(1). DOI: 10.1007/s11063-025-11732-2.

McGowan, T., Johnson, J.E., Kumar, P., Sajulga, R., Mehta, S. *et al*. (2020) Multi-omics visualization platform: An extensible galaxy plug-in for multi-omics data visualization and exploration. *GigaScience* 9(4), giaa025. DOI: 10.1093/gigascience/giaa025.

Michael, T.P. and VanBuren, R. (2020) Building near-complete plant genomes. *Current Opinion in Plant Biology* 54, 26–33. DOI: 10.1016/j.pbi.2019.12.009.

Min, S., Lee, B. and Yoon, S. (2017) Deep learning in bioinformatics. *Briefings in Bioinformatics* 18(5), 851–869. DOI: 10.1093/bib/bbw068.

Mishra, A.K., Sudalaimuthuasari, N., Hazzouri, K.M., Saeed, E.E., Shah, I. *et al*. (2022) Tapping into plant–microbiome interactions through the lens of multi-omics techniques. *Cells* 11(20), 3254. DOI: 10.3390/cells11203254.

Mohanty, S.P., Hughes, D.P. and Salathé, M. (2016) Using deep learning for image-based plant disease detection. *Frontiers in Plant Science* 7, 1419–1419. DOI: 10.3389/fpls.2016.01419.

Mukherjee, R., Ghosh, A., Chakraborty, C., De, J.N. and Mishra, D.P. (2025) Rice leaf disease identification and classification using machine learning techniques: A comprehensive review. *Engineering Applications of Artificial Intelligence* 139, 109639. DOI: 10.1016/j.engappai.2024.109639.

Müller, J.B., Geyer, P.E., Colaço, A.R., Treit, P.V., Strauss, M.T. *et al.* (2020) The proteome landscape of the kingdoms of life. *Nature* 582(7813), 592–596. DOI: 10.1038/s41586-020-2402-x.

Muthye, V., Kandoi, G. and Lavrov, D.V. (2020) MMPdb and MitoPredictor: Tools for facilitating comparative analysis of animal mitochondrial proteomes. *Mitochondrion* 51, 118–125. DOI: 10.1016/j.mito.2020.01.001.

Naqvi, R.Z., Siddiqui, H.A., Mahmood, M.A., Najeebullah, S., Ehsan, A. *et al.* (2022) Smart breeding approaches in post-genomics era for developing climate-resilient food crops. *Frontiers in Plant Science* 13, 972164. DOI: 10.3389/fpls.2022.972164.

Neethirajan, S. (2024) Net zero dairy farming—advancing climate goals with big data and artificial intelligence. *Climate* 12(2), 15. DOI: 10.3390/cli12020015.

Noor, Z., Ahn, S.B., Baker, M.S., Ranganathan, S. and Mohamedali, A. (2021) Mass spectrometry-based protein identification in proteomics—a review. *Briefings in Bioinformatics* 22(2), 1620–1638. DOI: 10.1093/bib/bbz163.

Oerke, E.-C. (2020) Remote sensing of diseases. *Annual Review of Phytopathology* 58(1), 225–252. DOI: 10.1146/annurev-phyto-010820-012832.

Orsburn, B.C. (2021) Proteome discoverer—a community enhanced data processing suite for protein informatics. *Proteomes* 9(1), 15. DOI: 10.3390/proteomes9010015.

Ozsolak, F. and Milos, P.M. (2011) RNA sequencing: Advances, challenges and opportunities. *Nature Reviews Genetics* 12(2), 87–98. DOI: 10.1038/nrg2934.

Pacal, I., Kunduracioglu, I., Alma, M.H., Deveci, M., Kadry, S. *et al.* (2024) A systematic review of deep learning techniques for plant diseases. *Artificial Intelligence Review* 57(11), 304. DOI: 10.1007/s10462-024-10944-7.

Panda, D., Kumar, M., Mahalingam, L. and Raveendran, M. (2022) Artificial intelligence-aided phenomics in high-throughput stress phenotyping of plants. In: Kose, U., Prasath, V.B.S., Mondal, M.R.H., Podder, P. and Bharati, S. (eds) *Artificial Intelligence and Smart Agriculture Applications*. Auerbach Publications, New York.

Pantazi, X.E., Moshou, D. and Tamouridou, A.A. (2019) Automated leaf disease detection in different crop species through image features analysis and one class classifiers. *Computers and Electronics in Agriculture* 156, 96–104. DOI: 10.1016/j.compag.2018.11.005.

Paul, K., Chatterjee, S.S., Pai, P., Varshney, A., Juikar, S. *et al.* (2022) Viable smart sensors and their application in data driven agriculture. *Computers and Electronics in Agriculture* 198, 107096. DOI: 10.1016/j.compag.2022.107096.

Peng, S. (2023) Ultrasound-assisted extraction of Chinese Propolis. *Frontiers in Sustainable Food Systems* 7, 1131959.

Peralbo-Molina, A., Solà-Santos, P., Perera-Lluna, A. and Chicano-Gálvez, E. (2023) Data processing and analysis in mass spectrometry-based metabolomics. In: González-Domínguez, R. (ed.) *Mass Spectrometry for Metabolomics*. Springer, New York, pp. 207–239. DOI: 10.1007/978-1-0716-2699-3_20.

Perrakis, A. and Sixma, T.K. (2021) AI revolutions in biology: The joys and perils of AlphaFold. *EMBO Reports* 22(11), e54046. DOI: 10.15252/embr.202154046.

Pomyen, Y., Wanichthanarak, K., Poungsombat, P., Fahrmann, J., Grapov, D. *et al.* (2020) Deep metabolome: Applications of deep learning in metabolomics. *Computational and Structural Biotechnology Journal* 18, 2818–2825. DOI: 10.1016/j.csbj.2020.09.033.

Prakash, V.S., Vanitha, R., Nikam, S.V., Athappan, V., Krishnamoorthy, R. *et al.* (2024) Revolutionizing agriculture: Artificial intelligence assisted plant leaf disease detection using deep learning principles. In: *2024 10th International Conference on Communication and Signal Processing (ICCSP)*, IEEE, Melmaruvathur, India, pp. 586–591. DOI: 10.1109/ICCSP60870.2024.10543984.

Raza, A. (2022) Metabolomics: A systems biology approach for enhancing heat stress tolerance in plants. *Plant Cell Reports* 41(3), 741–763. DOI: 10.1007/s00299-020-02635-8.

Ryu, K.H., Huang, L., Kang, H.M. and Schiefelbein, J. (2019) Single-cell RNA sequencing resolves molecular relationships among individual plant cells. *Plant Physiology* 179(4), 1444–1456. DOI: 10.1104/pp.18.01482.

Saleem, M.H., Potgieter, J. and Arif, K.M. (2021) Automation in agriculture by machine and deep learning techniques: A review of recent developments. *Precision Agriculture* 22(6), 2053–2091. DOI: 10.1007/s11119-021-09806-x.

Salman-Minkov, A., Sabath, N. and Mayrose, I. (2016) Whole-genome duplication as a key factor in crop domestication. *Nature Plants* 2(8), 16115. DOI: 10.1038/nplants.2016.115.

Sarkar, T.K., Roy, D.K., Kang, Y.S., Jun, S.R., Park, J.W. *et al.* (2024) Ensemble of machine learning algorithms for rice grain yield prediction using UAV-based remote sensing. *Journal of Biosystems Engineering* 49(1), 1–19. DOI: 10.1007/s42853-023-00209-6.

Satrio, R.D., Fendiyanto, M.H. and Miftahudin, M. (2024) Tools and techniques used at global scale through genomics, transcriptomics, proteomics, and metabolomics to investigate plant stress responses at the molecular level. In: Shahid, M. and Gaur, R. (eds) *Molecular Dynamics of Plant Stress and Its Management*. Springer Nature Singapore, Singapore, pp. 555–607. DOI: 10.1007/978-981-97-1699-9_25.

Schunck, D., Magistri, F., Rosu, R.A., Cornelißen, A., Chebrolu, N. *et al.* (2021) Pheno4D: A spatio-temporal dataset of maize and tomato plant point clouds for phenotyping and advanced plant analysis. *PLoS ONE* 16(8), e0256340. DOI: 10.1371/journal.pone.0256340.

Shahhosseini, M., Hu, G. and Archontoulis, S.V. (2020) Forecasting corn yield with machine learning ensembles. *Frontiers in Plant Science* 11. DOI: 10.3389/fpls.2020.01120.

Sharma, A. (2021) Smart agriculture services using deep learning, big data, and IoT (Internet of Things). In: *Advances in Environmental Engineering and Green Technologies. IGI Global*. IGI Global, pp. 166–202. DOI: 10.4018/978-1-7998-5003-8.ch010.

Sharma, M., Sudheer, S., Usmani, Z., Rani, R. and Gupta, P. (2020) Deciphering the omics of plant–microbe interaction: Perspectives and new insights. *Current Genomics* 21(5), 343–362. DOI: 10.2174/1389202921999200515140420.

Singh, R. (2016) Chemotaxonomy: A tool for plant classification. *Journal of Medicinal Plants Studies* 4(2), 90–93.

Shulse, C.N., Cole, B.J., Ciobanu, D., Lin, J., Yoshinaga, Y. *et al.* (2019) High-throughput single-cell transcriptome profiling of plant cell types. *Cell Reports* 27(7), 2241–2247. DOI: 10.1016/j.celrep.2019.04.054.

Singh, A., Ganapathysubramanian, B., Singh, A.K. and Sarkar, S. (2016) Machine learning for high-throughput stress phenotyping in plants. *Trends in Plant Science* 21(2), 110–124. DOI: 10.1016/j.tplants.2015.10.015.

Singh, A.K., Ganapathysubramanian, B., Sarkar, S. and Singh, A. (2018) Deep learning for plant stress phenotyping: Trends and future perspectives. *Trends in Plant Science* 23(10), 883–898. DOI: 10.1016/j.tplants.2018.07.004.

Singh, S., Singh, A.V., Singh, A. and Singh, S.P. (2021) Speed breeding: An innovative method for crop improvement. *EC Agriculture* 7(3), 1–8.

Sinitcyn, P., Hamzeiy, H., Soto, F.S., Itzhak, D., McCarthy, F. *et al.* (2021) MaxDIA enables library-based and library-free data-independent acquisition proteomics. *Nature Biotechnology* 39(12), 1563–1573. DOI: 10.1038/s41587-021-00968-7.

Sladojevic, S., Arsenovic, M., Anderla, A., Culibrk, D. and Stefanovic, D. (2016) Deep neural networks based recognition of plant diseases by leaf image classification. *Computational Intelligence and Neuroscience* 1–11. DOI: 10.1155/2016/3289801.

Sneath, P., Andrews, H. and Sokal, R. (1973) The principles and practice of numerical classification. *Numerical Taxonomy* 573, WH Freeman and Co. Available at: https://cir.nii.ac.jp/crid/1570572699884106752

Stephens, Z.D., Lee, S.Y., Faghri, F., Campbell, R.H., Zhai, C. *et al.* (2015) Big data: Astronomical or genomical? *PLoS Biology* 13(7), e1002195. DOI: 10.1371/journal.pbio.1002195.

Thakur, A. and Hota, C. (2013) Sustainable wireless internet connectivity for rural areas. In: *2013 International Conference on Advances in Computing, Communications and Informatics (ICACCI)*, pp. 1335–1340. DOI: 10.1109/ICACCI.2013.6637371.

Tirthyani, D., Kumar, S. and Vats, S. (2024) Clustering and unsupervised learning. In: *Advances in Systems Analysis, Software Engineering, and High Performance Computing*. IGI Global, pp. 119–139. DOI: 10.4018/979-8-3693-3609-0.ch005.

Upadhyay, A., Patel, A., Chandel, N.S., Chakraborty, S.K. and Bhalekar, D.G. (2025) Leveraging AI and ML in precision farming for pest and disease management: Benefits, challenges, and future prospects. In: Jatav, H.S., Raiput, V.D. and Minkina, T. (eds) *Ecologically Mediated Development, Sustainable Development and Biodiversity*, Vol. 41. Springer Nature Singapore, Singapore, pp. 511–528. DOI: 10.1007/978-981-96-2413-3_23.

Varadi, M., Anyango, S., Deshpande, M., Nair, S., Natassia, C. *et al.* (2022) AlphaFold protein structure database: Massively expanding the structural coverage of protein-sequence space with high-accuracy models. *Nucleic Acids Research* 50(D1), D439–444. DOI: 10.1093/nar/gkab1061.

Varshney, R.K., Bohra, A., Yu, J., Graner, A., Zhang, Q. *et al.* (2021) Designing future crops: Genomics-assisted breeding comes of age. *Trends in Plant Science* 26(6), 631–649. DOI: 10.1016/j.tplants.2021.03.010.

Vollert, H. (2009) Plant extract for the prophylaxis and treatment of hyperglycemic diseases. EP2226076A1. Munich, Germany: European Patent Office.

Wang, L.-S., Hashimoto, F., Shiraishi, A., Aoki, N., Li, J.-J. *et al.* (2001) Phenetics in tree peony species from China by flower pigment cluster analysis. *Journal of Plant Research* 114(3), 213–221. DOI: 10.1007/PL00013985.

Wang, Z., Gerstein, M. and Snyder, M. (2009) RNA-seq: A revolutionary tool for transcriptomics. *Nature Reviews Genetics* 10(1), 57–63. DOI: 10.1038/nrg2484.

Wen, B. and Zhang, B. (2023) PepQuery2 democratizes public MS proteomics data for rapid peptide searching. *Nature Communications* 14(1), 2213. DOI: 10.1038/s41467-023-37462-4.

Williamson, H.F., Brettschneider, J., Caccamo, M., Davey, R.P., Goble, C. *et al.* (2023) Data management challenges for artificial intelligence in plant and agricultural research. *F1000Research* 10, 324. DOI: 10.12688/f1000research.52204.2.

Wolfert, S., Ge, L., Verdouw, C. and Bogaardt, M.-J. (2017) Big data in smart farming – a review. *Agricultural Systems* 153, 69–80. DOI: 10.1016/j.agsy.2017.01.023.

Yang, Q., Liu, Y., Chen, T. and Tong, Y. (2019) Federated machine learning: Concept and applications. *ACM Transactions on Intelligent Systems and Technology* 10(2), 1–19. DOI: 10.1145/3298981.

Yan, T.-C., Yue, Z.-X., Xu, H.-Q., Liu, Y.H., Hong, Y.-F. *et al.* (2023) A systematic review of state-of-the-art strategies for machine learning-based protein function prediction. *Computers in Biology and Medicine* 154, 106446. DOI: 10.1016/j.compbiomed.2022.106446.

Yang, Z., Zeng, X., Zhao, Y. and Chen, R. (2023) AlphaFold2 and Its applications in the fields of biology and medicine. *Signal Transduction and Targeted Therapy* 8(1), 1–14. DOI: 10.1038/s41392-023-01381-z.

Younis, S., Weiland, C., Hoehndorf, R., Dressler, S., Hickler, T. *et al.* (2018) Taxon and trait recognition from digitized herbarium specimens using deep convolutional neural networks. *Botany Letters* 165(3–4), 377–383. DOI: 10.1080/23818107.2018.1446357.

Yousaf, A., Kayvanfar, V., Mazzoni, A. and Elomri, A. (2023) Artificial intelligence-based decision support systems in smart agriculture: Bibliometric analysis for operational insights and future directions. *Frontiers in Sustainable Food Systems* 6. DOI: 10.3389/fsufs.2022.1053921.

Zarco-Tejada, P.J., González-Dugo, V., Williams, L.E., Suárez, L., Berni, J.A.J. *et al.* (2013) A PRI-based water stress index combining structural and chlorophyll effects: Assessment using diurnal narrow-band airborne imagery and the CWSI thermal index. *Remote Sensing of Environment* 138, 38–50. DOI: 10.1016/j.rse.2013.07.024.

Zhang, J., An, T., Chi, F., Wei, D., Zhou, B. *et al.* (2019) Evolution over years of structural characteristics of humic acids in black soil as a function of various fertilization treatments. *Journal of Soils and Sediments* 19(4), 1959–1969. DOI: 10.1007/s11368-018-2212-z.

Zhang, M., Qin, Z., Liu, X. and Ustin, S.L. (2003) Detection of stress in tomatoes induced by late blight disease in California, USA, using hyperspectral remote sensing. *International Journal of Applied Earth Observation and Geoinformation* 4(4), 295–310. DOI: 10.1016/s0303-2434(03)00008-4.

Zhang, R., Zhang, C., Yu, C., Dong, J. and Hu, J. (2022) Integration of multi-omics technologies for crop improvement: Status and prospects. *Frontiers in Bioinformatics* 2, 1027457. DOI: 10.3389/fbinf.2022.1027457.

Zhang, Z., Zhang, R., Xiao, H., Bhattacharya, K., Bitounis, D. *et al.* (2019) Development of a standardized food model for studying the impact of food matrix effects on the gastrointestinal fate and toxicity of ingested nanomaterials. *NanoImpact* (13), 13–25. DOI: 10.1016/j.impact.2018.11.002.

Zhou, J. and Troyanskaya, O.G. (2015) Predicting effects of noncoding variants with deep learning–based sequence model. *Nature Methods* 12(10), 931–934. DOI: 10.1038/nmeth.3547.

Zhou, Y., Jia, E., Pan, M., Zhao, X. and Ge, Q. (2020) Encoding method of single-cell spatial transcriptomics sequencing. *International Journal of Biological Sciences* 16(14), 2663–2674. DOI: 10.7150/ijbs.43887.

Zhu, F., Saluja, M., Dharni, J.S., Paul, P., Sattler, S.E. *et al.* (2021) PhenoImage: An open-source graphical user interface for plant image analysis. *The Plant Phenome Journal* 4(1), e20015. DOI: 10.1002/ppj2.20015.

17 Deep Learning in Plant Pest Control

Battana Swapna*, Srinivasan Kameswaran and Mandala Ramakrishna
Department of Botany, Vikrama Simhapuri University College, Kavali, SPSR Nellore District, Andhra Pradesh, India

Abstract
Food quality control standards must include efficient control of pests since pests seriously degrade agricultural productivity and quality. Preventing significant crop damage requires early pest identification. Crop problems are usually detected by human specialists, but this method might be pricey and take a lot of time. As a result, new detection techniques are required. This chapter explores an outline of current research studies on the use of deep learning techniques for pest identification, including convolutional neural network (CNN), long short-term memory (LSTM), deep convolutional neural network (DCNN), and deep belief network (DBN). By automating field monitoring, these techniques increase crop protection and yield while lowering labor costs and human errors.

Keywords: Artificial intelligence, Biotic stress, Crop protection, Machine learning, Pest resistance

17.1 Introduction

Every year, pests cause a 20–40% reduction in the world's food production, making agricultural production extremely difficult (Karar *et al.*, 2021). Because these pests have similar features, it is challenging to identify and observe them, which makes classification challenging. Agricultural pests have historically put crop production at risk. Often, damage is severe by the time farmers discover it. A lot of fungicides need to be used immediately to fight pests. Classification of insects was difficult owing to their complex morphology and sharing a degree of resemblance among various groups (Thenmozhi and Reddy, 2019). The preservation of agricultural products and crop output has long been seriously threatened by agricultural pests (Cheng *et al.*, 2017). When farmers discover pest damage, it has frequently tripled in intensity and spread out of control. Numerous fungicides must be sprayed on the fields at that time in order to eradicate pests and reduce agricultural damage (Chen *et al.*, 2020). Although visual symptoms can be used to diagnose plant infections, precise diagnosis requires a specific understanding. It is difficult to diagnose in rural locations, particularly in poor countries. Diagnosis frequently entails extensive laboratory testing with chemicals or time-consuming professional visual inspections (Sethy *et al.*, 2020; Ahmad *et al.*, 2021).

Several strategies have been the main focus of crop leaf disease identification; for instance, fluorescence *in situ* hybridization (FISH), thermography, polymerase chain reaction (PCR), hyperspectral imaging, and enzyme-linked immunosorbent testing (ELISA) (Wang *et al.*, 2018; Hou *et al.*, 2021). To identify various pest species, features are directly gathered and

*Corresponding author: swapnaivsr@gmail.com

 Plant Genomic Prediction: AI-Enhanced Models for Plant Genomics and Crop Improvement (ed. Jen-Tsung Chen)
DOI: 10.1079/9781836991328.0017

a corresponding filter is applied. This process takes time when the collection is large (Li *et al.*, 2020b). Typical applications include the identification and classification of insects, computer-aided image classification systems that are associated with conventional machine learning techniques (Ayan *et al.*, 2020), and convolutional neural networks (CNNs), which have been in use for a long time (Lv *et al.*, 2020). A detection and classification system is used to classify various classes. Deep learning is utilized for classification (Shi *et al.*, 2020), and dual processes have been described as the models of Training and Testing (Nigam *et al.*, 2020). It appears that deep learning is more effective in identifying crop issues in agricultural output. A deep learning model can be used to track, diagnose, and stop crop growth (Ai *et al.*, 2020), and it can be adjusted to improve identification accuracy (Krishnamoorthy *et al.*, 2021). Ultimately, the study by Wang *et al.* (2020) provides a deep network model for agricultural pest identification and generates data for refining and enhancing the CNN model's structure and training parameters for pest picture features for common forestry and agricultural pests. Applications of deep learning in pest control are presented in Fig. 17.1.

17.2 Pest Detection

One of the major causes of the decline in crop quality and yield is the prevalence of agricultural pests and diseases. It is difficult for artificial intelligence methods to quickly and accurately identify pests, because of their great resemblance and quick movement.

Plant pests have a significant impact on plant quality and productivity. Digital image processing is a useful tool for identifying plant pests. Recent advancements in deep learning have led to its substantial superiority over traditional methods in the realm of digital image processing. Plant pathologists are becoming more focused on how to utilize deep learning technologies for the detection of plant diseases and pests. This study summarizes the research and process of contemporary deep learning-dependent plant pest detection (Fig. 17.2). Based on this, the chapter explores probable difficulties

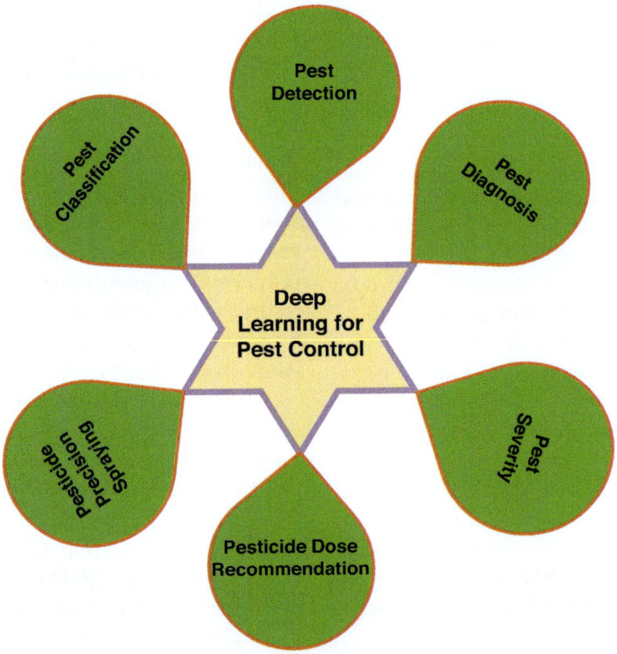

Fig. 17.1. Application of deep learning in pest control (Author's own image)

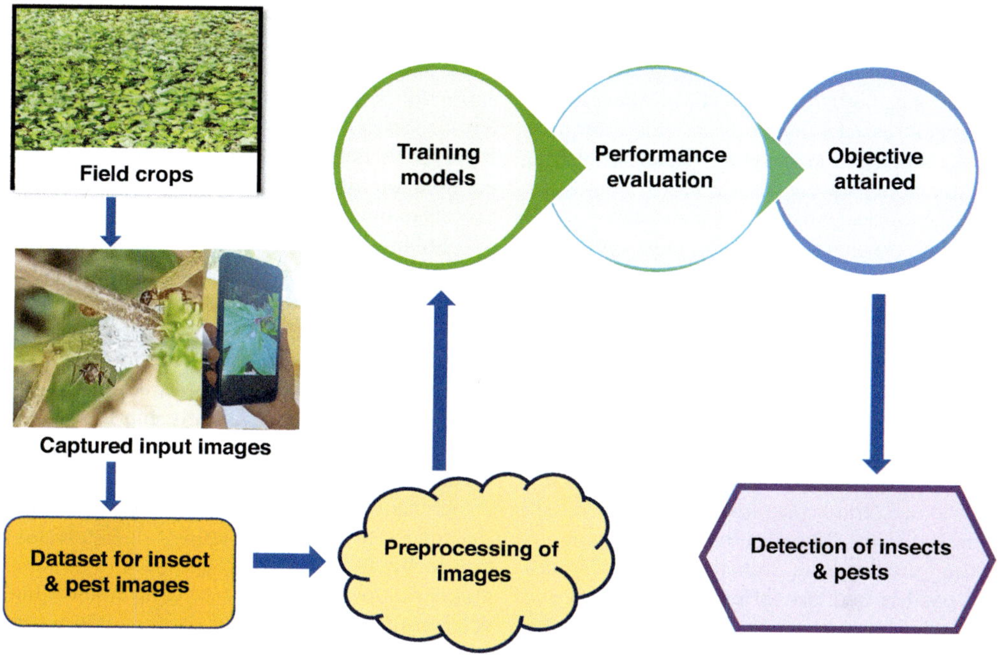

Fig. 17.2. Deep learning for the detection of pests (Author's own image)

in practical deep learning-powered plant disease and pest recognition applications. Additionally, a few commendations are made, along with potential research ideas and solutions to the problems. Lastly, this chapter provides an analysis and outlook for the future of deep learning-based plant disease and pest detection.

17.2.1 Pest detection in rice

The deadliest illness, according to a study by Tian *et al.* (2021), is rice blast, which endangers world output and results in enormous financial losses everywhere. The results of the study demonstrated that during plant–pathogen interactions, a notable pattern of variability persisted in rice plants. The great precision of approximately 80% at the initial infection phase demonstrates that consistent hyperspectral signals can be used to identify rice leaf blast (RLB) infection non-destructively.

A machine vision-based automated light-trap pest monitoring system in rice was developed by Yao *et al.* in 2020. That system includes a cloud server and a PC. When insects are killed and dispersed by the light trap, pictures of the insects are gathered and sent to a cloud server. The accuracy and effectiveness of the automated light trap for pest management were proven in field tests. Additionally, Jiang *et al.* (2020) found that support vector machines (SVMs) and deep learning may be utilized to identify leaf pathology in rice with pictures. Convolutional neural networks were used to identify the characteristics of disease on rice foliage images. The diseases were categorized and predicted using the SVM approach. According to the experimental data, 96.8% was the average detection rate of rice illness using SVM and deep learning. Compared with standard back-propagation neural networks, the accuracy was higher. It can be quite beneficial to use SVM and deep learning to forecast rice issues and offer technical plans for different agricultural areas.

Rice disease was observed in 2019 by Zhou *et al.* using FCM-KM as well as faster R-CNN fusion. In addition to addressing several problems using rice pathology photographs, FCM-KM also suggests faster R-CNN fusion.

These problems include blurred, vague images, excessive ambient noise, and reduced detectability. Different lengths of faster region CNNs were chosen using both the hybrid FCM-KM approach and region-based CNN for disease identification in rice. Toward identifying bacterial blight, sheath blight, and rice blast diseases, a database of images of rice pests was constructed.

The problems of rice blast artificial recognition were addressed in 2018 by Xiao *et al.*, who used a neural net and PCA in rice blast recognition. A photograph of the excised abscess was used to extract ten morphological features, six color features, and five texture qualities for each sickness. Then, using stepwise regression analysis, the relationship between the distinguishing parameters was examined. On average, the recognition rate was higher than that of the BP neural network recognition method. Additionally, Shiba *et al.* (2018) found that paddies had rice stripe disease. The virus known as rice stripe is carried by the tiny brown planthopper. The rapid decline in rice production in Japan due to diseases made the development of management plans imperative. According to research, sickness first manifests in diseased plants between mid-June and July, after which it spreads from the afflicted plants to nearby ones. Since the number of infected plants within the meadow can suddenly change, pest control measures should focus on either the first farmed rice paddies or both.

17.2.2 Pest detection in legumes

The effectiveness of the 'Resnet-50, Inception-v3, VGG-16, Xception, and VGG-19 architectures of deep learning' for categorizing photos of soybean pests was assessed by Tetila *et al.* in 2020. To separate the pests in the photos taken in real-world settings, visual content classification utilizing the superpixels generated by the Simple Linear Iterative Clustering method was taken into consideration. With accuracies of up to 93.82%, the performance of deep learning architectures for various 'fine-tuning and transfer learning' procedures was evaluated against alternative methods throughout the classification challenge. Furthermore, the semi-supervised learning technique OPFSEMImst, the supervised learning techniques for image classification like random forest (RF), k-NN, and SVM, and conventional techniques like feature extraction and Bag-of-features with SURF were all surpassed by deep learning architectures. In soybean fields, the findings show that the assessed architectures can help farmers and experts manage pests.

Subburaj *et al.* (2024) developed an instantaneous effectual pest-finding system using the transfer learning approach and the deep learning-based EfficientNetB7 architecture. With the help of its novel compound scaling technique, EfficientNetB7 has successfully classified pests from photos while using the fewest resources possible by striking a compromise between computation efficiency and accuracy. To efficiently create a pest detection system, the suggested method makes use of regularization techniques, including optimizers and data augmentation, as well as suitable fine-tuning of training parameters. Using CubeAI in STM32CubeIDE, the validated model is transferred to an STM32 embedded controller. The model achieves a classification accuracy of 93.5% when evaluated on a publicly accessible dataset. This technology supports precision agriculture by automating pest detection and providing a match percentage for the pests found. The dataset will be expanded in future research, and edge AI methods for decentralized decision making will be investigated. The revolutionary probability of using EfficientNetB7 in smart farming is demonstrated in this publication, which provides a scalable, affordable, and sustainable pest detection solution with pertinent wider implications for agricultural automation methods.

One of India's most important crops is the mung bean (*Vigna radiata*). CNNs were utilized in producing impressive results in image categorization thanks to the development of deep learning techniques. This holds great potential for the identification of diseases and pests through efficient picture classification. A revolutionary deep learning-based method in the identification of pests and diseases in mung bean was proposed by Mallick *et al.* in 2023. Transfer learning, which can produce a very promising outcome for rapid and cheap pest and disease diagnosis, was implemented to address the limitation resulting from a lack of mung bean crop photos for training. Six distinct kinds of mung bean disease and four pest types have been successfully identified using the created model, collecting healthy and afflicted leaves during various seasons. The suggested mobile deep learning model for disease and pest recognition in mung bean showed a typical accuracy of 93.65%, according to the experiments that were carried out.

Because traditional approaches to discover pest infestations are time-consuming, soybean farming is constantly at risk of significant crop losses. A state-of-the-art technique that uses particularly CNNs and RNNs to change initial soybean pest identification in farming areas was presented by Kim *et al.* in 2024 in order to bridge this crucial gap. Through the utilization of a wide-ranging dataset containing environmental factors and crop close-up photos, this approach provides valuable insights into the intricacies of soybean ecosystems. Images are interpreted using CNNs in the process, while temporal interactions of environmental factors are recorded using RNNs. As a result, a sophisticated deep learning model is produced. The collection is made up of 1050 images of soybean crop infested with pests such as Coccinellidae and *Anticarsia*, as well as pest-free, normal plants. The accuracy, sensitivity, and efficiency of the model in the quick detection of insects are confirmed after it has been meticulously trained over 20 epochs. The results show how effective the model is, consistently surpassing conventional techniques with an impressive 95% validity. The enhanced sensitivity lowers economic and environmental expenses while demonstrating its adaptability to various soybean growing environments. The proposed method is beneficial to the agricultural industry since it offers a state-of-the-art, environmentally responsible means of safeguarding soybean crops against pest-related issues.

The yields of soybeans are lowered by pests like moths, mites, beetles, and stink bugs. *Riptortus pedestris* infests soybean leaves and fruits throughout the plant's growth. Research by Park *et al.* (2023) created a surveillance system using an unmanned ground vehicle GoPro CAM to identify *R. pedestris* in accordance with distinct environmental factors at the seed development phase, seed maturation phase, and artificial cage. Through a web application, the models utilized in this experiment that are based on deep learning technologies (MRCNN, YOLOv3, Detectron2) may be easily challenged (i.e., constructed with lightweight parameters) right away. Prior to pre-processing by labeling the image for annotation, the dataset of images was dispersed via 'random selection for training, validation, and testing'. The deep learning model used a clipping area and segmented region in the image data to locate and categorize the *R. pedestris* individuals. With computed 'means of average precision' (mAP) values of 0.952, 0.716, and 0.873, respectively, the model demonstrated strong performance. The produced model can be a useful implementation for insect prediction in the early infestation in the agricultural process and will allow *R. pedestris* detection in farming conditions (Park *et al.*, 2023).

Insect pest misclassification can result in the incorrect use of pesticides, which can negatively impact agricultural productivity besides the environment. It requires creating an automatic structure that can identify and classify pests with greater accuracy. In 2022, Ullah and colleagues introduced a new end-to-end DeepPestNet for pest detection and classification. Three fully connected (FC) and eight convolutional layers make up the 11 learnable layers in the suggested model. Techniques for picture augmentation and rotation were employed to evaluate the broader relevance of the suggested DeepPestNet method and expand the dataset, respectively. To evaluate the DeepPestNet system, the well-known Deng's Crops dataset was utilized. Crop pests were identified and categorized using this method into ten classes: *Euproctis pseudoconspersa* strand, *Locusta migratoria*, *Spodoptera exigua*, *Empoasca flavescens*, *Chrysochus chinensis*, larvae of *Acrida cinerea*, *Parasa lepida*, *Laspeyresia pomonella*, larvae of *S. exigua*, and *L. pomonella*. With the suggested approach, the accuracy was 100%. The DeepPestNet method was contrasted with conventional deep learning (DL) models that had already been trained. The model proposed was evaluated on the Kaggle data collection 'Pest Dataset' in identifying various pests, including aphids, armyworms, beetles, bollworms, grasshoppers, mites, mosquitoes, sawflies, and stem borers. The accuracy of the model was 98.92%, confirming its wide adaptability. The suggested model has the potential to decrease crop productivity and economic losses by giving experts and farmers prompt and efficient assistance in identifying pests (Ullah *et al.*, 2022).

In 2023, Rajeshram *et al.* proposed a DNN method for pre-processing leaf pictures from a plant village dataset, disease classification, and pesticide recommendation. The large set of data is divided into a trained and a test dataset once the dataset for plant leaf images has been expanded through geometric manipulation. The ResNet-50 models are then trained on the extensive ImageNet picture dataset. The findings showed that the greater precision and competence for crop spring picture identification may be achieved by employing the transfer learning-based CNN system. It could increase agricultural productivity and quality, decrease the usage of pesticides and fertilizers, and diagnose crop illnesses quickly and accurately.

17.2.3 Pest detection in horticulture plants

Fuentes *et al.* (2017) used in-place photos taken by a photographic camera at varying levels of clarity to demonstrate a deep learning-based method for identifying tomato pests and illnesses. Additionally, the research recommended a method for adding descriptive information to data and data enhancement in order to increase accuracy and reduce the quantity of false positives throughout training. Deep learning meta-architectures and feature extraction methods were shown to work well. The extensive Tomato Illnesses and Pests Dataset, which includes difficult photos of illnesses and pests together with a number of variables, including disease state and plant position, was used to train and test the models. According to experimental results, our suggested approach can successfully identify nine distinct illnesses and pests and handle challenging situations from a plant's immediate surroundings (Fuentes *et al.*, 2017).

A knowledge graph about lychee- and longan-related pests and diseases was created by Zhu *et al.* in 2023. The camera was controlled by a Raspberry Pi to take pictures of the illnesses and pests. The camera's pictures were then processed and identified by the system. Lastly, the results were aired in real time via the Bluetooth speaker. Data gathering, data retrieval, knowledge integration, and storing were used to build the knowledge graph. When used on a Raspberry Pi, the trained model VGG-16 demonstrated 94.9% exactness in the pest identification test (Zhu *et al.*, 2023).

Mango trees are tropical and subtropical trees that thrive in warm areas. When illnesses damage mango trees, productivity diminishes; farmers find it difficult to market their goods. Expert methods for disease prevention and innovative technology for the detection and diagnosis of mango plant illnesses were developed as a result of this issue. Mango illnesses are detected using three machine learning algorithms. A dataset comprising 20 distinct classes of photographs of both healthy and sick mango leaves and fruit has been produced. With exactness, sensitivity, and F1 scores of 97%, 96%, and 96%, respectively, DenseNet169 achieves the greatest accuracy of 97.81% among various machine learning techniques. Together with the machine learning model, an Android app was shaped to support identifying mango illnesses and propose pesticides based on disease detection (Rahaman *et al.*, 2023).

Utilizing deep learning-powered image recognition, Kusrini *et al.* (2020) proposed a cutting-edge machine learning (ML) technique for evaluating extensive mango fields and diagnosing the outbreak of disease. It builds upon the VGG-16 and also considers the practical working circumstances that farmers of Indonesia frequently encounter when gathering and analyzing visual data from mango fields. The use of data augmentation systems that can faithfully replicate the circumstances farmers confront helps to overcome the scarcity of data available for efficiently training deep learning networks. On both the validation and testing datasets, the suggested training solution's total accuracy was 73 and 76%, respectively. The accuracy of the test data is improved by 13.43% when the augmentation transformation function is used.

One of the most harmful agricultural pests in the world, thrips can infect plants with viruses through oviposition, eating, and infection. Many pest species are invasive and present significant problems for growers of a wide range of crops, particularly mangoes, in greenhouses and open fields around the world. It has been common practice to employ sticky paper traps to detect a variety of pests. As part of an integrated pest management (IPM) procedure, sticky paper traps must be inspected. The procedure of identifying and counting the pests captured in the traps is time-consuming. So, a prompt and efficient plan is needed to fix the problem. An ultrahigh-resolution pest monitoring device and a deep learning-based object detection model were combined to create a comprehensive and workable integrated pest detection and monitoring system (IPDMS). The technology that was developed was able to gather 64-megapixel sticky paper photographs and store them in an AWS database for analysis. The model can detect thrips with 90% accuracy and an F1-score of 0.95, according to test findings from two mango greenhouses. Furthermore, a web-based user interface was created to offer thrips population analysis and visual picture identification findings. Thus, by offering real-time data on pest occurrences in greenhouses, the proposed IPDMS tackles the major issues with pest detection systems (PDSs). Consequently, it improves agricultural pest monitoring and detection (Wu *et al.*, 2024).

An image-based monitoring system was investigated by Rustia *et al.* in 2023 to automatically record environmental variables and insect pest numbers in a mango grove. The structure consists of sensing elements that run on solar power and can take and analyze pictures of sticky paper traps on a regular basis. A segmental algorithm was created with an average classification F1-score of 0.96 to identify and categorize key pests of mangoes, including thrips, mango leafhoppers, and oriental fruit flies. The system accurately categorizes insect pests based on various taxonomical hierarchies, even in field settings, in contrast to existing insect counting methods. Over two years were spent testing and implementing the monitoring system in a distant mango farm. In order to illustrate the advantages of utilizing the suggested technique and provide fresh approaches to integrated pest management, the gathered spatiotemporal data were examined. By employing statistical techniques to identify insect hotspots, data analysis showed a notable drop in the specific insect pest abundance following system use. Better IPM methods can be developed by utilizing the hardware and software innovations presented in this work to create more intelligent insect pest monitoring systems.

American-native guava is a major contributor to the Brazilian economy, utilized extensively in food preparation and herbal medicine, and exported to many nations. Guava is prone to illnesses and pests, though. The most troublesome of them is psyllid *Triozoida limbata*, which attacks young leaves, resulting in lower fruit harvests and stunted plant growth. In order to tackle this problem and help producers identify insect infestations, a new dataset of Brazilian pictures was presented. This dataset offers a starting point for creating pest detection systems by utilizing computer vision and artificial intelligence approaches. We tested several object detection architectures, such as variations of Vision Transformer (ViT) and YOLO. With a precision of 93.14%, the tests showed that YOLO V5 produced encouraging results. Furthermore, this dataset's availability creates new chances to improve crop security, strengthen pest control tactics, and boost guava industry productivity overall (de Mesquita *et al.*, 2025).

17.2.4 Pest detection in floriculture plants

It is suggested to utilize a wireless imaging device to automatically identify small greenhouse insect pests using many classes of photos of sticky paper traps. A cascaded technique is used in the developed algorithm, which employs CNN image classifiers and CNN object detectors independently. A deep learning classifier was used to remove things other than insects from the things that were primarily found after the object detector had been trained to recognize objects in an image. A multi-class image classifier was used in the next phase to further classify the gathered insect items into gnats, flies, thrips, and whiteflies. This method's benefits include sample control techniques to enhance classification performance and the ability to add extra programs to the multi-label classifier for insects. Images captured by Wi-Fi-enabled imaging positioned in various plant houses with normal and changing light conditions were used to design and test the algorithm. On the basis of findings from experiments conducted over an extended period in glasshouses, the algorithm was able to achieve mean counting accuracies of 0.91 and

0.90, also average F1-scores of 0.92, 0.90 when verified on a different dataset of images from a different greenhouse and a distinct dataset of images taken for 6 months, respectively. With its ability to solve significant issues utilizing smart pest monitoring and provide live updates on insect pest incidences in glasshouses, the suggested approach in this study holds great promise for creating more effective IPM strategies in agriculture (Rustia *et al.*, 2021).

Early recognition of crop pests and diseases is crucial in stopping their spread and reducing the harm they cause. Many ways have been created to employ deep learning techniques to detect indicators of pests and illnesses from plant photos; however, the lack of datasets to train for tiny changes in plants still makes it difficult to detect early signs. An automated approach for gathering a sizeable collection of plant photos was constructed in order to address these issues, and an ensemble model was trained to identify specific plant pests and illnesses. With an average AUPRC of 0.81, our ensemble model exhibits a respectable detection performance after acquiring 13,393 plant picture data. With more data being gathered, this data collection and detection method can also be used for other plant anomalies (Lee *et al.*, 2022).

The output and aesthetic value of roses are significantly impacted by diseases and insect infestations. Applying a target detection algorithm to the identification of insect pests and rose illnesses helps to increase detection efficiency and is crucial to the successful implementation of intelligent rose cultivation. Research by Zimao *et al.* (2021) developed a two-stage rose pest detection approach called TSDDP, which takes into account the complicated background that affects disease and insect recognition within an actual planting scenario. First, the YOLOv3 model's feature extraction and fusion capabilities were enhanced with the addition of the improved Inception module. To eliminate the shadow in the intricate background, multi-leaf photos of roses in their natural habitat were subjected to leaf detection. The anchor box generated through non-hierarchical clustering was utilized in optimizing 'faster R-CNN' to suit the requirements of insect pests and rose illnesses finding. The findings of comparing the detection effects of TSDDP, Faster R-CNN, and YOLOv3 in a natural setting revealed that TSDDP outperformed other algorithms in terms of both positioning and detection accuracy, with an overall average detection accuracy of 82.26% for diseases and pests. In addition to improving the detection and location effect of small-scale diseases and insect pests, this can successfully lower the false detection brought on by complicated backdrops (Zimao *et al.*, 2021).

Rapid as well as accurate detection of plant pathologies is crucial in agriculture to preserve the harvest in addition to agricultural goods standards. Using photos of *Rosa damascena* Mill. in its native habitat and season, a new dataset containing seven distinct diseases and pests was produced. Pre-training for this method was done using a combination of the CNN models DenseNet169, ResNet152, MobileNetV2, VGG19, and NasNet. The models with the best classification results among mixed-CNN models, DenseNet169 and MobileNetV2, were blended into the new model through fine-tuning using the ensemble learning technique. An accuracy of 95.17% was attained by the model in the performance tests. Furthermore, Duman (2025) presented a mobile application for Android that integrated these models, which sets it apart from prior studies of a similar nature. The top-performing models, DenseNet169 and MobileNetV2 in both quantized and flat buffered versions, were tested independently on an Android emulator, a real mobile device, and a computer. With the lowest average inference time (301 ms) on mobile devices, MobileNetV2 performed better than DenseNet169 (2271 ms). These findings show how mobile devices can be used to effectively identify pests and diseases in rose plants in their natural habitats (Duman, 2025).

17.2.5 Pest detection in medicinal plants

Both ancient and contemporary Ayurveda make extensive use of medicinal plants to treat a wide range of illnesses. However, keeping them healthy and productive is severely hampered by pest infestation. Shreelakshmi (2024) trained five deep learning models, including Basic CNN, VGG16, ResNet50, VGG32, and a recently suggested CFNET, using three datasets of medicinal plants: neem, aloe vera, and ashwagandha.

Metrics including accuracy, exactness, true positive rate, F-measure, and learning time are utilized by these models in processing plant image data and forecasting the presence of pests. Building on the EfficientNetB3 architecture, the CFNET model focused on minimizing false negatives and attained the greatest accuracy of 0.98 on the neem dataset, 0.94 on aloe vera, and 0.96 on ashwagandha. The outcomes demonstrate CFNET's higher efficiency and performance over alternative approaches. The promise of deep learning for long-term pest control in the production of medicinal plants is demonstrated by this method (Shreelakshmi, 2024).

One of the most important aspects of pest management is multi-class pest detection, which involves both location and classification. On account of differences in pest characteristics, multi-class pest detection is far more challenging than general object detection. A region-based end-to-end method called PestNet was presented by Liu *et al.* in 2019 for intelligent pest detection and categorization. There are three main components of PestNet. Initially, it was suggested that the CNN support be fused with attention mechanisms of computer vision in order to extract and improve features. Concerning feature representations derived from pictures, the other component, region proposal network, was used to offer probable pest placements. Positional score maps, fully connected (FC) layers, were substituted for bounding box regression and pest categorization in the third constituent. To increase identification accuracy, circumstantial regions of interest were also used as information about the features of pests. With 75.46% mean average precision, PestNet outperformed the advanced techniques in multi-class pest detection, according to an evaluation conducted on the recently gathered 'large-scale pests' image dataset, Multi-class Pests Dataset 2018 (MPD2018) (Liu *et al.*, 2019).

In 2024, Nirmaladevi *et al.* suggested a computer vision method for identifying pests and diseases and a deep learning-based categorization system for pests and illnesses. The input picture categories are automatically recognized and categorized by this type of machine vision-based system.

For traditional Chinese medicinal materials to thrive as best they can, produce more, and have a higher concentration of beneficial ingredients, precise and prompt pest monitoring is essential. In their 2024 study, Guo *et al.* highlighted the critical needs aimed at tracking pests and diseases in *Cinnamomum camphora* chvar. Borneol planting base and also offered an unmanned aerial vehicle (UAV) image-based solution in overcoming the drawbacks of on-site and real-time inspections. It presented a meta learning method for CCB tree pests and diseases observation that combines a number of machine learning algorithms with RGB-derived vegetation indices. It made large-scale, highly accurate, and reasonably priced pest and disease monitoring possible, which helped to increase the precision of traditional Chinese medicinal material cultivation management.

17.2.6 Pest detection in jute

DCNN and transfer learning were used to create an intelligent model for identifying jute pests. With the help of photographs, the proposed model of DCNN could identify jute pests quickly and accurately. In particular, TL trained the VGG19 CNN model using the ImageNet database. Also established is a well-organized photographic dataset of four common jute pests. The final accuracy of our model on the four most important jute pest classes is 95.86%. The results of the precision, recall, F1-score, and confusion matrix further illustrate the model's effectiveness. Applications for iOS and Android incorporate the suggested approach for real-world applications (Sourav and Wang, 2023).

17.3 Cotton Pest Categorization

Wang *et al.* (2022) suggested unmanned aerial vehicle (UAV) remote sensing, cotton territory estimation by assistants, and GIS-dependent farm-level findings. A geographic information system network analysis platform was created to identify potential volunteer growing locations of cotton. The results suggest that remote sensing UAVs combined with image processing techniques could help remove volunteer cotton from transportation routes by pinpointing its precise location.

Wei *et al.* (2020) found that cotton became tolerant to Verticillium wilt caused by *Verticillium dahliae* after the host suppressed the 'acetolactate synthase VdILV2 and VdILV6'. To define their involvement in *V. dahliae*, HIGS technology decreases the expression of VdILV2. Similarly, VdILV6 or HIGS prohibits a fungicide and effective control for Verticillium wilt induced by *V. dahliae*. Furthermore, bidirectional long short-term memory systems associated with wind, and temperature patterns were anticipated by Chen *et al.* (2020). It was explained how to use weather parameters to identify the emergence of cotton diseases and pests using a bi-directional RNN with LSTM units. In the beginning, the problematic foreseeing of the occurrence of pests and diseases was described as a future forecasting. A bi-directional LSTM network, which may depict extended reliance on the prior and subsequent frameworks of successive input, was then employed as a solution.

Plant diseases were identified by Barbedo *et al.* in 2019 using spots and wounds. The segmentation into little chunks indirectly took into account the natural variation inside each image, and the suggested method can significantly increase variety in addition to significantly increasing the quantity of image datasets.

Zhang *et al.* (2018a) identified automatic picture segmentation for sick cotton in a natural environment. Based on local data and global slope, the Active Contour model was introduced. The visual gradient was incorporated into the sectional tone as part of the global system, which reduced the role of the border compound. The suggested method can be used to segment images of cotton leaves with a single background. For photos of cotton leaves with fuzzy margins and staggered conditions, the recommended method performs more effectively in the major infected region than in the leaf margins.

17.4 Pest Classification in Citrus

For human health, in a number of products, an excellent source of ascorbic acid is citrus. Citrus harvests are low in quality and quantity because of diseases and pests. A deep learning-based intelligent strategy was used by Khanramaki *et al.* in 2021 to identify general citrus pests, sooty mold, and leaf miner. Three levels of variety are taken into account when constructing an ensemble.

Barman *et al.* (2020) suggested on-site CNN categorization of citrus leaf diseases using smartphone photos. Self-Structured classifiers and MobileNet are dual CNN forms utilized in classifying illnesses on citrus leaves throughout cultivation. Using photos from smartphones, the proposed study produced a dataset on citrus diseases. Both models were exposed to training and evaluation using the dataset of citrus. The method showed that the SSCNN was more accurate for classifying citrus leaves based on smartphone images. Additionally, citrus disease classification using deep metric learning and sparse data was identified by Janarthan *et al.* (2020). For citrus pest diagnosis, a consistent deep metric learning was presented. It provides a spot-based categorization network that consists of prototype-based clustering, a module for embedding, along with a simple classifier that uses neural networks in order to precisely identify citrus diseases. A collection of commonly available citrus fruits has been used to show the efficiency of the approach in accurately diagnosing a variety of diseases using leaf pictures.

Intelligent mobile detection of citrus disorders was presented by Pan *et al.* in 2019. An intelligent sensor called Densely Connected Convolutional Networks (DenseNet) can identify six distinct types of images in order to diagnose the condition. Utilizing the WeChat applet, which enables operators to upload photos and get observations, the implementation was run on a smartphone. Citrus disease recognition accuracy has improved to nearly 88%, and the estimated time consumption has dropped.

Sharif *et al.* (2018) introduced a few combinations for citrus tree pest classification and detection. The suggested procedure is separated into two stages. They are a classification of citrus diseases on leaves and the detection of infection markings on fresh fruit. An improved input image was utilized in recovering the spots using an efficient weighted segmentation approach. These databases were used to identify and categorize diseases on citrus, including anthracnose, melanose, citrus canker, and citrus scab.

Ali *et al.* (2017) discussed the use of various data descriptors to identify and categorize diseases in citrus trees. Both the image and disease levels of classification were used. Additionally employed are the ensemble classification techniques of Boosted Tree KNN, Bagged Tree, and Cubic SVM. When any color features are used, the bagged tree performs better than other classifiers.

The quality of agricultural products is significantly lowered by plant diseases and pests. Pests like Mediterranean fruit flies, which are among the most serious plant diseases, seriously harm crops, causing farmers to lose a lot of money on their products each year. Thus, it is especially crucial to employ contemporary, non-destructive techniques like deep learning and machine vision systems for early pest identification in agricultural harvests.

Three phases of citrus fruit photos were taken for the citrus pest study: (i) prior to pest infestation; (ii) during the onset of fruit infection; and (iii) eight days following the second stage, under natural light circumstances (7000–11,000 lux). For every class, 1519 photos in total were provided. Ten and 20% of the pictures were utilized for validating and evaluating phases, whereas 70% of the pictures were utilized for the network training stage. The SGDm, RMSProp, and Adam optimization techniques, together with ResNet-50, VGG-16, GoogleNet, and AlexNet, were utilized to differentiate fruit that was healthy and fruit that was contaminated with the Mediterranean fly. The VGG-16 model, with the aid of the SGDm algorithm, demonstrated maximum efficacy related to accurate detection and 98.33 and 98.36% F1 scores, respectively, according to the findings obtained in evaluating the models during the pest outbreak stage. The AlexNet model utilizing the SGDm algorithm produced the best results, according to the third-stage evaluation, yielding precise recognition along with 99.33%, 99.34%, and F1 scores, respectively. The duration to train the network for AlexNet, which used the SGDm optimization algorithm, was the shortest (323 seconds). The study's findings demonstrated the effectiveness of machine vision systems and convolutional neural networks in managing and controlling pests in groves and further farming goods (Hadipour-Rokni *et al.*, 2023).

17.5 Pest Detection by CNN

This part is made up of CNN, which has garnered a lot of interest in agricultural industries like plant pest detection. The CNN-based model is appealing because it systematically extracts pertinent features from the dataset. It uses basic healthy leaves and damaged plants to identify and diagnose plant illnesses.

A deep learning technique dependent on response connections was found in 2022 by Hassan *et al.* Factorized convolution was utilized in lowering the criterion. The model is evaluated and trained on separate datasets of plant ailments. The performance accuracy on the plant record was 99.39%; on the dataset of rice diseases it was 99.66%; and on the dataset of cassava it was 76.59%. By using separable convolution, the inception architecture decreases the number of variables and minimizes expenditure. Additionally, CNN for fine-grained categorization of diseases on plant leaves was presented by Wu *et al.* in 2022. The intricate network overburdens the system and memory storage, making it challenging to meet the need for inexpensive implementations. In order to address this difficulty, the study offered a network-based, fine-grained classification of illness rules. Included in the training process was the 'Reconstruction-Generation Model.' Inadequacies of deep neural networks in crop disease identification were addressed by a fine-grained disease identification technique developed by Visual DNS for tomato and peach disease leaflet detection.

Li *et al.* (2020a) found that SCNN-RF and SCNN-KSVM were methods addressing uncertainty. Comparison tests were conducted using different models of deep learning on various sets of data. The results showed that other pre-trained deep models are outperformed by SCNN-RF and SCNN-KSVM with less variation. An important attempt was made to identify plant diseases using a CNN and a conventional machine learning method. Li *et al.* (2019) also created an excellent methodology to augment data for CNN-dependent algorithms. To infer input pictures having varying quality independently, a model with multiple scales employed the Test Time Augmentation (TTA) approach. 'Non-Maximum Suppression' (NMS) was utilized in combining the detection outcomes from several image scales for the final outcome. Experiments on rice planthoppers, wheat mites, aphids, and sawflies using data pre-processing 'effective data augmentation technique' achieved a mean average precision (MAP) accuracy of 81.4%, outperforming approaches like DAG-CNN, HR, and FPN by 11.63%, 7.93%, and 4.73%.

In 2020, it was suggested by Ai *et al.* that crop bugs may be automatically detected using CNN. The Inception-ResNet-v2 model was utilized for development. The remaining design element contained a combination of multilayered direct edge and multi-scale convolution. Once the combined convolution process is finished, ReLu function is stimulated. The results of the trial presented that the model's general detection accuracy was 86.1%, demonstrating its effectiveness. Ai *et al.* (2020) created and implemented a WeChat applet for identifying insect pests and crop illnesses after the model was trained.

Ferentinos (2018) used photos of damaged and healthy leaves to discover deep learning techniques. CNN models were developed to diagnose and identify plant diseases. A total of 87,848 images were classified. There were 58 different classes of healthy and diseased plants of 25 different plant species.

A number of designs were created, and the one that found the connected sickness pair (or healthy plant) with the highest success rate was 99.53%. This strategy has demonstrated a lot of promise in terms of obtaining more and better data to advance the system and increase its dependability in practical settings.

A method for recognizing several plants in vibrant images using a deep CNN model was presented by Wang *et al.* in 2021. The findings of the experiment demonstrated that the CNN-based method can achieve an accuracy of around 96.02% on ten plant classes in just 32 epochs with a logarithmic loss of 0.01. Using more epochs could yield even better results, even if our proposed study produces superior results.

17.6 Pest Detection by LSTM

To identify pests like *Tessaratoma papillosa* and forecast pest outbreaks, LSTM analyzes environmental data from weather stations.

Liu *et al.* (2021) looked into the challenge of plant diseases that are observable in a methodical manner. To support studies on plant disease identification, a thorough crop illness database containing 271 classes of crop diseases and 220,592 photographs was created. Plant diseases were identified by utilizing information and modifying the weights of each visual region. The strengths of each separated component from each image could be calculated to ascertain the cluster organization of these portions. The effectiveness of the suggested approach was quantitatively evaluated using the PDD271 and PlantVillage datasets.

Chen *et al.* (2020) suggested that artificial intelligence-based picture recognition technologies be incorporated into the Internet of Things and landscape in order to identify pests. YOLOv3 was utilized for image segmentation and LSTM for evaluating weather sensor data to forecast the presence of pests in order to locate *Tessaratoma papillosa*. Before they become a significant issue, the recommended approach notifies farmers of the existence of specific pests. It provides effective pest management techniques that reduce agricultural failures, hence increasing the overall economic value of agriculture.

Turkoglu *et al.* (2019) suggested Multi-model LSTM-based Pre-trained Convolutional Neural Networks (MLP-CNNs). Trained LSTM networks and CNN models serve as the foundation for the hybrid model and obtain deep features from numerous deep models via transfer learning. AlexNet, DenseNet201, and GoogleNet were the feature extraction models used in the study.

17.7 Pest Detection by DCNN

The DCNN process uses a deep learning methodology to identify plant disease in high detail. In order to identify neural networks designed as dependent on R-FCN, Shi *et al.* (2020) identified eight common stored grain insects. The network's response time has grown, and it can now detect insects in challenging environments. To increase the detection accuracy of deep convolutional neural networks in identifying grain storage insects, a modified DenseNet-121 was proposed. Additionally, Vaishnnave *et al.* (2020) proposed a DCNN-based effective method aimed at identifying and categorizing groundnut diseases. This technology was primarily utilized in identifying foliage illness, and the DCNN classification was used to classify the diseases. The industry-best comparison accuracy of 94.96% was attained by DCNN. The accuracy rate of the sixth layer of the DCNN is 95.28% when compared with the other layers. In every class, DCNN's comprehensive precision was 99.88%.

Thenmozhi and Reddy (2019) presented a difficult issue that farmers face: accurately identifying and classifying various agricultural insects because they all have similar appearances in the early phases of crop growth. Given its ability to learn intricate high-level characteristics and extract elements fast, CNN with deep architectures has been utilized to solve the challenge of picture classification applications. A deep CNN model that is effective in classifying insect species was constructed using accessible insect datasets. The network is kept from generalizing by using information processors like translation, spin, and projection. When it came to categorizing different insect species in field crops, the results showed that the proposed CNN model outperformed pre-trained algorithms.

Zhang *et al.* (2018b) enhanced Cifar10 and GoogLeNet for foliar disease identification in maize, availing deep learning in 2018. Nine different kinds of maize leaf images were utilized to train and test two improved models that were produced by regulating variables, exploring varied fusions, and lowering classifiers number. The VGG and AlexNet structures possess a lot more parameters than the enhanced models. It might have decreased parallel iterations and increased the accuracy of the corn leaf sickness, which would have improved model training and identification efficiency.

17.8 Pest Identification by DBN

Plant diseases are identified and their occurrence is predicted using DBN. In 2020, Cristin *et al.* used the RiderCuckoo Search Algorithm, dependent on deep neural networks, to identify plant disease. In order to eliminate distortions, the subject image underwent preprocessing. The pre-processed image is captured, then goes through a classification step that employs piecewise fuzzy C-means clustering (piFCM). The texture feature was exposed to the classification step, which makes use of DBN for efficient plant disease identification. In this application, the DBN was trained using the recommended Rider-CSA.

17.9 Advanced Prediction Methods

The CSPResNeXt-50 and VoVGSCSP modules were inserted into the YOLOv7-based Maize-YOLO network to enable precise and instantaneous pest identification and detection in maize crops. This method obtained an acceptable presentation at 76.3% mAP and 77.3% recall, outperforming the current modern YOLO family of object identification (Yang *et al.*, 2023).

A novel approach to collecting event data, known as Reliability and Multi-path Encounter Routing (RMER), was proposed by Dong *et al.* (2016). Consequently, energy consumption might be significantly reduced, allowing for an extended network lifespan. The findings of theoretical and experimental simulations show that RMER works better than alternatives when used for event detection.

Ye *et al.* (2024) developed a model for pest detection: PestNAS, grounded on 'adaptive feature integration and evolutionary neural architecture search', which demonstrated higher detection accuracy, in order to address the issues of inefficient use of contextual information of plants as well as insufficient neural architecture design for pest detection in field conditions.

17.10 Conclusion

In this chapter, many deep learning-based methods for detecting diseases in various crops are examined. The machine learning-based methods, RF, SVM, DT, and NB, have a significant influence on the identification and forecasting of foliar diseases. Likewise, compared with the machine learning method, the deep learning-based model performs better. LSTM, CNN, DBN, and DCNN are some of the deep learning algorithms that are already being used to enhance the presentation of disease forecasts for different plants, including citrus, rice, and cotton pest detection. A better model must be developed, because the existing methods are still not as effective at producing predictions. Therefore, a novel hybrid deep learning-based approach to enhance identification of crop illness by using a metaheuristic method needs to be developed. Better classification results and enhanced prediction performance are possible with this model. In order to accomplish live tracking and pest detection in granaries, it will also be necessary in the future to integrate Internet technologies, such as mobile terminal processors and the agricultural Internet of Things. This will help to advance the modernization and intelligence of the agricultural sector.

The existence of several illnesses in a variety of crop types and the distinct criteria for each plant disease make determining the severity of a disease a difficult task that calls for the development of standardized techniques. The acquisition of large and varied datasets and the incorporation of the influence of climatic fluctuations on plant distress to increase model applicability and vigor are important research targets to further advance the use of deep learning. Looking ahead, there is a great deal of promise to transform precision agriculture and increase market garden crop cultivation's productivity, sustainability, and profitability by combining deep learning-driven crop health monitoring with IoT, robots, and data analysis.

References

Ahmad, M., Abdullah, M., Moon, H. and Han, D. (2021) Plant disease detection in imbalanced datasets using efficient convolutional neural networks with stepwise transfer learning. *IEEE Access* 9, 140565–140580. DOI: 10.1109/ACCESS.2021.3119655.

Ai, Y., Sun, C., Tie, J. and Cai, X. (2020) Research on recognition model of crop diseases and insect pests based on deep learning in harsh environments. *IEEE Access* 8, 171686–171693. DOI: 10.1109/ACCESS.2020.3025325.

Ali, H., Lali, M.I., Nawaz, M.Z., Sharif, M. and Saleem, B.A. (2017) Symptom based automated detection of citrus diseases using colour histogram and textural descriptors. *Computers and Electronics in Agriculture* 138, 92–104. DOI: 10.1016/j.compag.2017.04.008.

Ayan, E., Erbay, H. and Varçın, F. (2020) Crop pest classification with a genetic algorithm-based weighted ensemble of deep convolutional neural networks. *Computers and Electronics in Agriculture* 179, 105809. DOI: 10.1016/j.compag.2020.105809.

Barbedo, J.G.A. (2019) Plant disease identification from individual lesions and spots using deep learning. *Biosystems Engineering* 180, 96–107. DOI: 10.1016/j.biosystemseng.2019.02.002.

Barman, U., Choudhury, R.D., Sahu, D. and Barman, G.G. (2020) Comparison of convolution neural networks for smartphone image based real time classification of citrus leaf disease. *Computers and Electronics in Agriculture* 177, 105661. DOI: 10.1016/j.compag.2020.105661.

Chen, P., Xiao, Q., Zhang, J., Xie, C. and Wang, B. (2020) Occurrence prediction of cotton pests and diseases by bidirectional long short-term memory networks with climate and atmosphere circulation. *Computers and Electronics in Agriculture* 176, 105612. DOI: 10.1016/j.compag.2020.105612.

Chen, C.J., Huang, Y.Y., Li, Y.S., Chang, C.Y. and Huang, Y.M. (2020) An AIoT based smart agricultural system for pests detection. *IEEE Access* 8, 180750–180761. DOI: 10.1109/ACCESS.2020.3024891.

Cheng, X., Zhang, Y., Chen, Y., Wu, Y. and Yue, Y. (2017) Pest identification via deep residual learning in complex background. *Computers and Electronics in Agriculture* 141, 351–356. DOI: 10.1016/j.compag.2017.08.005.

Cristin, R., Kumar, B.S., Priya, C. and Karthick, K. (2020) Deep neural network-based Rider-Cuckoo search algorithm for plant disease detection. *Artificial Intelligence Review* 53(7), 4993–5018. DOI: 10.1007/s10462-020-09813-w.

de Mesquita, I.M., Moreira, L.F.R., Silva Nunes, R. and Paula Júnior, I.C. (2025) A novel Brazilian dataset for automatic detection of psyllid attacks in Guava. *Revista de Informática Teórica e Aplicada* 32(1), 151–157. DOI: 10.22456/2175-2745.143544.

Dong, M., Ota, K. and Liu, A. (2016) Reliable and energy-efficient data collection for large-scale wireless sensor networks. *IEEE Internet of Things Journal* 3(4), 511–519. DOI: 10.1109/JIOT.2016.2517405.

Duman, B. (2025) Mobile device-based detection system of diseases and pests in rose plants using deep convolutional neural networks and quantization. *Journal of Agricultural Sciences* 31(2), 302–318. DOI: 10.15832/ankutbd.1514972.

Ferentinos, K.P. (2018) Deep learning models for plant disease detection and diagnosis. *Computers and Electronics in Agriculture* 145, 311–318. DOI: 10.1016/j.compag.2018.01.009.

Fuentes, A., Yoon, S., Kim, S.C. and Park, D.S. (2017) A robust deep-learning-based detector for real-time tomato plant diseases and pests recognition. *Sensors* 17(9), 2022. DOI: 10.3390/s17092022.

Guo, H., Cheng, Y., Liu, J. and Wang, Z. (2024) Low-cost and precise traditional chinese medicinal tree pest and disease monitoring using UAV RGB image only. *Scientific Reports* 14(1), 25562. DOI: 10.1038/s41598-024-76502-x.

Hadipour-Rokni, R., Asli-Ardeh, E.A., Jahanbakhshi, A. and Sabzi, S. (2023) Intelligent detection of citrus fruit pests using machine vision system and convolutional neural network through transfer learning technique. *Computers in Biology and Medicine* 155, 106611. DOI: 10.1016/j.compbiomed.2023.106611.

Hassan, S.M. and Maji, A.K. (2022) Plant disease identification using a novel convolutional neural network. *IEEE Access* 10, 5390–5401. DOI: 10.1109/ACCESS.2022.3141371.

Hou, C., Zhuang, J., Tang, Y., He, Y., Miao, A. *et al.* (2021) Recognition of early blight and late blight diseases on potato leaves based on graph cut segmentation. *Journal of Agriculture and Food Research* 5, 100154. DOI: 10.1016/j.jafr.2021.100154.

Janarthan, S., Thuseethan, S., Rajasegarar, S., Lyu, Q., Zheng, Y. *et al.* (2020) Deep metric learning based citrus disease classification with sparse data. *IEEE Access* 8, 162588–149. DOI: 10.1109/ACCESS.2020.3021487.

Jiang, F., Lu, Y., Chen, Y., Cai, D. and Li, G. (2020) Image recognition of four rice leaf diseases based on deep learning and support vector machine. *Computers and Electronics in Agriculture* 179, 105824. DOI: 10.1016/j.compag.2020.105824.

Karar, M.E., Alsunaydi, F., Albusaymi, S. and Alotaibi, S. (2021) A new mobile application of agricultural pests recognition using deep learning in cloud computing systems. *Alexandria Engineering Journal* 60(5), 4423–4432. DOI: 10.1016/j.aej.2021.03.009.

Khanramaki, M., Asli-Ardeh, E.A. and Kozegar, E. (2021) Citrus pests classification using an ensemble of deep learning models. *Computers and Electronics in Agriculture* 186, 106192. DOI: 10.1016/j.compag.2021.106192.

Kim, B.H., Alamri, A.M. and Al-Qahtani, S.A. (2024) Leveraging machine learning for early detection of soybean crop pests. *Legume Research* 47(6), 1023–1031. DOI: 10.18805/LRF-794.

Krishnamoorthy, N., Prasad, L.N., Kumar, C.P., Subedi, B., Abraha, H.B. *et al.* (2021) Rice leaf diseases prediction using deep neural networks with transfer learning. *Environmental Research* 198, 111275. DOI: 10.1016/j.envres.2021.111275.

Kusrini, K., Suputa, S., Setyanto, A., Agastya, I.M.A., Priantoro, H. *et al.* (2020) Data augmentation for automated pest classification in mango farms. *Computers and Electronics in Agriculture* 179, 105842. DOI: 10.1016/j.compag.2020.105842.

Lee, S., Arora, A.S. and Yun, C.M. (2022) Detecting strawberry diseases and pest infections in the very early stage with an ensemble deep-learning model. *Frontiers in Plant Science* 13, 991134. DOI: 10.3389/fpls.2022.991134.

Li, R., Wang, R., Zhang, J., Xie, C., Liu, L. *et al.* (2019) An effective data augmentation strategy for CNN-based pest localization and recognition in the field. *IEEE Access* 7, 160274–160283. DOI: 10.1109/ACCESS.2019.2949852.

Li, Y., Nie, J. and Chao, X. (2020a) Do we really need deep CNN for plant diseases identification? *Computers and Electronics in Agriculture* 178, 105803. DOI: 10.1016/j.compag.2020.105803.

Li, Y., Wang, H., Dang, L.M., Sadeghi-Niaraki, A. and Moon, H. (2020b) Crop pest recognition in natural scenes using convolutional neural networks. *Computers and Electronics in Agriculture* 169, 105174. DOI: 10.1016/j.compag.2019.105174.

Liu, L., Wang, R., Xie, C., Yang, P., Wang, F. *et al.* (2019) PestNet: An end-to-end deep learning approach for large-scale multi-class pest detection and classification. *IEEE Access* 7, 45301–45312. DOI: 10.1109/access.2019.2909522.

Liu, X., Min, W., Mei, S., Wang, L. and Jiang, S. (2021) Plant disease recognition: A large-scale benchmark dataset and a visual region and loss reweighting approach. *IEEE Trans Image Process* 30, 2003–2015.

Lv, M., Zhou, G., He, M., Chen, A., Zhang, W. *et al.* (2020) Maize leaf disease identification based on feature enhancement and DMS-robust alexnet. *IEEE Access* 8, 57952–57966. DOI: 10.1109/ACCESS.2020.2982443.

Mallick, M.T., Biswas, S., Das, A.K., Saha, H.N., Chakrabarti, A. *et al.* (2023) Deep learning based automated disease detection and pest classification in Indian mung beans. *Multimedia Tools and Applications* 82(8), 12017–12041. DOI: 10.1007/s11042-022-13673-7.

Nigam, A., Tiwari, A.K. and Pandey, A. (2020) Paddy leaf diseases recognition and classification using PCA and BFO-DNN algorithm by image processing. *Materials Today: Proceedings* 33, 4856–4862. DOI: 10.1016/j.matpr.2020.08.397.

Nirmaladevi, K., Chaitanya Reddy, P., Tharun Kumar, P., Asha, S. and Lingaraja, D. (2024) Prediction of leaf disease and pest detection using deep learning. In: *AIP Conference Proceedings*, AIP Publishing (Vol. 2935). DOI: 10.1063/5.0198895.

Pan, W., Qin, J., Xiang, X., Wu, Y., Tan, Y. *et al.* (2019) A smart mobile diagnosis system for citrus diseases based on densely connected convolutional networks. *IEEE Access* 7, 87534–8754262600. DOI: 10.1109/ACCESS.2019.2924973.

Park, Y.H., Choi, S.H., Kwon, Y.J., Kwon, S.W., Kang, Y.J. *et al.* (2023) Detection of soybean insect pest and a forecasting platform using deep learning with unmanned ground vehicles. *Agronomy* 13(2), 477. DOI: 10.3390/agronomy13020477.

Rahaman, M., Chowdhury, M., Rahman, M.A., Ahmed, H., Hossain, M. *et al.* (2023) A deep learning based smartphone application for detecting mango diseases and pesticide suggestions. *International Journal of Computing and Digital Systems* 13(1), 1–1.

Rajeshram, V., Rithish, B., Karthikeyan, S. and Prathab, S. (2023) Leaf diseases prediction, pest detection and pesticides recommendation using deep learning techniques. In: *2023 International Conference on Sustainable Computing and Data Communication Systems (ICSCDS)*, IEEE, pp. 1633–1639. DOI: 10.1109/ICSCDS56580.2023.10104652.

Rustia, D.J.A., Chao, J.J., Chiu, L.Y., Wu, Y.F., Chung, J.Y. *et al.* (2021) Automatic greenhouse insect pest detection and recognition based on a cascaded deep learning classification method. *Journal of Applied Entomology* 145(3), 206–222. DOI: 10.1111/jen.12834.

Rustia, D.J.A., Lee, W.C., Lu, C.Y., Wu, Y.F., Shih, P.Y. *et al.* (2023) Edge-based wireless imaging system for continuous monitoring of insect pests in a remote outdoor mango orchard. *Computers and Electronics in Agriculture* 211, 108019. DOI: 10.1016/j.compag.2023.108019.

Sethy, P.K., Barpanda, N.K., Rath, A.K. and Behera, S.K. (2020) Deep feature based rice leaf disease identification using support vector machine. *Computers and Electronics in Agriculture* 175, 105527. DOI: 10.1016/j.compag.2020.105527.

Sharif, M., Khan, M.A., Iqbal, Z., Azam, M.F., Lali, M.I.U. *et al.* (2018) Detection and classification of citrus diseases in agriculture based on optimized weighted segmentation and feature selection. *Computers and Electronics in Agriculture* 150, 220–234. DOI: 10.1016/j.compag.2018.04.023.

Shi, Z., Dang, H., Liu, Z. and Zhou, X. (2020) Detection and identification of stored-grain insects using deep learning: A more effective neural network. *IEEE Access* 8, 163703–163714. DOI: 10.1109/ACCESS.2020.3021830.

Shiba, T., Hirae, M., Hayano-Saito, Y., Ohto, Y., Uematsu, H. *et al.* (2018) Spread and yield loss mechanisms of rice stripe disease in rice paddies. *Field Crops Research* 217, 211–217. DOI: 10.1016/j.fcr.2017.12.002.

Shreelakshmi, C.M. (2024) Advanced deep learning technique for accurate pest detection in medicinal plants. In: *2024 4th International Conference on Mobile Networks and Wireless Communications (ICMNWC)*, IEEE, pp. 1–8. DOI: 10.1109/ICMNWC63764.2024.10872027.

Sourav, M.S.U. and Wang, H. (2023) Intelligent identification of jute pests based on transfer learning and deep convolutional neural networks. *Neural Processing Letters* 55(3), 2193–2210. DOI: 10.1007/s11063-022-10978-4.

Subburaj, S.D.R., Eswaramoorthy, C., Latha, V.G. and Chinnasamy, R.K.P. (2024) Efficient pest detection through advanced machine learning technique. *Current Agriculture Research Journal* 12(3). DOI: 10.12944/CARJ.12.3.08.

Tetila, E.C., Machado, B.B., Astolfi, G., Souza Belete, N.A., Amorim, W.P. *et al.* (2020) Detection and classification of soybean pests using deep learning with UAV images. *Computers and Electronics in Agriculture* 179, 105836. DOI: 10.1016/j.compag.2020.105836.

Thenmozhi, K. and Reddy, U.S. (2019) Crop pest classification based on deep convolutional neural network and transfer learning. *Computers and Electronics in Agriculture* 164, 104906. DOI: 10.1016/j.compag.2019.104906.

Tian, L., Xue, B., Wang, Z., Li, D., Yao, X. *et al.* (2021) Spectroscopic detection of rice leaf blast infection from asymptomatic to mild stages with integrated machine learning and feature selection. *Remote Sensing of Environment* 257, 112350. DOI: 10.1016/j.rse.2021.112350.

Turkoglu, M., Hanbay, D. and Sengur, A. (2019) Multi-model LSTM-based convolutional neural networks for detection of apple diseases and pests. *Journal of Ambient Intelligence and Humanized Computing* 1–11. DOI: 10.1007/s12652-019-01591-w.

Ullah, N., Khan, J.A., Alharbi, L.A., Raza, A., Khan, W. *et al.* (2022) An efficient approach for crop pests recognition and classification based on novel deeppestnet deep learning model. *IEEE Access* 10, 73019–73032. DOI: 10.1109/ACCESS.2022.3189676.

Vaishnnave, M.P., Suganya Devi, K. and Ganeshkumar, P. (2020) Automatic method for classification of groundnut diseases using deep convolutional neural network. *Soft Computing* 24(21), 16347–16360. DOI: 10.1007/s00500-020-04946-0.

Wang, F., Wang, R., Xie, C., Zhang, J., Li, R. *et al.* (2021) Convolutional neural network based automatic pest monitoring system using hand-held mobile image analysis towards non-site-specific wild environments. *Computers and Electronics in Agriculture* 187, 106268. DOI: 10.1016/j.compag.2021.106268.

Wang, J., Li, Y., Feng, H., Ren, L., Du, X. *et al.* (2020) Common pests image recognition based on deep convolutional neural network. *Computers and Electronics in Agriculture* 179, 105834. DOI: 10.1016/j.compag.2020.105834.

Wang, Q., Bao, W., Yang, F., Yang, Y. and Lu, Y. (2018) A PCR-based analysis of plant DNA reveals the feeding preferences of *Apolyguslucorum* (Heteroptera: Miridae). *Arthropod-Plant Interact* 12(4), 567–574. DOI: 10.1007/s11829-018-9604-2.

Wang, T., Mei, X., Thomasson, J.A., Yang, C., Han, X. *et al.* (2022) GIS-based volunteer cotton habitat prediction and plant-level detection with UAV remote sensing. *Computers and Electronics in Agriculture* 193, 106629. DOI: 10.1016/j.compag.2021.106629.

Wei, C., Qin, T., Li, Y., Wang, W., Dong, T. *et al.* (2020) Host-induced gene silencing of the acetolactate synthases VdILV2 and VdILV6 confers resistance to Verticillium wilt in cotton (*Gossypium hirsutum* L). *Biochemical and Biophysical Research Communications* 524(2), 392–397. DOI: 10.1016/j.bbrc.2020.01.126.

Wu, G.H., Huang, N.F., Huang, Y.H., Chan, P.C. and Bhat, S.A. (2024) Thrips detection and monitoring system in mango greenhouses based on deep learning image recognition. *IEEE Transactions on Instrumentation and Measurement* 73, 1–18. DOI: 10.1109/TIM.2024.3403191.

Wu, Y., Feng, X. and Chen, G. (2022) Plant leaf diseases fine-grained categorization using convolutional neural networks. *IEEE Access* 10, 41087–41096. DOI: 10.1109/ACCESS.2022.3167513.

Xiao, M., Ma, Y., Feng, Z., Deng, Z., Hou, S. *et al.* (2018) Rice blast recognition based on principal component analysis and neural network. *Computers and Electronics in Agriculture* 154, 482–490. DOI: 10.1016/j.compag.2018.08.028.

Yang, S., Xing, Z., Wang, H., Dong, X., Gao, X. *et al.* (2023) Maize-YOLO: A new high-precision and real-time method for maize pest detection. *Insects* 14(3), 278. DOI: 10.3390/insects14030278.

Yao, Q., Feng, J., Tang, J., Xu, W.G., Zhu, X.H. *et al.* (2020) Development of an automatic monitoring system for rice light-trap pests based on machine vision. *Journal of Integrative Agriculture* 19(10), 2500–2513. DOI: 10.1016/S2095-3119(20)63168-9.

Ye, Y., Chen, Y. and Xiong, S. (2024) Field detection of pests based on adaptive feature fusion and evolutionary neural architecture search. *Computers and Electronics in Agriculture* 221, 108936. DOI: 10.1016/j.compag.2024.108936.

Zhang, J.H., Kong, F.T., Wu, J.Z., Han, S.Q. and Zhai, Z.F. (2018a) Automatic image segmentation method for cotton leaves with disease under natural environment. *Journal of Integrated Agriculture* 17(8), 1800–1814. DOI: 10.1016/S2095-3119(18)61915-X.

Zhang, X., Qiao, Y., Meng, F., Fan, C. and Zhang, M. (2018b) Identification of maize leaf diseases using improved deep convolutional neural networks. *IEEE Access* 6, 30370–30377. DOI: 10.1109/ACCESS.2018.2844405.

Zhou, G., Zhang, W., Chen, A., He, M. and Ma, X. (2019) Rapid detection of rice disease based on FCM-KM and faster r-CNN fusion. *IEEE Access* 7, 143190–143206. DOI: 10.1109/ACCESS.2019.2943454.

Zhu, D., Xie, L., Chen, B., Tan, J., Deng, R. *et al.* (2023) Knowledge graph and deep learning based pest detection and identification system for fruit quality. *Internet of Things* 21, 100649. DOI: 10.1016/j.iot.2022.100649.

Zimao, L., Liandong, L., Meng, X., Jun, T. and Yue, Z. (2021) Detection of rose diseases and insect pests based on deep learning. *Journal of Chinese Agricultural Mechanization* 42(8), 169. DOI: 10.13733/j.jcam.issn.2095-5553.2021.08.23.

18 AI-Assisted Genomic and Epigenomic Modulations in Plant Stress Adaptation

Umer Fayaz[1], Showkat A. Wazat[2] and A.K.M. Aminul Islam[3]*

[1]Accelerated Crop Improvement, Pigeonpea Breeding, International Crops Research Institute for the Semi-Arid Tropics, Hyderabad, India; [2]Department of Genetics and Plant Breeding, Sher-i-Kashmir University of Agricultural Sciences and Technology of Kashmir, India; [3]Department of Genetics and Plant Breeding, Gazipur Agricultural University, Gazipur, Bangladesh

Abstract

Plants utilize both genomic and epigenomic modulations to adapt to stresses due to environmental conditions. Genomic changes involve alteration of the DNA sequence itself and change gene expression. While epigenomic changes involve DNA methylation and histone modifications, they affect gene expression without changing DNA sequences. The epigenomic mechanisms play a crucial role in how plants respond to or adapt to various stresses, including biotic and abiotic stresses such as diseases, drought, salinity, extreme heat, etc. Both genomic and epigenomic changes can work together to influence plant stress adaptation. For example, DNA methylation can change gene expression in response to drought; histone modification changes gene expression related to heat-shock proteins; also, DNA methylation and other epigenetic mechanisms can be involved in a plant's response to salinity stress. Therefore, the integration of different omics approaches, including genomics, transcriptomics, proteomics, metabolomics, and epigenomics, is essential to understand plant adaptation to stress conditions. Artificial intelligence (AI) and machine learning (ML) algorithm helps to optimize omics technologies to steer more precise and effective prediction of a plant's response to stresses. This chapter reviews different aspects of AI-assisted genomic and epigenomic modulations in plants for the improvement of their yield and resilience to climate change by understanding the molecular mechanisms and pathways of stress adaptation.

Keywords: DNA methylation, Histone modification, Machine learning, Multiple omics, Stress response

18.1 Introduction

Epigenetics encompasses heritable molecular changes that influence gene function without altering the DNA sequences (Bender, 2002). The concept originated from Conrad Waddington in the 1940s as an explanation for gene–protein interactions during development. Since then, it has undergone significant changes. Jablonka and Lamb's, 1994 study identified how the term has evolved to describe the phenomenon of epigenetic inheritance, which entails passing on traits through mechanisms distinct from genetic code modifications. The broad definition of epigenetics includes any molecular or physical modifications in organisms that occur without changes to the gene's coding region or its nearby regulatory sequences, including promoters. Epigenetic

*Corresponding author: aminulgpb@bsmrau.edu.bd

 Plant Genomic Prediction: AI-Enhanced Models for Plant Genomics and Crop Improvement (ed. Jen-Tsung Chen)
DOI: 10.1079/9781836991328.0018

changes result from multiple interconnected mechanisms such as reversible DNA methylation and various histone protein modifications along with chromatin restructuring processes (Wagner, 2003; Vanyushin, 2006). Developmental signals, along with environmental factors and stress, specifically influence epigenetic processes. Internal and external factors can permanently modify the epigenetic expression of genes throughout an organism's lifespan, despite the DNA sequence remaining intact. Sometimes, these changes are transferred to the next generation, which results in epigenetic memories. According to several scientific studies, the changes in gene expression seen in both bacteria and in higher organisms like plants and animals are kept over many cell generations (Bender, 2004). Remarkably, alterations in DNA methylation appear to be passed down more frequently in plants than in animals (Takeda and Paszkowski, 2006). Some plants are able to keep the epigenetic memories of previous environmental challenges when they remain in the same spot and only produce reproductive cells later. Plant breeding and development strategies benefit from epigenetic regulation, as it governs plant development and enables stress responses and adaptive changes across generations. Environmental stresses such as heat and salinity produce overlapping genetic reactions, but each type of stress activates its own distinct set of genes (Sadder *et al.*, 2011). These different reactions are assumed to happen through specific epigenetic processes, one of the big ones being chromatin remodeling, which helps organize the genome and manage gene action (Wang and Chang, 2018). By using synchronized gene expression as a response to DNA methylation and other histone modifications, plants can handle changes in their environment (Kim *et al.*, 2015). *Arabidopsis* formed higher amounts of Swi-Independent3 when exposed to salt stress, pointing out its function in the response to stress (Perrella *et al.*, 2013). As a result, salt exposure increased the expression of genes related to cell wall architecture in maize roots as well. At the same time, it was found that increased acetylation of histone H3 on lysine 9 was present, associating histone modification with the activation of genes responding to stress (Li *et al.*, 2014). The degree of stress influences the magnitude of epigenetic changes. Genes that are active under drought, such as RD20 and RD29A, had higher levels of H3K4me3 and H3K9ac proteins than usual during prolonged drought (Kim *et al.*, 2012). Under osmotic stress, new modifications are made to the *Arabidopsis* pericentromeric areas, including the addition of threonine-3-phosphate to histone H3, suggesting that these regions are subject to different kinds of epigenetic changes than non-genic areas (Wang *et al.*, 2015).

By using machine learning (ML) approaches, artificial intelligence (AI) is greatly changing plant genomes and epigenomic research. Because of high-throughput phenotyping and emerging next-generation sequencing, the amount of plant genetic information has increased greatly. Therefore, powerful methods for computing data are now very much needed to deal with and interpret these large and complex sets of information. With the help of ML algorithms, plant genomics has gained in strength by improving its effectiveness through processing data. Multiple techniques are being designed for plant genomes and to look for patterns that create accurate estimates in biology. Deep learning approaches are included, along with supervised, semi-supervised, and unsupervised learning (Angermueller *et al.*, 2016). The methods involved have been effective in discovering regulatory DNA areas, measuring gene expression, and describing major regions of plant genomes, especially promoters and enhancers (Wu and Zhao, 2019; Liu *et al.*, 2020). Growing plants, managing plant stress, and identifying the genes that determine traits are all based on artificial intelligence. To improve crops, scientists look at how certain parts of the genome and epigenome differ, which tells them about gene regulation and function. Researchers who cultivate crops that grow well and stand up to climate change now find that studying complicated gene interactions is easier and quicker when they use artificial intelligence (Koumakis, 2020). This is needed for exact breeding methods, since it uses genetics to select certain traits (Leung *et al.*, 2016; Camacho *et al.*, 2018).

18.2 Fundamentals of Plant Epigenetics

18.2.1 DNA methylation

A form of epigenetics called DNA methylation puts a methyl group (CH_3) onto a DNA cytosine, which affects the way a gene works without altering the DNA itself (Villicaña and Bell, 2021). It works like 'cellular memory' and is vital for keeping the structure of chromatin steady (Alvarez-Venegas *et al.*, 2014). MET1 and chromomethyltransferases are part of the ways plants maintain CG, CHG, and CHH methylation with help from siRNA (Law and Jacobsen, 2010; Akhter *et al.*, 2021). Monocots generally show greater methylation than dicots (Rajewsky *et al.*, 2017). Some organisms have methylation near 30% in their cytosine. Epimutations, which involve changes in how genes act and look, can come about the same way as genetic mutations when there are disruptions in methylation (Ashapkin *et al.*, 2020). Transposons and repeats, genes that exist in multiple places, play a big role in helping plants evolve new genetic variations (Eichten *et al.*, 2013). Several scientific studies have proved that epigenetics occurs fast and is normally kept stable, most of the time when CHG and CG sequences are involved (Gutzat and Mittelsten Scheid, 2012; Lauria *et al.*, 2014; Zhang *et al.*, 2020).

18.2.2 Remodeling of chromatin structure

Chromatin, a highly condensed and organized configuration of DNA, is crucial for maintaining DNA structure, controlling gene expression, assuring precise replication, and facilitating effective organization of genetic material (Rajewsky *et al.*, 2017). It appears in two distinct structural forms: euchromatin, characterized by a loose organization and active transcription, and heterochromatin, characterized by being more densely packed and typically transcriptionally inactive (Santos *et al.*, 2017; Pecinka *et al.*, 2020). When transcription factors recognize and bind their cognate DNA sequences, during a plant cell cycle, the dynamic structural modification of the structure of the genome, chromatin remodeling, is permitted (Arya *et al.*, 2010; Weaver *et al.*, 2017; Bhadouriya *et al.*, 2021). Nucleosome repositioning is common in structural reorganization that is usually accompanied by regulatory protein–DNA interaction and gene expression alterations, sometimes causing inappropriate gene expression patterns (Secco *et al.*, 2017). Remodeling does not just cause the architecture of nucleosomes, the basic components of chromatin, to flinch (Goldstein *et al.*, 2013). Furthermore, some chromatin marks are heritable and act as cellular memory for adaptation of plants to external conditions (Lämke and Bäurle, 2017; Santos *et al.*, 2017). This process requires the SWI2/SNF2 protein family, which serves as an ATP-dependent remodeling complex for histone dimers (H3 and H4), RNA and DNA polymerases, enabling DNA accessibility by opening up chromatin organization (Clapier *et al.*, 2017; Bhadouriya *et al.*, 2021). During developmental progression post-embryogenesis, the BRAHMA (BRM) gene, a member of the SWI2/SNF2 family, controls reproductive phase switch, regulating mechanisms of gene expression and gene silencing (Yang *et al.*, 2015; Zhang *et al.*, 2017).

18.2.3 Non-coding RNA

Recent epigenomics research has shown that, aside from a small portion of the plant's genome that makes proteins, the rest of that genome is instead involved in controlling genes (Liu *et al.*, 2015). Non-coding RNAs (ncRNAs) control gene regulation and developmental processes and assist plants to adapt to stresses elicited by living entities and non-living factors (Sunkar *et al.*, 2012; Liu *et al.*, 2017). However, most of the research on non-coding RNAs focuses on small non-coding RNAs (sncRNAs), which regulate gene-level activity, and on long non-coding RNAs (lncRNAs) has been neglected so far. In plants, two species of small non-coding RNA are mainly found: microRNAs (miRNAs) and small interfering RNAs (siRNAs). Unlike siRNAs, miRNAs are formed inside the cell, while siRNAs usually form from exogenous double-stranded RNA (Guleria *et al.*, 2011). Both are degraded by dicer-like enzymes, which subsequently associate with Argonaute (AGO) proteins to form the RNA-induced silencing complex (RISC), thereby mediating the post-transcriptional gene silencing (Prathiba *et al.*, 2017). miRNAs of 21 nt, 22 nt, or 24 nt may profoundly affect such essential biological processes as hormone signaling, stress responses, and development through the modulation of target genes with significant sequence complementarity, even though the miRNAs are so short (Xu *et al.*, 2017; Zhao *et al.*, 2017). They also help with systemic regulation (Skopelitis *et al.*, 2018). In addition, siRNAs induce RNA-directed DNA methylation (RdDM), a mode of silencing genes both within the genome and throughout the latter that directs epigenetic modifications (Law and Jacobsen, 2010). Expression of short RNAs in plant hybrids is differentially expressed in hybrids compared with progenitor lines, linking short RNA expression to hybrid vigor (He *et al.*, 2010).

18.3 Whole-Genome Prediction

Whole-genome prediction (WGP) enables researchers to predict what a plant is worth in terms of its breeding value based on its DNA and its background information. Using the data from the reference populations, WGP will be able to identify those individuals that are the most likely to carry out the beneficial traits without ever having carried out field tests first (Heffner *et al.*, 2009; Lorenz *et al.*, 2011; Cooper *et al.*, 2014). WGP by genotyping and phenotyping has revolutionized plant breeding. While MET trials check out a couple of intriguing hybrids, WGP is employed heavily in commercial breeding to examine lots of possible hybrids all at once. By avoiding the expensive field tests, plant breeding becomes more efficient and genetic change happens faster using this technique (Heffner *et al.*, 2009; Cooper *et al.*, 2014). DNA markers have been studied using the whole genome with the emphasis on effects of these factors on the whole genome (Meuwissen *et al.*, 2001). The effects of markers are frequently figured out with Bayesian methods using both observed traits and prior information. To be able to capture all of the variety in the plant species and the plant habitats, these models need to be able to use wide datasets of intermittently available data from many years and many geographies (Cooper *et al.*, 2014).

18.4 Genomic Selection vs Epigenomic Selection

Genetic markers across the whole genome have allowed for a recent way of estimating an individual's full genetic worth called genomic selection (GS) (Meuwissen *et al.*, 2001). Genomic estimated breeding values (GEBVs) are calculated by adding up the estimated effect of all the genome-wide markers. This is especially helpful because it shortens the pace of breeding, speeds up the evaluation process, and therefore means less constant, long-term monitoring of phenotypes in the case of climate change (Heffner *et al.*, 2010; Yuan *et al.*, 2019). The model is trained on data containing, among other things, the genotype and the outcome and which produces the first step towards making a GEBV prediction. The genotypic data of each individual in the breeding population is used to get the GEBVs for each individual, using this model. The accuracy of GS also depends on many factors, among which are population size, the range of genes present, the heritability of the traits, gene density, and interactions with various environmental conditions (De Roos *et al.*, 2009; Clark *et al.*, 2011). Epistasis is not usually considered by models, which is important for cross-pollinated crops, and the techniques are more expensive than common pedigree-based alternatives (Heffner *et al.*, 2009). Genomic selection promises improved progress in crop enhancement.

Epigenomic selection includes heritable changes in epigenetic marks in its prediction models, which helps increase the chances of selecting beneficial traits in plants. Unlike regular genomic selection, which is based only on differences in DNA sequences, epigenomic selection also considers variations in phenotypes caused by non-sequence differences in regulatory systems. This method can be very helpful in learning about hybrid vigor and inbreeding depression, because similar genetic backgrounds can cause different trait expression due to epigenetics. Epigenomic selection has an edge over genomic selection, since it takes advantage of dynamic, stress-responsive, and sometimes reversible epigenetic changes, especially benefiting traits related to handling stress, development, and gene expression regulation. New high-resolution methods for epigenomic mapping will greatly increase the range of genetic variants available to plant breeders and lead to better crop improvements (Springer and Schmitz, 2017; Zenda *et al.*, 2021).

18.5 Machine Learning

AI aims to create software and machines that handle tasks similar to what is expected of human intellect, for example learning, thinking, and decision making. A major sector of AI is machine learning (ML), where algorithms program themselves to learn from information in datasets. Because of this flexibility, these algorithms continue to improve their results over time with more data (Ching *et al.*, 2018).

Many fields, such as ICT and agriculture, now make use of machine learning, an important area in AI (Falk *et al.*, 2020; Jafari and Shahsavar, 2020). ML is used in various crop science technologies, among which are smart watering systems (Yoosefzadeh-Najafabadi *et al.*, 2022). By using these, it becomes simpler to choose methods that call for reduced use of fertilizer and pesticides. Deep learning is typically sorted into three categories: supervised learning, unsupervised learning, and reinforcement learning (Hesami and Jones, 2020). Among the various ML methods, supervised learning is the one most often used and this involves giving the model labeled data. Trying different settings, it adapts its method to predict closer to how the results actually occur. People use this method for crop improvement because it allows them to predict complex properties ahead of time, discover areas in the genome related to particular traits, and choose the right breeds with the help of markers (Yoosefzadeh-Najafabadi *et al.*, 2022). Lack of labeled data makes unsupervised ML an option. With this method, the algorithm studies the data by itself to find out if there are any hidden patterns. It gets knowledge by understanding where the data naturally forms into clusters. In crop breeding, unsupervised learning methods are often applied to spot the genetic relationships among breeding individuals, group rough data, and make kinship matrices. This form of ML allows a model to interact with an environment that keeps changing. The model is encouraged or stopped by receiving rewards or sanctions based on what it does. After that, it changes its strategy to outperform the competitors over time. Reinforcement learning is a recent development in plant science, but it is very promising. Even though many of its uses are still untested, such as guiding carefully planned breeding schedules and fine-tuning breeding times, its potential is getting notable attention.

A number of key issues must be considered to effectively use data-driven models in crop improvement for forecasting traits with many genes. People go through data collection, primary data processing, creating attributes, constructing models, tuning hyperparameters, coding, and measuring performance on an ongoing basis (Vabalas *et al.*, 2019). The first step is to collect data from different samples in varied circumstances and genotypes, including noting observations, taking pictures, providing descriptions, and capturing videos (Xu *et al.*, 2022). The following step after collecting data is to make the data usable, which is done by sorting it, cleaning any errors, and adapting it. At this point, errors, whitespaces, and wrong matches in the dataset must be fixed. The next part is called 'feature extraction' and means grouping important traits and presenting them as the model's input. A model works correctly when its features are shown accurately. Application data that is not familiar to the model is then used to test how well the model works and how dependable it is. If the preliminary results are not good for the task, changing hyperparameters can reconfigure the program to make it better. Model implementation is the act of adding a certain model to a system or program to make it operational in breeding programs. Continuous evaluation helps ensure that the predictions are correct, because both data and environments change over time (Yoosefzadeh-Najafabadi *et al.*, 2022). Using ML technology can help manage growing crops. By considering crop growth data, and information on the weather and soil, models advise how to plant, water, and add fertilizer (Ansari *et al.*, 2020). People are using machine learning methods to analyze gene expression in plants. Genomic data improves our understanding of what determines plant features and how they change with changes in their environment (Najafabadi *et al.*, 2023). Genomic data can be analyzed by ML models to uncover what genes do and which ones are connected to certain traits (Wang *et al.*, 2020). An analysis of transcriptome data made possible by machine learning helps determine the responses of plants to a wide range of treatments and problems (Singh *et al.*, 2022; Cembrowska-Lech *et al.*, 2023).

18.6 AI in Genomics and DNA Sequencing

Biotechnology and genomics have been improved by AI primarily when it comes to genome sequencing and gene editing. AI technology has enabled Google AI to develop Google DeepVariant, a major tool in the genomics field. Using deep learning, DeepVariant scans sequencing data and detects additions, removals and changes in just one letter in genes. Deep Variant reviews the input data closely, spotting complex connections and having complete trust in its variant predictions. Because of this, researchers can make personalized treatment plans and biologists are able to find out the genes involved in different diseases more effectively. AI is also being used to improve the ways CRISPR/Cas genome editing works, which can enhance the accuracy and success of editing genes. When tested, these models ensure that only guide RNAs (gRNAs) with a low potential for off-target editing are selected. Because of DeepCRISPR, which uses deep learning to estimate the accuracy and side effects of CRISPR, there are fewer problems and more successes using this technology. They are allowing greater control in altering genes, which has helped researchers complete functional genomics studies faster. Due to AI, analyzing genomic data from single crops becomes quicker, helping researchers to find genes important for disease resistance, higher yields, and surviving stress. Modern crop-breeding fields, because of AI, are able to develop new breeding techniques that are tailored to each plant's genetic background. As a result, AI is benefiting future crop research by boosting targeted crop growth, selection, and the prediction of important features, as well as by revealing the molecular reasons behind complicated traits.

With the help of deep learning, GS in plant breeding has seen great improvements by being able to discover more complex patterns in detailed genotypic information. Advanced neural networks such as DeepGS and DNNGP (Deep Neural Network for Genomic Prediction) have been introduced to make trait prediction more precise. DeepGS uses convolution in neural networks, dropout and ensemble methods to cope with the complicated aspects in genomic data. Since it looks at interactions not covered by other methods, it can predict breeding results more accurately than rrBLUP can (Ma *et al.*, 2018). Combining multiple types of genomic data, DNNGP allows this approach to get better and learn about different biological systems. Evaluating genomic, transcriptomic, and phenotypic data has become efficient because of a well-designed deep neural network. The use of DNNGP helps GS pipelines by improving prediction and lowering the amount of calculation needed (Wang *et al.*, 2023). SoyDNGP, which is a recent addition, illustrates the promise of deep learning techniques in plant breeding. It transforms the data from genome variants to an image style, which is not possible by converting DNA sequences to binary using one-hot encoding. CNN handles genetic data by applying strategies from image processing, such as pooling and several convolutional layers. It is due to its unique design that SoyDNGP can find small patterns and relationships in the genome related to complex traits (Gao *et al.*, 2023a). New advancements in deep learning are enabling better prediction of plant genetics because they can find new patterns and improve how features of plants relate to their DNA. They make it possible to find and use the best cultivars more quickly and more precisely for breeding.

18.7 Machine Learning and Artificial Intelligence in Proteomics and Metabolomics

Crop improvement uses both metabolomics and proteomics because they closely examine plant metabolites and proteins. Using the techniques of mass spectrometry and nuclear magnetic resonance (NMR), scientists collect large sets of data that must be analyzed through detailed computations to find useful biological information. Plant metabolomics works to measure and study small molecules that impact the way metabolic activities happen in various parts or phases of plants. The number of molecules in metabolomics and the way they interact are quite difficult to handle analytically. Unsupervised learning algorithms in machine learning) are useful for revealing patterns in such complicated data. They allow scientists to sort metabolites, uncover potential biomarkers for stress or nutrient management, and explain complex paths taken by substances in plants.

In metabolomics, similar metabolites, rather than large amounts, are arranged together in ways like hierarchical clustering and k-means. It allows for finding markers of metabolism that separate different plant species, exhibit how they grow differently, and describe their responses to various pressures. Random forests and GBMs, which are supervised learning techniques, are applied to help produce accurate predictions of metabolic biomarkers and support plant disease prevention and trait selection. Plant proteomics must deal with the large amounts of data collected by mass spectrometry. Protein classification, function prediction, and protein interactions can be done through AI that analyzes different protein-related data. By using a CNN, scientists can detect changes in proteins that occur after synthesis, learn about protein interactions, and organize proteins by how their levels alter in different situations. Support vector machines (SVMs) and other classification algorithms are capable of uncovering new protein molecules or differences in plant function because they analyze mass spectrometry profiles. Besides, machine learning (ML) uses quantitative proteomics to monitor protein quantities in many samples and notice how changes in stress, development, or genetics influence protein expression. ML is helping find new biomarkers faster in plant metabolomics and proteomics, which benefits both crop breeding and our understanding of plants. Because of ML, it is now possible to detect and examine thousands of metabolites or proteins together, also uncovering small differences that other methods might not notice. Agricultural improvements and the use of precision agriculture rely on a comprehensive approach.

18.8 AI-Assisted Plant Epigenome Editing Using CRISPR/Cas

Altering the way genes are expressed using CRISPR/Cas, which does not modify the original DNA sequence, is known as epi-GED. By combining the CRISPR/Cas system with special epigenetic enzymes, the method targets selected points in the DNA and affects their epigenetic patterns (Goell and Hilton, 2021). Plant functional genomics and crop improvement rely on epi-GED, since it is able to control gene activity and differs from CRISPR in that it does not change DNA sequences. It is important to use both types of technology, such as machine learning (ML) and deep learning (DL) analysis, to study chromatin accessibility and histone and epigenetic methylation (Rauschert *et al.*, 2020; Gao *et al.*, 2023b). Different biological tools are used by plant scientists to investigate epigenetic rules. Epigenome editing differs from regular gene editing, so new AI technology has to be built for it. EpiCas-DL is a newly developed model introduced by Yang *et al.* (2023). Using deep learning, the model estimates the quality of single-guide RNAs (sgRNAs) for genome editing by CRISPR. EpiCas-DL takes into account four special epigenetic features not used by traditional sgRNA design tools: levels of gene expression, types of DNA methylation, chromatin structure, and TSS-sgRNA target site separation. Because decisions in the model are based on multiple layers, its prediction accuracy improves, and the AUC (area under curve) is better than any current alternatives at 0.87. It also helps to find factors that influence the proper functioning of sgRNA, as well as assist with planning sgRNA for the modifications sought in the study. Epigenetics particularly helps plants in activities such as development, reactions to stress, and passing traits to offspring. What we have learned about plant epigenome editing with AI is light compared with what is known about base or prime editing. Because the changes in plant epigenomics are affected by tissues, growth stages, and environmental factors, it is challenging and changes a lot depending on the context. EpiCas-DL and other AI tools like it are expected to make the process faster, since the availability of multi-omics plant data is increasing. With such information, it will be easier to understand gene regulation at the epigenetic level and develop ways to improve crops that support sustainability, stronger resistance, and better yield without altering the genome permanently.

18.9 AI and ML in Modern Phenomics and Plant Breeding

Artificial intelligence and machine learning have enabled breeders to switch from using old breeding methods to ones involving data and predictions. Using machine learning, part of AI, involves both defined outcomes and patterns that appear on their own. Many important areas of plant biology and breeding have come to rely heavily on these methods (Aasim *et al.*, 2023). Data scientists use their tools and big data to assist breeders in making fast and smart decisions about genetics and looks. Given that advanced systems in high-throughput phenotyping (HTP) collect large amounts of field, greenhouse, or lab data, analyzing this information is now especially significant (Bai *et al.*, 2016). When plants in various environments are observed by these technologies daily, a lot of useful information about them is uncovered. Data from digital camera photos and machine learning allow AI to study and identify characteristics of plants. By finding previously unknown connections in their data, machine learning offers new opportunities for breeding programs (Shakoor *et al.*, 2017). For example, AI rapidly distinguishes oilseed crops from weeds, which helps in crop management (Abdalla *et al.*, 2019). AI is involved in managing all the processes related to maize and wheat farming to ensure better harvests (Sadeghi-Tehran *et al.*, 2017; Jung *et al.*, 2021). Cyberinfrastructure (CI) has been key in helping with the changeover, allowing scientists to store, handle, and exchange large amounts of data via fast computer networks (Li *et al.*, 2022). Having the data exchange system, organizations can quickly advance phenomics. With the availability of open-source phenotyping tools, public organizations are able to study crops more effectively, and this has also promoted new ideas in related businesses (Nabwire *et al.*, 2021). FBP is seen as a strong way to observe and collect detailed data about plants outside. Sensors placed on vehicles, UAVs, or machines can monitor plant performance in real time out in the field (Bai *et al.*, 2016; Nabwire *et al.*, 2021). Such data will map out how the plant develops, which helps find genotypes suited to new challenges. AI systems have helped researchers learn more about plant architecture, the shape of their leaves, and the color of different species. With processed image and trait information, DL can increase accuracy and dependability, and reduce the amount of manual work required (Koh *et al.*, 2021). Plant scientists apply DL to predict gene expression (Zampieri *et al.*, 2019), estimate transcription factor binding (Lai *et al.*, 2019), and predict protein structures (Wang and Xu, 2017). AI is used in part to assess traits and find the genes that are linked to them. These studies combine DNA-related features with observable phenotypic characteristics. Using different data layers, models based on Bayesian networks have helped find QTLs, identify better traits, and create valuable varieties of crops more rapidly (Sharma *et al.*, 2021). Working with plants is becoming different because of advances in plant breeding and AI. Because of these advances, new crop varieties can be developed more easily, which helps deal with hunger, healthy diets, and climate challenges worldwide.

18.10 Role of Epigenetics in Stress Adaptation in Plants

By studying epigenetics, farming practices could make plants more resistant to climate change. Even though histone modifications, DNA methylation, and non-coding RNAs do not alter DNA sequence, they play a key role in controlling how genes act and deal with stress. It is unlike the usual genetic modifications done today. Because of epigenetics, plants can detect and respond faster to any prior challenges like heat, drought, or disease. Such experiences being passed down over the years can improve and advance methods of plant breeding (Rival *et al.*, 1999; Qi *et al.*, 2010; Richter *et al.*, 2010). Selective breeding and stress responses help breeders provide useful gene changes to plants, so they grow more and can respond better to changes in their environment. CRISPR/Cas9 and similar genetic tools allow researchers to easily alter certain sections of the epigenome and this helps manage the expression of stress-related genes (Russo *et al.*, 1996; Sable *et al.*, 2018). Managing disease and pest issues in plants is possible with elegant techniques because of epigenetics (Scharf *et al.*, 2012). In addition, things such as epimutations and epigenetic effects brought on by stress can be found in both wild populations and germplasm collections. Knowing this makes it possible to grow crops that withstand harsh conditions. (Schramm *et al.*, 2008; Sharma *et al.*,

2022, 2022). Therefore, it becomes simpler for plants to cope with new climates and change over time, for example in palms with somaclonal variation (Slotkin and Martienssen, 2007; Song *et al.*, 2012; Song *et al.*, 2015; Shen *et al.*, 2019) and tomatoes with changing fruit ripeness. By developing crops using epialleles and stable epimutations that happen under stress, biologists can bring about new ways to cope with changing temperatures (Springer, 2013; Sorenson and Bailey-Serres, 2014; Stief *et al.*, 2014).

18.11 AI-Powered Protein Engineering and Genome Editing to Enhance Crops

Conventional breeding through hybridization, mutagenesis, and genetic modification has played a big role in boosting crop productivity and efficiency. Even so, these techniques are mostly inefficient, since breeding takes a lot of time, the outcomes are not always clear, precision is low, and gene manipulation processes are ineffective (Labroo *et al.*, 2021; Yali and Mitiku, 2022). Gene editing technologies like CRISPR/Cas9, which have stepped forward with genome sequencing, have allowed scientists to improve crops faster and more precisely than before (Wang *et al.*, 2021). Genome editing is made possible in part by using site-directed nucleases (SDNs) such as TALENs, MNs, ZFNs, HEs, and CRISPR/Cas9 (Gaj *et al.*, 2013; Adil, 2018). Cas proteins like CRISPR-Cas9 are generally considered more accessible and easier to use for gene editing, while ZFNs and TALENs can be used for more accurate cutting or editing (Rasheed *et al.*, 2021).

Genome editing, which is used to modify DNA, is greatly improving because of the predictive and enhancing powers of AI (Christian *et al.*, 2010). AlphaFold2 and similar tools have enhanced our ability to accurately predict protein structures, which has become very important in molecular biology (Jumper *et al.*, 2021). By using AI to upgrade genetic engineering and gene editing, scientists might be able to create ideal crops (Huang *et al.*, 2022). An example of recent progress is OpenCRISPR-1, made by a machine learning system from research on thousands of CRISPR operons (Ruffolo *et al.*, 2024), and AlphaFold2, making sure the source is up to date, came up with new deaminase groups to enhance accuracy in base editors (Huang *et al.*, 2023). With AI, new tools for gene editing are precise, perform well, and are smaller. Also, the latest progress in plant breeding results from joining AI, crop science, and protein engineering skills.

Table 18.1 sets out epigenetic modifications in crops associated with agronomic traits.

18.12 Challenges and Limitations

Achieving the benefits of AI and ML in plant biotechnology means that several important problems have to be well resolved. Among other things, this section also looks at making sense of models, judging data accuracy and quality, and how to connect them to living biological systems in the plant sciences.

18.12.1 Data quality and bias

Plant biology studies using machine learning depend on accurate data, since accurate predictive models come from well-built training sets. When the data for forecasting is incomplete, irregular, or biased, it leads to poor choices in breeding and in managing farms. Plant biological datasets may have bias that comes from several different sources. Poor generalization by machine learning can happen if it is trained with data representative of a limited selection of plants or plant-growing environments. Because of this selection distortion, some models may not be helpful for genetic improvements needed for use in many agricultural settings. The presence of assessment biases can change what a genetic model finds, such as causing traits to be incorrectly estimated or leading to improper identification of genotypes. These biases are easy to notice in the way phenotypic data is regularly collected. Such prejudices can narrow the chances for undervalued crops or areas to receive improved varieties. Precise predictions from AI models about multiple plant populations and environments need access to diverse, fair, and accurate data. If datasets are not complete

Table 18.1. Epigenetic modifications in crops associated with agronomic traits (Author's own table)

Epigenetic modifications	Crops	Trait/changes induced	References
DNA methylation	Pigeon pea	Methylation in DNA is associated with heterosis	Sinha *et al.*, 2020
	Tomato	Fruit ripening is regulated by SlDML2 through DNA demethylation of ripening-related genes	Zhong *et al.*, 2013
	Rice	DNA methylation at OsMADS1 affects grain yield and floral development	Hu *et al.*, 2012
	Cotton	Demethylation of DNA activates the expression of *COL2* gene, which is responsible for photoperiodic flowering.	Hauben *et al.*, 2009
	Brassica	Epialleles of *FWA*-like gene control flowering time	Shea *et al.*, 2011
miRNA	Rice	miR393 targets auxin receptors, modulates tillering and drought response	Xia *et al.*, 2012
	Maize	miR156 and miR172 regulate phase change and inflorescence architecture	Chuck *et al.*, 2007
	Barley	miR172 targets APETALA2 (AP2)-like TF and is associated with grain density as well as cleistogamous flowering	Nair *et al.*, 2010; Houston *et al.*, 2013
	Soybean	GmMIR156b affects plant height and leaf size by repressing SPLs	Sun *et al.*, 2019
	Potato	miR156 targets SQUAMOSA-promoter binding protein-like (SPL) TFs and is associated with plant architecture as well as tuber yield	Bhogale *et al.*, 2014
		miR172 targets AP2-like TF and is associated with flowering time as well as tuberization time	Martin *et al.*, 2009
	Cotton	miR828 targets MYB TF and is associated with fiber development.	Guan *et al.*, 2014
	Wheat	miR408 targets Timing of CAB Expression 1 TF and is associated with heading time.	Zhao *et al.*, 2016
		miR159 targets MYB TF and is associated with the development of anthers.	Wang *et al.*, 2012
	Tomato	miR164 modulates NAC transcription factors involved in leaf development and stress response	Zhang *et al.*, 2018

Continued

Table 18.1. Continued

Epigenetic modifications	Crops	Trait/changes induced	References
Histone modification	Arabidopsis	FLOWERING LOCUS C (FLC) silencing via H3K27me3 results in vernalization-induced flowering	De Lucia *et al.*, 2008
	Banana	Binding of MaHDA6 to the MaERF11/15 promoters results in ripening	Fu *et al.*, 2018
		MaHDA1 recruitment to the target gene impeding ripening	Han et al., 2016
	Cotton	GhHDAs are associated with fiber initiation	Kumar *et al.*, 2018
	Wheat	Cuticular wax biosynthesis by TaGCN5 and attenuation of a fungal pathogen.	Kong *et al.*, 2020
		H3K9ac and H3K4me3 enrichment at vernalization genes affects flowering	Luo and He, 2020
	Rice	OsHDA705 (histone deacetylase) represses defense genes under normal conditions and is degraded under stress	Liu *et al.*, 2023

or dependable, the answers from the model will also be much less precise. Adopting AI in crop breeding and agronomy can be more difficult for those who depend on precise results for selection and management, because unreliable data or inaccurate entries could make the predictions less accurate.

18.12.2 Interpretability of ML models

A major problem with many AI and ML methods in plant science is that highly complex models are often difficult to interpret. Because of how these models work internally, it is hard for research teams to fully grasp them, which slows down checking and using the results of scientific experiments. Being able to interpret results is necessary for crop improvement and breeding to verify that the trait relationships and predictions are based on biology, not errors. If the decision-making process of the model is not well explained, people might not support the breeding process or instructions, especially when the project is aiming for approval from regulators. Plant scientists and breeders will rely on AI if they can interpret the results created by models. Models that are complicated can lead some people to question their accuracy, which reduces how helpful they are. Dealing with this problem, LIME and simpler methods like decision trees and rule-based models are finding increased use. They ensure AI can be used along with what breeders discover and what is studied about plants.

References

Aasim, M., Ali, S.A., Altaf, M.T., Ali, A., Nadeem, M.A. *et al.* (2023) Artificial neural network and decision tree facilitated prediction and validation of cytokinin-auxin induced *in vitro* organogenesis of sorghum (*Sorghum bicolor* L.). *Plant Cell, Tissue and Organ Culture* 153, 611–624.

Abdalla, A., Cen, H., Wan, L., Rashid, R., Weng, H. *et al.* (2019) Fine-tuning convolutional neural network with transfer learning for semantic segmentation of ground-level oilseed rape images in a field with high weed pressure. *Computers and Electronics in Agriculture* 167, 105091.

Adil, M. (2018) The CRISPR tool kit for genome editing and beyond. *Nature Communications* 9, 191.

Akhter, Z., Bi, Z., Ali, K., Sun, C., Fiaz, S. *et al.* (2021) In response to abiotic stress, DNA methylation confers epigenetic changes in plants. *Plants* 10, 1096. DOI: 10.3390/plants10061096.

Alvarez-Venegas, R., De la Peña, C. and Casas-Mollano, J.A. (eds) (2014) *Epigenetics in Plants of Agronomic Importance: Fundamentals and Applications: Transcriptional Regulation and Chromatin Remodelling in Plants*. Springer Cham, Switzerland. DOI: 10.1007/978-3-319-07971-4.

Angermueller, C., Pärnamaa, T., Parts, L. and Stegle, O. (2016) Deep learning for computational biology. *Molecular Systems Biology* 12(7), 878.

Ansari, M.H.D., Santosh, L., Raviraj, M.K., Srivastava, P.L., Pandit, V. *et al.* (2020) Recent advances in plant nanobionics and nanobiosensors for toxicology applications. *Current Nanoscience* 16(1), 27–41.

Arya, G., Maitra, A. and Grigoryev, S.A. (2010) A structural perspective on the where, how, why, and what of nucleosome positioning. *Journal of Biomolecular Structure and Dynamics* 27(6), 803–820.

Ashapkin, V.V., Kutueva, L.I., Aleksandrushkina, N.I. and Vanyushin, B.F. (2020) Epigenetic mechanisms of plant adaptation to biotic and abiotic stresses. *International Journal of Molecular Sciences* 21, e7457. DOI: 10.3390/ijms21207457.

Bai, G., Ge, Y., Hussain, W., Baenziger, P.S. and Graef, G. (2016) A multi-sensor system for high throughput field phenotyping in soybean and wheat breeding. *Computers and Electronics in Agriculture* 128, 181–192.

Bender, J. (2002) Plant epigenetics. *Current Biology* 12(13), R412–R414.

Bender, J. (2004) DNA methylation and epigenetics. *Annual Review of Plant Biology* 55, 41–68.

Bhadouriya, S.L., Mehrotra, S., Basantani, M.K., Loake, G.J. and Mehrotra, R. (2021) Role of chromatin architecture in plant stress responses: An update. *Frontiers in Plant Science* 11, 600246.

Bhogale, S., Mahajan, A.S., Natarajan, B., Rajabhoj, M., Thulasiram, H.V. *et al.* (2014) MicroRNA156: A potential graft-transmissible microRNA that modulates plant architecture and tuberization in *Solanum tuberosum* ssp. *andigena*. *Plant Physiology* 164, 1011–1027.

Camacho, D.M., Collins, K.M., Powers, R.K., Costello, J.C. and Collins, J.J. (2018) Next-generation machine learning for biological networks. *Cell* 173(7), 1581–1592. DOI: 10.1016/j.cell.2018.05.015.

Cembrowska-Lech, D., Krzemińska, A., Miller, T., Nowakowska, A., Adamski, C. *et al.* (2023) An integrated multi-omics and artificial intelligence framework for advanced plant phenotyping in horticulture. *Biology* 12(10), 1298.

Ching, T., Himmelstein, D.S., Beaulieu-Jones, B.K., Kalinin, A.A., Do, B.T. *et al.* (2018) Opportunities and obstacles for deep learning in biology and medicine. *Journal of the Royal Society Interface* 15(141), 20170387.

Christian, M., Cermak, T., Doyle, E.L., Schmidt, C., Zhang, F. *et al.* (2010) Targeting DNA double-strand breaks with TAL effector nucleases. *Genetics* 186(2), 757–761.

Chuck, G., Cigan, A.M., Saeteurn, K. and Hake, S. (2007) The temporal and spatial expression pattern of mir156 regulates shoot development in maize. *Developmental Cell* 13(6), 843–853.

Clapier, C.R., Iwasa, J., Cairns, B.R. and Peterson, C.L. (2017) Mechanisms of action and regulation of ATP-dependent chromatin-remodelling complexes. *Nature Reviews Molecular Cell Biology* 18(7), 407–422.

Clark, S.A., Hickey, J.M. and Van Der Werf, J.H. (2011) Different models of genetic variation and their effect on genomic evaluation. *Genetics Selection Evolution* 43, 18.

Cooper, A.J., Duke, É., Pickering, A.D. and Smillie, L.D. (2014) Individual differences in reward prediction error: Contrasting relations between feedback-related negativity and trait measures of reward sensitivity, impulsivity and extraversion. *Frontiers in Human Neuroscience* 8, 248. DOI: 10.3389/fnhum.2014.00248.

De Lucia, F., Crevillen, P., Jones, A.M., Greb, T. and Dean, C. (2008) A PHD-polycomb repressive complex 2 triggers the epigenetic silencing of FLC during vernalization. *Proceedings of the National Academy of Sciences* 105(44), 16831–16836.

De Roos, A.P.W., Hayes, B.J. and Goddard, M.E. (2009) Reliability of genomic predictions across multiple populations. *Genetics* 183(4), 1545–1553.

Eichten, S.R., Briskine, R., Song, J., Li, Q., Swanson-Wagner, R. *et al.* (2013) Epigenetic and genetic influences on DNA methylation variation in maize populations. *The Plant Cell* 25(8), 2783–2797. DOI: 10.1105/tpc.113.114793.

Falk, K.G., Jubery, T.Z., Mirnezami, S.V., Parmley, K.A., Sarkar, S. *et al.* (2020) Computer vision and machine learning enabled soybean root phenotyping pipeline. *Plant Methods* 16, 1–19.

Fu, C.C., Han, Y.C., Guo, Y.F., Kuang, J.F., Chen, J.Y. *et al.* (2018) Differential expression of histone deacetylases during banana ripening and identification of MaHDA6 in regulating ripening-associated genes. *Postharvest Biology and Technology* 141, 24–32.

Gaj, T., Gersbach, C.A. and Barbas, C.F. (2013) ZFN, TALEN, and CRISPR/Cas-based methods for genome engineering. *Trends in Biotechnology* 31(7), 397–405.

Gao, L., Ma, X., Lin, J. and Callan, J. (2023a) Precise zero-shot dense retrieval without relevance labels. In: *Proceedings of the 61st Annual Meeting of the Association for Computational Linguistics*, Association for Computational Linguistics, Toronto, Canada, pp. 1762–1777 (Vol. 1: Long Papers). Available at: https://aclanthology.org/2023.acl-long.99.pdf

Gao, Y., Marceau, A., Iqbal, V., Torres-Vázquez, J.A., Neupane, M. *et al.* (2023b) Genome-wide association analysis of heifer livability and early first calving in holstein cattle. *BMC Genomics* 24, 628. DOI: 10.1186/s12864-023-09736-0.

Goell, J.H. and Hilton, I.B. (2021) CRISPR/Cas-based epigenome editing: Advances, applications, and clinical utility. *Trends in Biotechnology* 39(7), 678–691.

Goldstein, M., Derheimer, F.A., Tait-Mulder, J. and Kastan, M.B. (2013) Nucleolin mediates nucleosome disruption critical for DNA double-strand break repair. *Proceedings of the National Academy of Sciences* 110(42), 16874–16879.

Guan, X., Pang, M., Nah, G., Shi, X., Ye, W. *et al.* (2014) miR828 and miR858 regulate homoeologous MYB2 gene functions in Arabidopsis trichome and cotton fibre development. *Nature Communication* 5, 3050.

Guleria, P., Mahajan, M., Bhardwaj, J. and Yadav, S.K. (2011) Plant small RNAs: Biogenesis, mode of action and their roles in abiotic stresses. *Genomics, Proteomics & Bioinformatics* 9(6), 183–199.

Gutzat, R. and Mittelsten Scheid, O. (2012) Epigenetic responses to stress: Triple defense? *Current Opinion in Plant Biology* 15(5), 568–573. DOI: 10.1016/j.pbi.2012.08.007.

Han, Y.C., Kuang, J.F., Chen, J.Y., Liu, X.C., Xiao, Y.Y. *et al.* (2016) Banana transcription factor MaERF11 recruit's histone deacetylase MaHDA1 and represses the expression of MaACO1 and expansions during fruit ripening. *Plant Physiology* 171, 1070–1084.

Hauben, M., Haesendonckx, B., Standaert, E., Kelen, K.V.D., Azmi, A. *et al.* (2009) Energy use efficiency is characterized by an epigenetic component that can be directed through artificial selection to increase yield. In: *Proceedings of the National Academy of Sciences USA*, pp. 20109–20114 (Vol. 106).

He, G., Zhu, X., Elling, A.A., Chen, L., Wang, X. *et al.* (2010) Global epigenetic and transcriptional trends among two rice subspecies and their reciprocal hybrids. *The Plant Cell* 22, 17–33.

Heffner, E.L., Sorrells, M.E. and Jannink, J.-L. (2009) Genomic selection for crop improvement. *Crop Science* 49(1), 1–12.

Heffner, E.L., Lorenz, A.J., Jannink, J.-L. and Sorrells, M.E. (2010) Plant breeding with genomic selection: Gain per unit time and cost. *Crop Science* 50, 1681–1690. DOI: 10.2135/cropsci2009.11.0662.

Hesami, M. and Jones, A.M.P. (2020) Application of artificial intelligence models and optimization algorithms in plant cell and tissue culture. *Applied Microbiology and Biotechnology* 104, 9449–9485.

Houston, K., McKim, S.M., Comadran, J., Bonar, N., Druka, I. *et al.* (2013) Variation in the interaction between alleles of HvAPETALA2 and microRNA172 determines the density of grains on the barley inflorescence. In: *Proceedings of the National Academy of Sciences USA*, pp. 16675–16680 (Vol. 110).

Huang, J., Lin, Q., Fei, H., He, Z., Xu, H. *et al.* (2023) Discovery of deaminase functions by structure-based protein clustering. *Cell* 186(15), 3182–3195.

Huang, X., Huang, S., Han, B. and Li, J. (2022) The integrated genomics of crop domestication and breeding. *Cell* 185(15), 2828–2839.

Hu, Y.F., Liu, D.N., Zhong, X.C., Zhang, C.J., Zhang, Q.F. *et al.* (2012) CHD3 protein recognizes and regulates methylated histone H3 lysines 4 and 27 over a subset of targets in the rice genome. *Proceedings of the National Academy of Sciences of the United States of America* 109, 5773–5778. DOI: 10.1073/pnas.1203148109.

Jablonka, E. and Lamb, M.J. (1994) The changing concept of epigenetics. *Annals of the New York Academy of Sciences* 981(1), 82–96.

Jafari, M. and Shahsavar, A. (2020) The application of artificial neural networks in modeling and predicting the effects of melatonin on morphological responses of citrus to drought stress. *PLoS ONE* 15(10), e0240427.

Jumper, J., Evans, R., Pritzel, A., Green, T., Figurnov, M. *et al.* (2021) Highly accurate protein structure prediction with AlphaFold. *Nature* 596(7873), 583–589.

Jung, J., Maeda, M., Chang, A., Bhandari, M., Ashapure, A. *et al.* (2021) The potential of remote sensing and artificial intelligence as tools to improve the resilience of agriculture production systems. *Current Opinion in Biotechnology* 70, 15–22.

Kim, J.M., To, T.K., Ishida, J., Matsui, A., Kimura, H. *et al.* (2012) Transition of chromatin status during the process of recovery from drought stress in *Arabidopsis thaliana*. *Plant Cell Physiology* 53(5), 847–856.

Kim, J.M., Sasaki, T., Ueda, M., Sako, K. and Seki, M. (2015) Chromatin changes in response to drought, salinity, heat, and cold stresses in plants. *Frontiers in Plant Science* 6, 114.

Koh, J.C.O., Spangenberg, G. and Kant, S. (2021) Automated machine learning for high-throughput image-based plant phenotyping. *Remote Sensing* 13(5), 858.

Kong, L., Zhi, P., Liu, J., Li, H., Zhang, X. *et al.* (2020) Epigenetic activation of Enoyl-CoA reductase by an acetyltransferase complex triggers wheat wax biosynthesis. *Plant Physiology* 183, 1250–1267.

Koumakis, L. (2020) Deep learning models in genomics: Are we there yet? *Computational and Structural Biotechnology Journal* 18, 1466–1473.

Kumar, V., Singh, B., Singh, S.K., Rai, K.M., Singh, S.P. *et al.* (2018) Role of GhHDA5 in H3K9 deacetylation and fiber initiation in *Gossypium hirsutum*. *Plant Journal* 95, 1069–1083.

Labroo, M.R., Studer, A.J. and Rutkoski, J.E. (2021) Heterosis and hybrid crop breeding: A multidisciplinary review. *Frontiers in Genetics* 12, 643761.

Lai, X., Stigliani, A., Vachon, G., Carles, C., Smaczniak, C. *et al.* (2019) Building transcription factor binding site models to understand gene regulation in plants. *Molecular Plant* 12(6), 743–763.

Lämke, J. and Bäurle, I. (2017) Epigenetic and chromatin-based mechanisms in environmental stress adaptation and stress memory in plants. *Genome Biology* 18, 124.

Lauria, M., Piccinini, S., Pirona, R., Lund, G., Viotti, A. *et al.* (2014) Epigenetic variation, inheritance, and parent-of-origin effects of cytosine methylation in maize (*Zea mays*). *Genetics* 196, 653–666. DOI: 10.1534/genetics.113.160515.

Law, J.A. and Jacobsen, S.E. (2010) Establishing, maintaining and modifying DNA methylation patterns in plants and animals. *Nature Reviews Genetics* 11(3), 204–220.

Leung, M.K.K., Delong, A., Alipanahi, B. and Frey, B.J. (2016) Machine learning in genomic medicine: A review of computational problems and data sets. *Proceedings of the IEEE* 104(1), 176–197.

Li, H., Yan, S., Zhao, L., Tan, J., Zhang, Q. *et al.* (2014) Histone acetylation associated up-regulation of the cell wall related genes is involved in salt stress induced maize root swelling. *BMC Plant Biology* 14, 105.

Li, Q., Feng, Q., Snouffer, A., Zhang, B., Rodríguez, G.R. *et al.* (2022) Increasing fruit weight by editing a cis-regulatory element in tomato KLUH promoter using CRISPR/Cas9. *Frontiers in Plant Science* 13, 879642.

Liu, D., Mewalal, R., Hu, R., Tuskan, G.A. and Yang, X. (2017) New technologies accelerate the exploration of non-coding RNAs in horticultural plants. *Horticulture Research* 4, 17031.

Liu, J., Wang, H. and Chua, N.H. (2015) Long noncoding RNA transcriptome of plants. *Plant Biotechnology Journal* 13(3), 319–328.

Liu, J., Li, J., Wang, H. and Yan, J. (2020) Application of deep learning in genomics. *Science China Life Sciences* 63(11), 1860–1878.

Liu, K., Chen, J., Sun, S., Chen, X., Zhao, X. *et al.* (2023) Histone deacetylase OsHDA706 increases salt tolerance via H4K5/K8 deacetylation of OsPP2C49 in rice. *Journal of Integrative Plant Biology* 65(6), 1394–1407.

Lorenz, J., Rauhut, H., Schweitzer, F. and Helbing, D. (2011) How social influence can undermine the wisdom of crowd effect. *Proceedings of the National Academy of Sciences USA* 108(22), 9020–9025. DOI: 10.1073/pnas.1008636108.

Luo, X. and He, Y. (2020) Experiencing winter for spring flowering: A molecular epigenetic perspective on vernalization. *Journal of Integrative Plant Biology* 62(1), 104–117.

Ma, W., Qiu, Z., Song, J., Li, J., Cheng, Q. *et al.* (2018) A deep convolutional neural network approach for predicting phenotypes from genotypes. *Planta* 248, 1307–1318. DOI: 10.1007/s00425-018-2976-9.

Martin, A., Adam, H., Díaz-Mendoza, M., Zurczak, M., Gonzalez-Schain, N.D. *et al.* (2009) Graft-transmissible induction of potato tuberization by the microRNA miR172. *Development* 136, 2873–2881.

Meuwissen, T.H.E., Hayes, B.J. and Goddard, M.E. (2001) Prediction of total genetic value using genome-wide dense marker maps. *Genetics* 157(4), 1819–1829.

Nabwire, S., Suh, H.-K., Kim, M.S., Baek, I. and Cho, B.-K. (2021) Review: Application of artificial intelligence in phenomics. *Sensors* 21(13), 4363.

Nair, S.K., Wang, N., Turuspekov, Y., Pourkheirandish, M., Sinsuwongwat, S. *et al.* (2010) Cleistogamous flowering in barley arises from the suppression of microRNA-guided HvAP2 mRNA cleavage. In: *Proceedings of the National Academy of Sciences USA*, pp. 490–495 (Vol. 107).

Najafabadi, M.Y., Hesami, M. and Eskandari, M. (2023) Machine learning-assisted approaches in modernized plant breeding programs. *Genes* 14(4), 777.

Pecinka, A., Chevalier, C., Colas, I., Kalantidis, K., Varotto, S. *et al.* (2020) Chromatin dynamics during interphase and cell division: Similarities and differences between model and crop plants. *Journal of Experimental Botany* 71(17), 5205–5222.

Perrella, G., Lopez-Vernaza, M.A., Carr, C., Sani, E., Grossele, V. *et al.* (2013) Histone deacetylase complex1 expression level titrates plant growth and abscisic acid sensitivity in *Arabidopsis*. *Plant Cell* 25(9), 3491–3505.

Prathiba, K.Y., Usha, S., Suchithra, B., Jyothi, M.N., Devaraj, V.R. *et al.* (2017) Computational identification of miRNAs and their targets from Niger (*Guizotia abyssinica*). *Journal of Applied Biology and Biotechnology* 5(2), 53–58.

Qi, H., Huang, Q., Gao, W. and Zhao, D. (2010) Error-resistance and low-complexity integer inverse discrete cosine transform. *Journal of Signal Processing Systems* 61, 231–239.

Rajewsky, N., Jurga, S. and Barciszewski, J. (eds) (2017) *Plant Epigenetics*. Springer, Cham, Switzerland.

Rasheed, A., Gill, R.A., Hassan, M.U., Mahmood, A., Qari, S. *et al.* (2021) A critical review: Recent advancements in the use of CRISPR/Cas9 technology to enhance crops and alleviate global food crises. *Current Issues in Molecular Biology* 43(3), 1950–1976.

Rauschert, S., Raubenheimer, K., Melton, P.E. and Huang, R.C. (2020) Machine learning and clinical epigenetics: A review of challenges for diagnosis and classification. *Clinical Epigenetics* 12(1), 51.

Richter, K., Haslbeck, M. and Buchner, J. (2010) The heat shock response: Life on the verge of death. *Molecular Cell* 40(2), 253–266.

Rival, A., Jaligot, E., Beulé, T., Verdeil, J.-L. and Tregear, J. (1999) DNA methylation and somaclonal variation in oil palm. In: *International Symposium on Methods and Markers for Quality Assurance in Micropropagation*, pp. 447–454 (Vol. 530).

Ruffolo, J.A., Nayfach, S., Gallagher, J., Bhatnagar, A., Beazer, J. *et al.* (2024) Design of highly functional genome editors by modeling the universe of CRISPR-Cas sequences. *bioRxiv*. DOI: 10.11011/2024.04.22.590591.

Russo, V.E.A., Martienssen, R.A. and Riggs, A.D. (1996) *Epigenetic Mechanisms of Gene Regulation*. Cold Spring Harbor Laboratory Press, New York.

Sable, A., Rai, K.M., Choudhary, A., Yadav, V.K., Agarwal, S.K. *et al.* (2018) Inhibition of heat shock proteins HSP90 and HSP70 induces oxidative stress, suppressing cotton fiber development. *Scientific Reports* 8, 3620.

Sadder, M.T., Alsadon, A., Al-Thamra, M., Zakri, A.M. and Al-Doss, A. (2011) Phylogenetic analysis of SET domain in Trithorax SlTX1 of *Solanum lycopersicum*. *Plant Omics* 4(2), 95–99.

Sadeghi-Tehran, P., Sabermanesh, K., Virlet, N. and Hawkesford, M.J. (2017) Automated method to determine two critical growth stages of wheat: Heading and flowering. *Frontiers in Plant Science* 8, 252.

Santos, A.P., Ferreira, L.J. and Oliveira, M.M. (2017) Concerted flexibility of chromatin structure, methylome, and histone modifications along with plant stress responses. *Biology (Basel)* 6(1), 3.

Scharf, K.-D., Berberich, T., Ebersberger, I. and Nover, L. (2012) The plant heat stress transcription factor (Hsf) family: Structure, function and evolution. *Biochimica et Biophysica Acta (BBA) – Gene Regulatory Mechanisms* 1819(2), 104–119.

Schramm, F., Larkindale, J., Kiehlmann, E., Ganguli, A., Englich, G. *et al.* (2008) A cascade of transcription factor DREB2A and heat stress transcription factor HsfA3 regulates the heat stress response of *Arabidopsis*. *The Plant Journal* 53(2), 264–274.

Secco, D., Whelan, J., Rouached, H. and Lister, R. (2017) Nutrient stress-induced chromatin changes in plants. *Current Opinion in Plant Biology* 39, 1–7.

Shakoor, N., Lee, S. and Mockler, T.C. (2017) High throughput phenotyping to accelerate crop breeding and monitoring of diseases in the field. *Current Opinion in Plant Biology* 38, 184–192.

Sharma, M., Kumar, P., Verma, V., Sharma, R., Bhargava, B. *et al.* (2022) Understanding plant stress memory response for abiotic stress resilience: Molecular insights and prospects. *Plant Physiology and Biochemistry* 179, 10–24.

Sharma, N., Thakur, M., Suryakumar, P., Mukherjee, P., Raza, A. *et al.* (2022) Breathing out' under heat stress—respiratory control of crop yield under high temperature. *Agronomy* 12, 806.

Sharma, S., Pinson, S.R.M., Gealy, D.R. and Edwards, J.D. (2021) Genomic prediction and QTL mapping of root system architecture and above-ground agronomic traits in rice (*Oryza sativa L.*) with a multi-trait index and Bayesian networks. *G3: Genes, Genomes, Genetics* 11, jkab178.

Shea, D.J., Penfield, S. and Grant-Downton, R.T. (2011) Epigenetic variation and the control of flowering time in *Brassica rapa*. *The Plant Journal* 68(4), 543–553.

Shen, Y., Lei, T., Cui, X., Liu, X., Zhou, S. *et al.* (2019) Arabidopsis histone deacetylase HDA15 directly represses plant response to elevated ambient temperature. *The Plant Journal* 100(5), 991–1006.

Singh, A.V., Chandrasekar, V., Laux, P., Luch, A., Dakua, S.P. *et al.* (2022) Micropatterned neurovascular interface to mimic the blood–brain barrier's neurophysiology and micromechanical function: A BBB-on-CHIP model. *Cells* 11(18), 2801.

Sinha, P., Singh, V.K., Saxena, R.K., Kale, S.M., Li, Y. *et al.* (2020) Genome-wide analysis of epigenetic and transcriptional changes associated with heterosis in pigeonpea. *Plant Biotechnology Journal* 18, 1697–1710.

Skopelitis, D.S., Hill, K., Klesen, S., Marco, C.F., von Born, P. *et al.* (2018) Gating of miRNA movement at defined cell–cell interfaces governs their impact as positional signals. *Nature Communications* 9(1), 3107.

Slotkin, R.K. and Martienssen, R. (2007) Transposable elements and the epigenetic regulation of the genome. *Nature Reviews Genetics* 8(4), 272–285.

Song, L., Jiang, Y., Zhao, H. and Hou, M. (2012) Acquired thermotolerance in plants. *Plant Cell, Tissue and Organ Culture* 111(3), 265–276.

Song, Z.-T., Sun, L., Lu, S.-J., Tian, Y., Ding, Y. *et al.* (2015) Transcription factor interaction with COMPASS-like complex regulates histone H3K4 trimethylation for specific gene expression in plants. *Proceedings of the National Academy of Sciences* 112(9), 2900–2905.

Sorenson, R. and Bailey-Serres, J. (2014) Selective mRNA sequestration by OLIGOURIDYLATE-BINDING PROTEIN 1 contributes to translational control during hypoxia in *Arabidopsis*. *Proceedings of the National Academy of Sciences* 111(6), 2373–2378.

Springer, N.M. (2013) Epigenetics and crop improvement. *Trends in Genetics* 29(4), 241–247.

Springer, N.M. and Schmitz, R.J. (2017) Exploiting induced and natural epigenetic variation for crop improvement. *Nature Reviews Genetics* 18, 563–575.

Stief, A., Altmann, S., Hoffmann, K., Pant, B.D., Scheible, W.-R. *et al.* (2014) Arabidopsis miR156 regulates tolerance to recurring environmental stress through SPL transcription factors. *The Plant Cell* 26(4), 1792–1807.

Sun, Z., Su, C., Yun, J., Jiang, Q., Wang, L. *et al.* (2019) Genetic improvement of the shoot architecture and yield in soya bean plants via the manipulation of GmmiR156b. *Plant Biotechnology Journal* 17(1), 50–62.

Sunkar, R., Li, Y.F. and Jagadeeswaran, G. (2012) Functions of microRNAs in plant stress responses. *Trends in Plant Science* 17(4), 196–203.

Takeda, S. and Paszkowski, J. (2006) DNA methylation and epigenetic inheritance during plant gametogenesis. *Chromosoma* 115(1), 27–35.

Vabalas, A., Gowen, E., Poliakoff, E. and Casson, A.J. (2019) Machine learning algorithm validation with a limited sample size. *PLoS ONE* 14(11), e0224365. DOI: 10.1371/journal.pone.0224365.

Vanyushin, B.F. (2006) DNA methylation in plants. *Current Topics in Microbiology and Immunology* 301, 67–122.

Villicaña, S. and Bell, J.T. (2021) Genetic impacts on DNA methylation: Research findings and future perspectives. *Genome Biology* 22(1), 127. DOI: 10.1186/s13059-021-02347-6.

Wagner, D. (2003) Chromatin regulation of plant development. *Current Opinion in Plant Biology* 6(1), 20–28.

Wang, K.C. and Chang, H.Y. (2018) Epigenomics—technologies and applications. *Circulation Research* 122(9), 1191–1199.

Wang, S. and Xu, J. (2017) De novo protein structure prediction by big data and deep learning. *Biophysical Journal* 112(3), 55a.

Wang, K., Abid, M.A., Rasheed, A., Crossa, J., Hearne, S. *et al.* (2023) DNNGP, a deep neural network-based method for genomic prediction using multi-omics data in plants. *Molecular Plant* 16, 279–293. DOI: 10.1016/j.molp.2022.11.004.

Wang, T., Zhang, C., Zhang, H. and Zhu, H. (2021) CRISPR/Cas9-mediated gene editing revolutionizes the improvement of horticulture food crops. *Journal of Agricultural and Food Chemistry* 69(45), 13260–13269.

Wang, X., Li, N., Li, W., Gao, X., Cha, M. *et al.* (2020) Advances in transcriptomics in the response to stress in plants. *Global Medical Genetics* 7(1), 30–34.

Wang, Y., Sun, F., Cao, H., Peng, H., Ni, Z. *et al.* (2012) *TamiR159* directed wheat *TaGAMYB* cleavage and its involvement in anther development and heat response. *PLoS ONE* 7(11), e48445. DOI: 10.1371/journal.pone.0048445.

Wang, Z., Casas-Mollano, J.A., Xu, J., Riethoven, J.J.M., Zhang, C. *et al.* (2015) Osmotic stress induces phosphorylation of histone H3 at threonine 3 in pericentromeric regions of *Arabidopsis thaliana*. *Proceedings of the National Academy of Sciences* 112(27), 8487–8492.

Weaver, I.C.G., Korgan, A.C., Kristen, L., Wheeler, R.V., Hundert, A.S. *et al.* (2017) Stress and the emerging roles of chromatin remodeling in signal integration and stable transmission of reversible phenotypes. *Frontiers in Behavioral Neuroscience* 11, 41.

Wu, J. and Zhao, Y. (2019) Machine learning technology in the application of genome analysis: A systematic review. *Gene* 705, 149–156.

Xia, K., Wang, R., Ou, X., Fang, Z., Tian, C. *et al.* (2012) MicroRNA393-mediated regulation of the auxin signaling pathway in rice confers drought tolerance. *PLoS ONE* 7(10), e30231.

Xu, J., Li, Y., Wang, Y., Liu, X. and Zhu, X.-G. (2017) Altered expression profiles of microRNA families during de-etiolation of maize and rice leaves. *BMC Research Notes* 10, 108.

Xu, Y., Zhang, X., Li, H., Zheng, H., Zhang, J. *et al.* (2022) Smart breeding driven by big data, artificial intelligence, and integrated genomic-enviromic prediction. *Molecular Plant* 15(11), 1664–1695. DOI: 10.1016/j.molp.2022.09.001.

Yali, W. and Mitiku, T. (2022) Mutation breeding and its importance in modern plant breeding. *Journal of Plant Sciences* 10(3), 64–70.

Yang, Q., Wu, L., Meng, J., Ma, L., Zuo, E. *et al.* (2023) EpiCas-DL: Predicting sgRNA activity for CRISPR-mediated epigenome editing by deep learning. *Computational and Structural Biotechnology Journal* 21, 202–211.

Yang, S., Li, C., Zhao, L., Gao, S., Lu, J. *et al.* (2015) The Arabidopsis SWI2/SNF2 chromatin remodeling ATPase BRAHMA targets directly to PINs and is required for root stem cell niche maintenance. *Plant Cell* 27, 1670–1680.

Yoosefzadeh-Najafabadi, M., Rajcan, I. and Vazin, M. (2022) High-throughput plant breeding approaches: Moving along with plant-based food demands for pet food industries. *Frontiers in Veterinary Science* 9, 1467.

Yuan, Y., Cairns, J.E., Babu, R., Gowda, M., Makumbi, D. *et al.* (2019) Genome-wide association mapping and genomic prediction analyses reveal the genetic architecture of grain yield and flowering time under drought and heat stress conditions in maize. *Frontiers in Plant Science* 9, 1919.

Zampieri, G., Vijayakumar, S., Yaneske, E. and Angione, C. (2019) Machine and deep learning meet genome-scale metabolic modeling. *PLoS Computational Biology* 15(7), e1007084. DOI: 10.1371/journal.pcbi.1007084.

Zenda, T., Liu, S., Dong, A., Yang, Y., Wang, N. *et al.* (2021) In response to abiotic stress, DNA methylation confers epigenetic changes in plants. *International Journal of Molecular Sciences* 22(11), 5996.

Zhang, D., Li, Y., Zhang, X., Zha, P. and Lin, R. (2017) The SWI2/SNF2 chromatin remodeling ATPase BRAHMA regulates chlorophyll biosynthesis in Arabidopsis. *Molecular Plant* 10, 155–167.

Zhang, L., Yao, L., Zhang, N., Yang, J., Zhu, X. *et al.* (2018) Lateral root development in potato is mediated by stu-mi164 regulation of NAC transcription factor. *Frontiers in Plant Science* 9, 383.

Zhang, S., Shen, S., Yang, Z., Kong, X., Liu, F. *et al.* (2020) Coding and non-coding RNAs: Molecular basis of forest-insect outbreaks. *Frontiers in Cell and Developmental Biology* 8, 369. DOI: 10.3389/fcell.2020.00369.

Zhao, D., Chong, K. and Palanivelu, R. (2017) Editorial: Molecular and cellular plant reproduction. *Frontiers in Plant Science* 8, 199.

Zhao, X.Y., Hong, P., Wu, J.Y., Chen, X.B., Ye, X.G. *et al.* (2016) The tae-miR408-mediated control of TaTOC1 genes transcription is required for the regulation of heading time in wheat. *Plant Physiology* 170, 1578–1594.

Zhong, S., Fei, Z., Chen, Y.R., Zheng, Y., Huang, M. *et al.* (2013) Single-base resolution methylomes of tomato fruit development reveal epigenome modifications associated with ripening. *Nature Biotechnology* 31(2), 154–159.

19 Plant Genomic Prediction: Fundamental Models and Recent Applications in Crop Breeding

Aditya Pratap Singh[1]*, Anamika Das[1], Arup Kr. Saha[2], Sudip Bhattacharya[3] and Jaishree Tanwar[4]

[1]Department of Genetics and Plant Breeding, Bidhan Chandra Krishi Viswavidyalaya, Mohanpur, Nadia, West Bengal, India; [2]Department of Genetics and Plant Breeding, School of Agriculture, Seacom Skills University, Kendradangal, Birbhum, West Bengal, India; [3]Department of Biological Science, Midnapore City College, West Bengal, India; [4]Department of Genetics and Plant Breeding, Agriculture University, Jodhpur, Rajasthan, India

Abstract

Genomic prediction (GP) has revolutionized plant breeding by leveraging genome-wide marker data to estimate breeding values, enabling faster and more accurate selection of superior crop genotypes. This chapter explores the evolution, methodologies, and applications of GP in modern agriculture. Beginning with the historical context of plant breeding, it discusses the transition from phenotypic selection to molecular marker-based approaches and the emergence of GP as a transformative tool. Key statistical models, including genomic best linear unbiased prediction (GBLUP), ridge regression BLUP (RR-BLUP), Bayesian approaches, and machine learning techniques like deep learning, are detailed, emphasizing their ability to handle complex, polygenic traits and genotype-by-environment interactions. The integration of high-throughput phenotyping, multi-omics, and enviromics enhances prediction accuracy, supporting the development of climate-resilient, high-yielding cultivars. Applications in major cereals and minor crops highlight GP's role in accelerating genetic gain and optimizing breeding pipelines. Emerging trends, such as AI-driven models, practical haplotype graphs, and real-time prediction platforms, point to a future of personalized breeding. Despite challenges like data requirements and computational complexity, GP offers immense opportunities to transform global agriculture.

Keywords: Artificial intelligence, Genomic prediction, Machine learning, Plant breeding, Polygenic traits, Statistical models

19.1 Introduction to Genomic Prediction in Plants

Plant breeding aims to develop high-yielding, adapted cultivars. Over time, it evolved into a more sophisticated, scientific approach grounded in genetic principles. The early 20th century marked a significant milestone with the rediscovery of Mendel's laws of inheritance, which facilitated a systematic and experimental

*Corresponding author: adityapratapbckv@gmail.com

DOI: 10.1079/9781836991328.0019

framework for crop improvement through techniques such as hybridization, selective breeding, and cytogenetics (Hickey *et al.*, 2017). The latter half of the 20th century witnessed remarkable advances driven by molecular biology and biotechnology, which enhanced the precision of breeding efforts. The introduction of molecular markers and recombinant DNA technology enabled scientists to transfer specific genes across species boundaries, thereby expanding the potential for developing improved crop varieties with desired traits (Chen *et al.*, 2019; Varshney *et al.*, 2021). Although phenotypic selection has been effective in the past, it has notable drawbacks. It is a lengthy process that requires multi-year and multi-location experiments. Moreover, improving complex traits such as yield, stress tolerance, and nutritional quality remains challenging, because they are polygenic and heavily influenced by environmental factors (Cooper *et al.*, 2014). Conventional field assessments are laborious, expensive, and often imprecise, because they can only analyze a limited number of genotypes (Araus and Cairns, 2014). Therefore, early and accurate prediction methods using genomic models, machine learning, and high-throughput phenotyping are crucial for shortening breeding cycles, increasing genetic gains, and addressing issues like food security and climate change (Xu *et al.*, 2020; Montesinos-López *et al.*, 2021a).

Nowadays, a sophisticated breeding technique called genomic prediction (GP) estimates an individual's genetic potential, or genomic estimated breeding values (GEBVs), prior to phenotypic assessment using statistical models and genome-wide molecular markers. Ever since Meuwissen *et al.* (2001) first proposed the concept, GP has become a vital component of modern breeding pipelines. In contrast to marker-assisted selection (MAS), which depends on a small number of loci linked to traits of interest, or traditional phenotypic selection, which depends on observable traits across various environments and seasons, GP uses genome-wide information to capture both major and minor effect loci, allowing for more precise predictions of complex traits (Crossa *et al.*, 2017; Xu *et al.*, 2020). Conversely, genomic selection (GS), which estimates breeding values utilizing genome-wide data without the necessity for extensive phenotyping, has become practicable owing to advancements in genomics, sequencing, and bioinformatics. Although GP predominantly pertains to modeling and prediction methodologies, GS constitutes the practical application of GP within breeding programs. In order to pick superior individuals without requiring extensive field testing, GS uses GP models that have been trained on reference populations that have both genotype and phenotypic data (Cooper *et al.*, 2014; Hickey *et al.*, 2017). High-throughput phenotyping and genotyping (e.g. SNP arrays, sequencing-based platforms) have made GP possible in breeding workflows. By offering comprehensive genomic data at a lower cost, these technologies enable the identification of both major and minor effect loci that underlie quantitative traits. GP speeds up genetic gain by facilitating early and precise prediction, which improves selection accuracy, reduces breeding cycles, and permits more intense selection (Montesinos-López *et al.*, 2021a; Crossa *et al.*, 2023). The development of crop varieties that are resource-efficient, high-yielding, and climate-resilient has made GP a key component of contemporary breeding techniques.

19.2 Fundamentals of Genomic Prediction Models

Genetic mapping determines genetic distances between gene pairs and maps them to specific sites in the genome. Mapping also enables the identification of genomic regions associated with complex traits, which is the basis for genomic selection strategies (Huang and Han, 2014).

Genetic markers are effective in analyzing the molecular basis of traits. In recent years, single nucleotide polymorphisms (SNPs) have emerged as the marker of choice across diverse crop species, due to their abundance, widespread distribution across coding and non-coding genomic regions, and bi-allelic nature, which aligns well with population genetics models and supports co-dominant inheritance (Khlestkina and Salina, 2006). SNP markers offer significant logistical advantages, including reduced genotyping error rates and ease of automation for high-throughput applications (Elshire *et al.*, 2011).

High dimensionality is a condition where the analyzed loci (p) exceed the number of plant genotypes (n). This leads to the curse of dimensionality, where markers represent causal sites as quantitative trait loci (QTLs), allowing their participation in phenotypic variability to be identified specifically. Due to ascertainment bias (sampling bias of markers during their development), genetic markers generally do not assay the whole pool of DNA polymorphisms in any given population (Heslot *et al.*, 2013).

To resolve this, the effective and universal method for genomic prediction is a linear model in which markers are supposed to have an additive contribution to the phenotypic expression. The criteria that need to be defined are the effect of markers, which are determined by fitting the best phenotypic observations, subject to some side conditions. Techniques such as ridge regression – which is analogous to the genomic best linear unbiased prediction (GBLUP), the Bayesian alphabet (Gianola *et al.*, 2009; Gianola, 2013), or the least absolute shrinkage and selection operator (LASSO) (Santosa and Symes, 1986; Tibshirani, 1996) regression – fall into this class of linear models, among others. These models use genome-wide marker information for finding genomic estimated breeding values (GEBVs), which can speed up selection cycles in breeding programs (Meuwissen *et al.*, 2001). Regularization is beneficial as the genotype number is much less than the explanatory variables. A basic linear common least squares regression, therefore, would not lead to a distinct solution, and any of the probable outcomes would fit the training data ideally, but provide only less predictive capacity for new data (Crossa *et al.*, 2025b).

Total genetic variance is the $\sigma^2G = \sigma^2A + \sigma^2D$, where the first one is the additive genetic variance and the second relates to the dominance genetic variance. The genetic variance is further classified into additive (VA), dominance (VD), and epistatic (VI) interaction. Additive variance is the most useful for selection because it is transmitted reliably across generations, whereas epistasis and dominance may also play an important role in trait expression but are more complex to exploit in breeding. In genomic prediction, additive models are commonly used, though inclusion of non-additive components can enhance prediction for traits with significant gene interactions (Xu, 2013).

Generally, the precision of GEBV prediction increases almost directly with an increase in population size. Therefore, the training population should be adequately large, particularly when it is not related to the breeding population (Singh and Singh, 2015). Similarly, the genetic markers should be sufficiently more in numbers so that the maximum QTLs affecting the trait are in strong linkage disequilibrium (LD) with at least one genomic marker. Therefore, marker density will also depend on the extent of LD in the concerned species; in general, density should be much higher in cross-pollinated crops than in self-pollinated ones (Singh and Singh, 2015).

19.3 Classical Statistical Models in Genomic Prediction

Classical models are summarized in Table 19.1. BLUP allowed genetic advancement, which is to be achieved in many crop improvement programs. This methodology depends on pedigree records to explain the covariance present between relatives. Further, covariance can also be explained by using information on the number of DNA markers, frequently SNP markers. This model is referred to as the genomic relationship matrix (GRM). Here, two methods are explained to encompass genomic information into BLUP: ridge regression BLUP (RR-BLUP) and genomic BLUP (GBLUP), and how they are equivalent (Clark and Werf, 2013).

19.3.1 Genomic best linear unbiased prediction (GBLUP)

The GBLUP model is the most important statistical method that predicts the genotypic value of genotypes using GRM. This matrix is studied for multi-trait analysis. GBLUP is as follows:

$$Y = X\beta + g + e$$

where, $Y = [Y_1 \ldots Y_T]$ is a value of the phenotypes for all trait involved in the matrix, of magnitude $n \times T$, where n is the number of observations and T is the traits number; X is a fixed effects design model, of magnitude $n \times p$, where p is fixed effects number, $\beta = [\beta_1 \ldots \beta_t]$, such that β_t is fixed effects vector for trait $t = 1 \ldots T$; $g = [g'_1 \ldots g'_T]'$ is a matrix of genetic values of dimension $n \times T$, such that gt is the

Table 19.1. Classical statistical models in genomic prediction

Model	Key assumptions	Strengths	Limitations	Best suited for	Key references
BLUP	Linear, additive and normal distribution of random effects	Simple, effective, pedigree-based	No marker information is provided	Identifying pedigree-based breeding values	Henderson (1975)
GBLUP	Additive and Gaussian effects with GRM	Records genome-wide additive variance	Uniform shrinkage of effects ignores non-additive effects	Associating polygenic traits with dense markers	VanRaden (2008)
RR-BLUP	Equivalent to GBLUP via ridge penalty	More computationally efficient and marker-based	Shrinks large effects effectively	High-dimensional genomic datasets	Endelman (2011)
BayesA	Heavy-tailed prior for marker variances	Records large-effect loci	Computationally intensive	Traits with few large + many small effects	Meuwissen *et al.* (2001)
BayesB / BayesCπ	Many markers with zero effect (sparse priors)	Variable selection, good for sparse traits	Computational complexity	Oligogenic traits	Meuwissen *et al.* (2001) and Habier *et al.* (2011)
MAP estimators	Prior probability distribution, no full distribution	Fast, scalable	No uncertainty intervals	Large datasets, quick inference	Xu (2013)
LASSO / Elastic Net	Linear with sparsity (L1/L1 + L2 penalties)	Sparse, interpretable, efficient	Unstable under LD, biased effects	Useful for traits with major QTL	Tibshirani (1996); Zou and Hastie (2005)

vector of genetic values for trait t, with distribution g ~N : (0, $\Sigma_g \otimes G$), in which Σ_g is the additive genetic (co) variance matrix of the traits; G is the genomic relationship matrix (VanRaden, 2008); ⊗ represents their Kronecker product; and e = $[e'_1 \ldots e'_T]'$ is the matrix of normally distributed random errors with e~N (0, R ⊗ I), in which R is the error (co)variance matrix between the traits (Shokor *et al.*, 2025).

19.3.2 Ridge regression BLUP

Proposed by Whittaker *et al.* (2000) for MAS in biparental populations, this model simultaneously selects all marker effects rather than dividing into significant or having no effect. This method was analyzed by Meuwissen *et al.* (2001) and Habier *et al.* (2007), who assumed the model

$$y = 1_n\mu + \Sigma_i W q_i + e$$

where μ is the mean, W is a matrix containing genotypes coded as 0, 1, or 2, and q_i is the effect of each SNP. The components in W per column j contain $2p_j$ (where p_j is the minor allelic frequency of marker j), reduced from the genotype constitution to achieve zero as the sum of coefficients in every column. Here, the SNP effects are treated as random and summed over all segments (Clark and Werf, 2013).

The GBLUP procedure that uses a matrix is similar to the RR-BLUP method, provided that the markers in WW' in the ridge matrix are measured the same, as those used to compute G, i.e. $XX' = X_{ij}/2 \Sigma p_i (1-p_i)$. Furthermore, GBLUP and RR-BLUP are only equivalent when $G\sigma^2_g = WW'\sigma^2_q$ (Clark and Werf, 2013).

19.3.3 Bayesian approaches, including the Bayesian alphabet and MAP estimators

This approach eases the presumption of common variance and equal marker effects; it determines an individual variance for each one and incorporates different sizes of marker effects (more realistically by assuming a specified prior distribution). Meuwissen *et al.* (2001) proposed two Bayesian models: BayesA and BayesB.

BayesA

The marker variance distribution is an inverted chi-square distribution. The degree of freedom and parameters of distribution are chosen in such a way that the mean and variance for the distribution match the mean and variance of the marker (Singh and Singh, 2015).

BayesB

This approach enables a few markers to have zero variances and effects, whereas others may have greater than zero effects with a variance of an inverted chi-square distribution pattern. It is more accurate than BayesA and RR-BLUP, and its accuracy does not decrease with an increase in the number of markers (Singh and Singh, 2015). Bayesian methods provide a better choice than other methods when there is a high density of markers and fewer phenotypic records (Singh and Singh, 2015).

19.3.4 LASSO for sparse effect estimation

LASSO (least absolute shrinkage and selection operator) is an additive straight regression model. Similar to RR-BLUP, LASSO is a regularization method; however, in LASSO, parameters can be reduced to zero, whereas in ridge regression, they cannot. It identifies a small number of QTLs with large effects (Singh and Singh, 2015).

19.4 Advanced Machine Learning Approaches for Genomic Prediction

Machine learning (ML), a subfield of artificial intelligence, performs specific tasks in scientific research on agriculture (Falk *et al.*, 2020; Jafari *et al.*, 2020) and industrial and field-based applications based on large datasets using different algorithms. ML can be actively used in different agricultural areas such as plant yield prediction, plant growth and development monitoring, water requirement, climate effect on crop, and precision agriculture (Yoosefzadeh-Najafabadi *et al.*, 2022a). Hesami *et al.* (2022) extensively studied different subgroups of ML algorithms, which are mainly of three types: supervised learning, unsupervised learning, and reinforcement learning. Supervised learning tends to predict an output either in a discrete level (in the case of classification) or through a numerical value (in the case of regression) for a labeled dataset (Hesami *et al.*, 2020a). It performs by optimizing predictive models as corresponding output by fitting parameters from labeled training data comprising inputs. Precautions should be taken against overfitting of the data or in such areas where the model can generalize the output to new unseen data. Supervised learning is most frequently used to predict complex traits in the early crop-growth stage (Yoosefzadeh-Najafabadi *et al.*, 2021), detect genomic regions associated with different characters (Yoosefzadeh-Najafabadi *et al.*, 2022a) and thereby helps to select superior genotypes through genomic selection in early generation (Yoosefzadeh-Najafabadi *et al.*, 2022a). Unsupervised learning performs on unlabeled data. It is very hard to estimate if an unsupervised approach can perform well or not, as it does not consider outputs and it does not work on training data. However, the model performs by finding the pattern in input data, such as groups/clusters. It is used to find the possible association among different genotypes in breeding populations. Reinforcement learning is quite new in plant breeding and yet to be explored effectively; however, it works on the feedback in the form of prizes and penalties received on exposure of the model to an environment (Lee *et al.*, 2012).

Over the years, various ML algorithms have been used in the plant breeding field depending on their advantages and disadvantages. Here, we will discuss some common supervised learning-based ML algorithms such as support vector machines (SVMs), random forest, gradient boosting and ensemble learning. SVMs, frequently used for classification and regression-related problems, are based on the concept of decision planes defining decision boundaries (hyperplane) represented by y = mx + c in linear classification. Support vectors are the nearest data points to the hyperplane defining the class boundaries. The distance between the hyperplane and the support vector is called the margin. A larger margin signifies better performance of models on new and unseen data. The decision plane separates a group of objects by maximizing the margin between them, having different classes such as tolerants vs susceptibles. Several researchers have shown the performance superiority of the SVM model over other models in their different successful research findings in maize (Zhao *et al.*, 2020), and potato (Griffel *et al.*, 2018). Random forest (RF) is another common machine learning algorithm typically used for both classification and regression. It performs by analyzing several decision trees to produce an accurate and stable prediction. It has the benefits of judging different complex datasets with many features, which makes it a strong predictive analyst. RF was used to predict meaningful relationships among several phenomic traits and seed yield from a diverse panel of 292 soybean accessions tested across six variable environments (Parmley *et al.*, 2019). This RF method is found to be the best in predicting seed yield, with 84% accuracy. However, RF is slow to predict if they are fully trained (Yoosefzadeh-Najafabadi *et al.*, 2021). Additionally, the RF algorithms are not suited for real-time prediction as the result can vary greatly, depending on the dataset type (Belgiu and Drăguţ, 2016).

Gradient boosting, another ML algorithm, is thought to be the best predictive modelling method when the genetic makeup of a plant includes non-additive effects, as suggested by Abdollahi-Arpanahi *et al.* (2020) on the basis of their experimental results obtained on both real animal and stimulated datasets. But care should be taken for an optimization of hyperparameters during the training phase to prevent overfitting of new unseen datasets. Moreover, these new machine models can identify important environmental parameters affecting different phenological traits while analyzing large crop datasets coupled with environmental data. Much research has been done on maize for yield prediction under adverse climate conditions, such as high temperature at the reproductive phase (Cicchino *et al.*, 2010; Butler and Huybers, 2015; Lizaso *et al.*, 2018).

Ensemble learning is an ML technique that combines multiple algorithms to produce a more stable and powerful model (Hesami and Jones, 2020; Hesami *et al.*, 2020b). It is specifically used in such fields where a single algorithm may not be as accurate as a combined model (Polikar, 2012). Alongside, it is also computationally fast, as it needs less training data (Zhou *et al.*, 2021). Different types of ensemble learning, such as bagging, boosting, and stacking, are followed, depending on different functions and sampling methods. Although the application of ensemble learning in the plant breeding field is limited, some research has been conducted in the field of GWAS (genome-wide association study), genomic selection (Yoosefzadeh-Najafabadi *et al.*, 2022b), and yield prediction of soybean genotypes across four different environments (Yoosefzadeh-Najafabadi *et al.*, 2021).

The choice of a particular ML algorithm on a given dataset is one of the most important steps in plant breeding. It starts with the choice of grouping the entire datasets into training, validation, and testing with the help of the following methods:

- Splitting the data into two parts: one for testing, another one for training (Libbrecht and Noble, 2015). It is useful in analyzing genetic markers and RNA sequencing technology.
- Cross-validation, which includes splitting the whole dataset into k subsets, followed by training and evaluating the model *k* times (Browne, 2000), and each time, using one of the k subsets as the test set and the remaining k-1 as the training set. This is used in case of smaller data as for development of phenotypic algorithm in plant breeding.

- Random sampling with replacement from the original data followed by training and evaluating the model – bootstrap method
- LOOCV (leave-one-out cross-validation), which comprises splitting the whole data into *n* folds and after that training and evaluating the model *n* times (Cawley *et al.*, 2024), typically used for single-location trial, greenhouse experiment, plant tissue culture, etc. Currently, we are in the era of 'Breeding 4.0' where analyzing large datasets using computational technology has allowed plant breeders to frame their breeding goal globally (Kuriakose and Xiao, 2020).

19.5 Deep Learning in Genomic Prediction

Deep learning (DL), a subfield of machine learning, has recently emerged as a powerful approach in genomic prediction due to its ability to capture complex, nonlinear relationships and high-order interactions in large datasets. Unlike classical statistical models, which often rely on assumptions of linearity and additivity, DL algorithms are capable of extracting hierarchical feature representations directly from raw data. This enables them to better model the genetic architectures of complex traits, including dominance, epistasis, and genotype-by-environment (G×E) interactions (Montesinos-López *et al.*, 2021b).

The basic architecture of deep neural networks (DNNs) consists of multiple interconnected layers of computational nodes, with each layer transforming the input into increasingly abstract representations. Typically, the input layer receives genomic, phenotypic, or environmental data, which are then processed through hidden layers that perform nonlinear transformations, before finally producing trait predictions or genomic estimated breeding values (GEBVs) at the output layer. Training these networks involves iterative optimization of weights using algorithms such as stochastic gradient descent, while regularization strategies like dropout and early stopping are applied to avoid overfitting and ensure generalizability to new datasets.

Several DL architectures have been explored in the context of plant breeding. Multi-layer perceptrons (MLPs), which are fully connected feed-forward networks, represent one of the earliest applications, capable of modeling both additive and certain nonlinear effects but requiring large datasets to achieve reliable accuracy. Convolutional neural networks (CNNs), originally designed for image recognition, have shown promise in genomic prediction by identifying local dependencies within SNP data as well as in integrating phenomic images derived from UAV-based or ground-based sensors. Recurrent neural networks (RNNs), particularly suited to sequential or time-series data, have been used to model multi-seasonal or longitudinal datasets such as repeated phenotypic measurements across environments. More recently, transformer-based architectures, which employ attention mechanisms to model long-range dependencies, have demonstrated considerable potential for integrating genomic sequences with multi-omics and enviromics data.

The advantages of DL over classical approaches include its capacity to model nonlinear genetic architectures, its ability to integrate heterogeneous data layers such as genomics, transcriptomics, metabolomics, phenomics, and enviromics, and its effectiveness in explicitly capturing G×E interactions. These strengths allow for improved prediction accuracies across diverse environments, particularly when combined with high-throughput phenotyping platforms. For example, UAV-based imaging integrated with CNNs has been successfully applied to predict yield and stress responses in crops like maize and wheat, while transformer models incorporating genomic and environmental data have shown superior performance in predicting climate-resilient genotypes.

Despite these advances, DL also faces important challenges in practical breeding applications. Training effective models requires large, high-quality datasets, which are not always available in breeding programs. Computational demands are high, often necessitating access to advanced hardware such as GPUs or TPUs. Moreover, the interpretability of DL models remains limited compared with classical linear approaches, leading to their frequent characterization as 'black-box' systems. Without careful use of regularization techniques, DL models are also prone to overfitting, particularly when applied to smaller datasets.

Nevertheless, the integration of DL with multi-omics and phenomics is rapidly advancing. By combining dense genomic marker data with transcriptomic and metabolomic profiles, DL models can capture regulatory networks underlying trait expression, while phenomics data derived from imaging platforms provide dynamic measurements of plant performance under field conditions. Comparative studies suggest that DL does not always outperform classical methods when datasets are small, but in large and heterogeneous datasets characterized by non-additive effects, DL often achieves significantly higher predictive accuracy (Montesinos-López *et al.*, 2021b; Shokor *et al.*, 2025).

19.6 Prediction for Non-Gaussian Traits

Many agronomically important traits are significantly non-normal and thus genetic prediction of non-Gaussian phenotypes, such as binary, ordinal, and count phenotypes from pedigree information, is required for breeding and genetics purposes. However, conventional Gaussian-based models cannot adequately fit the discrete or skewed nature of these characteristics and underperform in obtaining unbiased parameter estimates and prediction accuracy. Alternatively, recent methods based on generalized linear mixed models (GLMMs), Bayesian threshold models (e.g. TGBLUP), and Poisson or negative binomial regressions for count data provide powerful alternatives that accommodate the structure of the model with the distribution of traits. More recently, Bayesian logistic ordinal regression and Bayesian recurrent neural networks have been introduced to provide additional flexibility in modeling and improved prediction for complex, nonlinear traits (Meyer, 2016; Mbebi *et al.*, 2025).

19.6.1 Importance of binary, ordinal and/ or count traits in breeding

Frequently, phenotypic values used in breeding programs do not follow a continuous normal distribution. Instead, they are often binary (e.g. resistant/susceptible to disease), ordinal (e.g. severity of plant disease), or count data (e.g. number of tillers or seeds). These traits have biological and agronomic value, determining survival, reproduction, and yield potential. Recording the underlying genetic variation for these traits allows selection to be made more accurately and genetic gain to be greatly increased. However, their discrete, skewed, or bounded nature creates problems for traditional quantitative genetic prediction (Meyer, 2016; de los Campos *et al.*, 2025).

19.6.2 Inadequacy of normality-based models

Traditional genetic prediction models using linear mixed models are based on normally distributed residuals and homogeneous variance at the levels of the response. Direct application of these methods to non-Gaussian data generally leads to biased parameter estimation, smaller variability estimates, and less accurate predictive abilities. However, this limit affects the effective usage of selection decisions in binary, ordinal, and count traits, thus calling for more appropriate modeling frameworks (Zhou *et al.*, 2014).

19.6.3 Application of generalized linear mixed models (GLMMs)

GLMMs are generalizations of classical mixed-effects models, in which the linear predictor is related to the mean of the response variable by suitable link functions. In the case of binary and ordinal outcomes, observed non-normal data are converted into a latent scale, which makes the data normally distributed using binomial distributions and links such as logistic or probit. Count characteristics are modeled with Poisson or negative binary (count) distribution. This model can account for genetic and environmental effects and respect distributional characteristics of each trait (Meyer, 2016).

19.6.4 Bayesian threshold models (BTM) for ordinal characteristics (TGBLUP)

For ordinal traits, Bayesian threshold genomic best linear unbiased predictor (TGBLUP), which assumes an underlying continuous latent trait from a normal distribution, seems to be a good choice. Observed ordinal groups are defined by thresholds on this latent variable. Bayesian inference is used to obtain posterior distributions for the parameters, improving the estimation precision and permitting an interval-level interpretation of ordinal phenotypes (de los Campos *et al.*, 2025).

19.6.5 Negative binomial and Poisson regression for count data

Count variables such as tiller counts are often modeled using Poisson regression, which is valid for equidispersed data. However, biological time series are often over-dispersed (variance > mean). Negative binomial regression (Nb) includes an extra dispersion parameter to account for this, thus improving the model fit and the prediction accuracy for count traits (Montesinos-López *et al.*, 2017).

19.6.6 Models for zero-inflated characteristics

Traits where zero observations are excessive (i.e. lack of disease symptoms resulting in zero scores) are best fitted by zero-inflated models that separate a binary component specification on the probability of zero outcome, and a count model on positive outcomes. Zero-inflated Poisson and zero-inflated negative binomial models have been shown to adequately represent the underlying biology and are more effective for predictions in breeding settings (Montesinos-López *et al.*, 2017).

19.6.7 Bayesian logistic ordinal regression (BLOR)

Bayesian logistic ordinal regression is a flexible foundation for analyzing ordinal characteristics by making connections to Bayesian inference on cumulative logit functions. These models allow for the possibility of heterogeneous genetic and environmental effects across threshold categories, providing posterior estimates of model parameters and improved predictive ability (de los Campos *et al.*, 2025).

19.6.8 Bayesian recurrent neural networks (BRNN) for modeling nonlinear traits

Given nonlinear genotype–phenotype relationships that govern the many complex traits, BRNNs are a powerful tool to capture interactions and temporal dependencies. By combining recurrent neural network architectures with Bayesian models, BRNNs effectively model high-dimensional data, improving prediction accuracy over what is achievable with linear or threshold models (Montesinos-López *et al.*, 2017).

19.6.9 Practical case examples

Ordinally scaled disease resistance obtained from animal studies has been successfully modeled with TGBLUP and BLOR, providing better predictions of the estimated level of resistance and better selection efficiency. Moreover, tiller counts simulated using negative binomial regression and zero-inflated models accurately represent the true distribution of counts in the population, which is important for making more accurate genetic inference and breeding decisions (Meyer, 2016; Montesinos-López *et al.*, 2017)

19.7 Multi-trait and Multi-environment Genomic Prediction

Taking advantage of the genetic correlations and the genotype-by-environment (G × E) interaction, multi-trait and multi-environment genomic prediction provides better prediction accuracies and selection abilities than conventional single-trait or single-environment models (Burgueño *et al.*, 2025; de los Campos *et al.*, 2025). The integration of data from varied traits and environments allows breeders to better capture complex genetic architecture and the reflection of environment to identify genotypes with improved levels of performance and stability, or with special adaptation to target environments. Recent advances in computation have made models increasingly scalable and predictive, such as MegaLMM and deep learning-based approaches that enable the integration of phenomics and environics data layers. These advances aid breeding choices under ever-changing and challenging environmental conditions while optimizing genetic improvement by minimizing phenotyping cost (Runcie *et al.*, 2021).

19.7.1 Predicting multiple traits at once to leverage genetic correlations

Predicting for multiple traits at once takes advantage of correlations in the genes for the different traits to improve genetic prediction accuracy. Multi-trait genomic prediction models make use of a group of traits correlated in order to increase the accuracy of traits that are hard to measure or have low heritability by integrating phenotypic and genotypic datasets from related traits (de los Campos *et al.*, 2025). Contrary to single-trait methods, multi-trait models model covariance structures between traits, allowing for a more complete understanding of the genetic architecture. Contemporary improvements also include local genetic correlations from particular regions of the genome, which enable greater precision and improvement in prediction performance, taking heterogeneity of trait relationships across the genome into account (Teng *et al.*, 2025). This integrated approach is very beneficial for complex, multigenic phenotypes in breeding programs.

19.7.2 Modeling genotype-by-environment (G×E) interactions

The modeling of G×E interactions is indispensable in genomic prediction because G×E occurs when genotypes behave differently in different environments, thus influencing trait expression and ranking (Burgueño *et al.*, 2025). Such interactions complicate breeding efforts because they reduce the predictability of genotype performance in untested environments and force modeling requiring differential genotype performances. Incorporation of G×E into prediction models enhances the discovery of genotypes that exhibit stable performance or specific adaptation that is critical for the development of cultivars that show optimal performance across diverse locales or targeting of specific conditions (Crossa *et al.*, 2025a). Quantitatively, G×E accounts for a significant portion of phenotypic variation, and including it improves the usefulness and strength of genomic predictions (Hudson *et al.*, 2022).

19.7.3 Application of MegaLMM to large-scale multi-trait data

MegaLMM is a state-of-the-art Bayesian multi-trait linear mixed model framework for efficient analysis of large-scale multi-trait genomic data with hundreds to thousands of correlated traits simultaneously (Runcie *et al.*, 2021). It overcomes the computational and statistical shortcomings that are inherent in traditional multi-variate models by making use of Bayesian factor analysis and Markov chain Monte Carlo methods for robust parameter estimation despite missing data. MegaLMM leverages the sharing of genetic covariance structures across traits and environments and leads to significant improvements in prediction accuracy in a variety of breeding scenarios, including for wheat and maize (Runcie *et al.*, 2021). Its scalability and flexibility make MegaLMM essential for the integration of high-dimensional phenomics data and therefore advance multi-trait genomic prediction for complex breeding objectives. Traditional multi-trait models often face computational and convergence issues in scaling with trait number; these issues are overcome in MegaLMM using Bayesian factor analysis and advanced sampling algorithms that enable reliable estimation of genetic covariance structures even in the presence of incomplete data (Runcie *et al.*, 2021). An added beneficial effect of the capacity of MegaLMM to incorporate phenomics data derived from high-throughput platforms is the further increase in predictive precision, as demonstrated by its use in wheat and maize breeding programs, establishing it as a key tool for modern multi-trait genomic prediction.

19.7.4 Phenomics and enviromics data integration

The incorporation of phenomics and enviromics data in genomic prediction models dramatically increases the accuracy and robustness of predictions for breeding across environmental spectra. Phenomics data, gained through high-throughput phenotyping platforms, such as multi-spectral imaging and remote sensing, provides granular data-based measurements of plant physiological states over time and space during plant development. Enviromics data refers to data on environmental parameters (e.g. soil properties, temperature, moisture) that affect the genotype performance. The integration of these rich streams of data enables the modeling of complex interactions between traits and environments, and genotype-specific responses, in greater detail than genomic data alone (Jung *et al.*, 2024). Empirical evidence has shown that phenomics data can be more accurate for predicting yield than genomic data, and both of these sources of data achieved even greater yield gains when combined (Jackson *et al.*, 2023). Thus, integration of phenomics and enviromics information promises to speed up genetic improvement and facilitate environment-conscious selection strategies in modern breeding programs (Kaushal *et al.*, 2024).

19.7.5 Application of kernels and DL in modeling complex trait–environment relationships

Kernel methods and deep learning (DL) models have been shown to be instrumental in modeling complex, nonlinear interactions of genotype, phenotype, and environment in genomic prediction. Kernel-based approaches encode raw input data as inputs to high-dimensional feature spaces that allow machine learning algorithms to model complex genotype-by-environment interactions and trait dependencies that may not be captured in linear methods (Cuevas *et al.*, 2016). Deep learning, convolutional, and recurrent neural networks are especially good at joining heterogeneous types of data, i.e. genomics, phenomics, and enviromics data, together and the extraction of hierarchical features and temporal patterns that are essential for prediction under different environmental conditions (Jubair and Domaratzki, 2023; Montesinos-López *et al.*, 2025a, b). These techniques enable breeders to model genotype–environment relationships with increased accuracy and thereby help to predict complex traits with increasing accuracy and to select for stability and adaptation to various environments.

19.7.6 Role of training data structure and cohort effects

The architecture of training data has a critical impact on fidelity as well as generalizability of genomic prediction models. Training cohorts should include a large amount of genetic heterogeneity and mimic the makeup of the proposed target populations for use in breeding programs to maximize predictive power. Population structure, kinship, and cohort-related phenomena, like familial clustering or stratification, can also cause a bias in the predicted accuracy if they are not sufficiently controlled for, which might lead to overstatement of performance (or overestimation of the predictive accuracy) or undermining the transferability of the model into other different cohorts (Isidro *et al.*, 2014). Boosting the training set with stratified sampling or with maximum possible coverage of phenotypic variance helps increase robustness (Isidro *et al.*, 2014). Temporal or geographically based sampling artefacts that are cohort effects need to be corrected in order to prevent erroneous genetic evaluations and to ensure stable predictions for a year-to-year basis and environmental context (Crossa *et al.*, 2025a). Competent stewardship of training data architecture is therefore essential for reliable multi-trait and multi-environment genomic predictions.

19.7.7 Cross-year and cross-location predictions

Cross-year and cross-location genomic prediction exercises are indispensable for evaluating the robustness and generalizability of predictive frameworks under temporal and spatial heterogeneity of the environment. Such predictions mimic real-world breeding situations where genotypes are being questioned across different years and locations, which in turn allows for the identification of germplasm with steady performance, or of specific environmental adaptation (Barreto *et al.*, 2024). Successful cross-year and cross-location inference requires the explicit modeling of genotype-by-environment (G×E) interactions and integration of exhaustive phenotypic and environmental data. The attainment of correct forecasts finesse breeding efficiency by allowing for pre-emptive, reliable selection before the completion of extensive multi-site phenotyping efforts, reducing the expenses and accelerating genetic improvement (Alemu *et al.*, 2024; Crossa *et al.*, 2025a).

19.8 Applications of Genomic Prediction in Crop Improvement

As a highly versatile technology, GP is applicable across a wide range of crops and breeding objectives. GP has significantly contributed to major cereals such as wheat, maize, and rice, as well as to minor crops. In wheat, empirical research indicates that rapid-cycle genomic selection enhances grain yield by enabling earlier and more frequent selection of superior genotypes (Rutkoski *et al.*, 2015). Maize breeding programs have used GP to predict hybrid performance and leverage genotype–environment interactions, resulting in improved yield and flowering-time predictions under various conditions. The accuracy of yield projections under stress conditions has also been increased through the integration of meteorological and genetic data in rice breeding, particularly for drought-resistance traits (Crossa *et al.*, 2023). Beyond cereals, GP has opened new opportunities for pre-breeding and expanding the genetic base in minor crops such as peanut, chickpea, and cassava, where traits like provitamin A content, yield stability, and seed quality are being precisely predicted (Yu *et al.*, 2016; Roorkiwal *et al.*, 2020). Breeding efficiency has been improved by GP, which makes fast cycling and early-generation selection possible. In contrast to conventional phenotypic selection, which necessitates many years of multi-location trials, GP enables precise selection before comprehensive field testing, reducing cycle time. GP allows for numerous selection cycles annually when combined with speed breeding or doubled haploid technologies, significantly increasing yearly genetic gain, especially in crops with lengthy generation durations (Gaynor *et al.*, 2021; Nannuru *et al.*, 2025). The use of parental prediction and optimal crossing strategy is another crucial application. In order to maximize genetic variety and long-term benefit, breeders can create optimal mating schemes by forecasting both individual performance and the utility of crosses. Given that methods like expected cross-value estimate and genomic mate selection have been demonstrated to enhance progeny merit while controlling inbreeding, GP has been found to be a useful decision-support tool for crossing programs (Allier *et al.*, 2019; Simeão *et al.*, 2021). Genomic prediction (GP) is highly effective for improving complex traits such as yield, drought tolerance, and nutrient-use efficiency, which are polygenic and environment-dependent. Unlike marker-assisted selection that targets a few major QTLs, GP captures genome-wide effects from many loci. Prediction accuracy is further enhanced by integrating enviromics, high-throughput phenotyping, and multi-omics data (Gorjanc *et al.*, 2017; Juliana *et al.*, 2019; Xu *et al.*, 2020). There is ample evidence of GP's overall effects on genetic gain and breeding efficiency. The results of modeling and empirical investigations clearly demonstrate that GP works better than traditional phenotypic pipelines by lowering cycle duration, improving selection accuracy, and sustaining greater rates of gain per unit time (Hickey *et al.*, 2017). Lastly, one of the most important factors for adoption is how cost-effective GP is. Despite the high initial cost of genotyping and computing infrastructure, GP has become more and more accessible thanks to advancements like pooled training populations, sparse testing designs, and low-density SNP panels with imputation (Crossa *et al.*, 2017). The cost-effectiveness of GP is increased for large breeding projects due to its scalability, whereas hybrid approaches that integrate phenotypic and genomic selection guarantee long-term use for smaller operations (Norman *et al.*, 2017). GP has several uses in crop improvement, ranging from increasing

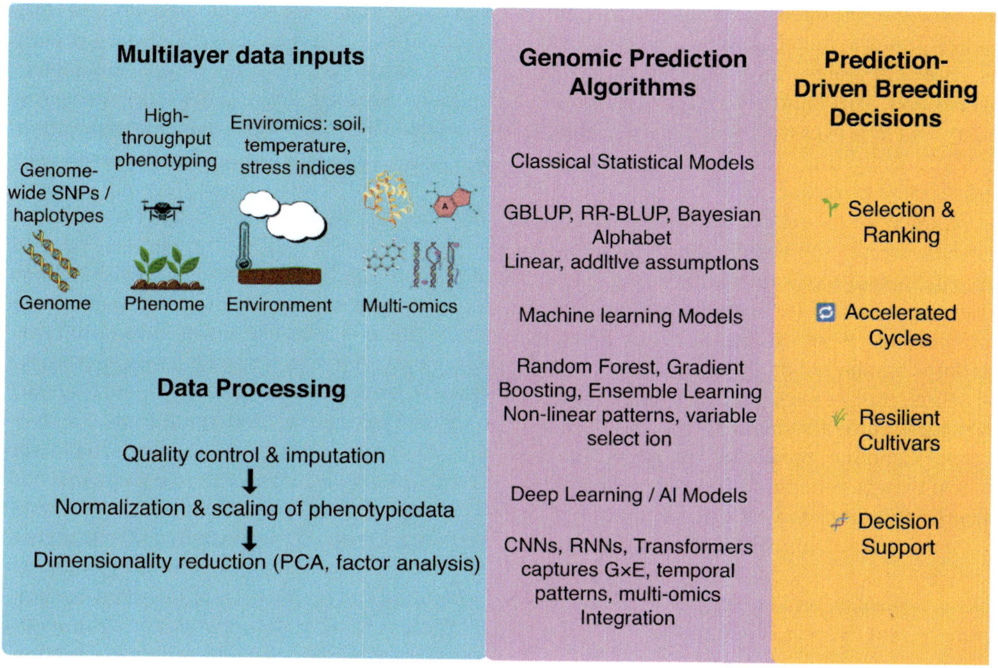

Fig. 19.1. Integrated genomic prediction framework for modern plant breeding (Author's own image)

complex characteristics and cost-effectiveness to optimizing crossing and speeding up genetic gain. Its widespread use in breeding contexts and across crops demonstrates its potential to revolutionize global agriculture by more effectively producing high-yielding, climate-resilient variants (Fig. 19.1).

19.9 Emerging Trends and Future Prospects

Beyond the capabilities of linear models, the inclusion of AI and machine learning techniques, such as deep neural networks, ensemble approaches, and transfer learning, has enhanced the ability to model nonlinear genetic architectures and G×E interactions (Montesinos-López *et al.*, 2021a, b). Meanwhile, developments in high-throughput phenotyping (HTP) such as Internet of Things (IoT)-enabled ground sensors, LiDAR, thermal imaging, and unmanned aerial vehicle (UAV)-based multispectral and hyperspectral imaging, have revolutionized GP by offering temporally dense and high-resolution phenotypic and environmental datasets that improve accuracy and allow for almost real-time trait prediction (Araus and Kefauver, 2018). Practical haplotype graphs (PHGs) and graph-based pangenomes are significant advances in genomics because they improve imputation from low-coverage sequencing, capture structural and haplotypic diversity that is not present in linear references, and lower genotyping costs. These attributes expand the applicability of GP across diverse germplasm and breeding programs with limited resources (Jensen *et al.*, 2020; Bradbury *et al.*, 2022). These developments are combining to create real-time prediction platforms that incorporate genomic and environmental data, AI-driven models, and UAV and IoT phenomics. These platforms offer interactive tools for breeders that can forecast yield dynamically and recommend genotypes during the growing season, moving GP away from retrospective analyses and toward proactive, decision-support systems (Xu *et al.*, 2022). Furthermore, GP is positioned as a predictive and mechanistic tool when multi-omics data layers, including transcriptomics, metabolomics, and epigenomics, are integrated into network-based GP frameworks. This improves biological interpretability and improves the prediction of complex, low-heritability traits (Montesinos-Lopez *et al.*, 2024). Despite advances, GP raises ethical and governance issues related to data ownership, consent, access, and benefit sharing, especially for smallholders and public breeding. Strong governance is vital for fairness and inclusivity. These trends point to personalized breeding pipelines similar to precision medicine, using AI, PHGs, multi-omics, and real-time analytics to suggest cultivars suited to specific environments. Genomic prediction is poised to be central to next-generation breeding by combining predictive models with scalable phenotyping and data stewardship, offering new opportunities to boost genetic gains while ensuring adaptability, resilience, and inclusivity in global crop improvement. (Crossa *et al.*, 2017).

19.10 Conclusion

Genomic prediction stands as a cornerstone of modern plant breeding, bridging traditional selection methods with cutting-edge genomic and computational technologies. By harnessing genome-wide marker data, GP enables breeders to predict genetic potential with unprecedented accuracy, significantly reducing the time and cost associated with conventional phenotyping. The integration of classical statistical models, such as GBLUP and Bayesian approaches, with advanced machine learning techniques, including deep learning and ensemble methods, has expanded GP's capacity to model complex traits and genotype-by-environment interactions. The incorporation of multi-omics, high-throughput phenotyping, and enviromics further enhances predictive power, facilitating the development of resilient, high-yielding crop varieties tailored to diverse environments. Practical applications across major cereals like wheat, maize, and rice, as well as minor crops such as chickpea and cassava, underscore GP's versatility and impact on global agriculture. Emerging trends, including AI-driven platforms, practical haplotype graphs, and real-time prediction systems, herald a new era of data-driven breeding, often termed Breeding 4.0 and 5.0. However, challenges such as the need for large, high-quality datasets, computational demands, and ethical considerations surrounding data ownership and access must be addressed to ensure equitable adoption, particularly in resource-limited settings. Future prospects lie in scalable, cloud-based pipelines and federated learning systems that democratize access to GP technologies. By combining predictive accuracy with biological interpretability and inclusive governance, GP is poised to accelerate genetic gains, enhance food security, and support sustainable agriculture. As breeding programs worldwide embrace these advancements, GP will continue to redefine crop improvement, fostering resilient and productive agricultural systems for a rapidly changing world.

References

Abdollahi-Arpanahi, R., Gianola, D. and Peñagaricano, F. (2020) Deep learning versus parametric and ensemble methods for genomic prediction of complex phenotypes. *Genetics Selection Evolution* 52(1), 12.

Alemu, S.G., Crossa, J., Montesinos-López, O.A., y Sanchez, J.I., Fernandez-Gonzalez, J. *et al.* (2024) Genomic selection in plant breeding: Key factors shaping prediction accuracy across environments. *Theoretical and Applied Genetics* 137(3), 409–423.

Allier, A., Lehermeier, C., Charcosset, A., Moreau, L. and Teyssèdre, S. (2019) Improving short- and long-term genetic gain by accounting for within-family variance in optimal cross-selection. *Frontiers in Genetics* 10, 1006.

Araus, J.L. and Cairns, J.E. (2014) Field high-throughput phenotyping: The new crop breeding frontier. *Trends in Plant Science* 19(1), 52–61.

Araus, J.L. and Kefauver, S.C. (2018) Breeding to adapt agriculture to climate change: Affordable phenotyping solutions. *Current Opinion in Plant Biology* 45, 237–247.

Barreto, C.A.V., das Graças Dias, K.O., de Sousa, I.C., Azevedo, C.F., Nascimento, A.C.C. *et al.* (2024). Genomic prediction in multi-environment trials in maize using statistical and machine learning methods. *Scientific Reports* 14(1), 1062.

Belgiu, M. and Drăguț, L. (2016) Random forest in remote sensing: A review of applications and future directions. *ISPRS Journal of Photogrammetry and Remote Sensing* 114, 24–31.

Bradbury, P.J., Casstevens, T., Jensen, S.E., Johnson, L.C., Miller, Z.R. *et al.* (2022) The practical haplotype graph, a platform for storing and using pangenomes for imputation. *Bioinformatics* 38(15), 3698–3702.

Browne, M.W. (2000). Cross-validation methods. *Journal of Mathematical Psychology* 44(1), 108–132. DOI: 10.1006/jmps.1999.1279.

Burgueño, J., de los Campos, G., Weigel, K. and Crossa, J. (2012). Genomic prediction of breeding values when modeling genotype × environment interaction using pedigree and dense molecular markers. *Crop Science* 52(2), 707–719.

Butler, E.E. and Huybers, P. (2015) Variations in the sensitivity of US maize yield to extreme temperatures by region and growth phase. *Environmental Research Letters* 10(3), 034009.

Cawley, S., Bekiranov, S., Ng, H.H., Kapranov, P., Sekinger, E.A. *et al.* (2024) Unbiased mapping of transcription factor binding sites along human chromosomes 21 and 22 points to widespread regulation of noncoding RNAs. *Cell* 116(4), 499–509.

Chen, K., Wang, Y., Zhang, R., Zhang, H. and Gao, C. (2019) CRISPR/Cas genome editing and precision plant breeding in agriculture. *Annual Review of Plant Biology* 70, 667–697.

Cicchino, M., Edreira, J.R., Uribelarrea, M. and Otegui, M.E. (2010) Heat stress in field-grown maize: Response of physiological determinants of grain yield. *Crop Science* 50(4), 1438–1448.

Clark, S.A. and Werf, J.V. (2013) Genomic best linear unbiased prediction (gBLUP) for the estimation of genomic breeding values. In: Gondro, C., Werf, J. and Hayes, B. (eds) *Genome-Wide Association Studies and Genomic Prediction*. Humana Press, New York, pp. 321–330.

Cooper, M., Messina, C.D., Podlich, D., Totir, L.R., Baumgarten, A. *et al.* (2014) Predicting the future of plant breeding: Complementing empirical evaluation with genetic prediction. *Crop and Pasture Science* 65(4), 311–336.

Crossa, J., Pérez-Rodríguez, P., Cuevas, J., Montesinos-López, O., Jarquín, D. *et al.* (2017) Genomic selection in plant breeding: Methods, models, and perspectives. *Trends in Plant Science* 22(11), 961–975.

Crossa, J., Zhang, X., Pérez-Rodríguez, P., Jarquin, D., de Los Campos, G. *et al.* (2023) Enviromics and genomic prediction of complex traits in plant breeding. *Current Opinion in Plant Biology* 73, 102360.

Crossa, J., Martini, J.W.R., Vitale, P., Pérez-Rodríguez, P., Costa-Neto, G. *et al.* (2025a) Expanding genomic prediction in plant breeding: Harnessing big data, machine learning, and advanced software. *Trends in Plant Science* 30, 756–774.

Crossa, J., Montesinos-Lopez, O.A., Costa-Neto, G., Vitale, P., Martini, J.W. *et al.* (2025b) Machine learning algorithms translate big data into predictive breeding accuracy. *Trends in Plant Science* 30(2), 167–184.

Cuevas, J., Crossa, J., Soberanis, V., Pérez-Elizalde, S., Pérez-Rodríguez, P. *et al.* (2016). Genomic prediction of genotype × environment interaction kernel regression models. *The plant genome* 9(3), plantgenome2016-03.

Elshire, R.J., Glaubitz, J.C., Sun, Q., Poland, J.A., Kawamoto, K. *et al.* (2011) A robust, simple genotyping-by-sequencing (GBS) approach for high diversity species. *PloS ONE* 6, e19379.

Endelman, J.B. (2011) Ridge regression and other kernels for genomic selection with R package rrBLUP. *The Plant Genome* 4(3), 250–255.

Falk, K.G., Jubery, T.Z., Mirnezami, S.V., Parmley, K.A., Sarkar, S. *et al.* (2020) Computer vision and machine learning enabled soybean root phenotyping pipeline. *Plant Methods* 16(1), 5.

Gaynor, R.C., Gorjanc, G. and Hickey, J.M. (2021) AlphaSimR: An R package for breeding program simulations. *G3: Genes, Genomes, Genetics* 11(2), jkaa017.

Gianola, D., de Los Campos, G., Hill, W.G., Manfredi, E. and Fernando, R. (2009) Additive genetic variability and the Bayesian alphabet. *Genetics* 183, 347–363.

Gianola, D. (2013) Priors in whole-genome regression: The Bayesian alphabet returns. *Genetics* 194, 573–596.

Gorjanc, G., Dumasy, J.F., Gonen, S., Gaynor, R.C., Antolin, R. *et al.* (2017) Potential of low-coverage genotyping-by-sequencing and imputation for cost-effective genomic selection in biparental segregating populations. *Crop Science* 57(3), 1404–1420.

Griffel, L.M., Delparte, D. and Edwards, J. (2018) Using support vector machines classification to differentiate spectral signatures of potato plants infected with potato virus Y. *Computers and Electronics in Agriculture* 153, 318–324.

Habier, D., Fernando, R.L. and Dekkers, J.C.M. (2007) The impact of genetic relationship information on genome-assisted breeding values. *Genetics* 177, 2389–2397.

Habier, D., Fernando, R.L., Kizilkaya, K. and Garrick, D.J. (2011) Extension of the Bayesian alphabet for genomic selection. *BMC Bioinformatics* 12, 186.

Henderson, C.R. (1975) Best linear unbiased estimation and prediction under a selection model. *Biometrics* 31(2), 423–447.

Hesami, M. and Jones, A.M.P. (2020) Application of artificial intelligence models and optimization algorithms in plant cell and tissue culture. *Applied Microbiology and Biotechnology* 104(3), 1031–1044.

Hesami, M., Naderi, R., Tohidfar, M. and Yoosefzadeh-Najafabadi, M. (2020a) Development of support vector machine-based model and comparative analysis with artificial neural network for modeling the plant tissue culture procedures: Effect of plant growth regulators on somatic embryogenesis of chrysanthemum, as a case study. *Plant Methods* 16(1), 112.

Hesami, M., Alizadeh, M., Naderi, R. and Tohidfar, M. (2020b) Forecasting and optimizing Agrobacterium-mediated genetic transformation via ensemble model-fruit fly optimization algorithm: A data mining approach using chrysanthemum databases. *PLoS ONE* 15(9), e0239901.

Hesami, M., Alizadeh, M., Jones, A.M.P. and Torkamaneh, D. (2022) Machine learning: Its challenges and opportunities in plant system biology. *Applied Microbiology and Biotechnology* 106(9), 3507–3530.

Heslot, N., Rutkoski, J. and Poland, J. (2013) Impact of marker ascertainment bias on genomic selection accuracy and estimates of genetic diversity. *PLoS ONE* 8, e74612.

Hickey, J.M., Chiurugwi, T., Mackay, I. and Powell, W. (2017) Genomic prediction unifies animal and plant breeding programs to form platforms for biological discovery. *Nature Genetics* 49(9), 1297–1303.

Huang, X. and Han, B. (2014) Natural variations and genome-wide association studies in crop plants. *Annual Review of Plant Biology* 65, 531–551.

Hudson, A.I., Odell, S.G., Dubreuil, P., Tixier, M.H., Praud, S. *et al.* (2022). Analysis of genotype-by-environment interactions in a maize mapping population. *G3* 12(3), jkac013.

Isidro, J., Jannink, J.L., Akdemir, D., Poland, J., Heslot, N. *et al.* (2014) Training set optimization under population structure in genomic selection. *Theoretical and Applied Genetics* 127(1), 749–757.

Jafari, M., Marquez, G., Selberg, J., Jia, M., Dechiraju, H. *et al.* (2020) Feedback control of bioelectronic devices using machine learning. *IEEE Control Systems Letters* 5(4), 1133–1138.

Jensen, S.E., Charles, J.R., Muleta, K., Bradbury, P.J., Casstevens, T. *et al.* (2020) A sorghum practical haplotype graph facilitates genome-wide imputation and cost-effective genomic prediction. *The Plant Genome* 13(1), e20009.

Jubair, S. and Domaratzki, M. (2023) Crop genomic selection with deep learning and environmental data: A survey. *Frontiers in Artificial Intelligence* 5, 1040295.

Juliana, P., Montesinos-López, O.A., Crossa, J., Mondal, S., González Pérez, L. *et al.* (2019) Integrating genomic-enabled prediction and high-throughput phenotyping in breeding for climate-resilient bread wheat. *Theoretical and Applied Genetics* 132(1), 177–194.

Jung, M., Muranty, H., Guerra, W., García-Gómez, B.E., Rymenants, M. *et al.* (2024). Modelling genotype by environment interaction using genomic and environmental data in apple. *Acta Horticulturae* 1412, 145–150. DOI: 10.17660/ActaHortic.2024.1412.23.

Kaushal, S., Gill, H.S., Billah, M.M., Khan, S.N., Halder, J. *et al.* (2024) Enhancing the potential of phenomic and genomic prediction in winter wheat breeding using high-throughput phenotyping and deep learning. *Frontiers in Plant Science* 15, 1410249.

Khlestkina, E.K. and Salina, E.A. (2006) SNP markers: Methods of analysis, ways of development, and comparison on an example of common wheat. *Russian Journal of Genetics* 42, 585–594.

Kuriakose, D. and Xiao, Z. (2020) Pathophysiology and treatment of stroke: Present status and future perspectives. *International Journal of Molecular Sciences* 21(20), 7609.

Lee, D., Seo, H. and Jung, M.W. (2012) Neural basis of reinforcement learning and decision making. *Annual Review of Neuroscience* 35(1), 287–308.

Libbrecht, M.W. and Noble, W.S. (2015) Machine learning applications in genetics and genomics. *Nature Reviews Genetics* 16(6), 321–332.

Lizaso, J.I., Ruiz-Ramos, M., Rodríguez, L., Gabaldon-Leal, C., Oliveira, J.A. *et al.* (2018) Impact of high temperatures in maize: Phenology and yield components. *Field Crops Research* 216, 129–140.

Mbebi, A.J., Mercado, F., Hobby, D., Tong, H. and Nikoloski, Z. (2025). Advances in multi-trait genomic prediction approaches: Classification, comparative analysis, and perspectives. *Briefings in bioinformatics* 26(3), bbaf211.

Meuwissen, T.H.E., Hayes, B.J. and Goddard, M.E. (2001) Prediction of total genetic value using genome-wide dense marker maps. *Genetics* 157(4), 1819–1829.

Meyer, K. (2016) General methods for evolutionary quantitative genetic inference from generalized mixed models. *Genetics* 204(4), 1281–1294.

Montesinos-Lopez, A., Crespo-Herrera, L., Dreisigacker, S., Gerard, G., Vitale, P. *et al.* (2024) Deep learning methods improve genomic prediction of wheat breeding. *Frontiers in Plant Science* 15, 1324090.

Montesinos-López, O.A., Montesinos-López, A., Gianola, D., Crossa, J. and Hernández-Suárez, C.M. (2021a) Multi-trait machine learning models for genomic selection. *Frontiers in Genetics* 12, 606843.

Montesinos-López, O.A., Montesinos-López, A., Pérez-Rodríguez, P., Barrón-López, J.A., Martini, J.W. *et al.* (2021b) A review of deep learning applications for genomic selection. *BMC Genomics* 22(1), 19.

Montesinos-López, O.A., Montesinos-López, A., Brandon, A., Delgado-Enciso, I., Chavira-Flores, M. *et al.* (2025a) Genomic prediction powered by multi-omics data. *Frontiers in Genetics* 16, 1636438.

Montesinos-López, O.A., Montesinos-López, A., Crossa, J., Toledo, F.H., Montesinos-López, J.C. *et al.* (2017). A Bayesian Poisson-lognormal model for count data for multiple-trait multiple-environment genomic-enabled prediction. *G3: Genes, Genomes, Genetics* 7(5), 1595–1606.

Nannuru, V.K.R., Dieseth, J.A., Lillemo, M. and Meuwissen, T.H. (2025) Evaluating genomic selection and speed breeding for Fusarium head blight resistance in wheat using stochastic simulations. *Molecular Breeding* 45(1), 14.

Norman, A., Taylor, J., Tanaka, E., Telfer, P., Edwards, J. *et al.* (2017) Increased genomic prediction accuracy in wheat breeding using a large Australian panel. *Theoretical and Applied Genetics* 130(12), 2543–2555.

Parmley, K., Nagasubramanian, K., Sarkar, S., Ganapathysubramanian, B. and Singh, A.K. (2019) Development of optimized phenomic predictors for efficient plant breeding decisions using phenomic-assisted selection in soybean. *Plant Phenomics*.

Polikar, R. (2012) Ensemble learning. In: *Ensemble Machine Learning*. Springer, New York, pp. 1–34.

Roorkiwal, M., Bharadwaj, C., Barmukh, R., Dixit, G.P., Thudi, M. *et al.* (2020) Integrating genomics for chickpea improvement: Achievements and opportunities. *Theoretical and Applied Genetics* 133(5), 1703–1720.

Runcie, D. E., Qu, J., Cheng, H. and Crawford, L. (2021). MegaLMM: Mega-scale linear mixed models for genomic predictions with thousands of traits. *Genome biology* 22(1), 213.

Rutkoski, J., Singh, R.P., Huerta-Espino, J., Bhavani, S., Poland, J. *et al.* (2015) Genetic gain from phenotypic and genomic selection for quantitative resistance to stem rust of wheat. *The Plant Genome* 8(2), 1–10.

Santosa, F. and Symes, W.W. (1986) Linear inversion of band limited reflection seismograms. *SIAM Journal on Scientific & Statistical Computing* 7, 1307–1330.

Shokor, F., Croiseau, P., Gangloff, H., Saintilan, R., Tribout, T. *et al.* (2025) Deep learning and genomic best linear unbiased prediction integration: An approach to identify potential nonlinear genetic relationships between traits. *Journal of Dairy Science* 108, 6174–6189.

Simeão, R.M., Resende, M.D., Alves, R.S., Pessoa-Filho, M., Azevedo, A.L.S. *et al.* (2021) Genomic selection in tropical forage grasses: Current status and future applications. *Frontiers in Plant Science* 12, 665195.

Singh, B.D. and Singh, A.K. (2015) Genomic selection. In: *Marker Assisted Plant Breeding: Principles and Practices*. Springer, New Delhi, pp. 295–312.

Teng, J., Zhai, T., Zhang, X., Zhao, C., Wang, W. *et al.* (2025). Improving multi-trait genomic prediction by incorporating local genetic correlations. *Communications Biology* 8(1), 307.

Tibshirani, R. (1996) Regression shrinkage and selection via the lasso. *Journal of the Royal Statistical Society* 56, 267–288.

VanRaden, P.M. (2008) Efficient methods to compute genomic predictions. *Journal of Dairy Science* 91, 4414–4423.

Varshney, R.K., Sinha, P., Singh, V.K., Kumar, A., Zhang, Q. *et al.* (2021) 5Gs for crop genetic improvement. *Current Opinion in Plant Biology* 60, 101997.

Whittaker, J.C., Thompson, R. and Denham, M.C. (2000) Marker-assisted selection using ridge regression. *Genetics Research* 75(2), 249–252.

Xu, S. (2013) Mapping quantitative trait loci by controlling polygenic background effects. *Genetics* 195(4), 1209–1222.

Xu, Y., Liu, X., Fu, J., Wang, H., Wang, J. *et al.* (2020) Enhancing genetic gain through genomic selection: From livestock to plants. *Plant Communications* 1(5), 100005.

Xu, Y., Zhang, X., Li, H., Zheng, H., Zhang, J. *et al.* (2022) Smart breeding driven by big data, artificial intelligence, and integrated genomic-enviromic prediction. *Molecular Plant* 15(11), 1664–1695.

Yoosefzadeh-Najafabadi, M., Earl, H.J., Tulpan, D., Sulik, J. and Eskandari, M. (2021) Application of machine learning algorithms in plant breeding: Predicting yield from hyperspectral reflectance in soybean. *Frontiers in Plant Science* 11, 624273. DOI: 10.3389/fpls.2020.624273.

Yoosefzadeh-Najafabadi, M., Eskandari, M., Torabi, S., Torkamaneh, D., Tulpan, D. *et al.* (2022a) Machine-learning-based genome-wide association studies for uncovering QTL underlying soybean yield and its components. *International Journal of Molecular Sciences* 23(10), 5538.

Yoosefzadeh-Najafabadi, M., Rajcan, I. and Eskandari, M. (2022b) Optimizing genomic selection in soybean. *Heliyon* 8(11), e11873.

Yu, X., Li, X., Guo, T., Zhu, C., Wu, Y. *et al.* (2016) Genomic prediction contributing to a promising global strategy to turbocharge gene banks. *Nature Plants* 2(10), 1–7.

Zhao, W., Lai, X., Liu, D., Zhang, Z., Ma, P. *et al.* (2020). Applications of Support Vector Machine in Genomic Prediction in Pig and Maize Populations. *Frontiers in Genetics* 11, 598318. DOI: 10.3389/fgene.2020.598318

Zhou, J., Gandomi, A.H., Chen, F. and Holzinger, A. (2021) Evaluating the quality of machine learning explanations: A survey on methods and metrics. *Electronics* 10(5), 593.

Zhou, X., Carbonetto, P. and Stephens, M. (2014) Linear-time algorithms for Gaussian and non-Gaussian trait evolution models. *Systematic Biology* 63(3), 397–408.

Zou, H. and Hastie, T. (2005) Regularization and variable selection via the elastic net. *Journal of the Royal Statistical Society: Series B* 67(2), 301–320.

Index

Note: Page numbers in **bold** and *italics* refer to tables and figures, respectively

Printed and bound by CPI Group (UK) Ltd, Croydon, CR0 4YY

13/09/2026

14961085-0006